D0214143

Mining and its Impact on the Environment

Also available from Taylor & Francis

Mining and its Impact on the Environment

Fred G. Bell and
Laurance J. Donnelly

LONDON AND NEW YORK

First published 2006
by Taylor & Francis
2 Park Square, Milton Park, Abingdon, Oxon OX14 4RN

Simultaneously published in the USA and Canada
by Taylor & Francis
270 Madison Ave, New York, NY 10016

Taylor & Francis is an imprint of the Taylor & Francis Group, an informa business

Typeset in Sabon by
Newgen Imaging Systems (P) Ltd, Chennai, India
Printed and bound in Great Britain by
TJ International Ltd, Padstow, Cornwall

British Library Cataloguing in Publication Data
A catalogue record for this book is available from the British Library

Library of Congress Cataloging in Publication Data
Bell, Fred G.
Mining and its impact on the environment / Fred G. Bell & Laurance J. Donnelly.
p. cm.
Includes bibliographical references and index.
1. Mineral industries – Environmental aspects. 2. Environmental management. I. Donnelly, Laurance J. II. Title.

TD195.M5B45 2006
622.028'6–dc22 2005026719

ISBN10: 0–415–28644–1 (hbk)
ISBN10: 0–203–96951–0 (ebk)

ISBN13: 978–0–415–28644–2 (hbk)
ISBN13: 978–0–203–96951–9 (ebk)

Contents

Preface

Most individuals today are more conscious of their environment than in the past and they are interested especially in those factors that degrade the environment. Mining is one of those factors. Mining refers to the process of extraction or abstraction of mineral deposits from either the surface of the Earth or from beneath the surface. In this context it also involves the abstraction of oil, gas, water and brine from the ground. Furthermore, mining represents one of man's earliest activities reaching back into Palaeolithic times and, as such, has played an important role in the development of civilization. It is also an activity that has and is going on more or less worldwide. With time, the use of minerals has increased in both volume and variety in order to meet a greater range of purposes and demand by society. Accordingly, present day society is more dependent on the minerals industry than in the past. Indeed, the exploitation of minerals is fundamental to society now and will continue to be in the future. In other words, the mining of minerals contributes to the sustained economic progress of developed nations and helps to alleviate poverty and improve the quality of life of people in developing countries.

First, mineral deposits have to be located, they then are worked and subsequently processed. It is the working and processing of mineral deposits that can give rise to environmental damage. This can mean that land is disturbed, that the topography is changed and that the hydrogeological conditions are affected adversely. The degree of impact that mining has on the environment varies depending on the mineral worked, the method of working, and the location and size of the working. However, land that has become spoiled by past mining activity generally can be rehabilitated but at a cost. This cost may be recovered indirectly by the benefit that a more attractive environment brings to the area so affected. In order to rehabilitate land that has been affected adversely by mining activity, the site or area concerned has to be investigated before it can be treated.

Consequently, the primary objective of this book is to provide an overview of various aspects of mining and how they affect the surrounding environment, and, just as importantly, how the arising environmental problems can be investigated and subsequently dealt with. Hopefully, this should enable the hazard liabilities from mining and their associated risks to be assessed more readily. As such, it is hoped that this book will be of value to those who are involved with the development or redevelopment of mining areas. These include civil engineers, mining engineers, geotechnical engineers, engineering geologists, mining geologists, environmental scientists and managers, hydrologists and hydrogeologists, utilities engineers, builders, mineral surveyors, conveyance lawyers, insurance officers, land owners and planners. Although it has been attempted to give a broad overview of the subject this, no doubt, has been influenced by the experience of the authors so that it may reflect a certain bias. Case histories have been provided at the end of each chapter, except that of the Introduction. These generally illustrate the content of the individual chapter but in some instances go further and introduce new material. Numerous references have been provided to help allow the reader to pursue his/her interests further.

As far as the authors are aware, there is no book available today that is similar to this in content. In other words, no book provides a survey, in particular, of various aspects of subsidence, of waste disposal, of pollution and of dereliction as caused by different types of mining, and their investigation and treatment. In addition, topics such as spontaneous combustion, gases and induced seismicity also are included within the text. Nonetheless, not every aspect of environmental degradation such as noise and dust is covered. This primarily is because of the limitations of the author's experience.

The authors would like to express their thanks to various individuals and organizations for the help provided during the production of this book. In particular, they would like to thank Professor Dick Stacey, Mr Howard Siddle, Mr John Sinclair, Dr John Cripps, Dr David Reddish, Dr Dave Buttrick, Mr Roy Piggott, Mr Ross Parry-Davies, Mrs Susan Mills, the Halcrow Group Limited, the British Geological Survey, International Mining Consultants Limited, the Coal Authority, Anglo American Plc, Infoterra Limited, White Young Green Consulting Limited, the Canadian Geotechnical Society, the United States Department of the Interior, the International Association of Hydrological Sciences, the United States Geological Survey, the United States Bureau of Mines, SRK Consulting South Africa, Thomas Telford Publishing, the Institution of Civil Engineers, the International Society for Soil Mechanics and Geotechnical Engineering, the International Institution for Environment and Development, the South African Institution of Civil Engineering, Geological and Geotechnical Engineering, the International Mine Water Association, Nottingham University, Sheffield University, Robertson Geoconsultants Inc, Mining and Environmental Management, the Institute of Materials, Minerals and Mining, Elsevier, the National Museum and Gallery of Wales, the American Society of Civil Engineers, Richards Bay Minerals Limited, A.A. Balkema, the Geological Society of London, the Journal of Environmental Geochemistry and Health, Geologie en Mijnbouw, Kluwer Academic Publishers and the World Health Organization for providing illustrative material.

Chapter 1

Introduction

1.1. Introduction

Mining, alongside agriculture, represents one of man's earliest activities, the two being fundamental to the development and continuation of civilization. In fact, the oldest known mine in the archaeological record is the Lion Cave in Swaziland, which has a radio carbon age of 43 000 years. There Palaeolithic humans mined hematite, which they presumably ground to produce the red pigment ochre. Moreover, the dependence of primitive societies on mined products is illustrated by the terms Stone Age, Bronze Age and Iron Age, a sequence of ages that indicate the increasing complexity of the relationship between mining and society. In order to produce a flint axe meant that a suitable deposit of flint had to be located, that a method of mining be developed that would allow the flint to be worked, and that a means of processing be evolved to fashion the flint into a finished product. These requirements are more or less the same for mineral production at the present time. This general sequence applied in the Bronze and Iron Ages but the processing in particular became more sophisticated. With time, the use of minerals has increased in both volume and variety in order to meet a greater range of purposes and demand by society, and the means of locating, working and processing minerals has increased in complexity. Today, society is even more dependent on the minerals industry than in the past. In fact, it can be claimed that every material thing in society is either directly derived from a mineral product or is produced with the aid of mineral derivatives such as steel, energy or fertilizers. The importance of the minerals industry therefore cannot be overstressed.

However, unlike agriculture where there is some choice in where and what to grow, mining can only take place where minerals are present and economically viable to work. The localized nature of mineral deposits, as opposed to the widespread demand for them, led to the development of trade between peoples from prehistoric times onwards. The desire to acquire minerals, notably precious minerals, has led to exploration of far off lands, to wars and to conquest. Another important aspect of mining is that it is a robber economy, in that a mineral deposit is a finite resource and so mining comes to an end when the mineral deposit is exhausted or becomes uneconomic to mine or abstract. Perhaps groundwater abstraction is the exception in that with proper management it should be sustainable. Deposits that are abandoned when they become uneconomic to work may be reworked at some time in the future when mining technology or demand makes their exploitation once again worthwhile.

Mining, and associated mineral processing and beneficiation, does impact on the environment. Unfortunately, this frequently has led to serious consequences. The impact of mining depends on many factors, especially the type of mining and the size of the operation. It can mean that land is disturbed, that the topography is changed and that the hydrogeological conditions are affected adversely. Mining also has social impacts on the environment. Communities grow up around mines. When the mines close, the communities can suffer and even may die, becoming 'ghost towns'.

In the past the mining industry frequently showed a lack of concern for the environment. This does not necessarily imply that society was not aware of the environmental problems that could be associated with mining. For instance, in 1306 a royal proclamation prohibited the use of coal in London for domestic and industrial purposes because of the nuisance caused by smoke but it proved impossible to enforce. In addition, Agricola (1556) referred to environmental problems created by mining such as the devastation of fields and the pollution of streams. What is more, as technology has developed, so mining has had an increasing impact on the environment. The disposal of waste, in particular, has led to unsightly spoils being left to disfigure the landscape, and to surface streams and groundwater being polluted. Urban areas have suffered serious subsidence damage by undermining. Today, however, the greater awareness of the importance of the environment has led to tighter legislation being imposed by many countries to lessen the impact of mining. This is especially the case in the developed, more affluent nations. Unfortunately, many poorer countries, in which the primary minerals industry is proportionately of greater economic importance, are reluctant to impose non-essential restrictions on their principal earners of wealth and foreign exchange. What is more, the concept of reclamation of a site after mining operations have ceased, has become entrenched in law in the developed countries. An environmental impact assessment is necessary prior to the development of any new mine and an environmental management programme has to be produced to show how the mine will operate. Plans for reclamation of the mine site have to be made.

Although the adverse impacts on the environment should be minimized some environmental degradation due to mining is inescapable. Mines, however, are a local phenomena, although they may impact beyond mine boundaries. They also account for only a small part of the land area of a country (e.g. the mining industry accounts for less than 1% of the total area of South Africa). Land that has become derelict by past mining activity can be reclaimed. Rehabilitated spoil heaps frequently become centres of social amenity such as parklands, golf courses and even artificial ski slopes. Open-pits, when they fill with water, can be used as marinas, for fishing or as wildlife reserves. Even some underground mines can be used, such as those in limestone at Kansas City, Missouri, which are used as warehouses, cold storage facilities and offices (Fig. 1.1). Mining therefore can be looked upon as one of the stages in the sequential use of land.

Figure 1.1 An entrance to old workings in the Bethaney Falls Limestone, Kansas City, United States, which are now used for warehousing, cold storage etc.

1.2. Environmental problems and impacts

Mining may be regarded as the removal of minerals, in solid, liquid or gaseous form, from the Earth's crust for use in the service of society. It therefore involves exploration for and exploitation of mineral deposits by surface or underground methods. Accordingly, mining inevitably brings about change to the environment and so has an environmental impact. The degree of impact can vary from more or less imperceptible to highly intrusive and depends on the mineral worked, the method of working, and the location and size of the mine. Unfortunately, there is no consistent means by which to quantitatively assess some environmental impacts, notably those of a visual nature, so that any assessment of such impact tends to be qualitative with an appreciable element of subjectivity. On the other hand, some impacts can be quantitatively assessed such as the degree to which groundwater has been polluted. Obviously, the most severe environmental impacts occur where mining is going or has gone on, the effect of the impact reducing with increasing distance from the mine responsible.

Mining can pose a direct hazard to humans, apart from mine employees. Fortunately, mining hazards only infrequently lead to death or serious injury. Nonetheless, there are instances when, for example, old partially sealed shafts have opened beneath people who then have fallen to their death, where tailings dams have burst and killed people or where spoil heaps have failed causing death. Effluents from mines and mine wastes may be toxic and kill vegetation, and injure animals. Water pollution can continue for a long period after the mine causing the problem has ceased operation and as such can pose a health hazard. Many aspects of mining have the potential to cause damage to land and property, and thereby represent a direct financial loss to those affected. Admittedly, compensation can be paid to those affected but this may involve contentious legal disputes.

Loss of amenity or nuisance frequently is associated with mining but this does not necessarily threaten human life, although adversely affecting the quality of life. In some instances, this may be difficult to quantity whilst in others such as blasting it is possible to recognize and so assess levels of nuisance below the threshold of damage. Even so, the nuisance caused by blasting depends upon a number of factors such as intensity, frequency, duration, time of day, human susceptibility and location. Mining also is associated with dereliction, although derelict areas can be rehabilitated either during or more usually after mining has ceased, but at a cost. This cost may be recovered indirectly by the benefit that a more attractive environment brings to the area so affected.

As remarked, there is a wide variation in both the severity and type of environmental impact caused by mining, this being influenced by the size of the operation, the location of the operation, the type of mineral or minerals worked and the method of mining used to win these minerals.

Obviously, the larger the size of a mining operation, the larger the impact is likely to be since it will produce more waste, occupy more land and have a greater number of buildings and plant.

The extent to which mining becomes an environmental impact depends to a large extent upon the number of people that a mine affects. Frequently, communities grow up around mines and so have to live with many of the environmental problems that are consequent on mining. A good example is provided by coal mining. For instance, many of the industrial and urban areas of Britain developed on coal fields (Fig. 1.2). That is not to say that mineral exploitation in rural areas does not make an environmental impact. For example, restrictions are placed on quarry or mine development in national parks in Britain because of their amenity and recreational value. However, the opposition to such development tends to come from those who are not directly involved with mining. Those to whom mining represents a source of employment are much less likely to offer objections to the expansion and continuance of their means of livelihood. In developing nations even less opposition is forthcoming against the development of mining because it represents a means of employment and hopefully a way of raising living standards. The exception to this at the present day is likely to arise when mining is perceived to be a hazard to health.

Landscaping can be undertaken about mineral workings to reduce their visual impact. For instance, a mine may be screened from view to some extent by the construction of embankments around it that are subsequently planted with grass and trees, or a mine may be surrounded by trees, climate and soil permitting. Embankments can be used to dispose of some coarse discard but this may need to be covered with topsoil prior to the establishment of vegetation. They also can help to baffle noise and reduce dust problems. The shape, height and location of embankments need to be taken into account in their design. Certain information is required in order to carry out a landscaping plan. This generally includes details of the minerals worked, the method of working; the nature and location of the fixed surface plant; the likely development of a mine throughout its operational life; the amount, character and method of disposal of the waste produced; and the transport and service facilities required. In addition, it may be possible, for instance, in the case of quarries or pits, to orientate the working face in a direction that minimizes the visual impact from a critical viewpoint (Fig. 1.3(a)). At times an area of the mineral deposit can remain unworked in order to act as a screen (Fig. 1.3(b)). The avoidance of breaches in the skyline about a quarry usually contributes towards the preservation of visual amenity. It may be possible to limit visual impact by increasing the depth of a quarry but this may be limited by the position of the water table and will increase the difficulties of subsequent rehabilitation. In some instances, with increasing depth of overburden it may be better to commence subsurface mining of a deposit rather than continue with quarrying. This happened at

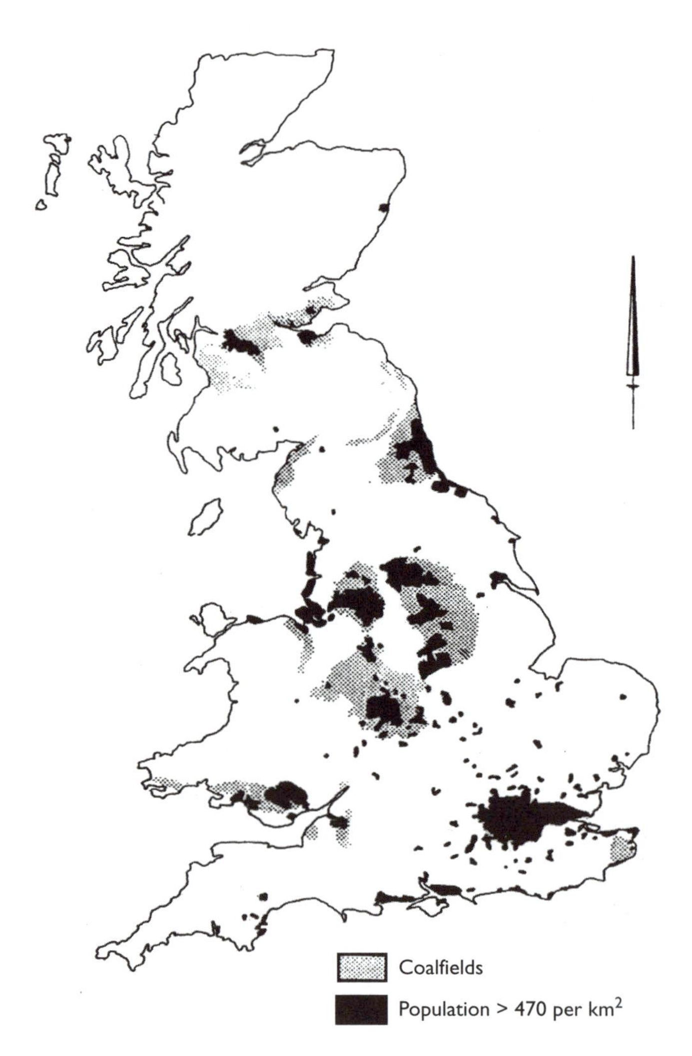

Figure 1.2 Location of former coalfields of Britain along with urban areas. Stippled areas represent coalfields, black areas represent urban centres with a population of over 475 per km^2.

a limestone quarry at Middleton, Derbyshire, England, where a mine was opened, leading off the quarry, when the overburden became too great (Fig. 1.4). Partial regrading, seeding and planting of spoil heaps at times can take place as their construction proceeds.

1.3. Examples of environmental impacts

One of the most notable environmental problems associated with past and present mining, and one that has an impact on development and construction,

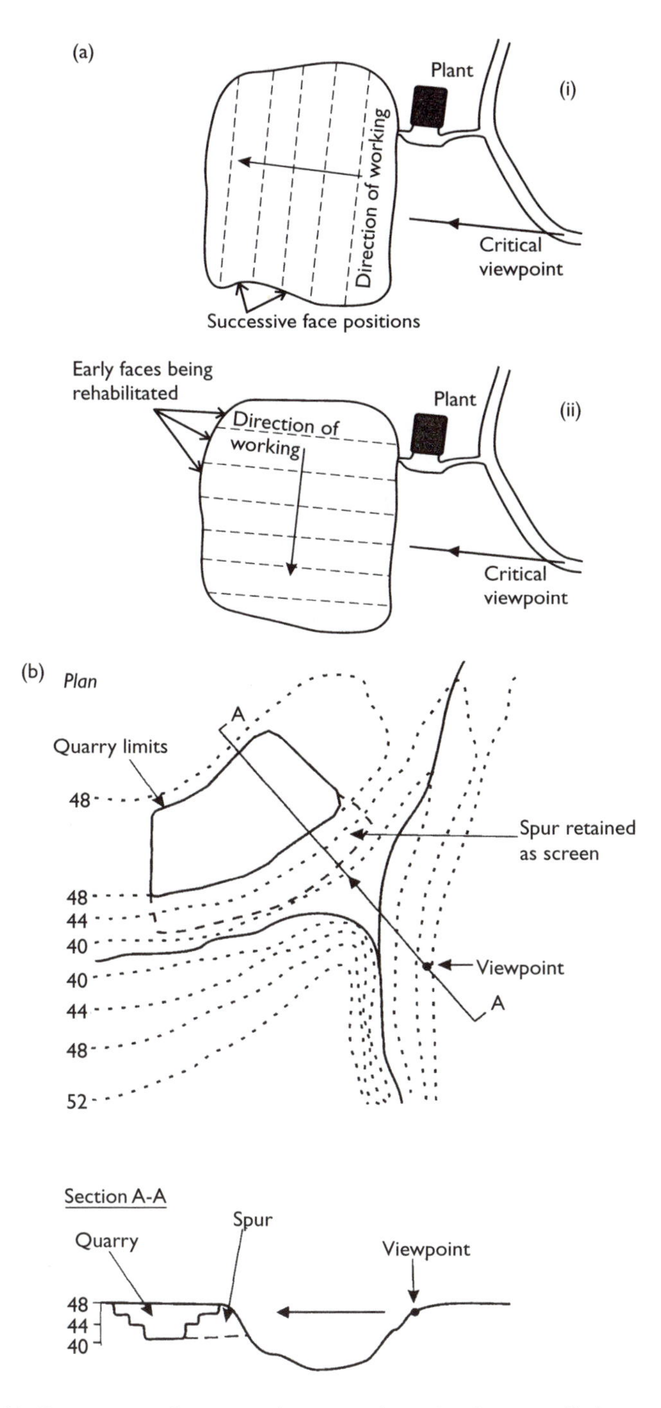

Figure 1.3 (a) Orientation of a quarry face to reduce visual impact (i) the quarry face is worked throughout its life in a direction that can be seen from a critical viewpoint. (ii) the direction of the face is orientated normal to a critical viewpoint and disused faces can be rehabilitated. (b) Retention of an area of unworked mineral in order to provide screening of a quarry, contours given in metres.

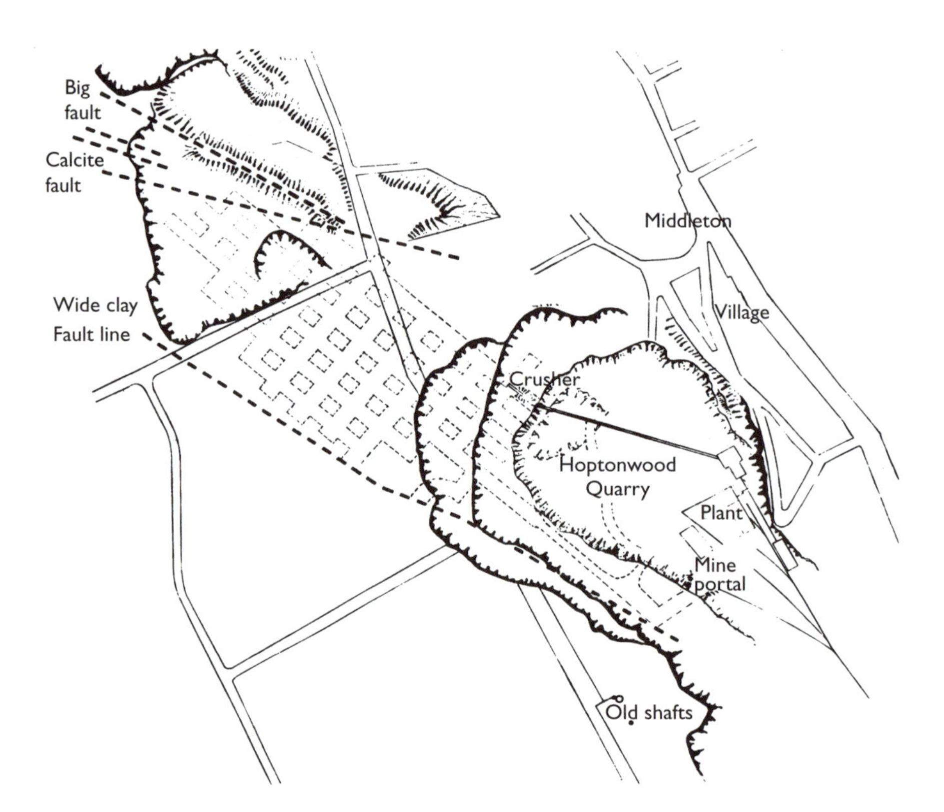

Figure 1.4 Middleton Mine, a limestone mine, at Middleton-by-Wirksworth, Derbyshire, England, that was developed when the overburden in Hoptonwood Quarry became excessive.

is subsidence. The type of subsidence that develops is largely dependent upon the method of mining employed. Also, the time when the subsidence occurs depends upon the type of mining, as does the reliability of subsidence prediction. Abandoned mines, especially when their presence is unknown, present a problem for developers. In the case of coal, burning seams may be difficult to extinguish and may lead to additional subsidence. Waste from mines in the form of spoil heaps and tailings lagoons represent unsightly blemishes on the landscape and after closure of a mine often require rehabilitation. Then there are the effluents, including acid mine drainage, associated with many past and present coal and metalliferous mines. These can have a serious effect upon the ecology of an area. The emission of gas from mines generally has not given rise to major problems at the ground surface but it still can give rise to fatalities or destruction of property and adversely affect land. Fortunately, seismic events associated with mining rarely give rise to damage of any consequence but individuals may find them psychologically disturbing. Some of the problems that are having to be faced, and presumably will become increasingly noteworthy, are attributable to the closure of mines and the subsequent rise of water tables. In Britain most coalfield areas have been closed and mine water pumping, which was necessary to dewater the mines, has ceased. Consequently, one of the most notable effects of coal mine closure is the emission of ferruginous effluents and acid mine drainage from the various exits to mines due to groundwater rebound. In addition, faults have been reactivated and subsidence of pillared workings may be brought about.

Subsidence of the ground surface can be regarded as ground movement that takes place due to the removal of mineral resources from within the ground, whether they be solid, liquid or gas, in other words due to mining of metals or stratiform deposits or the abstraction of water, oil or natural gas. It commonly is an inevitable consequence of such activities and reflects the movements that occur in the area so affected. The subsidence effects of working minerals depend on the type of deposit, the geological conditions, in particular the nature and structure of the overlying rocks or soils, the mining method and any mitigative action. In addition to subsidence due to the removal of support, mining often entails the lowering of groundwater levels that, since the vertical effective stress is raised, can cause consolidation and settlement of some of the overburden. Furthermore, groundwater lowering in carbonate areas can lead to the occurrence of sinkholes at the surface. Unfortunately, subsidence can have serious effects on surface structures, buildings, services and communications and can damage agricultural land through the disruption of drainage and alteration of gradient. It also can be responsible for flooding, can lead to the sterilization of land or call for extensive remedial measures, or special constructional design in site development. Hence, a number of questions concerning the stability of the surface in areas where extraction/abstraction has or is occurring require answers, these include:

- Will subsidence occur and if so, what will be its magnitude?
- When will subsidence happen and how will it develop?
- What form will the subsidence take?
- Is it practical and economical to prevent or reduce its effect?

Obviously, the answers to these questions may be difficult or impossible to obtain as they frequently depend on the assessment of a large number of factors and the nature of any answers that are obtained tend to be general rather than precise.

In some parts of the world mining has gone on for centuries, if not thousands of years, and methods of mining have changed with time. The oldest examples of mining in England date back to Neolithic times when flints were mined in the Chalk. The largest group of Neolithic flint mines in England are represented by Grimes Graves in the Brecklands near Thetford, Norfolk. Similar mines occur in many areas, including Blackpatch, Assbury, Harrow Hill and Stoke Down in Sussex, and Easton Down in Wiltshire, as well as at Buckenham Tofts, Lynford, Mass and Whitlingham in Norfolk. Probably some 700 to 800 pits are present at the Grimes Graves site indicating the significance of mining at that time. Three types of pits have been identified (Bell *et al.*, 1992). Primitive pits occur in a confined area around the northern limits of the site. These are quite small (e.g. about 2 m in diameter and 3 or 4 m in depth) simple opencast workings. The pits are in close proximity and commonly merge with each other. Intermediate sized pits have been identified along the western boundary of the site. These also are opencast workings but have larger diameters (e.g. around 6 m). Galleries were excavated from the bottom of some of these pits. Some 366 deep galleried shafts have been found over the remaining area of the site (i.e. 6.6 ha out of a total mined area of 13.7 ha). The shafts usually are funnel-shaped with galleries radiating from the bottom of each shaft (e.g. Pit 2 is 12.8 m in diameter at the surface reducing to 4.2 m at its base). The shafts vary from 6.1 to 12.2 m in depth and they were sunk to the depth of the best quality flint. The galleries spread out from the shafts to allow working of the flint; the main galleries are up to 2.1 m in width and 1.5 m in height, with smaller galleries between 0.6 and 0.9 m high. Galleries are connected by small holes through which a miner could crawl. Smaller holes between galleries presumably were for ventilation only. The galleries from one shaft sometimes communicate with another. This shows that mining was quite sophisticated and this type of mining carried on for centuries, similar mining methods being used for early working of coal.

The developments that took place in the coal mining industry in Britain, from workings at surface outcrops and in bell pits, to the development of pillared workings and longwall mining are outlined in Chapter 2. Bell pitting began to be overtaken by pillared working as a method of mining in the sixteenth century because of concern about a shortage of timber, coal becoming an increasingly important source of fuel. Longwall mining evolved in the eighteenth century with the use of coke for iron smelting. The latter created an increasing demand for coal, which fuelled the new industrial age.

One of the problems associated with old abandoned subsurface mineral workings is that there may be no record of their existence. For example, coal mining in Britain has gone on for centuries but the first statutory obligation to keep mine records only dates from 1850 and it was not until 1872 that the production and retention of mine plans became compulsory for mines over a certain size. Then in 1911, mine owners were required to maintain accurate plans of a specified scale and revise them every three months. Plans had to show the position of the mine workings in relation to surface features, including mine shafts and faults. However, even if old records exist, they frequently may be inaccurate.

Old abandoned workings occur at shallow depth beneath the surface of many urban areas of the European Community and North America. Such old workings can represent a hazard, especially during any subsequent redevelopment. The need for redevelopment in urban areas, together with the increasing scarcity of suitable sites means that sites formerly regarded as unsuitable are now considered for building purposes. In particular, many of the large industrial centres in the European Community and North America, where redevelopment has and is going on, are underlain by rocks of Coal Measures age, coal and other minerals present in the Coal Measures being one of the reasons for the initial development of industry. Therefore, an added factor as far as redevelopment in such areas is concerned is the problem of past or existing mine workings. It, however, must not be assumed that the frequent problems associated with mining in these areas are only related to the extraction of coal for other materials have been and are extracted from the Coal Measures. These include fireclay; gannister; ironstones; clays, shales and mudstones for bricks; sandstones for building purposes etc. Such materials have been both mined and quarried.

The increased demand of society for mineral products has led to the growth of the minerals industry and the demand has been met by developments in technology that have enabled new less accessible resources to be mined. In the case of metalliferous deposits, at the present day many of these are worked at much lower concentrations than formerly because of the increasing efficiency of extraction processes. The consequence of this is that larger mines are required and that much greater volumes of waste are produced. In fact, mine wastes represent the highest proportion of waste produced by industrial activity, billions of tonnes being produced annually. The fine ground down waste from metalliferous mining usually is contaminated by chemical reagents used in processing the ore. Coarse waste from mines is deposited on the surface in spoil heaps whilst fine waste or slurry is disposed of in tailings lagoons. Hence, mining waste can be inert or contain hazardous constituents but generally is of low toxicity. Nevertheless, in some areas where metals were mined in the past, because little regard was given to the disposal of waste, relatively high concentrations of heavy metals can represent an environmental problem. The chemical characteristics of mine waste and waters arising from them are governed by the type of mineral being mined and the chemicals that are used in the extraction or beneficiation processes. A special case of visual impact from waste disposal is that associated with leach heaps that contain acid-bleached rock spoil devoid of vegetation. Because of its high volume, mine waste in the past has been disposed of at the lowest cost, generally without regard for safety and frequently with significant environmental impacts. Worse still, catastrophic failures of tailings dams and spoil heaps, although fortunately not common events, have led to the loss of lives and destruction of property.

Spoil heaps that are to be rehabilitated need to be subject to geotechnical investigations to determine their soil texture, pH value, potential leachable contaminants and acid generation. It is particularly important that they are analysed to determine their content and distribution of toxic and combustible substances. Spoil heaps may be covered by rock and soil imported to the site, then fertilized, if necessary, and grasses and trees planted on them. Spoil at times may be reprocessed as extractive technology improves and the value of the mineral concerned increases, or to reclaim minerals that were initially considered as by-products. For example, some of the spoil heaps around Johannesburg, South Africa, have been reworked on a number of occasions to extract gold. Spoil in the former lead producing areas of Derbyshire, England, contain much calcite, barite and fluorspar. This often is reworked for fluorspar and barite. Similarly, many old colliery spoil heaps have been reprocessed to extract coal following advancements in coal processing technology.

The problems that are commonly associated with tailings lagoons include the generation of dust, low organic content and low levels of plant nutrients for vegetation to be established, likely toxicity (especially if pyrite or other sulphides are present), and radioactivity and radon gas emissions in the case of uranium tailings (see Chapter 7). A cover, using natural clays or synthetic materials such as geosynthetics or asphalt, may help in reducing the environmental concerns. Other factors that need to be considered include dust suppression, water erosion and the stability of the tailings dam.

In the case of open-pit mining, these are often on a large scale and in themselves make a significant visual impact on the environment, although some large open-pit mines are located in out of the way areas of low population density such as the arid west of the United States or northern Chile (Fig. 1.5). The most significant impact occurs where the excavation intersects with the skyline. In addition, many environmental impacts are more notable in open-pit mining such as blasting, vibrations, noise, dust, surface pollution and dereliction. Large quantities of waste normally have to be disposed of. Nevertheless, the nature of a surface excavation and its relationship with surrounding landforms influences the impact that an excavation makes. Opencast workings also give rise to significant visual environmental impact. However, opencast coal mining in Britain has tended to represent a limited phase of land use, partial site restoration going on at the same time as other parts of the site are being exploited. The whole site is restored after mining comes to an end.

Groundwaters contain dissolved salts derived from the rocks that they have been in contact with. They can be characterized according to the concentrations and proportions of combinations of ions that they contain. For instance, groundwaters derived from near-surface Coal Measures

Figure 1.5 Bingham Canyon open-pit copper mine, Utah. This is the largest excavation of its type in the world and it is one of the most important tourist attractions in Utah.

contain approximately 60–80% magnesium and calcium bicarbonates with some sodium sulphate and chloride. The dissolved salt concentration usually is less than 1000 mg l^{-1}. Groundwater from deeper seams contains higher proportions of chlorides. However, groundwater that has been in contact with coal mine workings may have its chemistry changed by reactions with iron pyrite, found in most coal seams and associated argillaceous rocks. Iron pyrite also is associated with many metalliferous deposits. Due to the solution of pyrite oxidation products groundwater is commonly highly ferruginous and often has a low pH value. Such groundwater is referred to as acid mines drainage. Acid mine drainage commonly contains high values of sulphate and elevated levels of heavy metals so that it can be toxic. Where this groundwater flows into surface watercourses the latter may become grossly polluted, with extensive precipitation of ochreous ferric hydroxide on stream beds. Acid mine drainage also can be formed when pyrite in surface waste or stockpiles is oxidized and comes in contact with water. As noted, the closure of a mine with the cessation of pumping results in the ultimate flooding of the old workings and subsequent rebound of groundwater levels until equilibrium is reached. If it reaches the surface this groundwater rebound can lead to streams being polluted. It also can give rise to other problems such as faults and subsidence being reactivated or the displacement and emergence of mine gases into the environment. On the other hand, many abandoned mines subsequently may function as aquifers after mining has ended (Wood *et al.*, 1999).

1.4. Closure and reclamation of mine workings

The closure of mines at the present day in developed countries normally requires the reclamation and restoration of the land in and around the mine concerned. In this regard, the design and phasing of the closure should seek to identify those factors necessary to ensure that the mine is left in a secure and safe state. If mine closure is carried out correctly, then this should mean that the maximum value for that land is secured for future redevelopment, as well the minimizing future liabilities. Furthermore, mine closure in developed nations in recent years has been associated with rapid developments in environmental legislation and regulations, and an increased awareness of the responsibilities of the mine operators, stakeholders and general public. As a result, many mineral operators now seek to eliminate any future liabilities following mine closure (Foster, 1998). However, for many developing countries planning for mine closure is a relatively new concept.

In some circumstances, the potential environmental impacts associated with the closure of mines may be more severe than those that occurred during mining operations. It therefore is crucial that prior to the closure, demolition and surface restoration, that an environmental audit should be commissioned (Ricks, 1995). The mine infrastructure should be assessed including roads, buildings, processing plant, offices, spoil heaps, tailings lagoons, concealed pipework, and utilities to determine whether any hazardous materials could cause problems. Some environmental concerns may be obvious whilst others will appear during demolition. Hazardous substances require identification, monitoring and disposal in a controlled manner according to hazardous waste management regulations. These substances include, for example, explosives, fuels, oils, greases, asbestos, polychlorinated biphenyls (PCBs), and boiler and water treatment chemicals. Monitoring also may be required for methane emissions and/or other gases, odourization plant, effluent and sewage discharge, mineral processing, water abstraction, noise and dust. Spoil heaps, dams and tailings lagoons may require specialist inspection prior to some form of rehabilitation. Such areas may provide future sites for public open space, parkland, forestry or agriculture.

Planning for mine closure involves the consideration of a number of complex interrelating technical, legal, financial and socio-economic issues. A mine closure plan ideally should be implemented before a mine is closed. Indeed, many modern mines now devise mine closure procedures before mining has begun and dedicate sufficient financial resources for this to reduce the liabilities and risks. For instance, opencast coal mines in Britain usually have closure and remediation plans as part of the mine design, before any mining has taken place. Each underground mine requires a unique mine closure treatment plan to deal with the mine complex including mine entries (shafts, drifts and adits), roadways and mine areas, surface and underground infrastructure, mine gas (if present) and groundwater regimes (Table 1.1). If the mine closure plan begins in the final stages of a mining operation and before the cessation of production, then this can facilitate the salvage of equipment, machinery and infrastructure. When the closure and rehabilitation of a deep mine has not been considered or finalized, then care and maintenance of the mine infrastructure is required before the cessation of mining operations.

The mine closure design plan in this case should follow the period of care and maintenance. This should identify actual and potential problems that affect pre-closure, closure and post-closure. Any necessary preventative measures also should be included in the design plan. This plan should be flexible and updated throughout the phased closure of the mine so as to take consideration of any new observations resulting from any subsequent investigations. If an abandoned mine requires closure treatment this will include controlled re-entry of the mine to determine the conditions of the mine entries and main roadways, and to verify that the ventilation is still operational, as well as the presence or absence of mine water and mine gasses (Table 1.1).

Table 1.1 Typical engineering investigations that might be considered in a mine closure plan

Engineering investigation	Detailed requirements
Rock mechanics and stability	Long-term stabilization of mine roadways and support pillars to prevent collapse, or the controlled collapse of part of the mine complex to reduce the risks of subsidence
Water and fire dams	To inhibit the inflow of mine waters
Seals and barriers	To control gas migration and airflow
Mine entries	Treatment of shafts, construction of shaft caps
Inclined roadways and adits	Treatment, stabilization to maintain the long-term integrity and stability of mine entries
Mine gas and mine water pumping	Post-closure control of gaseous emissions and mine water. This may involve the retention of shafts
Mechanical and electrical components	Inspection of type and conditions of roadway furnishings, surface mine winder system, haulage systems and electrical network. These may have an influence on how the closure is carried out and will eventually require decommissioning and salvage
Asset verification	Including all underground and surface equipment, structures and buildings. This will establish proof of ownership, saleable and recoverable assets and identify potential buyers

Following a mine closure treatment design, detailed closure specifications are prepared and made available to prospective contractors for competitive tendering. The way in which a mine is demolished is determined by specialist contractors and specified in their tender proposal. This also should include a detailed method statement and risk assessments. Demolition will include the controlled explosion of structures, excavation of services in the ground, capping of shafts etc. (Fig. 1.6). Although care and maintenance, and mine closure costs are high, these are relatively small in comparison with any longer term costs resulting from the risk of major mining induced hazards such as the collapse of abandoned mine workings, ground deformation, mine water discharge, mine gas emissions and environmental contamination.

Mine workings are commonly associated with derelict land, especially after a mine is closed. In this context, derelict land may be regarded as land that has been spoiled by the extraction or abstraction of minerals and so is not capable of beneficial use without treatment. Derelict land is a wasted resource, has a blighting effect on the surrounding area and can deter new development. Hence, its restoration or rehabilitation is highly desirable not only by improving the appearance of an area but also by making a significant contribution to its economy by bringing derelict land back into worthwhile use. Accordingly, there is both economic and environmental advantage in the regeneration of derelict land. Moreover, its regeneration should help prevent the exploitation of greenfield sites and encourage the use of brownfield sites. The term brownfield normally is applied in a broad

Figure 1.6 Controlled demolition of the head gear at Hem Heath (Trentham) Colliery, Staffordshire, England. (Reproduced by kind permission of White Young Green Consulting Limited and International Mining Consultants Limited.)

sense to land that has been developed previously and as such includes derelict land. The regeneration of brownfield sites is linked with the process of sustainable development. Sustainable land use is essential if present day development needs are to be met without compromising the ability of future generations to meet their needs. A sustainable land use therefore requires derelict sites to be re-used, thereby recovering such sites as a land resource. The use to which derelict land is put should suit the needs of the surrounding area and ideally be compatible with other forms of land use. Restoring or rehabilitating a site to a condition that is well integrated into its surroundings upgrades the character of the environment beyond the confines of the site.

Restoration, rehabilitation and reclamation have slightly different meanings. As far as restoration is concerned, it commonly refers to returning a mined area to its previous condition and land use such as when an opencast mine is filled and the restored land returned to agriculture. The meaning of rehabilitation and reclamation in the context of mining are not as widely accepted as the meaning of restoration. Nevertheless, rehabilitation (also referred to as regeneration) can be regarded as the development of a mine site for a substantially different land use from that which had existed prior to mining. Reclamation has been used in a general way simply to mean returning a mine site to some other land use, whether it be the same as before mining began or different.

As remarked above, when a new mine opens the conditions under which it is allowed to operate are likely to include provision for reclamation. This should mean that the impact of dereliction is reduced, indeed some reclamation can be undertaken whilst a mine is in operation. The plans to establish some form of reclamation obviously require certain information such as the confines of the mine and whether any further land will need obtaining in order to carry out the reclamation plan. In addition, information on buildings and plant, services, surface streams, hydrogeology and the effects of groundwater rebound, the character and volume of the waste and the manner in which it has been disposed of, and the location of any potential hazards such as shafts also need to be known. Hence, an investigation is needed to determine the input data for the design of the reclamation measures. The site investigation response to derelict land is no different from that for a greenfield site in that volumes of ground (including man-made structures) that will impact upon the foundation design of any future development must be identified and their properties quantified. The nature of the reclamation subsequently carried out will be influenced by the type of mining that was undertaken, be it surface or underground, and the size of the operation.

Abandoned quarries and pits, where ground conditions are suitable such as quarries in impermeable rock or old bricks pits located in thick formations of clay, frequently have been used for the disposal of domestic waste and then subsequently reclaimed. Some large old quarries have been used for storage or for the location of industrial units. When an attempt is made to restore a quarry, then landform replication represents the best method of restoration, especially of limestone quarries (Brashaw *et al.*, 2001). Landform replication involves the construction of landforms and associated habitats similar to those of the surrounding environment (see Chapter 6). Sand and gravel pits often have a high water table, indeed some are worked by dredging, and so become flooded when closed. Such workings frequently have been converted to marinas and wetlands that act as nature reserves. Restoration of an opencast area usually begins before the site is closed. Hence, worked out areas behind the excavation front are filled with rock waste. This means that the final contours can be designed with less movement of spoil than if the two operations were undertaken separately. In strip mining the hill and dale formation is regraded and then spread with subsoil and topsoil. The after uses of strip mined land are normally agriculture or forestry. Former spoil heaps tend to be regraded and may be returned to agriculture, used for recreational purposes or, in part, built over.

Derelict land caused by some earlier mining or associated industrial operation may contain structures or machinery of historical interest and worthy of preservation. Any assessment of the industrial archaeological value of a site must consider its scientific and engineering interest, its state of preservation and completeness and its representiveness and rarity. The amenity, recreation and tourism aspects also need to be considered. Some old mines have been turned into museums and some others now are tourist attractions such as the famous Big Hole at Kimberley, South Africa (Fig. 1.7).

Derelict mining sites often are contaminated and as such can give rise to polluted soils and water. Consequently, there is good reason to attempt to reclaim such sites. When hazardous substances are present in contaminated land, then this implies a degree of risk, but the degree of risk varies according to what is being risked. It depends, for example, upon the mobility of the contaminant(s) within the ground and different types of soils have different degrees of reactivity to compounds that are introduced. It also is influenced by the future use of a site. Consequently, the concept of risk analysis for contaminated land assessment is complex and needs a great deal of investigation and assessment of data before it can be used with effect. A fundamental objective of risk assessment and risk management is the need to define whether risks are real or are perceived. The nature of the relationship between the source of the contaminants, the pathway(s) and the receptor(s) determines the degree of risk. Adoption of a risk management framework allows proper characterization and evaluation of a site, selection of appropriate remedial strategies, effective reduction or control of defined risks, and thereby effective technical and financial control of a project. Risk assessment requires the acquisition of data by a site investigation

Figure 1.7 The Big Hole, Kimberley, South Africa. This old diamond working is now a popular tourist attraction.

so that hazards are identified and evaluated. The assessment should ensure that any unacceptable health or environmental risks are identified and dealt with appropriately.

Some countries make financial incentives available to encourage and assist the reclamation of derelict land, whether it is contaminated or not. For instance, the United States was the first country to establish a national fund for the reclamation of contaminated sites. Under the Comprehensive Environmental Response, Compensation and Liability Act (CERCLA) of 1980 a federally controlled fund, known as the Superfund, was established. A national priority listing of sites based on risk assessment criteria specified in CERCLA was drawn up. If an investigation of a former mining site that is contaminated is going to achieve its purpose, then its objectives must be defined and the level of data required determined, the investigation being designed to meet the specific needs of the reclamation project concerned. As in a normal site investigation, one that is involved with the exploration for contamination needs to determine the nature of the ground. In addition, it needs to assess the ability of the ground to transmit any contaminants either laterally or upward by capillary action. Sampling procedures are of particular importance and the value of the data obtained therefrom is related to how representative the samples are. A wide range of technologies are available for the remediation of contaminated sites and the applicability of a particular method depends on the site conditions, the type and extent of contamination be it soil or groundwater contamination, and the extent of the remediation required (Bell, 2004). In some instances, the remedial operation requires the use of more than one method.

1.5. Alternative uses of old mine workings

Old mine workings usually are regarded as a liability, but under favourable circumstances may actually represent an asset and provide commercially viable opportunities. The conversion of an abandoned mine for practical commercial purposes depends upon the geological and hydrogeological conditions, as well as the nature and geometry of the mining that took place. For instance, old pillared mines in limestone that are dry and have large stalls can be utilized for various purposes. Probably the most noteworthy example of the use of abandoned mines is that found in Kansas City, United States, where the workings in the Bethaney Falls and Argentine Limestones provide space for commercial operations. The more or less constant temperature in such mines often proves advantageous for storage facilities. The oil, gas and chemical industries, in particular, have used abandoned mines for economic storage for their products. In addition, use has been made of abandoned mines for the storage of waste, a good example being provided by the salt industry in Cheshire, England. Many countries have considered the use of abandoned mines for the long-term disposal of radioactive waste materials. At Hasse, near Hannover, Germany, approximately 25 000 m^3 of such low and intermediate level waste were deposited in caverns in former salt mines, which subsequently were backfilled with salt and sealed.

The following represent other possible uses of abandoned mine workings:

- *Mining and power plant waste storage.* Disposal of these types of waste has taken place in abandoned mine workings in Romania, Hungary and the Czech Republic, by backfilling and stowing. The materials used for backfill were tested and classified according to their safety for such use. The wastes included slags, ashes from coal fired power stations (i.e. fly-ash and bottom ash) and waste incinerators, foundry waste, sandblasting waste and construction waste.

- *Storage of consumables and food-stuffs.* In highly populated urban areas where surface space is in short supply and therefore valuable, the use of underground space helps conserve such surface areas. Old mines can be used to store a range of food-stuffs, deep frozen products and even drinking water. Yet another example is the use of abandoned mines as mushroom growing facilities, as illustrated by the Tennants Mine at Carrickfergus, Northern Ireland, and as nurseries as occurs in some old chalk mines in Kent, England.
- *Storage of documents.* Storage of paper documents may be possible in some mines due to the low humidity.
- *Secure storage.* Storage of sensitive and valuable documents, artwork or military data may be more secure in old mines than in surface structures. In particular, some abandoned mines in Europe were used for these purposes during the First and Second World Wars. The largest such storage facility in the United States is Iron Mountain, some 72 km north of Pittsburg. This is a huge former limestone mine, which was in operation from 1902 to 1950, supplying limestone to the steel industry of the area. It initially was used as an atomic shelter before being used to store government, commercial and industrial documents. The facility is dry because of the overlying shales, is located in a stable area and has a temperature of around 13°C.
- *Research.* Scientific research that requires protection from cosmic radiation and weather fluctuations may be suited to abandoned mine workings, where the temperature and humidity remain relatively constant. Oil well research involving sensors used in oil wells, is practised in some abandoned gypsum and anhydrite mines in the English Midlands, where the underground mine provides a shield from radiowave interference. It has been suggested that some medical research, such as research into bronchitis, may benefit from an atmosphere like that found in anhydrite mines. Research facilities for drilling, cutting and digging machines (i.e. excavating equipment), for testing rock stabilization methods such as rock bolts, for support systems in tunnelling and for ventilation equipment may take place in abandoned mines.
- *Museums.* Museums, and sites of industrial archaeological and historical value often are based in abandoned mine workings. In Britain, for example, mining museums exist for metalliferous deposits, coal and evaporitic rocks.
- *Sites of Special Scientific Interest (SSSI) and regionally important geological sites (RIGS).* Abandoned mines, in favourable circumstances, may be designated as sites of special scientific interest, for example, where rare ore minerals or unique geological conditions occur.
- *Storm water channels.* Storm water channels and water reservoirs in areas prone to flooding or at sites where the surface storage of water is difficult may be constructed in some abandoned mine workings that

are located near the coast, for storage and then subsequent discharge to the sea.

- *Munitions storage*. Pillar and stall mines have been used for the storage of munitions, an example being provided by the limestone mines of Wiltshire, England.

References

Agricola, G. 1556. *De Re Metallica*. Translated by Hoover, H.C. and Hoover, L.H. (1950), Dover Publications, New York.

Bell, F.G. 2004. *Engineering Geology and Construction*. E. & F.N. Spon, London.

Bell, F.G., Culshaw, M.G., Moorlock, B.S.P. and Cripps, J.C. 1992. Subsidence and ground movements in chalk. *Bulletin International Association of Engineering Geology*, **45**, 75–82.

Brashaw, P., Cripps, J.C., Czerewko, M.A., Elison, M. and Bradley, P. 2001. Limestone landform simulation in quarry restoration. *Quarry Management*, December, 41–46.

Foster, S.M. 1998. *Mining and Environmental Management*. Financial Times Energy, Financial Times Business Ltd, London.

Ricks, G. 1995. Closure considerations in environmental impact statements. *Minerals Industry*, Institution Mining and Metallurgy, Bulletin. No. 1022, January, 5–10.

Wood, S.C., Younger, P.L. and Robins, N.S. 1999. Long term changes in the quality of polluted mine water discharges from abandoned underground coal mines in Scotland. *Quarterly Journal Engineering Geology*, **28**, S101–S113.

Chapter 2

Subsidence due to the partial extraction of stratiform mineral deposits

2.1. Introduction

Stratified deposits such as coal, limestone, chalk, gypsum, salt, sedimentary iron ore, sandstone, slate etc. have been and are worked by partial extraction methods. In this method of mining only part of the deposit is removed the remainder being left in place to support the roof of the workings.

Coal mining has a long history. Taking Britain as an example, coal has been the most important mineral mined and in areas of the Productive Coal Measures if no record of mining activity is available, then it should not be assumed that no mining has taken place. Indeed, old shallow workings close to outcrop should be expected in any coalfield where exploitable beds are not covered by thick superficial deposits. Although mining has gone on in many British coalfields for several centuries, the first statutory obligation to keep mine records only dates from 1850 and it was not until 1872 that the production and retention of mine plans became compulsory for coal mines over a certain size. Because many of these old workings were unrecorded they can represent a potential hazard to those engaged in urban redevelopment, in particular they can give rise to ground subsidence. Even if records exist, they often are inaccurate. Matters are often complicated further by the fact that many old workings were subsequently built over. Old abandoned coal workings also occur at shallow depth beneath the surface of many urban areas, for example, in western Europe and North America (Fig. 2.1(a)). Indeed, as in Britain, the presence of coal was frequently one of the major reasons why urban development took place in the first instance.

Limestone has been mined for building stone, aggregate or lime. For example, limestone was mined extensively by the pillar and stall method around Kansas City in the United States (Fig. 2.1(b)). According to Hasan (1996), the mining began around 1875 and is still going on, although prior to 1950 mines were not designed for long-term stability. The entrances to such mines tend to be by adits driven into the hillsides so that the cover above most mines is shallow, ranging from 12 to 61 m. In the older mines the pillar shape varied and pillars were randomly spaced, and the extraction ratio (i.e. the amount of material removed) may have been 85% or higher. Furthermore, pillars often were robbed as a mine came towards the end of its working life. These factors have contributed towards subsidence. However, much of the underground space created by mining, especially that formed since 1950, is now used for warehouses, offices, light manufacturing and retail businesses. Fortunately, karstic features are essentially non-existent in the limestone that was or is worked. Rock bolting of the roofs and pillar reinforcement are used to help ensure the stability of the excavated areas that are used for commercial purposes. Indeed, these mines represent the

Figure 2.1a Old pillared workings in coal exposed at the Tinsley Park opencast site, Sheffield, England.

Figure 2.1b Old pillar working in the Bethaney Falls Limestone, greater Kansas City, United States.

Figure 2.1c Spalling and partial collapse of pillars in the Bath Stone, old workings near Corsham, Wiltshire, England.

Figure 2.1d A pillar from the workings above punching through the roof of the workings below, in strong limestone of Carboniferous age, Middleton-by-Wirksworth, Derbyshire, England.

largest area of utilization of underground space in the world. A number of case histories of subsidence in the greater Kansas City area have been outlined by Hasan.

In Britain the Bath Stone (Great Oolite, Jurassic) in Wiltshire and Somerset would appear to have been mined since Roman times for building stone. For instance, in the Box-Corsham district of Wiltshire there are some 96.6 km^2 of pillar and stall workings in the limestone. The extraction ratio varied from 65% to as high as 85% in places. Many of the mines are in a quasi-stable condition due to pillars spalling and subsidences have occurred (Fig. 2.1(c)). An example of subsidence associated with a working limestone mine took place at Middleton-by-Wirksworth, Derbyshire, where strong competent limestone of Carboniferous age is worked for lime, the extraction ratio being around 50%. In one part of the mine pillared workings were developed at two levels separated by a few metres of limestone, pillars from the upper workings punched through into those below, with subsidence at the surface being manifested in the form of tension gashes, crown holes or surface lowering (Fig. 2.1(d)). Limestone of Silurian age has been worked extensively in the West Midlands of England (see Section 2.8).

Historically sedimentary ironstone has been mined in several parts of England, for example, in Northamptonshire, Lincolnshire and the Cleveland Hills, over a period spanning several centuries. The mines were worked by the pillar and stall method. The precise locations of many of these workings are not known since the mines were unmapped. Left behind is a legacy of mining relics, which include abandoned shafts and underground workings at shallow depth. These forgotten mine workings occasionally have collapsed.

Subsidence due to the room and pillar mining of salt has been fairly well documented in parts of the Britain, notably the Cheshire Basin. However, other mined regions are less well known. For instance, the Permo-Triassic evaporite deposits around Stockton-on-Tees in north-eastern England and in Northern Ireland have resulted in occasional dramatic subsidence (see Section 2.9).

2.2. Resumé of past working methods

Taking coal as an illustration of past working methods, it has been mined in Britain from at least Roman times but mining did not begin to be carried out on a significant scale until the thirteenth century. Drifts and adits into shallow workings usually were situated at the base of quarries and open-pits or along the coal outcrops in hilly country. The workings extended as far as natural drainage and ventilation permitted.

By the fourteenth century such outcrop workings had largely given way to bell pits (Fig. 2.2(a)). The shafts of bell pits rarely exceeded 12.2 m in depth and their diameter was usually 1.3 m. Because bell pits were very shallow workings, they tended to be concentrated near coal outcrops where the coal was more or less horizontal or had a small amount of dip. Indeed, the outcrop of a seam in rural areas may be delineated by a series of bell pits and associated mounds of spoil. Obviously, bell pitting is a feature of coalfield areas where the superficial cover is thin. Extraction was carried on around the shaft until such times as roof support became impossible, then another shaft was sunk nearby. Hence, where such mining went on, the number of bell pits may be very numerous (Fig. 2.2(b)). If bell pits were backfilled, then the state of compaction of the fill is generally unsatisfactory.

It is generally impossible to predict the distribution or amount of void space in the ground associated with the old bell pits. This can present problems in terms of development or redevelopment of a site. For instance, at a site in Sheffield, England, where the presence of old workings had been proved by drilling, the distribution of voids was irregular and no estimate could be made of the amount of void space. As a consequence, a grouting contractor was only prepared to quote a minimum figure for filling the voids. In fact, this amounted to 28% of the contract price. The client decided that as the amount of void space that required filling was unknown he could be signing an 'open cheque' if he sanctioned such ground treatment. He therefore abandoned development of the site.

When a coal seam occurred at more than about 7 m below the surface, bell pit mining tended to be replaced by headings that radiated into the coal seam for short distances around the shaft. The pillars of coal between the headings generally represented the only type of support to the overlying strata. The layout of a mine was unplanned and simply consisted of a complex of interconnected headings. Hence, the support pillars were irregular in shape and size.

Because of the scarcity of timber, the demand for coal increased in the sixteenth century and led to the development of the pillar and stall method of extraction. Underground workings were shallow and not extensive, for example, they rarely penetrated more than 40 m from the shaft. Indeed, when such limits were reached, it was usually less costly to abandon a pit and sink another shaft nearby. Workings extending 200 m from the shaft were exceptional even at the end of the seventeenth century, the shaft itself usually being less than 60 m deep (Bell, 1988a). In very early mining the remnant pillars and voids (variously referred to as stalls, rooms, bords) usually differed in size and arrangement, but with time mining became

Figure 2.2a A bell pit working in an ironstone band revealed by opencasting at Sproats site, Northumberland, England.

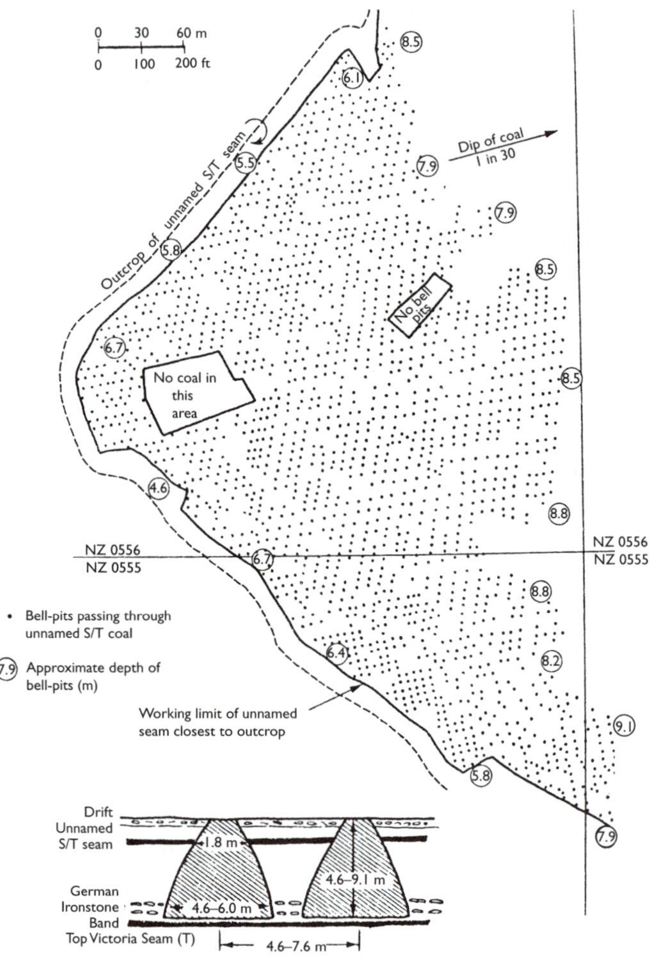

Figure 2.2b A large number of bell pits revealed in close proximity at Sproats opencast site, Northumberland, England.

more systematic and pillars of more or less uniform shape were formed by driving intersecting roadways in the seam. Also, there was a general tendency for the size of stalls to increase. Wardell and Wood (1965) noted that in the nineteenth century the normal width of stalls varied from 1.83 to 4.57 m, the extraction ratio varying from 30% to 70%. Several variations of the pillar and stall method were devised, for example, Staffordshire squarework was developed to work the Ten Yards Coal seam and in the Sheffield area ribs of coal generally were left to support the roof rather than

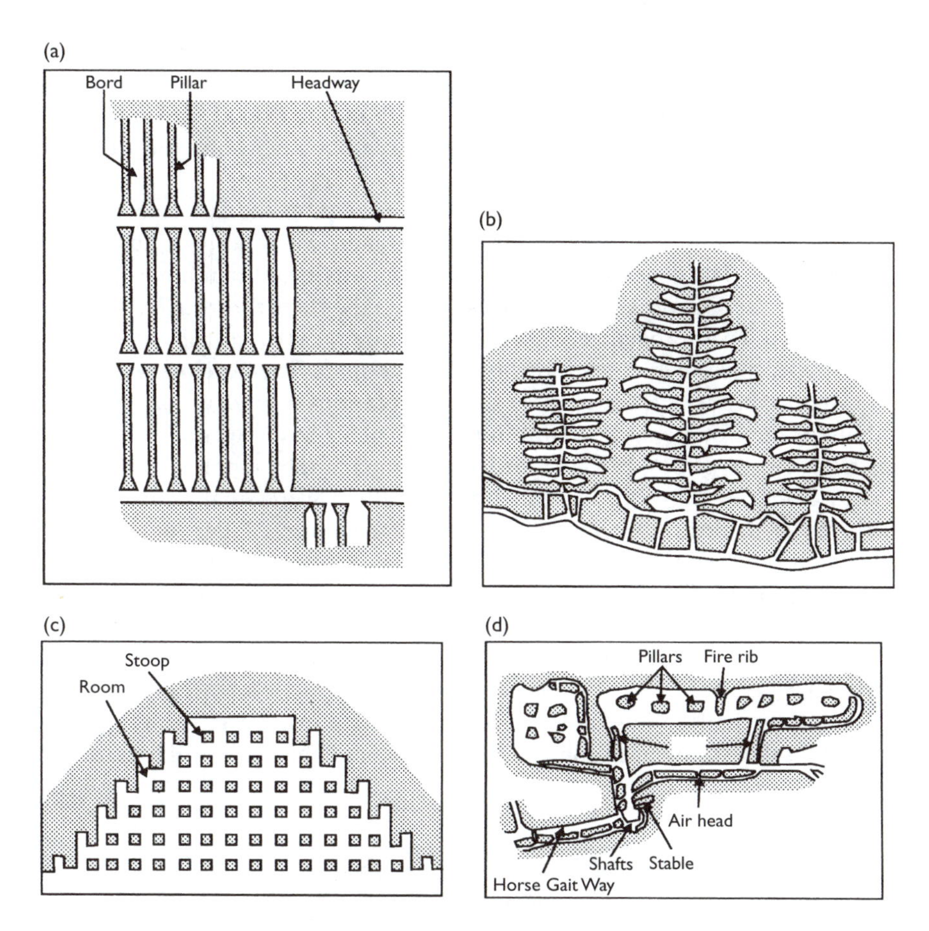

Figure 2.3 Different configurations of pillared workings. (a) Bord and pillar workings, Newcastle upon Tyne, seventeenth century. (b) Post and stall workings, South Wales, seventeenth century. (c) Stoop and room workings, Scotland, seventeenth century. (d) Staffordshire squarework, developed for conditions involving spontaneous combustion, airtight stoppings could be placed in narrow access ways to prevent the spread of fire.

pillars (Fig. 2.3). In present day workings the pillar support system can be designed for long-term stability if it is important to protect the surface or to minimize damage to the roof, as the miner can control the pillar dimensions and the percentage extraction (Orchard, 1964). For example, partial extraction by leaving pillars equivalent to approximately 0.25 depth of the workings has been used successfully. This enables significant areas of coal to be extracted in part (usually around 50% being removed) that otherwise would be left to protect the surface against subsidence. A similar method has been used in the Pittsburg area of the Pennsylvania Coalfield (Gray and Meyers, 1970). There an angle of 15° projected from the vertical outside a safety area, 5 m in width, around the perimeter of a foundation has been used to define the zone of support of a building (Fig. 2.4). Up to 50% of the coal in the support zone can be removed without any adverse subsidence effects occurring at the surface.

The expansion of mining, especially coal mining, in the latter part of the eighteenth and the nineteenth centuries in Britain was due to a number of developments. The development of the steam pump meant that mine drainage became more effective. Ventilation was improved with the introduction of the Buddle Fan. These, together with the wire rope meant that mines could be sunk deeper and could extend further from the shaft. The development of the safety lamp meant that the number of gas explosions was reduced.

Similar developments in coal mining took place in the United States. For instance, Bruhn *et al.* (1981) recorded that during the eighteenth and nineteenth centuries mines in the Pittsburg Coalfield were small and shallow, with rather arbitrarily spaced pillars of differing sizes. Pillar arrangement became more systematic during the second half of the nineteenth century. The stalls were 1.5 to 1.8 m wide and alternated with pillars of equal width so that the extraction ratio did not exceed 50%. In the 1890s the width of the stalls increased to between 6 and 7.3 m and pillar dimensions to between 3 and 5.4 m.

Mining of chalk in England also has a very long history and consequently has seen a development in mining methodologies. The chalk in England has

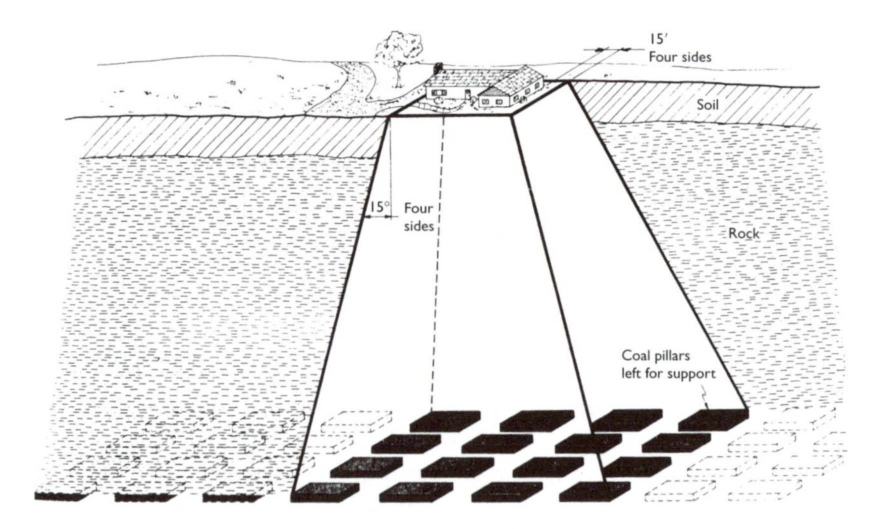

Figure 2.4 Pillars of coal left in place to prevent surface subsidence. (After Gray and Meyers, 1970; reproduced by kind permission of the American Society of Civil Engineers.)

been, and still is, extensively worked for lime. In addition to being dug for lime, the chalk also has been worked as a source of flint. In the past many lime workings took the form of surface quarries and pits often associated with subsurface mining. These mine workings were located at shallow depths and, because they lay above the water table, were dry. Subsurface workings took the form of cavities, galleries and chambers and relied on the strength of the chalk to form self-supporting roofs. Old mine workings are entered by shafts, drifts or adits that run directly from the surface. Each mine could have more than one entrance. Most of these workings are unrecorded and can represent a hazard to surface development, as well as to existing property, and indeed lives have been lost owing to roof collapse into the workings (Fig. 2.5). The timing of collapse is generally difficult or impossible to predict. A review of the geological conditions that may aid the collapse of mine workings in chalk has been provided by Smith and Rosenbaum (1993).

Old mine workings in the chalk in south-east England are referred to as deneholes. Some of the deneholes have ancient origins and are believed

Figure 2.5 A double-decker bus caught in a crown hole that opened in chalk, Norwich, England. (Reproduced by kind permission of the British Geological Survey. © NERC. All rights reserved. IPR/61–05C.)

to have been excavated during pre-Roman times. Others are presumed to be of Roman origin whilst the remainder could have been excavated as late as the fourteenth century (Bell *et al.*, 1992). Commonly, a circular shaft, 10 to 20 m or so in depth, leads to two or more chambers. The most commonly occurring form is the six chambered or double trefoil type of denehole (Fig. 2.6). Subsequent improvement in mining practice led to the development of the pillared denehole in which the walls between the chambers were partially removed to create pillars. Other forms of workings in chalk include chalkwells, bell pits, chalkangles and pillared workings (Edmonds *et al.*, 1987). Pillared workings in chalk are the most recent and generally date from the beginning of the nineteenth century. The pattern of the workings varies, in particular the size of the pillars. The joint pattern and the presence of any solution pipes tended to influence the layout of the mines. The stalls were typically 2 to 3 m wide and about 9 m in height and were worked by benching.

An additional factor that must be considered when dealing with subsidence of workings in chalk is that it subject to dissolution. In fact, many chalk subsidence incidents have been man-induced where flows of water

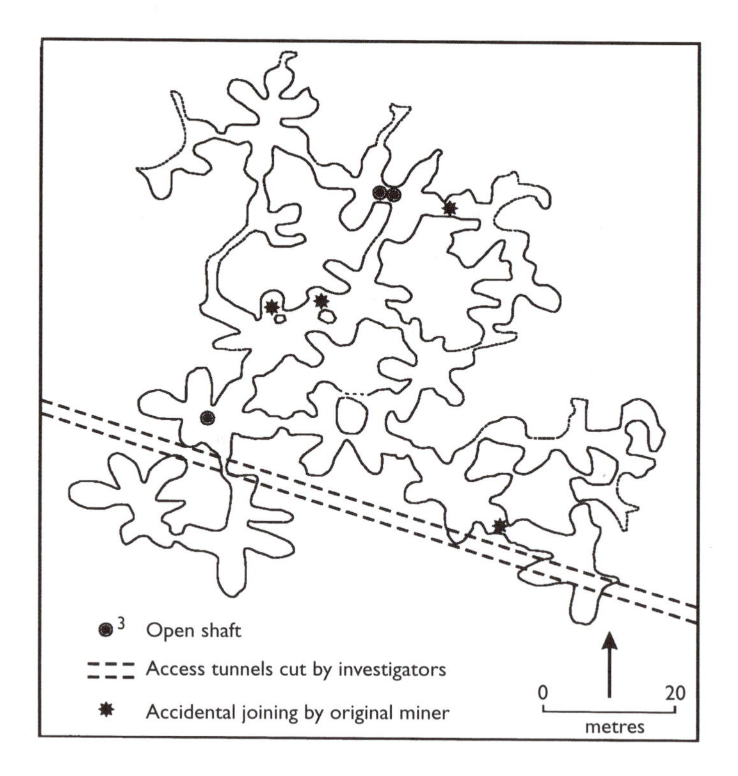

Figure 2.6 Linked deneholes in chalk in Essex, England. (From Bell *et al.*, 1992.)

Figure 2.7 Collapse of shallow workings in chalk at Jacqueline Close, Bury St Edmunds, Suffolk, England, the Close ultimately was abandoned.

into the ground have been concentrated into small areas by soakaways, leaking drains and water supply pipes (Fig. 2.7). Such concentrated flows accelerate dissolution, as well as reducing the strength of chalk. Urbanization can make a significant contribution to causing subsidence in that it radically changes the way in which water enters the ground compared with an open undeveloped site by rendering 50–80% of the surface plan area relatively impermeable.

2.3. Pillar and stall workings

In pillared workings the pillars sustain the redistributed weight of the overburden, which means that they and the rocks immediately above and below are subjected to added compression. From Fig. 2.8 it can be seen that a is the pillar width whilst b is the width of the stall. The extraction ratio, r, that is, the percentage amount of mineral removed, therefore can be derived from the following expression:

$$\text{extraction ratio } (r) = \frac{2ab + b^2}{(a + b)^2} \tag{2.1}$$

Stress concentrations tend to be located at the edges of pillars and the intervening roof rocks tend to sag. The effects at ground level normally are insignificant. Although stresses acting on the corners and sides of pillars can give rise to fracturing, at depths of less than 50 m the stresses are usually

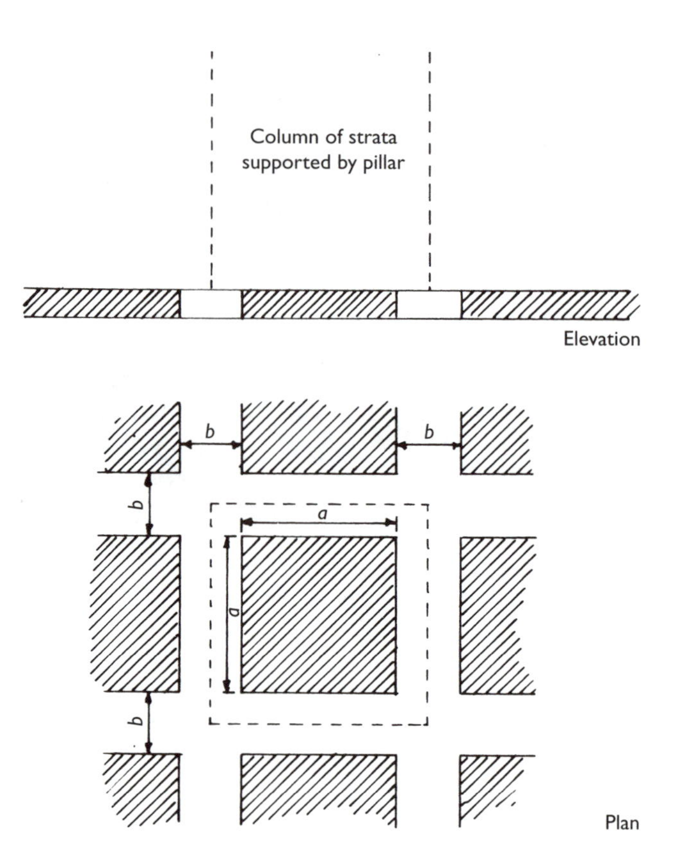

Figure 2.8 An illustration of extraction ratio and pillar loading.

too small to cause significant fracture development. Conversely, at depths greater than 120 m, the theoretical stress values commonly exceed the unconfined compressive strength as far as coal is concerned. Although the intrinsic strength of a stratified deposit varies, the important factor in the case of pillars is that their ultimate behaviour is a function of the height of the workings to pillar width, the depth below ground and the size of the extraction area. For example, Madden and Hardman (1992) showed that 60% of the pillar collapses in coal mines in South Africa had a pillar width to mine height ratio of less than 2 and that no pillar failure was recorded with a pillar width to height ratio above 4. They also showed that 65% of the failures occurred at less than 70 m depth with 35% occurring at less than 40 m depth, and that 65% of the failures occurred where extraction exceeded 75%. Previously, MacCourt *et al.* (1986) had noted that subsidence as a proportion of mined height decreased from around 0.8 at very shallow depths to between 0.1 and 0.5 at depths exceeding 100 m. The mode of failure also involves the character of the roof and floor rocks. Pillars in the centre of a mined out area are subjected to greater stress than those at the periphery. Individual pillars in dipping seams tend to be less stable than those in horizontal seams since the overburden produces a shear force on the pillar. In addition, the greatest stress occurs at the edges of pillars, between pillar and roof or between pillar and floor.

Because of roof failure between the pillars, their height may increase over time and hence their structural stability is reduced. Fallen roof material can provide confinement to pillars, as well as surcharging weak floor materials that otherwise may heave.

Pillars often experience local failures whilst mining is taking place (Fig. 2.1(b)). If a pillar is highly jointed, then its margin may fail and fall away under relatively low stress. Such action reduces and ultimately removes the constraint from the core thereby subjecting it to increasing stress. This can lead to pillar failure. Collapse in one pillar can bring about collapse in others in a sort of chain reaction because increasing loads are placed on those remaining. The effective pillar width can be reduced by blast damage or local geological weaknesses. Old pillars at shallow depth have occasionally failed near faults and they may fail if they are subjected to the effects of subsequent mining, notably longwall mining of coal. Furthermore, the collapse of roof strata around pillars alters their geometry and so can affect their stability. Yielding of a large number of pillars can bring about a shallow broad subsidence over a large surface area that Marino and Gamble (1986) referred to as sag (Fig. 2.9). The ground surface in a sag displaces radially inwards towards the area of maximum subsidence. This inward radial movement generates tangential compressive strain and circumferential tension fractures frequently develope. Sag movements depend on the mine layout, in particular the extraction ratio,

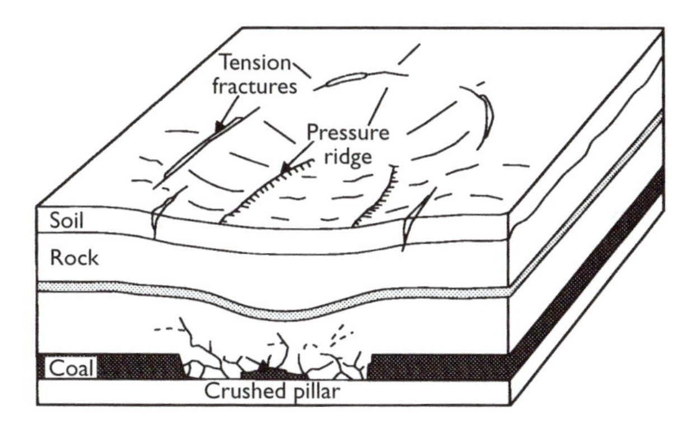

Figure 2.9 The development of a sag due to the multiple failure of pillars in the workings.

and geology, as well as the topographic conditions at the surface (Speck and Bruhn, 1995). They tend to develop rather suddenly, the major initial movements lasting, in some instances, for about a week, with subsequent displacements occurring over varying periods of time. The initial movements can produce a relatively steep-sided bowl-shaped area. Nonetheless, the shape of a sag profile can vary appreciably and because it varies with mine layout and geological conditions, it can be difficult to predict accurately. Normally, the greater the maximum subsidence, the greater is the likelihood of variation in the profile. Maximum profile slopes and curvatures frequently increase with increasing subsidence. The magnitude of surface tensile and compressive strains can range from slight to severe.

Carter *et al.* (1981) described subsidence in Bathgate near Edinburgh, Scotland, which was attributable to failure in old workings in coal with a high extraction ratio, located at shallow depth. Thirty per cent of the subsidence occurred immediately and 50% had occurred after the first two weeks. After one year, probably 95% of the subsidence had occurred. The subsidence took the form of a broad trough affecting over 7500 m^2. The maximum subsidence exceeded 0.3 m. This collapse of pillars threw extra loading onto the up-dip pillars, which eventually failed about two years later. The local newspaper reported that the initial ground movement was like an earthquake, buildings were cracked and window and door frames very severely distorted. Eventually, some buildings had to be demolished.

The most notable example of pillar failure occurred in a working coal mine at Coalbrook, South Africa, where collapse occurred in a matter of minutes over a whole pillar district, involving an area of 300 ha. The workings occurred in a seam approximately 7.5 m thick at an average depth of around 140 m. In this case the ground surface suffered substantial rapid subsidence, severe local strain and tilt damage, and a fracture zone developed. Over 400 miners lost their lives. The associated seismic activity produced shock waves that were recorded over 720 km from Coalbrook. Bryan *et al.* (1964) examined a number of possible causes and concluded that the particular geological conditions had a significant influence on the collapse, especially the occurrence of a dolerite sill in the roof immediately above the seam. This tended to concentrate the stresses on the pillars.

Slow deterioration and failure of pillars may take place years after mining operations have ceased, although observations at shallow depth in coal mines and the resistance of coal to weathering in pillared workings suggests that this is a relatively uncommon feature at depths of less than 30 m. However, old workings affect the pattern of groundwater drainage, which in turn may influence pillar deterioration. On the other hand, small pillars in coal mines of earlier workings may be crushed out once the overburden exceeds 50 or 60 m (Piggott and Eynon, 1978).

Very frequently, when a mine neared the end of its life, the pillars were robbed (Thorburn and Reid, 1978). Extraction of pillars during the retreat phase simulates the longwall surface condition although it can never be assumed that all pillars have been removed. As noted earlier, at moderate depths pillars, particularly pillar remnants, are probably crushed and the goaf (i.e. the worked out area) compacted, but at shallow depths lower crushing pressures may mean the closure is variable. This is likely to cause foundation problems when large or sensitive structures are to be erected above. Bullock and Bell (1997) noted that subsidence at a mine in the Witbank Coalfield, South Africa, had coincided with the onset of pillar robbing. This meant that the extraction ratio increased from some 60% to 85% or more, thereby increasing the load that the pillars had to carry by as much as one-third. Robbing also changed the shape of pillars by reducing their width significantly while the pillar height remained constant. The result was a reduction in the pillar carrying capacity to the extent that many pillars could no longer support the overburden stress, collapsing to produce subsidence of the ground surface. The surface subsidences, in this instance caused by multiple pillar failure, are generally several hundred square metres in extent and the collapsed areas are often bounded by near vertical sides (Fig. 2.10(a)). Surface tension cracks occur around the outer edges of the collapsed areas, and are typically 200–800 mm in width and can extend in length for up to 100 m (Fig. 2.10(b)).

A more interesting example of subsidence associated with the reduction of pillar size is provided by the failure at the salt mine at Retsof in Livingston county, New York. The mine had operated for 109 years by extracting approximately 3.7 m of salt with an extraction ratio of approximately 50–60% from a gently dipping layer within the Vernon Formation.

Figure 2.10a Subsidence at the surface caused by pillar failure in shallow abandoned workings, Witbank Coalfield, South Africa.

Figure 2.10b A tension crack running from an area that has subsided as a result of pillar failure, Witbank Coalfield, South Africa.

However, in 1993 the yield pillar panel method was introduced in certain selected areas in the mine, supposedly so that the roof could settle and thereby become more stable. This method uses pillars that are small enough so that the load is transferred onto larger abutment pillars that are relatively stiff and unyielding (Gowan and Trader, 2000). Unfortunately, in March 1994 an area of the roof, some 150 m by 150 m, collapsed allowing brine, which subsequently gave way to fresh groundwater, and methane to flow into the mine. It also has been suggested that the brine and gas, which occurred in a pocket above the roof were under pressure, this representing a further reason for the failure. This failure was subsequently followed by another collapse leading ultimately to the mine being flooded in December 1995. The initial inflow of was around 400 $l\ s^{-1}$ rising to approximately 1500 $l\ s^{-1}$ after the second collapse (Yager *et al.*, 2001). Land subsidence caused by the collapse led to the appearance of two large sinkholes (213 m and 274 m in diameter) at the surface, the failure of a bridge on State Route 20A, damage to property and a decline in the water levels of several domestic wells in the district. In addition, the decline in groundwater level in the basal aquifer above the mine allowed hydrogen sulphide and methane to escape from the depressurized groundwater. These gases appeared in some wells and in some cases had to be flared off.

In coal mining the floor of an extraction area frequently heaves, this being especially characteristic of areas where argillaceous rocks, notably fireclays, form the floor. Floor heave can occur at shallow depth, for example, old roadways at less than 20 m depth have at times been completely closed. Obviously, significant floor heave places a constraint on pillars and reduces the void volume thus pillar spalling and void migration (see later) are reduced.

Squeezes or crushes sometimes occur in a coal mine as a result of the pillars being punched into either the roof or floor beds. In such cases the roof or floor may have been weakened or altered by the action of groundwater or weathering (Piggott and Eynon, 1978). Once again any resulting surface subsidence adopts a trough-like or basin form, and minor strain and tilt problems occur around the periphery of the basin thereby produced (Bruhn *et al.*, 1981). Briggs (1929) reported that when pillars are forced into a yielding pavement subsidence may restart. This could take place many years after mining had ceased and could damage property. He quoted an example in Wallsend, England, where the surface subsidence amounted to 1.2 m. Such action becomes more significant with increasing depth or high extraction ratio since these give rise to greater loads upon the pillars. Sizer and Gill (2000) documented a case of subsidence of terraced houses at Bedlington, Northumberland, England, where it is possible that pillars punched into the weak roof and floor rocks. These consisted of fireclay and mudstone respectively. Progressive subsidence at the surface led to the development of a subsidence trough in which the average subsidence of the ground surface amounted to 250 mm. Mining records indicated that the workings were at a depth of 44 m and drillhole evidence showed that some of the pillars were much smaller than indicated on the mine plans.

Although much subsidence is often slow it may be quickened by the ingress of water into the workings. In addition, Marino and Choi (1999) reported that a major cause of subsidence associated with coal mines in the Illinois Coalfield is softening of mine floors. In particular, softening of argillaceous material once coal has been removed can give rise to bearing capacity failure of pillars. Softening is brought about by slaking or swelling on the one hand or creep or strain softening due to sustained loading on the other. The time taken for such pillar failure and associated subsidence to occur varies.

The potential for pillar failure should not be ignored, particular warning signs being strong roofs and floors allied with high extraction ratios and moderately to steeply dipping seams. Prediction of subsidence as a result of pillar failure requires accurate data regarding the layout of a mine. Such information frequently does not exist in the case of abandoned mines. On the other hand, when accurate mine plans are available or in the case of a working mine, the method outlined by Goodman *et al.* (1980) may be used to evaluate collapse potential. Basically, the method involves calculating the vertical stress based upon the tributary area load concept that assumes that each pillar supports a column of rock with an area bounded by room centres and a height equal to the depth from the surface, and comparing this with the strength of the pillar. Previously, Wardell and Wood (1965) had suggested more or less the same, that is, that determination of the loading on pillars was best approximated by averaging the load on a given pillar due to the weight of overburden. The latter is equal to the weight of the column of strata over an area equal to $(a + b)^2$ (see Fig. 2.8). It is assumed that the load acts vertically and is distributed uniformly over the cross sectional area of the pillars. It can be shown that the average loading, P, on a pillar is equal to:

$$P = \frac{\gamma z}{1 - r} \qquad (2.2)$$

where γ is the unit weight, z is the depth and r is the extraction ratio. When a structure is to be built over an area of old pillared workings the additional load on the pillars can be estimated simply by adding the weight of the appropriate part of the structure to the weight of the column of strata supported by a given pillar. This method is very conservative except when used for large concentrated loads where old workings are located at shallow depth. Bell (1992) provided a number of other means of calculating the strength of or stresses on pillars (Table 2.1). More recently, Sheorey *et al.* (2000), after a survey of subsidence at some mines in India, suggested that a modification of the influence function method (see Chapter 3) could be used to predict the differential subsidence associated with pillar and stall mines located at shallow depth.

Even if pillars are relatively stable the strata above a worked out area (i.e. a room or stall) may be affected by void migration (Fig. 2.11). This can take place within a few months of or a very long period of years after mining and in Britain is a much more serious problem than pillar collapse. Void migration develops when roof rock falls into the worked out areas. When this occurs the material involved in the fall bulks, which means that migration is eventually arrested, alternatively upward migration may be halted by a strong competent bed of rock that acts as a beam spanning the void. Nevertheless, the process can, at shallow depth, continue upwards to the ground surface leading to the sudden appearance of a crown hole (Fig. 2.12). The geometry of crown holes is influenced by the nature and thickness of the overlying strata, the state of stress existing in the roof rocks and the height of the extraction but generally the surface shape of crown holes is circular or elliptical. Singh and Dhar (1997) suggested that the maximum diameter of crown holes in areas of shallow coal mining in India

Table 2.1 Some expressions for determining the strength of and the stress on coal pillars (After Bell, 1992a)

Author(s)	*Expression*	*Remarks*
1 Greenwald *et al.* (1941)	Pillar strength $(\tau_\rho) = 2800 \times \frac{w^{0.5}}{h^{0.85}}$ psi w = width of pillar, in inches h = height of seam, in inches	Based on the results of *in situ* tests on square pillars in Pittsburg bed. Pillars varied in width from 1 to 5.3 ft and height from 1.4 to 5.3 ft
2 Terzaghi (1943)	$\sigma_v = \frac{\rho^{0.5B}}{\kappa \tan \phi}(1 - \exp^{-\kappa \tan \sigma/B}) + q \exp^{-\kappa \tan z/B}$ σ_v = vertical pillar loading pressure κ = constant representing ratio between horizontal and vertical pressure B = breadth of room ρ = density or unit weight of overburden z = depth ϕ = angle of internal friction q = uniform surcharge carried by soil per unit area	This expression was developed in relation to theories of arching and more particularly to the vertical load above a tunnel. It has been suggested that it could be used to calculate the total vertical load, including any proposed surface structure, above old mine workings
3 Denkaus (1962)	$\sigma_v = wz \frac{1}{1 - r}$ r = extraction ratio $= \frac{2ab + b^2}{(a + b)^2}$ (see Fig. 2.8)	This expression is based on two assumptions, namely, that each pillar supports the column of rock over an area that is the sum of the cross sectional area of the pillar plus a part of the room area; and that the load is vertical only and uniformly distributed across the cross sectional area. Steart (1954) has pointed out, however, that the pressure is not evenly distributed, it frequently assuming a parabolic distribution with maximum pressure exerted on the central pillars of the mined out area. He further stated that the pressure gradient gradually increased with increasing area of development to a maximum given by Equation 3, and that this maximum was reached when the development, if roughly circular, attained a radius equal to the depth, divided by the ratio $(a + b)^2/a^2$ (Fig. 2.8)

(*Continued*)

Table 2.1 Continued

Author(s)	*Expression*	*Remarks*
4 Coates (1965)	$\sigma_v = \dfrac{1.1z}{1 - \dfrac{r}{100}}$ psi r = extraction ratio	According to Coates this expression gives valid results where the depth of the workings is less than half their extent, as is the case with most old mine workings
5 Wardell and Wood (1965)	$\sigma_v = \dfrac{z}{1 - r}$ psi	When a structure is to be built over old pillared workings the additional weight on the pillars can be estimated simply by adding the weight of the appropriate part of the structure to the over-burden pressure supported by the pillar. The total load on a pillar determined in this manner is probably greater than the true load. Moreover, the distribution of load over the area of the pillar is not uniform for stress is concentrated at the pillar edges. The transfer of the weight of a surface structure to residual pillars is normally calculated on the assumption that the additional weight acts vertically downwards and that there is no lateral spreading of the load. This is based on the fact that void migration causes dilation above the seam as well as joint opening so reducing or preventing such lateral distribution
6 Salamon and Munro (1967) and Salamon (1967)	(a) $\tau_p = 1320 \dfrac{w^{0.46}}{h^{0.66}}$ psi (dimensions in ft) (b) $\sigma_v = 1.1z \left(\dfrac{w + B}{w}\right)^2$ psi (c) factor of safety (FS) against pillar failure $FS = \dfrac{\text{strength of pillar}}{\text{pillar loading pressure}}$	These expressions are based upon a statistical analysis of a survey of pillar dimensions in both stable (98 cases) and collapsed (27 cases) mining areas in South Africa

Summary of data	*Stable*	*Collapsed*
1 Depth (m)	20–220	21–192
2 Room height (m)	1.2–4.9	1.5–5.5
3 Pillar width (m)	2.7–21.3	3.4–15.9
4 Extraction ratio	37–89	49–91
5 Pillar width:height ratio	1.2–8.8	0.9–3.6

		Salamon and Munro concluded that 99% of collapses occurred at safety factors lower than 1.48. Holland (1964), however, had previously suggested that a factor of 1.8 was generally necessary and that in critically important areas it was 2.2
7 Bieniawski (1967, 1970)	$\tau_p = 400 + 200 \frac{w}{h}$ psi	Based on tests carried out on coal pillars in South Africa, the pillars varying in height and width from 2 to 6.7 ft
8 Wilson (1972)	(a) Wide pillars, $w > 0.003$ mz ft (i) Square pillars $\sigma_v = 4\rho z(w^2 - 3wmz \times 10^{-3} + 3m^2z^2 \times 10^{-5})$ tons (ii) Rectangular pillars $\sigma_v = 4\rho z\ (wl - 1.5(w + l mz \times 10^{-3} + 3\ m^2z^2 \times 10^{-6})$ tons (iii) Long pillars $\sigma_v = 4\rho z(w - 1.5\ mz \times 10^{-3})$ tons/ft run (b) Narrow pillars, $w \leqslant 0.003$ mz ft (i) Square pillars $\sigma_v = 444\rho \frac{w^3}{m}$ tons (ii) Rectangular pillars $\sigma_v = 1333\rho \frac{w^2}{m}\left(1 - \frac{w + l}{2} + \frac{w}{3}\right)$ tons (iii) Long pillars $\sigma_v = 667\rho \frac{w^2}{m}$ tons/ft run where σ_v = load which pillar will carry, in tons ρ = average density of rock in tons/ft^3 l = length of pillar in ft m = height of roadway in ft w = width of pillar in ft z = depth of overburden in ft	Wilson suggested that a pillar was surrounded by an outer yield zone in which the stress distribution varied in linear fashion from zero at the surface to the point of failure at the pillar core. The constraint given to coal in the pillar core can increase its strength appreciably. The six expressions take no account of the strength of the coal and it has been suggested that although they may be admissible for deep workings they could give underestimates of strength of pillars at shallow depths

(*Continued*)

Table 2.1 Continued

Author(s)	*Expression*	*Remarks*
9 Goodman *et al.* (1980)	(a) $\tau_\rho = \frac{\text{UCS} \times \text{N-shape} \times \text{N-size}}{\text{FS}}$ UCS = unconfined compressive strength N-shape = $(0.875 \pm 0.250\, w/h_p)$ h_p = height of pillar FS = factor of safety (b) $\sigma_v = \frac{\sigma_{vi} \times A_t}{A_p - A_w}$ σ_{vi} = initial vertical stress at roof level A_t = area tributary to each pillar A_p = cross sectional area of pillar A_w = area of pillar lost from load carrying (c) FS $= \tau_p/\sigma_v$	Allows zones of minimum stability in abandoned pillar and stall workings to be located. Locations and dimensions of pillars plotted on mine plan. Factor of safety for each pillar determined and plotted on plan. Pillars which have factors of safety of less than one are considered potentially unstable and removed from plan. Their tributary areas are reassigned to adjacent pillars and the calculation repeated. In second calculation more pillars may fail. They are removed for further iteration of the calculation, if so required. Although method requires knowledge of exact shape of pillars, as well as strength of rocks involved, it offers practical approach to recognizing where the most potentially unstable areas exist so that appropriate measures may be taken optimally
10 MacCourt *et al.* (1986)	(a) $\tau_p = 7176\, w^{0.459}\, h^{-0.66}$ kPa (b) $\sigma_v = \frac{\rho q z_f c^2}{w^2}$ ρ = density of overburden g = acceleration due to gravity z_f = depth to floor of workings c = pillar centre distance (c) FS $= \tau_p/\sigma_v$	MacCourt *et al.* (1986) extended the work of Salamon and Munro (1967). Like Salamon and Munro, MacCourt *et al.* adopted tributary area theory, assuming that each pillar carried the mass of superincumbent strata immediately above it. This theory is valid when pillars have reasonably uniform geometry and are mined over an area where the width of mining exceeds the depth of the seam. After a survey of recent pillar collapses in South Africa collieries, MacCourt *et al.* concluded that the majority occurred when the depth was less than 60 m, the pillar width was less than 6 m, the pillar width to height ratio was less than 1.75 and the extraction ratio exceeded 70%. They found that subsidence as a proportion of mined height decreased from around 0.8 at very shallow depths to between 0.1 and 0.5 at depths exceeding 100 m

Figure 2.11 Void migration above a coal seam exposed by opencast workings in Staffordshire, England.

Figure 2.12 A crown hole formed at the surface due to void migration from shallow workings in coal at Springs, South Africa.

usually was less than 10 m. Site investigations frequently reveal partially choked voids in abandoned pillar and stall workings. For example, the remnant voids may be less than a metre in height and the bulked material may have suffered some amount of breakdown (Thorburn and Reid, 1978).

The factors that influence whether or not void migration will take place include the width of the unsupported span; the height of the workings; the nature of the cover rocks particularly their shear strength, and the incidence and geometry of discontinuities; the thickness and dip of the seam; the depth of overburden; and the groundwater regime. According to Wardell and Eynon (1968) the maximum width of a self-supporting span, S, can be derived from the following expression:

$$S = \sqrt{2t\sigma_t} \tag{2.3}$$

where t is the thickness of the seam and σ_t the ultimate tensile strength of the roof rock beam (this should make due allowance for discontinuities). Subsequently, Palchik (2000) outlined a statistical method of predicting the presence of voids in abandoned shallow underground workings in coal, which is based upon the thickness and type of the beds in the mass of rock above the workings and the strength of the immediate roof rock.

Obviously, when rock is removed by mining, the strata immediately above is distressed, the void remaining stable as long as the stresses do not exceed the strength of the roof rock. However, roof stability over a period of time may be compromised by changes in the strength of the roof rock and in the stress. Such changes may be brought about by changes in the groundwater regime or by creep deformation. Singh and Dhar (1997), for instance, reported that crown holes started to appear at Niga Mahabir Colliery in India after dewatering of shallow abandoned workings, which removed hydrostatic support.

High compressive stresses acting in the vicinity of the pillars can cause failure of roof strata. Stresses tend to develop at an angle to the maximum compressive stress. In incompetent roof strata this tends to be at a high angle (e.g. 75°). These stress fractures in the mine in the Witbank Coalfield, South Africa, referred to by Bullock and Bell (1997), combined with inter-pillar tensional areas, caused collapse, resulting in upward void migration through overlying strata. Crown holes that appeared at the surface have diameters between 5 and 10 m (Fig. 2.13). Tension cracks are found around the perimeters of many of the crown holes.

Garrard and Taylor (1988) summarized four methods that have been used to predict the collapse of roof strata above stalls. These are clamped beam analysis that considers the tensile strength of the immediate roof rocks (Wardell and Eynon, 1968; Hoek and Brown, 1980); bulking equations that consider the maximum height of collapse before a void is choked (Tincelin, 1958; Price *et al.*, 1969; Piggott and Eynon, 1978); arching theories that estimate the height to which a collapse will occur before a stable arch develops (Terzaghi, 1946; Szechy, 1970); and coefficients based on experience and field observations that act as multipliers of either seam thickness or span width (Walton and Cobb, 1984). For example, the

Figure 2.13 A succession of crown holes in the Witbank Coalfield, South Africa.

Tincelin expression, which was developed in relation to ironstone mines in France, has been used to determine the height to which a void can migrate above the roof of a working, the expression being:

$$D_c = t\left[\frac{\rho_1/\rho}{1 - (\rho_1/\rho)}\right] \tag{2.4}$$

where t is the thickness of the seam (or height of the workings if not the same), ρ is the bulk density of the roof rocks, ρ_1 is the bulk density of collapsed roof materials and D_c is the height of migration. However, if as Piggott and Eynon (1978) conceived, the void can adopt various geometrical forms such as due to conical, wedge and rectangular collapses, then different expressions must be used to calculate void migration. They showed that for a particular width of mine opening, B, then the height of collapse or migration, D_c, is a function of the original height of the mine opening, h, and the bulking factor, b_f, of the overlying strata. They derived the following simple expressions for obtaining the height of void migration for the three geometrical forms mentioned (Fig. 2.14):

1 Conical collapse

$$\text{Volume of intact beds } (V_o) = \frac{\pi B^2}{4} \times \frac{D_c}{3} \tag{2.5a}$$

Total volume of collapse zone (V_c)

$$V_c = V_o + \frac{\pi B^2}{4} \times h \qquad (2.5b)$$

But

$$b_f = \frac{V_c - V_o}{V_o} \qquad (2.5c)$$

Hence

$$b_f = \frac{3h}{4} \quad \text{or} \quad D_c = \frac{3h}{b_f} \qquad (2.5d)$$

2 Wedge collapse

$$V_o = \frac{BD_c l}{2} \qquad (2.5e)$$

where l = length of collapsed roadway and

$$V_c = V_o + Blh \qquad (2.5f)$$

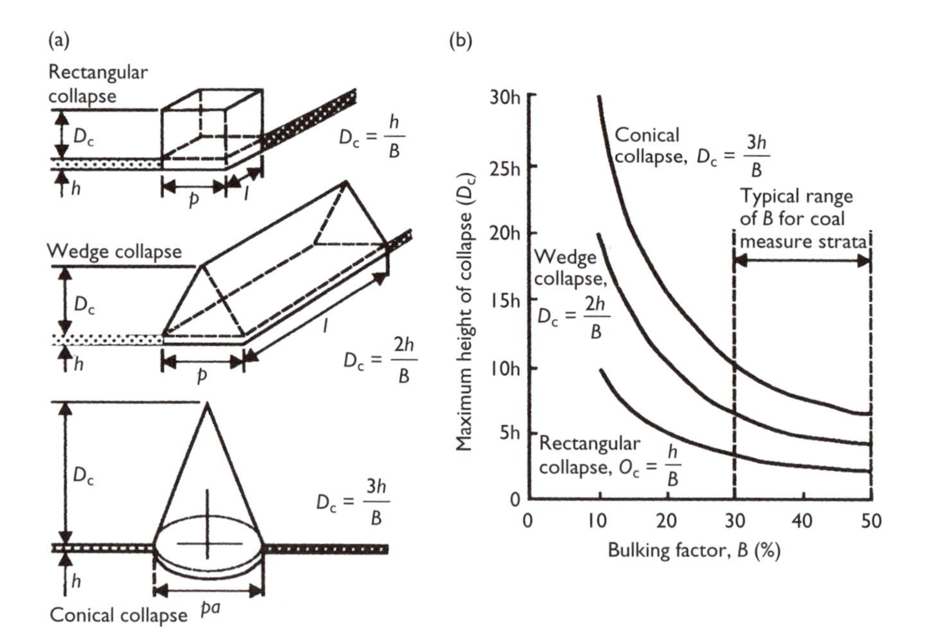

Figure 2.14 Type and estimation of the amount of void migration. (a) Diagram showing notations relating to maximum height of collapse and geometry. (b) Postulated variations in maximum height of collapse for different modes of failure and bulking factors. (After Piggott and Eynon, 1978.)

Hence

$$b_f = \frac{2h}{D_c} \quad \text{or} \quad D_c = \frac{2h}{b_f} \tag{2.5g}$$

3 Rectangular collapse

$$V_o = Blh \tag{2.5h}$$
$$V_c = V_o + Blh \tag{2.5i}$$

Hence

$$b_f = \frac{h}{D_c} \quad \text{or} \quad D_c = \frac{h}{b_f} \tag{2.5j}$$

It frequently is maintained that the maximum height of void migration is directly proportional to the thickness of seam mined, assuming that the total thickness is worked, and inversely proportional to the change in volume of the collapsed material. Garrard and Taylor (1988) also showed that the height of collapse in pillared workings in coal was frequently proportional to the width of the excavation and that the larger the span, the more likely is collapse to occur. The maximum height of migration in exceptional cases might extend to 10 times the height of the original stall, however, it is generally 3 to 5 times the stall height. Depth of cover should not include superficial deposits or made-ground since low bulking factors are characteristic of these materials. Weak superficial deposits may flow into voids that have reached rockhead, thereby forming features that may vary from a gentle dishing of the surface to inverted cone-like depressions of large diameter.

In a sequence of differing rock types, if a competent rock beam is to span an opening, then its thickness should be equal to twice the span width in order to allow for arching to develop. A thick bed of sandstone usually will arrest a void, especially if it is located some distance above the immediate roof of the working (Van Besian and Rockaway, 1988). Sandstones apart, however, most voids are bridged when the span decreases through corbelling to an acceptable width, rather than when a more competent bed is encountered. Chimney-type collapses can occur to abnormally high levels of migration in massive strata in which the joints diverge downwards.

Exceptionally, void migrations in excess of 20 times the worked height of a coal seam have been recorded. The self-choking process may not be fulfilled in dipping seams, especially if they are affected by copious quantities of groundwater that can redistribute the fallen material (Statham *et al.*, 1987). The redistribution of collapsed material can lead to the formation of supervoids. In addition, the migration of a void from a coal seam into a worked seam directly above can lead to pillar collapse in that seam with the formation of voids that are larger than the original stalls. Migration of super voids up to rockhead, then produces large scale subsidence at ground level. Under such circumstances simple analysis according to bulking factors proves inadequate (Carter, 1985).

2.4. Investigations in areas of abandoned mine workings

A site investigation for an important structure(s) requires the exploration and sampling of all strata likely to be significantly affected by the structural loading, that is, to determine the geology and physical properties of the soils and rocks that have an influence on the structures erected, and the long-term stability of the ground (Anon., 1999a). Where a mineral (e.g. a coal seam) of workable thickness occurs at shallow depth (i.e. within about 30–50 m of rockhead), it should be assumed that it has been worked at some time, although this may not have been in a systematic way. Even if there are no mine plans or records of past workings, the ground investigation should be planned on the assumption that mining has taken place until it is possible to prove otherwise (Taylor, 1968). In particular, the location of subsurface voids due to mineral extraction is of prime importance in this context. In other words, an attempt should be made to determine the number and depth of mined horizons, the extraction ratio, the geometry of the layout and the condition of any old pillar and stall workings (Bell, 1986). The sequence and type of roof rocks may provide some clue as to whether void migration has taken place and if so, its possible extent. Of particular importance is the state of the old workings, careful note should be taken of whether they are open, partially collapsed or collapsed, and the degree of fracturing, joint dilation and bed separation in the roof rocks should be recorded, if possible. This helps to provide an assessment of past and future collapse that is obviously very important.

Each investigation should be designed to meet the requirements of the construction operations to be carried out. The first stage of a site investigation involves a desk study and a reconnaissance survey, which are then followed by the necessary field exploration. The desk study should include a survey of appropriate maps, documents, records and literature. The presence on geological maps of mineral deposits that could have been mined suggests the possibility of past mining unless there is evidence to the contrary, and geological and topographic maps may show evidence of past workings such as old shafts, adits and spoil heaps. All the geological and topographic maps of the area in question, going back to the first editions, should be examined. The presence of any faults should be noted. Abandoned mines record offices, when they exist, represent primary sources of information relating to past mining activity such as old mine plans. Other sources include public record offices, museums, libraries, specialist contractors and consultants, private collections and geological surveys. Unfortunately, however, it is often the case that the extent, age and condition of old abandoned workings are poorly documented or else unknown. Even when former mine workings have been recorded the plans are often incomplete and inaccurate, therefore it may not be possible to determine their precise locations accurately in all situations. Furthermore, old abandonment plans rarely show information about the condition of the workings such as whether the mine openings have been stowed to reduce the risk of collapse, or the type of support systems used. In fact, sometimes in Britain mine plans may be misleading showing shafts and adits in areas where a mineral was not mined. In yet other cases, old mine plans may show numerous faults that do not exist. This practice was to avoid the payment of royalties, the faulted mineral being tax-free. Hence, the prediction of instability problems associated with abandoned mine workings cannot always be aided by old mine plans.

The use of remote sensing imagery and aerial photography for the detection of surface features caused by subsidence to a large extent is restricted to rural areas and scale is a critical factor. Nonetheless, according to Russell *et al.* (1979) Landsat imagery, Side Looking Airborne Radar (SLAR) imagery and high-altitude aerial photography may provide data on fractures and lineaments that could be used to locate zones of potential subsidence. High resolution airborne geophysical surveys involving gamma spectrometry, magnetic and very low frequency electromagnetic sensors are improving the ability to produce maps. These techniques help to map high conductivities that might be related to abandoned mine sites. Laser and radar sensors on airborne platforms are being used to produce high resolution (centimetre to metre) digital terrain models. The Light Detecting And Ranging (LIDAR) system sends a laser pulse from an airborne platform to the ground and measures the speed and intensity of the returning signal. From this changes in ground elevation can be mapped. Similarly, the Wide-Angle Airborne Laser Ranging System (WA-ALRS) provides a means of measuring land subsidence with millimetre accuracy (Bock and Thom, 2001). Radar systems use radar rather than lasers to achieve the same end. Satellite radar measurements are becoming increasingly sophisticated, with the potential to measure millimetric movements in urban areas where hard surfaces act as 'permanent' reflectors. Synthetic Aperture Radar Interferometry (InSAR), for instance, provides a means of mapping from a satellite continuous displacements, over large areas (100 km × 100 km), with a spatial resolution of 10 m and an accuracy of a few centimetres. Recently, Carnec and Delacourt (2000) indicated that differential interferogram images derived from repeat pass spaceborne InSAR systems afford the possibility of mapping surface deformations of small extent, as well as providing a method of monitoring their development. A similar satellite technique known as Permanent Scatterer Interferometry (PSInSAR) also uses radar and produces maps showing rates of displacement, accurate to a few millimetres per year, over extensive time periods, currently up to a decade long. The process provides the millimetric displacement histories for each reflector point across the entire time period analysed, as calculated at every individual radar scene acquisition. Small incremental ground movements therefore can be detected that might be caused by subsidence.

The resolution necessary for the detection of relatively small subsidence features (1.5–3 m across) is provided by aerial photographs with scales between 1:25 000 and 1:10 000. Colour photographs may be more useful than black and white ones in the detection of past workings since they can reveal subtle changes in vegetation related to subsidence and, if there are differences in thermal emission, then infra-red (false colour) photographs should show these differences. The details obtained from aerial photographs should be represented on a site plan at a scale of 1:2500 or larger.

The reconnaissance survey involves a walk-over visit of the site to allow familiarization, to find any new evidence of former mining activity and perhaps to confirm any anecdotal and documentary evidence of past mining. Subtle variations in the topography may be observed together with evidence of past land use. If sufficient information has been gathered at this stage, it may be possible to pass straight into a field investigation involving direct exploration of the ground by drilling. In other words, the information derived from the desk study and reconnaissance survey can hopefully be used to help plan the field exploration programme. If this is not the case, then indirect subsurface exploration using geophysical techniques, may be undertaken. One of the main advantages of geophysical methods of investigation over intrusive methods is that information is obtained for much larger volumes of ground at lower cost (McDowell *et al.*, 2002). This is an important consideration because the probability of finding a small target within a large volume of ground is very low using point-sampling methods. Be that as it may, geophysical methods are not a substitute for direct methods but should be regarded as complimenting them. Because the geophysical data gathered relates to the variation of physical properties of a volume of ground, then the data must be processed and interpreted based upon a conceptual ground model and some ground control data. Both Anon. (1999a) and Anon. (1999b) provide guidance relating to the usefulness of different, commonly available, geophysical methods for the detection of a range of applications.

Over the past thirty or so years attempts have been made to develop geophysical methods for the location and delineation of abandoned mine workings. Unfortunately, no one geophysical method has yet been developed that completely resolves all problems of this nature. A variety of surface traversing techniques are available that provide readings at close station intervals for the location of shallow voids where the lateral dimensions of the void are at least of the same order as the depth of burial.

Considerable care should be exercised at the planning stage of a geophysical survey for the location of subsurface voids because of the variable nature of the target (McCann *et al.*, 1987; Cripps *et al.*, 1988). The selection of the most appropriate technique necessitates consideration of four parameters, namely, penetration, resolution, signal to noise ratio and contrast in physical properties. The size and depth of the workings and the character of any infill control the likelihood of the workings being detected as an anomaly. With the information obtained from the desk study and the reconnaissance survey, many of the available geophysical methods can be assessed at the selection stage, using a model study, and accepted, or rejected, without any requirement for field trials. Generally, it is possible to detect a cavity whose depth of burial is less than twice its effective diameter. However, since the presence of a void is likely to affect the physical properties and drainage pattern of the surrounding rock mass, this can give rise to a larger anomalous zone than that produced by the void alone.

The nature of the environment around a site affects the success of geophysical surveys. For instance, traffic vibrations adversely affect the results obtained from seismic surveys, as do power lines and electricity cables in the case of electromagnetic and magnetic techniques. Of particular importance is that there should be sufficient physical property contrast between the void and the surrounding rock mass so that an anomaly can be detected.

Seismic refraction has not been used particularly often in searching for voids at shallow depth created by previous mining since such voids are often too small to be detected by this method because of attenuation of seismic waves in the rock mass. Furthermore, if the workings are dry, then high attenuation of the energy from the seismic source occurs so that penetration in to the rock mass is poor. McCann *et al.* (1982) concluded that there was little likelihood of a cavity producing a measurable anomaly where it is buried at a depth greater than its diameter. However, the disturbed zone around an anomaly may increase its effective size. Voids in coal seams above a competent sandstone sometimes can be detected by localized increase in travel time and decrease in amplitude of the seismic event, provided that the anomalous region is comparable in dimensions with the seismic wavelengths. For instance, a seismic wave having a dominant frequency of 100 Hz and a velocity of 2000 m s^{-1} has a wavelength of 20 m. As a consequence voids less than 20 m in width probably will not be recognized (Anon., 1988). Recent advances in seismic reflection techniques have improved the potential for detection of shallow voids.

Generally, where dry pillar and stall workings occur at a depth greater than 5 m, it is unlikely that resistivity profiling will detect their presence. In addition, where the depth of a cavity is equal to its diameter, the maximum disturbance in the resistivity profile is only about 10%, in which case, cavities at depths greater than twice their average dimension usually are not recorded. Electrical resistivity depth sounding can be applied to the location of voids where the width to depth ratio is large. Mine workings that produce an air-filled layer often can be identified on the sounding curve as an increase in apparent resistivity.

Terrain conductivity meters have several advantages over conventional electrical resistivity equipment when used for locating small, near-surface anomalous features. For example, in the depth range up to 30 m, terrain conductivity surveys are more effective than resistivity traversing. Conductivity values are taken at positions set out on a grid pattern and the results can be contoured to indicate the presence of any anomalies. Penetration into the ground achieved by electromagnetic radiation can be limited by excessive attenuation in ground of high conductivity. However, problems with detecting an extremely small secondary field in the presence of a strong primary signal can be overcome by the use of pulsed radiation.

Ground probing radar is capable of detecting small subsurface cavities directly and appears to be one of the most promising methods for the future, especially for surveys in urban areas. The method is based upon the transmission of pulsed electromagnetic waves, the travel time of the waves reflected from subsurface interfaces being recorded as they arrive at the surface so allowing the depth to an interface to be obtained (White, 1992). The high frequency of the system provides high resolution and characteristic arcuate traces are produced by air-filled voids (Fig. 2.15). Depths to voids can be determined from the two-way travel times of reflected events if velocity values can be assigned to the strata above the void. The conductivity of the ground imposes the greatest limitation on the use of radar probing in site investigation (Leggo and Leach, 1982). In other words, the depth to which radar energy can penetrate depends upon the effective conductivity of the strata being probed. This, in turn, is governed chiefly by the water

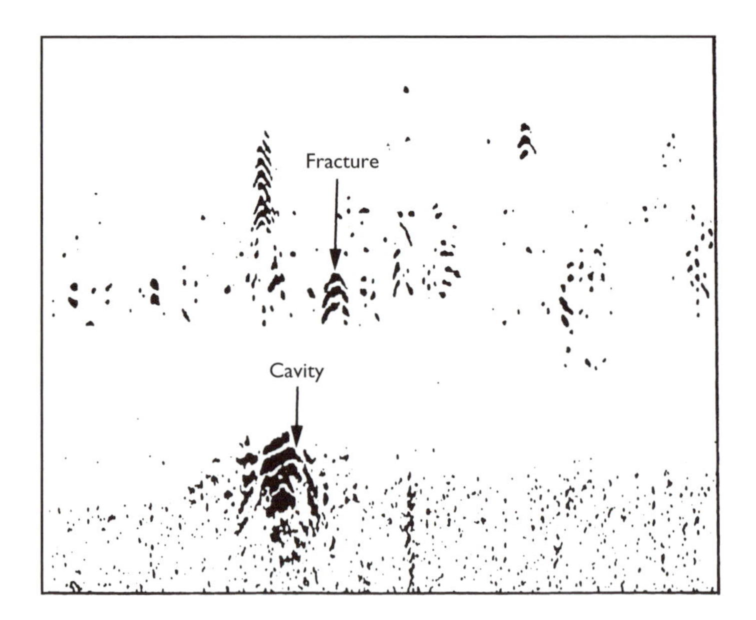

Figure 2.15 Ground probing radar scan providing an indication of a small subsurface cavity, depth of cavity is about 1.5 m.

content and its salinity. Furthermore, the value of effective conductivity is a function of temperature and density, as well as the frequency of the electromagnetic waves being propogated. The least penetration occurs in saturated clayey materials or when the pore water is saline. Useful data can be obtained from sites where clayey topsoil is more or less absent because wet clay and silt, in particular, cause great attenuation of electromagnetic energy frequently giving penetration depths of less than 1 m. On the other hand, the technique appears to be reasonably successful in sandy soils and rocks in which the pore water is non-saline. The penetration of radar energy can be increased by using a lower frequency but unfortunately this reduces its resolution, which means that subsurface anomalies have to be correspondingly larger if they are to be detected.

Koerner *et al.* (1982) investigated a microwave method as a means of locating subsurface voids. There are two microwave techniques, namely, the pulsed type and the continuous wave type. In each technique an electromagnetic wave is sent through the rock material under investigation. The wave is partly reflected back to the surface by any interface with different electrical properties, and the amplitude of the reflected wave is recorded. It would appear that the continuous wave microwave technique is the more successful.

Generally speaking, voids in shallow abandoned mine workings are too small and located at depths too great to be detected by normal magnetic or gravity surveys. However, the fluxgate magnetic field gradiometer permits surveys of shallow depths to be carried out. It provides a continuous recording of lateral variations in the vertical gradient of the Earth's magnetic field rather than giving the total field strength. The gradiometer tends to give better definition of shallow anomalies by automatically removing the regional magnetic gradient. Quantitative analysis of the depth, size and shape of an anomaly generally can be made more readily for near-surface features, using the gradiometer, than from total field measurements, obtained with a proton-magnetometer. The sensor of a gradiometer is small (0.5 m or 1 m) in order to record features within 2–3 m of the ground surface. On the other hand, a proton-magnetometer can more easily detect larger and deeper features, and yields results that are more suitable for contouring.

Micro-gravity meters, accurate to 2 μgal (1 mgal = 10 g.u., that is, gravity units), may be successful when the voids have a significant lateral extent, as in some pillared workings (Bishop *et al.*, 1997). A series of traverses can be used to map the lateral extent of such features but there are still size/depth constraints and an inherent ambiguity in the interpretation of results.

Seismic tomography uses two or more drillholes, and possibly the ground surface, for the location of sources and detectors, the object being to derive one or more two-dimensional images (Jackson and McCann, 1997). In this way it utilizes a multitude of wave paths to enable the location, shape and velocity contrast relating to an anomaly, such as a void, between drillholes to be delineated. The quality of the results obtained is a function of the nature of the ground, in particular the physical property contrasts, the number of drillhole pairs, the distances apart of the drillholes and their location in relation to the target. In seismic tomographic surveys, the wavelength of the seismic event needs to be less than the average dimensions of the target. Electrical resistance tomography is a relatively new geophysical imaging technique that uses a number of electrodes in drillholes, and sometimes at the ground surface, to image the resistivity of the subsurface. Electromagnetic techniques also have been used to produce tomographic imagery. Corin *et al.* (1997) used drillhole radar tomography to assess foundation conditions in which voids occurred and concluded that crosshole methods are probably the best tools available at present to provide the detailed information for good foundation design.

The location of old workings generally has been done by exploratory drilling, the locations of drillholes being influenced by data obtained from the desk study and/or the data gathered by indirect methods. However, it must be admitted that although frequently successful in locating the presence of old mine workings, exploratory drilling is not necessarily able to establish their layout. Nevertheless, when the results from drilling are combined with a study of old mine plans, if they exist, then it should be possible to obtain a better understanding of methods of working, sizes of voids and directions of roadways and galleries.

Drilling to prove the existence of old mine workings is usually done by open holes, which allow relatively quick probe drilling (Bell, 1986). Rotary percussion drilling with a cruciform bit may be used (Fig. 2.16). The drillholes should be taken to a depth where any voids present are not likely to influence the performance of the structures to be erected. For sites where there is little data available Littlejohn (1979) suggested that drillholes should be sunk to a depth of 60 m, especially if substantial structures such as multi-storey buildings are to be erected. If a grid pattern of drillholes is used some irregularity should be introduced into the pattern to avoid holes coinciding with pillar positions. The presence of old voids is indicated by the free-fall of the drill string and the loss of flush.

One of the principal objectives of investigations of abandoned mine workings is to determine their extent and condition. Accordingly, core material may need to be obtained. In addition, the stratigraphic sequence should be established by taking cores in at least three drillholes. Double barrel sampling tubes with inner plastic liners can be used to obtain core, which then can be photographed and logged, and the rock quality designation (RQD) or fracture spacing index recorded. Drilling penetration rates, water flush returns and *in situ* permeability tests may be used to assess the degree of fracturing. The degree of fracturing is important in that it tends to increase as old workings are approached. Determination of groundwater

Figure 2.16 Probing for shallow old mine workings in coal with a rotary percussion rig, Sheffield, England.

conditions is necessary, especially where a grouting programme is required to treat the workings.

Detailed mapping of galleries is best made by driving a heading from an outcrop if this is close at hand or by sinking a shaft to the level of the mineral deposit to obtain access to the workings (Thorburn and Reid, 1978). Sometimes access can be gained via old shafts. Radial holes may be drilled from a shaft to establish the dimensions of pillars and stalls. As old workings may prove dangerous, exploration should be undertaken with the advice and aid of experts.

Below surface workings may be examined by using drillhole cameras or closed circuit television, information being recorded photographically, or on videotape, and used to assess the geometry of voids and, possibly, the percentage extraction. However, their use in flooded old workings has not proved very satisfactory. Occasionally, smoke tests or dyes have been used to aid the exploration of subsurface cavities.

It is possible to study the interior of large abandoned mine workings that are flooded by using a rotating ultrasonic scanner (Braithwaite and Cole, 1986). The ultrasonic survey can be carried out within the void, the probe being lowered down a drillhole to mine level. Horizontal scans are made at 1 m intervals over the height of the old workings and then tilting scans at 15° intervals are taken from vertical to horizontal. The echoes received during the horizontal scan are processed by computer to provide a plan view of the mine. Vertical sections are produced from the vertical scan. Hence, it is possible to determine the positions of pillars and the extent of the workings.

2.5. Old mine workings and hazard zoning

A hazard involves a degree of risk, the elements at risk being life, property, possessions and the environment. Risk involves quantification of the probability that a hazard will be harmful and the tolerable degree of risk depends upon what is being risked, life being much more important than property. The frequency of a particular hazard event can be regarded as the number of events of a given magnitude in a particular period of time at a certain location. The risk to society can be regarded as the magnitude of a hazard multiplied by the probability of its occurrence. Risk, as mentioned, should take account of the magnitude of the hazard and the probability of its occurrence. Risk arises out of uncertainty due to insufficient information being available about a hazard and to incomplete understanding of the mechanisms involved. The uncertainties prevent accurate predictions of hazard occurrence. Risk analysis involves identifying the degree of risk, then estimating and evaluating it.

Assessments of mining hazards usually have been on a site basis, regional assessments being much less common. Nonetheless, regional assessments can offer planners an overview of the problems involved and can help them avoid imposing unnecessarily rigorous conditions in areas where they really are not warranted.

Any spacial aspect of a particular hazard can be mapped provided there is sufficient information on its distribution. Hence, when hazard assessment is made of an area, the results can be expressed in the form of hazard maps. An ideal hazard map should provide information relating to the spacial and temporal probabilities of the hazard mapped. However, hazard maps, like other maps, do have disadvantages. For instance, they are highly generalized and represent a static view of reality. They therefore need to be updated periodically as new data becomes available. Hazard maps of areas where old mine workings are present should ideally represent a source of clear and useful information for planners and developers. In this respect, they should realistically present the degree of risk (i.e. the probability of the occurrence of a hazard event). Descriptive terms such as high, medium and low risk must be defined, and there are social and economic dangers in overstating the degree of risk in the terms used to describe it. Ideally, numerical values should be assigned to the degree of risk. However, this is by no means an easy matter, for numerical values can only be derived from a comprehensive record of events, which unfortunately in the case of abandoned mine workings is available only infrequently. Furthermore, many events in the past may not have been recorded, which throws into question the reliability of any statistical analysis of data. Indeed, subsidence is affected by so many mining and geological variables that many regard it as, to all intents and purposes, a random process. In addition, the matter of risk assessment is complicated further in terms of its tolerance, for example, people are less tolerant in relation to loss of life than to loss of property. Hence, the likelihood of an event causing loss of life would be assigned a higher value than loss of property (e.g. the probability of a 1 in 100 loss of life presumably would be regarded as very risky whilst very risky in terms of loss of property may be accepted as 1 in 10). Nonetheless, if planners can be provided with some form of numerical assessment of the degree of risk, then they are better able to make sensible financial decisions and to provide more effective solutions to problems arising. For example, Merad *et al.* (2004) developed a methodology for risk zoning of subsidence that is occurring above old iron mining districts in Lorraine, France, which is based on a multi-criteria decision-aid approach that assigns zones at risk to predefined classes.

An attempt at hazard zoning in an area of old mine workings was made by Price (1971). After a detailed investigation at a site in Airdrie, Scotland, underlain by shallow abandoned mine workings, he was able to propose safe and unsafe zones (Fig. 2.17). In the safe zones the cover rock was regarded as thick enough to preclude subsidence hazards (about 10 m of rock or 15 m of till was regarded as sufficient to ensure that crown holes did not appear at the surface) and normal foundations could be used for the two-storey dwellings that were to be erected. On the boundaries between the safe and unsafe zones, the dwellings were constructed with reinforced foundations, or rafts, as an added precaution against unforeseen problems. Development was prohibited in zones designated unsafe. In effect, Price produced a thematic mining information plan of the site to facilitate its development.

In recent years thematic maps have been produced in Britain of both urban and rural areas with a view to benefiting planners and civil engineers concerned about ground stability and associated land use. Early thematic maps produced by the British Geological Survey depicted areas of undermining assumed to be within 30 m of the surface on the one hand and at depths exceeding 30 m below the surface on the other (McMillan and Browne, 1987). This 30 m depth is based on limited information and therefore is subject to interpretation. It assumes that bulking factors of 10–20% will affect the strata involved in void migration. Initially, known and

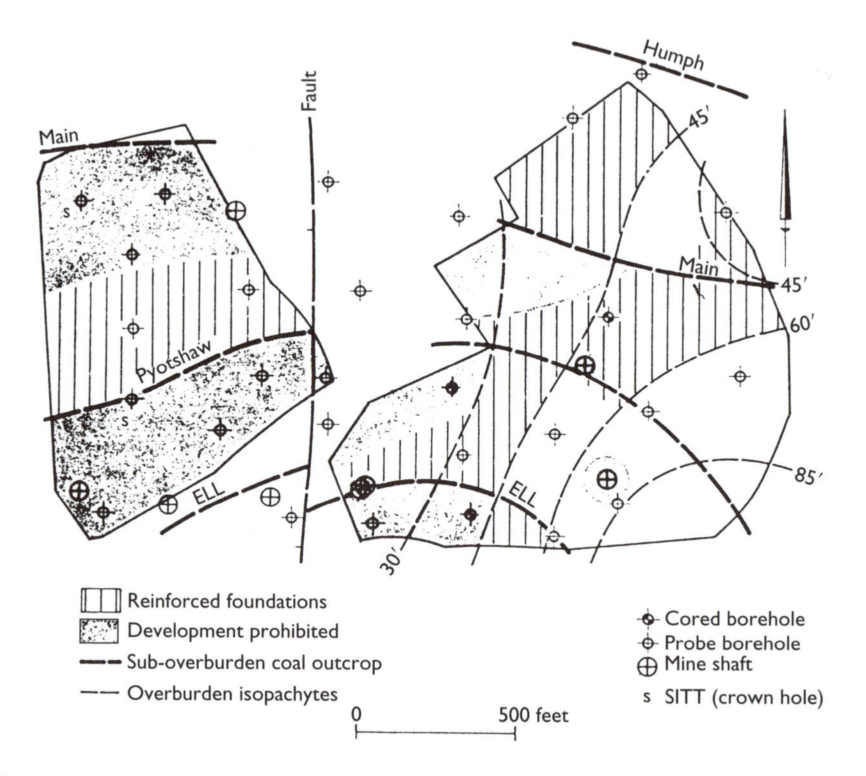

Figure 2.17 A plan of different foundation zones at a site in Airdrie, Scotland. Normal strip foundations were used in the unshaded area, that is, the zone deemed safe. (After Price, 1971; reproduced by kind permission of the Geological Society of London.)

suspected mining areas were not differentiated on these maps. However, only areas of mining shown on mine plans were represented on the map of the Glasgow district and no areas of suspected mining were shown (Fig. 2.18). No attempt was made to infer the extent of working beyond the limits defined by mine plans other than to plot relevant drillhole data. The recognition of single and multiple seam working led to the requirement that areas of shallow working (i.e. less than 30 m below rockhead) should be identified in terms of seams worked. Separate maps were prepared illustrating areas of total known mining; current mining; known mining within 30 m of rockhead together with the locations of shafts and drillholes encountering old shallow workings; and mining for minerals other than coal and ironstone. In Fife, known and inferred old shallow mining were differentiated on the same map. Each modification has reflected an attempt to clarify the presentation of known and inferred past mining. An indication of the area in which old mine workings might be expected can be

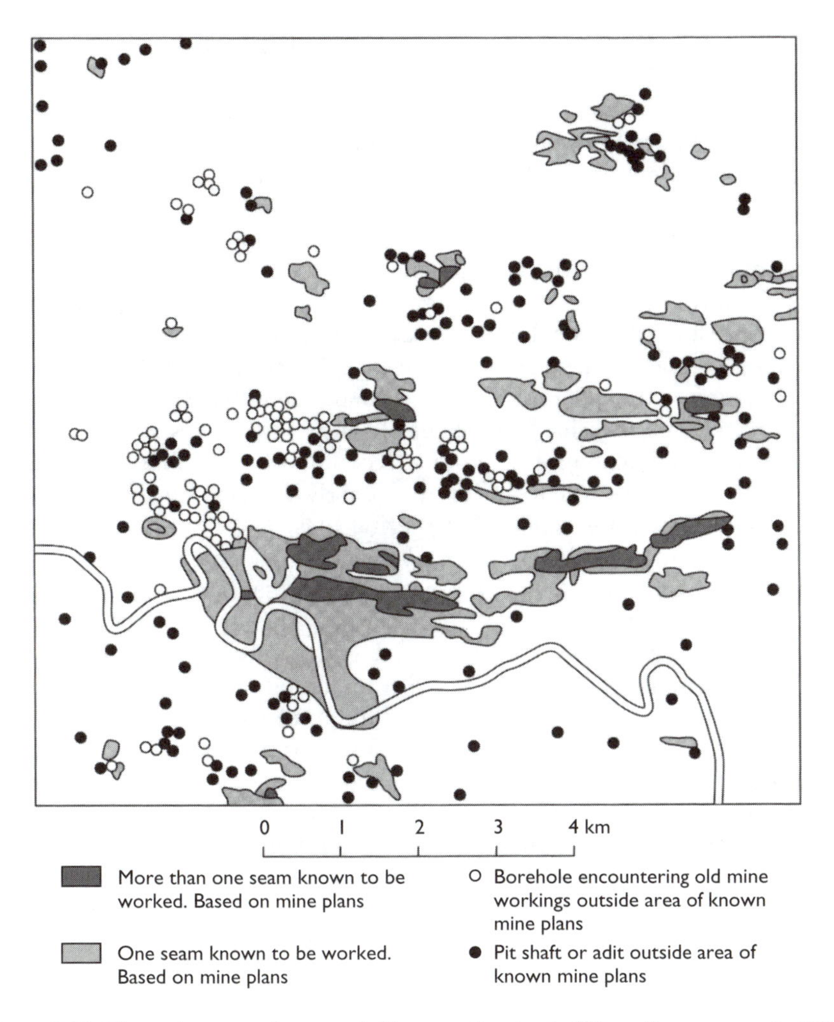

Figure 2.18 Thematic map of part of Glasgow, Scotland. (After Browne *et al.*, 1986; reproduced by kind permission of the Geological Society of London.)

obtained by plotting all drillholes that encounter spoil outside areas of workings known from abandonment plans. In areas where mineral outcrops are reasonably well known, areas of suspected workings can be mapped as a separate category, although it is then necessary to assume that all workable beds have been exploited, at least in the near surface area.

In an assessment of the degree of risk due to subsidence incidents associated with abandoned mine workings in South Wales, Statham *et al.* (1987) found that of the 388 events recorded, 64% occurred in open land and so posed no threat to person or property. Twenty-one per cent had occurred when people were nearby or property threatened. The remainder caused damage to highways, buildings or other property and only one of these events resulted in minor injuries. In the context of the South Wales Coalfield this represents a low level of hazard. Assuming that a typical incident affects an area of 5 m^2, then the probability of collapse occurring on any 25 m^2 plot is of the order of 10^{-7} per year. Even if the number of subsidence incidents that have remained undiscovered increased the above total figure by a factor of 3, then the overall risk would still be low. Statham *et al.* noted that over 90% of the incidents occurred within 100 m of the outcrop of the seam concerned. They produced a development advice map for South Wales that showed two zones inside the outcrops of worked seams corresponding to migration ratio (thickness of rock cover/extracted thickness) values of 6 and 10, which were expected to contain 90% and 100% respectively of relevant subsidence incidents. The map makes a contribution towards regional planning by taking account of a possible development constraint at an early stage and offers an early warning on the likely scale of ground investigation required at specific sites.

However, it must be borne in mind that thematic maps that attempt to portray the degree of risk of a hazard event represent generalized interpretations of the data available at the time of compilation. Therefore, they cannot be interpreted too literally and areas outlined as 'undermined' should not automatically be subjected to planning blight. Obviously, there is a tendency to assume that the limits of old mine workings represented on a map indicate the full extent of the workings but the interpretation of their location is based on scanty information and includes assumptions, some of which may be unfounded. It should be recognized that engineering problems in areas of past mining only occur if buildings are not properly planned, designed and constructed with reference to the state of undermining. Also, zoning based entirely upon depth of cover above workings cannot be relied on completely, since occasionally subsidences have occurred in zones labelled 'safe' (Carter, 1985).

Thematic geological maps of past mining areas therefore have their limitations due to factors such as scale of presentation, reliability and availability of data. They are transitory and require amendment as new data becomes available. These maps must not be used as a substitute for site investigation.

2.6. Measures to reduce or avoid subsidence effects due to old mine workings

Where a site that is proposed for development is underlain by shallow old mine workings there are a number of ways in which the problem can be dealt with (Healy and Head, 1984). However, any decision regarding the most appropriate method should be based upon an investigation that has evaluated the site conditions carefully and examined the economic implications of the alternatives. One of the most difficult assessments to make is related to the possible effects of progressive deterioration of the workings and associated potential subsidence risk. The placement of any new structure must be carefully considered to ensure that the ground is not adversely affected by the additional load and/or that the ground is sufficiently stabilized that it will not suffer distress during the anticipated life of the structure.

The first and most obvious measure is to locate the proposed structure on sound ground away from old workings or over workings proved to be stable. It is not generally sufficient to locate immediately outside the area undermined as the area of influence should be considered. In such cases the angle of influence usually is taken as 25°, in other words the area of influence is defined by projecting an angle of 25° to the vertical from the periphery and depth of the workings to the ground surface. For example, following a site investigation for a hotel near Newcastle upon Tyne, England, it was found that part of the site was underlain by old pillar and stall workings, and that coal would be extracted from beneath the site by longwall mining within the next five years. It was recommended that the hotel complex should be relocated and the design of the building altered. Such relocation, of course, is not always possible. Alternatively, it may be possible to redesign the layout of a site. For example, Price *et al.* (1969) referred to a site in Edinburgh, Scotland, where the proposed layout of buildings was changed to accommodate subsurface workings revealed by the subsurface investigation. The initial proposed layout and new layout of the site are shown in Figure 2.19 from which it can be seen that the nature of the buildings also were changed but they nonetheless provided a similar number of dwellings.

If old mine workings are at very shallow depth, then it might be feasible, by means of bulk excavation, to found on the strata beneath. This is an economic solution, at depths of up to 7 m or on sloping sites and is well suited to areas that were worked by means of bell pits. Such excavations may be carried out rapidly if the overburden consists of clays, shales or fragmented and weathered rocks. For example, apartment blocks in Leeds, England, were founded below old workings in the Beeston seam, with 21 740 tonnes of coal being opencasted, which proved a profitable enterprise.

Where the allowable bearing capacity of the foundation materials has been reduced by mining, it may be possible to use a raft. Rafts can consist of massive concrete slabs, stiff slabs and beam or cellular rafts (Fig. 2.20). The latter are suitable for the provision of jacking sockets in the upstand beams to permit the columns or walls to be relevelled if subsidence distorts the raft. A raft can span weaker and more deformable zones in the foundation, thus spreading the weight of a building well outside the limits of the building. However, rafts are expensive and therefore tend to be used where

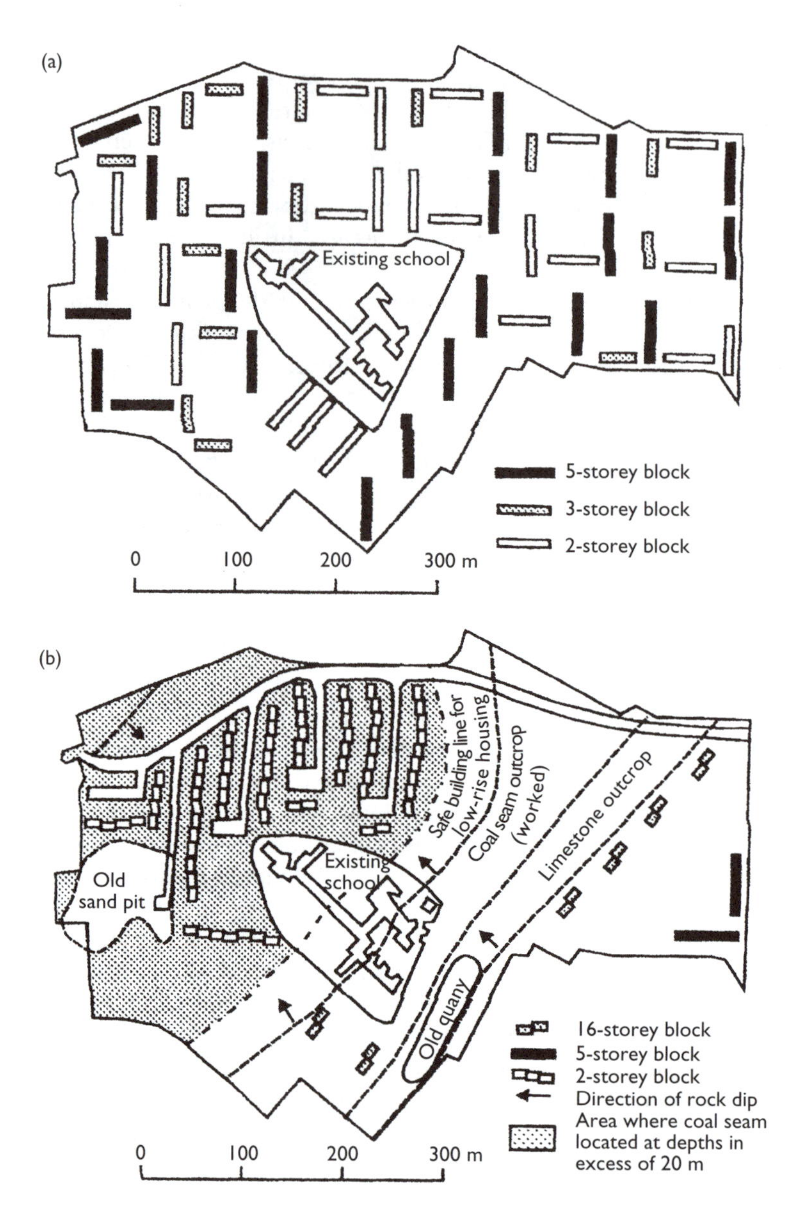

Figure 2.19 Redesign of a site layout in Edinburgh, Scotland, to accommodate subsurface workings (a) Proposed layout prior to site investigation. (b) Redesign of a site layout to accommodate subsurface workings in Edinburgh, Scotland. (After Price *et al.*, 1969; reproduced by kind permission of the Geological Society of London.)

Figure 2.20 A cellular raft.

no alternative exists. For low buildings, that is, up to four storeys in height, it is occasionally possible to use an external reinforced ring beam with a central lightly reinforced raft as a practical and more economic foundation.

Reinforced bored pile foundations also have been resorted to in areas of abandoned mine workings. In such instances the piles bear on a competent stratum beneath the workings. They also should be sleeved so that concrete is not lost into voids, and to avoid the development of negative skin friction if overlying strata collapse. Some authorities, however, have suggested that piling through old mine workings seems inadvisable because, first, their emplacement may precipitate collapse and, second, subsequent collapse at seam level could possibly lead to piles being either buckled or sheared (Price *et al.*, 1969). There also may be a problem with lateral stability of piles passing through collapsed zones above mine workings or through large remnant voids.

Where old mine workings are believed to pose an unacceptable hazard to development and it is impractical to use adequate measures in design or to found below their level, then the ground itself can be treated. Such treatment involves filling the voids in order to prevent void migration and pillar collapse. In exceptional cases where, for example, the mine workings are readily accessible, barriers can be constructed underground and the workings filled hydraulically with sand, crushed rock, pulverized fuel ash (PFA), fluidized-bed combustion ash (FBC) or coarse colliery discard; or pneumatically with some suitable material (Siriwardane *et al.*, 2003). Hydraulic stowing also may take place from the surface via drillholes of sufficient diameter. For example, Karfakis (1993) referred to the stabilization of abandoned pillar and stall workings beneath Hanna, Wyoming, by hydraulic backfilling with sand. Michalski and Gray (2001) suggested that because many power stations in India are located on or near coalfields, then ash could be used to backfill mines. Gravity or pneumatic stowing (where a compressed jet of air is used to maintain the suspension of sand) are often considered where large subsurface voids have to be filled. The Bureau of Mines in the United States pioneered pumped slurry injection (Waite and Allen, 1975). This involves drilling slurry injection wells into the mine workings. The injected slurry acts initially as a passive support and begins to take on load with continuing deformation of the roof and/or pillars. Slurry can be pumped into wet and dry mines but the technique appears to be more successful when the mine is flooded. Pumped slurry injection and hydraulic backfilling can be used to stabilize relatively large areas, that is, areas larger than 1 ha. If the fill material, once drained, is likely to suffer adversely from any inflow of groundwater, for instance, if it may become liquefied and flow from the treated area or if fines may be washed out so causing consolidation of the fill, then it may be necessary to prevent this by constructing barriers around the area to be treated.

Grouting may be undertaken to stabilize old mine workings beneath projected future surface structures. It is carried out by drilling holes from the surface into the mine workings, on a systematic basis, and filling the remnant voids with an appropriate grout mix (Lloyd *et al.*, 1995; Fig. 2.21).

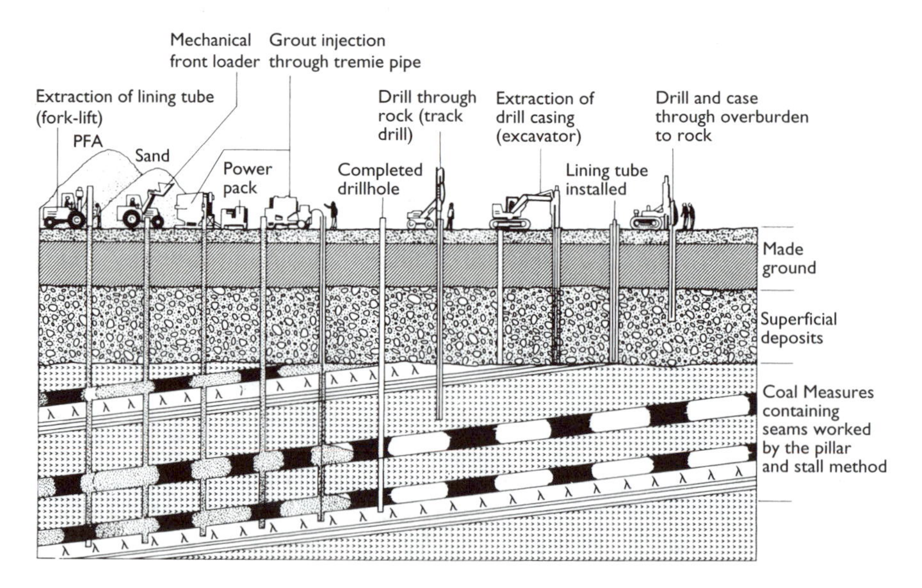

Figure 2.21 Schematic layout of boreholes and plant for filling and grouting abandoned mine workings.

If it has been impossible to obtain accurate details of the layout and extent of the workings, then the zone beneath the intended structure can be subjected to consolidation grouting. The grouts used in these operations commonly consist of cement, fly-ash and sand mixes, economy and bulk being their important features. Gravel may be used as a bulk filler where a large amount of grout is required for treatment. Alternatively, foam grouts can be used. If the workings are still more or less continuous, then there is a risk that grout will penetrate the bounds of the zone requiring treatment. In such instances, dams can be built by placing gravel down large diameter drillholes around the periphery of the site (Littlejohn, 1979). When the gravel mound has been formed it is grouted. Then, the area within this barrier is grouted (Fig. 2.22). If the old workings contain water, then a gap should be left in the dam through which water can drain as the grout is emplaced. This minimizes the risk of trapped water preventing the voids being filled. As bulk grouting results in a significant change in mass permeability, consideration must be given to the implications of such treatment on groundwater flow. The adequacy of the infilling can be assessed by the quantities of grout injected, the use of water absorption tests and/or downhole cameras.

An example of a grout treated area is provided by the A1 trunk road approach to the Tyne Bridge in Gateshead in north-east England. The road is carried on a viaduct, which is a prestressed concrete structure with support piers founded in sandstone, each pier carrying some 2000 tonnes. Generally, the sandstone was recorded at 0.6–4.9 m depth but was not present in some areas where it was believed that quarrying had taken place in the past. Five coal seams were present beneath the sandstone (Table 2.2). After the site investigation, it was considered that the risk of surface subsidence due to void migration from the High Main seam was low but that the stress on the pillars and the seatearth beneath could result in some subsidence damage. In view of the relatively high risk structure involved, it was decided to pressure grout the workings in the High Main seam, although the seams below were not treated. A perimeter wall some 1675 m in length was formed to contain the grout. This barrier consisted of approximately

Table 2.2 Depth and thickness of seams at the site of viaduct leading up to Tyne Bridge, Gateshead, England

Seam	*Average depth below g.l. (m)*	*Seam thickness (m)*	*Nature of workings*
High Main	16.8	1.67	Pillar and stall
Top Main	41.2	1.70	Presumed old workings
Bottom Main	71.7	0.66	No record of workings
Main	126.5	0.45–0.91	Thicker part worked
Maudlin	209.0	1.07	Presumed old workings

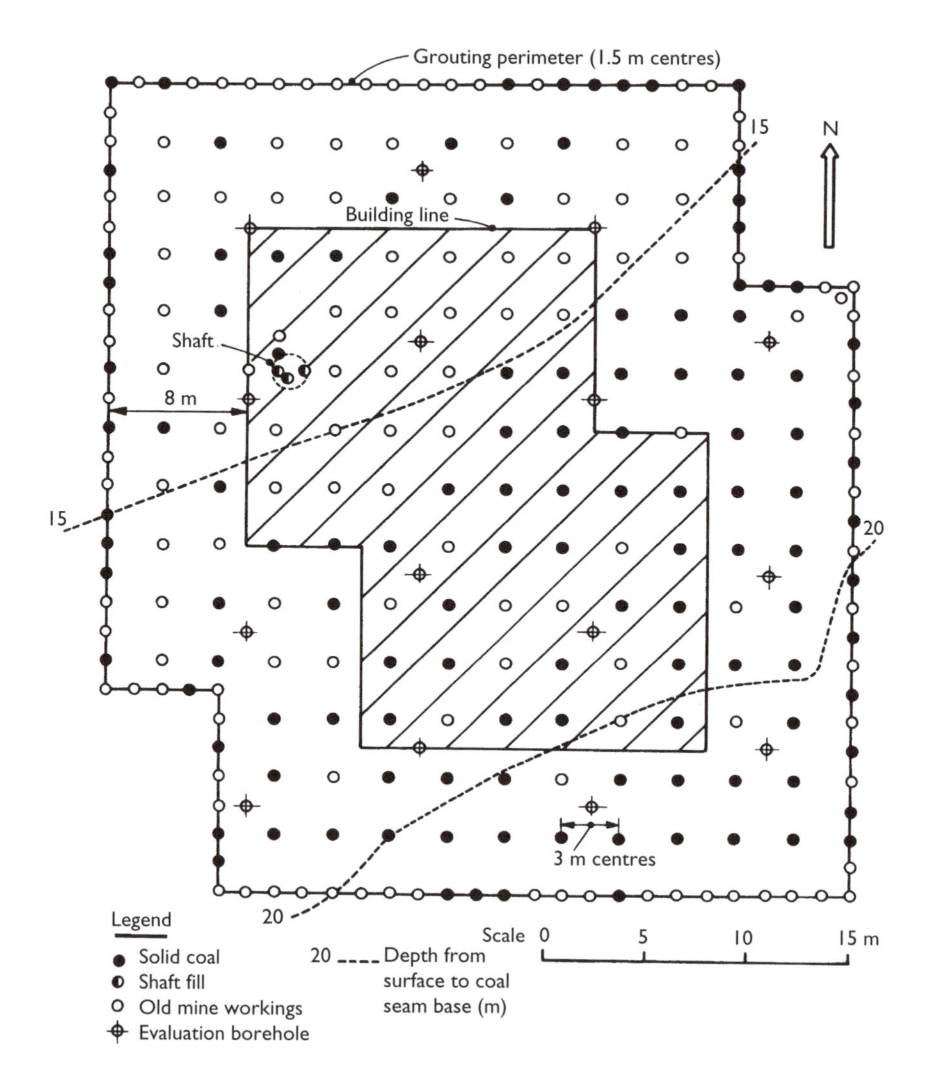

Figure 2.22 Layout of boreholes for grouting shallow abandoned workings in coal at Newcastle upon Tyne, England. (After Price *et al.*, 1969; reproduced by kind permission of the Geological Society of London.)

2300 m^3 of grout, the mix being 1 part cement, 2.5 parts fly-ash, 10.5 parts sand and 0.1 parts bentonite. Infilling within the perimeter wall took 2850 m^3 of grout using a mix of 1 part cement, 5 parts fly-ash and 18 parts sand. The groutholes were set out at approximately 3 m intervals and provided an 18 m wide treated area beneath each pier. After completion of the infill grouting, the discontinuities in the overlying sandstone were grouted with a 1:1 mix of cement and fly-ash at a pressure of 275 kPa. The adequacy of the grouting was determined by injecting water into the last holes. As full returns resulted, it was assumed that the grouting exercise had been satisfactory.

Karfakis (1993) referred to the injection of sand-grout mixtures to fill fractured rock. Zones of fractured rock and small voids formed by roof collapse need to be stabilized to provide support to the remaining roof and prevent further void migration.

A method of ground treatment used in the Pennsylvania Coalfield has been described by Gray and Meyers (1970). It involves sinking 150 mm drillholes down to the workings. Gravel is added if a void is encountered, thus a mound is formed that supports the roof, the gravel then is grouted. Next, grouting is carried out from the drillhole. The holes are simply grouted if no voids are encountered, the basic concept being to create a series of grouted columns, thereby strengthening the ground (Fig. 2.23). The technique is applied most successfully where roof subsidence has not occurred to any significant degree. The published costs indicate that there is little economic difference between this process and consolidation grouting based on closely spaced drillholes.

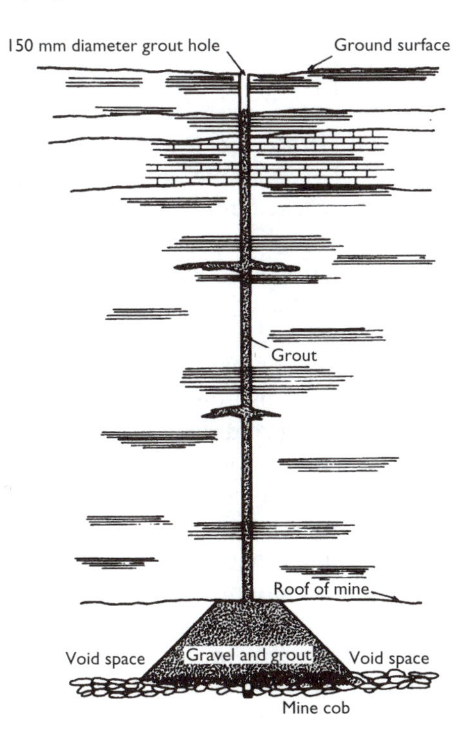

Figure 2.23 The formation of a grout column to minimize subsidence. (After Gray and Meyers, 1970; reproduced by kind permission of the American Society of Civil Engineers.)

Anon. (1977) suggested that if the roofs of old pillar and stall workings occur within 5 m of the formation level of a projected road, then the whole of the underlying ground should be excavated to the floor level of the workings. Where the stalls are occupied by debris that has collapsed as a result of void migration, then the depth of excavation can be reduced and the debris compacted. Any galleries that open into a cutting should be sealed with a concrete wall and clay plug to prevent water subsequently draining onto the carriageway. If the roofs of workings occur at a greater depth than 5 m below formation level of a road, then a reinforced concrete slab, some 200 mm in thickness, can be placed immediately below the base, replacing sub-base material. The slab should extend approximately 1.2 m beyond the carriageway. Alternatively, the workings may be grouted. More recently, steel mesh reinforcement or geonets have been used in road construction over areas of potential mining subsidence (Cameron-Clarke *et al.*, 1992). If a void should develop under a road, then the reinforcing layer is meant to prevent the road from collapsing into the void. Any surface depression that occurs in the road suggests the presence of a void. It can be investigated and, if required, remedial measures taken.

2.7. Old mine shafts

Centuries of mining in many countries have left behind a legacy of old shafts. Unfortunately, many, if not most, are unrecorded or are recorded inaccurately (Dean, 1967). For example, as many as 10 000 unrecorded coal mine shafts are believed to exist in Britain. In addition, there can be no guarantee of the effectiveness of shaft treatment unless it has been carried out in recent years. Shafts represent an environmental hazard in that they may collapse or emit noxious gases or acid mine drainage waters. The location of a shaft is of importance as far as the safety of an existing or future structure is concerned for although shaft collapse is fortunately an infrequent event, its occurrence can prove disastrous (Bell, 1988b). One of the worst disasters in England occurred in 1945 at Abram, near Wigan, Lancashire. A train with several wagons was being shunted across a goods yard when a shaft opened up and swallowed it. Neither crew, train nor wagons were recovered.

The shape of the shafts and mode of lining the walls were governed mainly by mining custom. For example, in most coalfields in England the shafts were circular, while in Wales they were elliptical and where necessary the walls of the shafts were lined with stone or brickwork. Iron tubbing at times was used for lining shafts. In Scotland the shafts were usually rectangular and often lined with wood. Shaft diameters ranged from 2 to 5 m, and the maximum side of rectangular shafts ranged from 2 to 6 m. Shafts that were used solely for ventilation and pumping were usually smaller in cross-section than in winding shafts.

Shafts often were capped with wrought iron domes and turfed over, or trees were dropped into a shaft to form a bridge, on top of which fill was placed. More usually, a wooden platform was laid across the buntons some 3–15 m below the surface and topped up with fill. With time the wood decayed to expose an open shaft. In the nineteenth century stronger wooden rafts were laid across the shaft void and projected outside its perimeter.

If shafts were filled, it frequently was undertaken in a haphazard manner without regard for the future. Many shafts were filled with unsuitable material, that at hand being the most obvious, for instance, rails, timbers, bogies, scrap metal and mine waste were used. This rarely meant that the shaft was filled properly, bridging and the formation of voids usually occurring. Such fills are capable of sudden collapse. Generally, no attempt was made to seal shafts from the workings, which after a mine was abandoned frequently became waterlogged. In such instances, fine fill is likely to flow into the workings with the result that a cavity is produced in the shaft. With time the plug above the cavity is denuded until eventually the remnant collapses.

The location of a shaft, as mentioned, is of great importance as far as the safety of structures is concerned. Moreover, from the economic point of view the sterilization of land due to the suspected presence of a mine shaft is unrealistic. The number of shafts per mine varied, indeed some shallower coal mines had as many as 10 shafts giving easy access to all parts of the workings. On the other hand, some deep mines had only one shaft. If the function of a shaft is known, it may be possible to guess its proximity to other shafts. For example, pumping shafts sometimes occur within a metre or so of each other. Again in Britain, in the latter half of the eighteenth century ventilation shafts for coal mines were connected to the main shaft and at the surface could be from 2 to 8 m away from the latter. After 1863 the legal minimum distance between two shafts had to be 10 ft (3 m), and this was increased to 41 ft 6 in. (13.6 m) in 1887.

The ground about a shaft may subside or, worse still, collapse suddenly (Fig. 2.24). Collapse of filled shafts can be brought about by deterioration of the fill, which is usually due to adverse changes in groundwater conditions, or to the surrounding ground being subjected to vibration or overloading. In addition, ground movements attributable to more recent mining, notably longwall mining in the case of coal, may trigger shaft failure. Collapse at an unfilled shaft usually is due to the deterioration and failure of the shaft lining, the older the shaft, the more likely it is to fail. Shaft collapse may manifest itself at the surface as a hole roughly equal to the diameter of the shaft if the lining remains intact or if the ground around the shaft consists of solid rock. More frequently, however, shaft linings deteriorate with age to a point at which they are no longer capable of retaining the surrounding material. As far as the geological conditions surrounding a shaft are concerned, the geomechanical properties of the cover rocks, the groundwater conditions and possibly the geological structures influence the stability of shafts. In particular,

Figure 2.24 Collapse of an old shaft, trapping a car, Cardiff, South Wales.

if superficial deposits surround a shaft that is open at the top, the deposits are likely to collapse into the shaft at some point in time and thereby form a crown hole at the surface. The thicker these superficial deposits are, the greater will be the dimensions of the crown hole (Fig. 2.25). Such collapses may affect adjacent shafts if they are interconnected.

Abandoned mine shafts may or may not have a surface expression, in the former instance this may be a circular depression. In the latter case an investigation is needed to locate the shaft(s). The search for old shafts on land that is about to be developed should not be confined to the site itself but should extend beyond it for a sufficient distance to find any shafts that could affect the site if a collapse occurred. An investigation should include a desk study involving a survey of maps, literature and remote sensing imagery and aerial photographs (Anon., 1976). The principal sources of information include plans of abandoned mines, geological records of shaft sinking, all available editions of topographic and geological maps, imagery, aerial photographs, archival and other official records. Mooijman *et al.* (1998), for example, described the use of aerial photographs to help locate abandoned shafts in

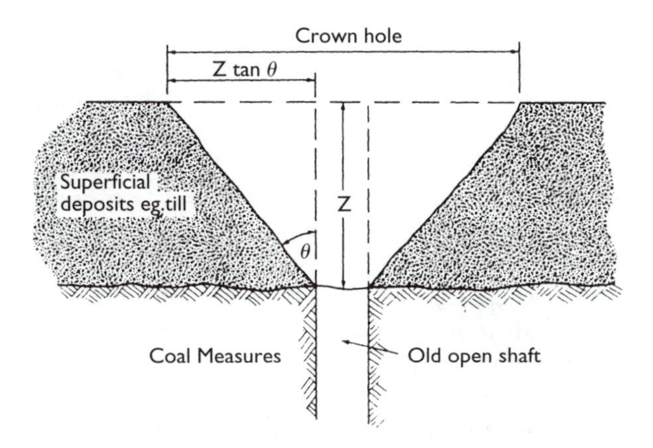

Figure 2.25 Indication of possible diameter of a crown hole developed in superficial deposits due to the collapse of a mine shaft. $\theta = 90^\circ - \phi$ (ϕ = angle of friction of superficial deposits).

the coal mining district of East Limburg, the Netherlands. Fortunately, aerial photographs were available from 1935 onwards and the older photographs proved more useful since much land has been built over in more recent years. They found that digital enhancement of the photographs helped identification and exact determination of shaft location.

Because of their different heat capacities, mine shafts and the surrounding ground are likely to exhibit temperature differences, mine shafts generally being cooler for much of the year but warmer during winter. Consequently, airborne thermal images (e.g. Airborne Thematic Mapper, ATM, imagery) may be used to detect the presence of an anomaly caused by a mine shaft (Gunn *et al.*, 2004). However, the situation is not always straightforward and may be complicated by a number of factors (Ager *et al.*, 2004). Obviously, the moisture content of the ground around a shaft affects its temperature, as does the nature of the surface vegetation, which tends to act as a thermal blanket. Wetter ground tends to be cooler than drier ground and mining disturbed ground often gives rise to changes in vegetation. Old shafts can either drain surface waters or provide spring sources for rising mine water and thereby be responsible for localized changes in groundwater flow. Any discard from a mine around a shaft or discard mixed with soil has an affect on soil chemistry that, in turn, can affect vegetation. Changes in the density of soil, caused by soil mixing or soil disturbance, affects the thermal inertia of the soil, that is, its resistance to temperature change. Furthermore, the oxidation of mine gasses as they escape from a mine shaft leads to an increase in temperature of a degree or so. However, barometric pressure has an influence on the emission of gas in that high pressure inhibits its escape. Direct observation of surface temperature differences can be made with thermometers such as the digital thermometer and the precision radiation thermometer. Both have a resolution of around 0.1°C. A review of thermal techniques used in the location of shafts has been provided by Donnelly and McCann (2000).

The success of geophysical methods in locating old shafts depends on the existence of a sufficient contrast between the physical properties of the shaft and its surroundings to produce an anomaly. If no contrast exists a shaft cannot be detected. Moreover, a shaft will frequently remain undetected when the top of the shaft is covered by more than 3 m of fill. The size, especially the diameter, of a shaft influences whether or not it is likely to be detected.

A resistivity survey may be successful where there is a significant contrast in electrical resistance between a shaft and its surroundings. Anomalies are detectable if the depth of cover around a shaft is less than its radius and success generally has been achieved where the cover was less than 1 m in thickness. Both sounding and profiling surveys should be undertaken, the latter are used to detect vertical anomalies that could be attributable to shafts. An isoresistivity map can be drawn from the results of profiling and may indicate the presence of a shaft or shafts.

Electromagnetic methods can be effective in the detection of mine shafts when shafts contain a mixture of materials with highly contrasting conductivities. They are used along profiles or grids.

Mooijman *et al.* (1998) used ground probing radar to help detect old coal mine shafts in the Netherlands. They pointed out that the use of shielded antennae during a survey provided a clearer subsurface image than unshielded antennae. In addition, they emphasized that a detailed site description was a prime requirement in order to avoid misinterpretation of apparent subsurface reflections and other anomalies.

Magnetic surveys, especially those using the proton-magnetometer and, more recently, the fluxgate magnetic field gradiometer, have had some success in locating old mine shafts. Abandoned mine shafts may be detected where a shaft is lined by bricks or iron tubbing, bricks containing a certain amount of iron. Similarly, if any shaft filling contains metallic objects such as rails or boggies, then they may be detected by a magnetic survey. An isomagnetic contour map is produced from the results (Fig. 2.26).

An open shaft offers a gravity contrast but unfortunately most old shafts are too small to produce a difference in density that is sufficient to yield a gravity anomaly. The maximum depth at which an open shaft can be located by a gravity survey is about 1.5 times the diameter of the shaft. If the shaft has been backfilled the depth is less. The production of a gravity contour map hopefully allows any anomalies to be recognized.

Seismic surveys are generally not used to locate old mine shafts. However, crosshole seismic testing, which involves sending both compressional and shear waves between drillholes, can be used to detect shafts (see earlier).

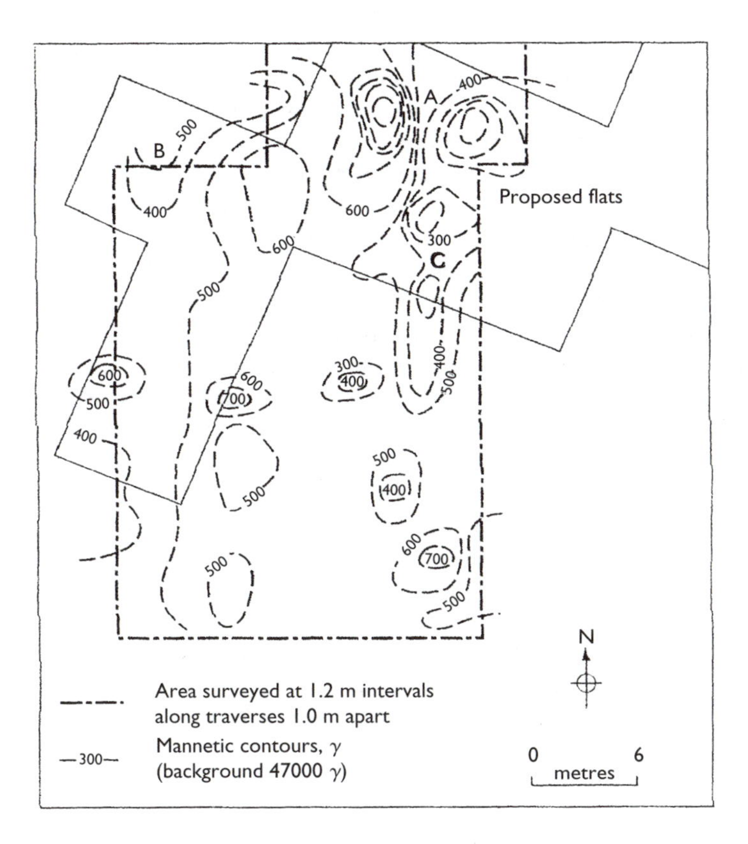

Figure 2.26 Isomagnetic map of a site containing old mine shafts at A, B and C.

Crosshole techniques can be used when the depth of burial of a void is more than 2 or 3 times the diameter of the void. Generally, the source and receiver are at the same level in the two drillholes and are moved up and down together. Drillholes must be spaced closely enough to achieve the required resolution of detail. The method can be used to detect subsurface cavities if the cavity is directly in line between two drillholes and has at least one-tenth of the drillhole separation as its smallest dimension. Air-filled cavities are detectable more readily than those filled with water.

Geochemical exploration depends on identifying chemical changes such as changes in mineral content or in the chemical character of the moisture content in the soil associated with old mine workings. In addition, gases such as carbon dioxide, carbon monoxide, methane and nitrogen may accumulate in open or partially filled shafts. Indeed, the most effective geochemical method yet used in the location of abandoned coal mine shafts is that of methane detection. Methane, being a light gas, may escape from old shafts and methane detectors can record concentrations as low as 1 ppm. Anomalies associated with old coal mine shafts generally have ranged between 10 and 100 ppm. The detector is carried over the site near the ground and should not be used on a windy day. A contour map showing methane concentration is produced.

The confirmation of the existence of a shaft, is accomplished by excavation. As this may be a hazardous task, necessary safety precautions should be taken. Gregory (1982) suggested that when a possible position of a shaft has been determined, a mechanical boom-type digger can be used to reveal its presence. The excavator is anchored outside the search area, and harness and lifelines should be used by the operatives. A series of parallel trenches are dug at intervals reflecting the possible diameter of the shaft. If excavation to greater depth than can be achieved by an excavator is required, resort may be made to a dragline. Again the dragline should be anchored safely outside the search area.

Where a site is considered potentially dangerous, in that shaft collapse may occur, or where obstructions prevent the use of earthmoving equipment, a light mobile rig can be used to drill exploratory holes. The rig should be placed on a platform or girders long enough to give protection against shaft collapse. Since many old shafts have diameters around 2 m or even less, drilling should be undertaken on a closely spaced grid. Rotary percussion drilling can be done quickly and the holes can be angled. Changes in the rate of penetration may indicate the presence of a shaft or the flush may be lost. Significant differences in the depth of unconsolidated material may indicate the presence of a filled shaft or the fill may differ from the surrounding material.

Once a shaft has been located, the character of any fill occurring within it needs to be determined. A drillhole alongside the shaft to determine the thickness of the overburden and the stratal succession, especially if the latter is not available from mine plans, proves very valuable. In particular, in old coal mines the positions of the seams may mean that mouthings open into the shaft at those levels. It also is important to record the position of the water table and, if possible, the condition of the shaft lining.

If the depth is not excessive and the shaft is open, it can be filled with suitable granular material. The lower part, approximately five times the shaft diameter, should be filled with non-degradable, open graded hardcore or boulders that do not exceed 330 mm in diameter. This will allow for drainage of mine water or groundwater without the loss of fill material. The remainder of the shaft usually is filled with uniform granular crushed rock or gravel. Excess fill material may be required during the filling process to use for any subsequent consolidation and settlement. The top of the fill should be compacted. Where more than one shaft is to be filled, then the filling should be carried out simultaneously so as to reduce the chance of inducing collapse. Alternatively, Dean (1967) suggested that if the exact positions of the mouthings in an abandoned open shaft are known, then

Figure 2.27 Heavy duty reinforced concrete slab acting as a cover for an old mine shaft. (Reproduced by kind permission of White Young Green Consulting Limited, International Mining Consultants Limited and the Coal Authority.)

these areas should be filled with gravel and grouted, the rest of the shaft being filled with mine waste. However, the latter will tend to consolidate much more than will gravel. If, as is more usual, the shaft is filled with debris in which there are voids, then these should be filled with pea gravel and grouted. The types of grout may vary but usually consist of cement, sand and pulverized fly-ash (PFA), or polymer mixes. In general, shafts that are up to 2 m in diameter can be grouted successfully with one central grout hole whereas shafts that have larger diameters up to 6 m may require five or more holes, grouting being carried out in stages. Grouting is necessary when structures are to be located over or in close proximity to abandoned shafts, or when it is not possible to cap a shaft at rockhead due to thick overburden.

A reinforced concrete capping commonly is used to cover a shaft. The concrete capping should take the form of an inverted cone. The zone immediately beneath the capping is grouted. The cap width is normally twice the internal diameter of the shaft and is located at or below rockhead. Cap thicknesses are variable but should not be less than 0.5 m. Anon. (1982) supplied details concerning the concrete covers and cappings needed to seal mine shafts (Fig. 2.27). A cap should be marked suitably by a permanent marker so that it can be easily located and identified. Hazards such as subsidence, mine water and gas emissions also need to be considered, and it may be necessary to vent a cap to prevent concentrations of gas.

Obviously, the easiest way to deal with old mine shafts is to avoid locating structures in their immediate vicinity. Nevertheless, for practical purposes it is necessary to define a safe limit around a mine shaft, that is, a distance from the shaft, beyond which damage to a property will not occur in the event of shaft collapse. This is not an easy task since the stability of an individual shaft is controlled by a number of factors including the type of mining that went on and the geological conditions around the shaft. In addition, a consideration of the long-term stability of shaft linings is required, as well as consideration of the effects of increased lateral pressure due to the erection of structures nearby. Because of the possible danger of collapse and cratering it was recommended in Britain that the minimum distance for siting buildings from open or poorly filled shafts should be twice the thickness of the superficial deposits up to a depth of 15 m, unless they are exceptionally weak. Alternatively, Anon. (1982) recommended that a safety zone around a shaft should have dimensions that may be subtended by an angle of 45° to the surface from the sides of the shaft where it intersects rockhead. Protective sheet piles or concrete walls can be constructed around shafts to counteract cratering, but such operations should be carried out with due care to avoid shaft disturbance or collapse.

2.8. Case history 1

From around 1780 to 1920 large quantities of limestone were extracted from mines in the West Midlands of England, giving rise to about 250 ha of derelict land at the present day (Bell *et al.*, 1989). Unfortunately, the existence of these workings were not fully recorded and so the potential subsidence risk has adversely affected attempts to redevelop the area. The limestone that was worked is Silurian in age and several extensive mines were developed at depths of up to 250 m below the surface. Pillar and stall workings occur where the dip of the limestone is less than 30° but where steeper dips occur, the mines take the form of long galleries that run parallel to the strike. The limestone mines were left unfilled once abandoned. Although some old shafts were filled with debris, most simply had a stage constructed near the top that then was covered with earth.

Since the existence of limestone mines was known, it was tacitly accepted that shallow workings (i.e. at less than 60 m depth beneath the surface) could give rise to instability problems and so development did not take place in such areas. Where mines existed at greater depth they were not considered as likely to affect the surface, hence no restrictions were placed on development. However, in 1978 partial collapse occurred at Cow Pasture Mine, Wednesbury, and this mine is located at around 145 m depth. The resulting surface depression was approximately 200 m by 300 m and the maximum subsidence was 1.5 m. The initial rate of subsidence was slow but increased to a maximum of around 50 mm per day and then gradually tailed off over a period of several months.

This event caused an intensive investigation of these limestone mines to be put into operation, the object being to assess the risk of collapse and the need for remedial work. The desk study collected all information onto one database and some 30 mines requiring investigation were identified. Although some mines were dry and could be entered and inspected, most were flooded. Inspection of Castlefields Mine at Dudley revealed that the pillars were in good condition, showing no evidence of spalling or crushing due to overburden load. However, many roof collapses above the stalls have in the past led to the appearance of crown holes at the surface and major roof falls were observed within the mine. Many of the voids in the roof formed by falls were bounded by joint faces and were rectangular in outline. Forster (1988) showed that the falls were due to the deterioration of stiff clay in unevenly dilated master joints, roof falls occurring most frequently where joint dilation was greatest and clay infill was thickest. Microseismic monitoring of roof fall within Castlefields Mine indicated that rates of collapse were greater than expected (Miller *et al.*, 1988).

An influence zone for each mine was identified within which damage could be caused if subsidence occurred. The boundaries of such zones were taken as the limit of 0.2% horizontal ground strain, outside of which normal structures are unlikely to undergo more than negligible damage if affected by subsidence. Any new development inside the zones was frozen until the conditions of the mines concerned were ascertained and any necessary remedial action undertaken. The subsidence risk, either in terms of extensive trough subsidence as a result of pillar collapse, or the appearance of crown holes at the surface due to void migration, was evaluated. Each mine was ranked in terms of its relative potential instability and land use within its zone.

The principal objective of the investigations was to determine the extent and condition of the mine workings. To this end, an extensive drilling programme was carried out. Double barrel sampling tubes with inner plastic liners were used to obtain core. Core was photographed and logged, and the rock quality designation (RQD) and fracture spacing index recorded. Drilling penetration rates, water flush returns and *in situ* permeability tests also were used to assess the degree of fracturing. The degree of fracturing is important in that it tends to increase as old workings are approached. In fact, the shale above the top limestone is intensely fractured for several metres above voids that have migrated into the shale.

Geophysical logging, namely, gamma-ray, gamma-gamma, neutron, caliper, multi-channel sonic, resistivity and temperature logging, were carried out in drillholes. This was done to identify changes in lithology and to assess the degree of fracturing. In addition, a dipmeter was run in conjunction with a verticality log (which records the tilt and azimuth of a drillhole) so that the dip direction of the strata within the drillhole could be evaluated. Correlation between dipmeter dip and *in situ* dip proved to be very good. An ultrasonic survey was carried out within each drillhole that entered a void exceeding 1 m in height at mine level. Such surveys were undertaken in flooded mines. Horizontal scans were made at 1 m intervals over the height of the old workings and then tilting scans at 15° intervals were made from vertical to horizontal. Thus, old workings within approximately 60 m of the drillhole were mapped and the shapes of voids measured.

Initially, the use of closed circuit television to explore old workings was of little value since illumination only allowed a clear view to a distance of some 2–3 m. However, the use of stronger lighting, image intensifiers and computer enhancement allowed the field of view to be increased to around 20 m.

It would appear that for given rock conditions and mine layout, as revealed at Cow Pasture Mine, there is a critical height for the development of 'pillars' in the shale overlying the limestone at which the overburden pressure causes the shale to be crushed. These 'pillars' in the shale are an upward extension of the pillars in the limestone and have been formed by the collapse of roof rock in the stalls. In such cases, a factor of safety can be regarded as the difference in height of a pillar between its present height and that at which failure by crushing will occur. Obviously, the rate at which roof fall occurs, thereby extending pillar height, is important since when one pillar fails its load is transferred onto those surrounding it, thus causing them to fail. There is no way of accurately predicting when and where such failures will occur, nor the form they will adopt.

In such circumstances the question is how to deal with the old workings. One option is to do nothing other than maintain periodic checks on surface levels and underground events and then to take action if a subsidence event is forecasted. This can be acceptable over agricultural land but is not satisfactory in areas of urban development. Here mines have to be treated, usually by filling, to prevent the possibility of future collapse. Filling these large limestone mines (e.g. some may occupy a volume of approximately 500 000 m^3) requires a cheap material that is locally available in large quantities and that can be placed at a high rate. It also needs to be capable of spreading a long way in a mine before it stiffens, so that the number of injection holes can be limited. Fortunately, in the West Midlands there is an abundance of colliery spoil that can be used to make rock paste. Rock paste is waste rock debris, especially colliery spoil, mixed with water. However, it should contain enough fines (i.e. silt and clay sized material) for it to flow under pressure through a pipeline and to spread through the mine workings as a plastic material at constant yield stress without segregation or drainage occurring during flow. Rock paste may have small percentages of PFA and lime (less than 5 and 3% respectively) added to it. This is to ensure that it will achieve a design strength of 20 kPa by pozzolanic reaction between the constituents.

2.9. Case history 2

Tennant Mine is 1 of 8 mines that extracted rock salt from Triassic strata near Carrickfergus, Northern Ireland, at various times from the mid-nineteenth century until 1958. Both underground mining and brine extraction methods were practised. On 19 October 1990, the national seismic monitoring network of the British Geological Survey recorded a seismic event measuring 2.5 on the Richter scale that was associated with collapse at the disused Tennant Mine. Until the collapse, there was no evidence to indicate that the mine was liable to subside. However, an underground rumbling noise had been heard on 2 September, and was followed by 'creaking and groaning' noises on 5 October 1990, these being noted by residents in houses nearby. The collapse of Tennant Mine generated signals on seismographs in Northern Ireland, the Isle of Skye, south-west Scotland and north-west England, all of which were used to locate the event. The seismic wave form was of low frequency with large surface waves, consistent with mine collapse (Donnelly, 1998). On the morning of the collapse an electricity substation subsided into the ground by approximately 8.0 m just off Trailcock Road, and crown holes developed within a major subsidence depression defined by fractures and tension gashes (Fig. 2.28). The latter ruptured Trailcock Road.

Following closure and abandonment of a salt mine, groundwater, if it gains access to the mine, will dissolve remnant support pillars, which can lead to collapse. However, once the mine becomes flooded and groundwater has dissolved sufficient salt, the groundwater becomes saturated. At this point, no further dissolution of the pillars occurs unless groundwater is moving so that fresh supplies are supplied continuously. The subsidence depression at the Tennant Mine site is now flooded, and the tension gashes and

Figure 2.28 Flooded subsidence depression at Carrickfergus, Northern Ireland.

Figure 2.29 Collapsed mine shaft at Carrickfergus, Northern Ireland.

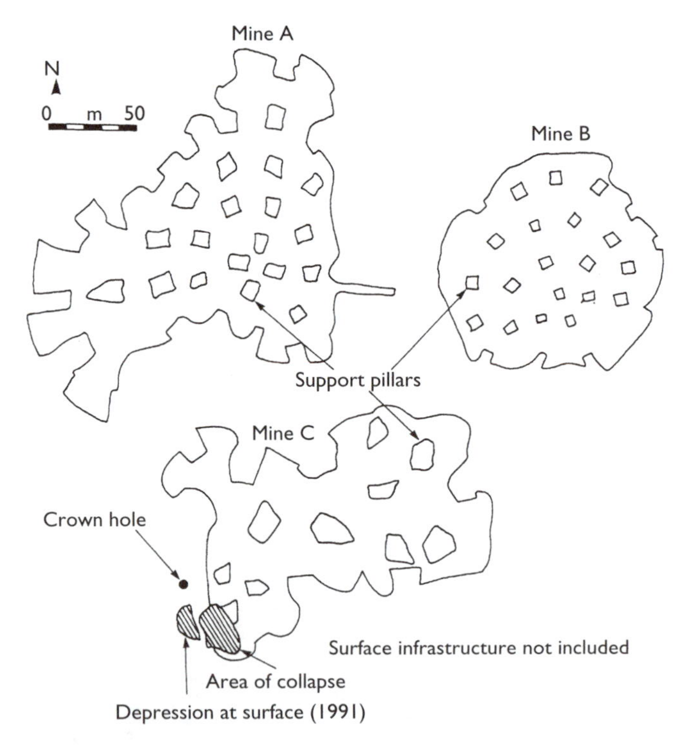

Figure 2.30 Plan of abandoned salt mine workings and the 1991 collapse at Carrickfergus, Northern Ireland. (Reproduced by kind permission of the British Geological Survey. © NERC. All rights reserved. IPR/61–05C.)

fractures on the north-eastern part of the subsidence depression are located approximately 5 m above the level of the water. These fractures are extensive and expose superficial deposits. Consequently, they probably afford access of water to the mine. Other likely means of access to the mine are crown holes developed above shaft positions (Fig. 2.29). Although many of these shaft sites have been secured and fenced, and some are monitored and maintained on a regular basis, nonetheless large areas of poorly supported, or unsupported stalls are likely to pose a risk of future collapse (Fig. 2.30). As such, this could present problems for future planning and development.

References

Ager, G.J., Gibson, A.D., Gunn, D.A., Raines, M.G., Marsh, S.H., Farrant, A.R., Lowe, D.R., Forster, A. and Culshaw, M.G. 2004. *Thermal Imaging Techniques to Map Ground Disturbances in a Former Nottinghamshire Coal Mining Area.* British Geological Survey, Internal Report IR/02/007, Keyworth, Nottinghamshire.

Anon. 1976. *Reclamation of Derelict Land: Procedure for Locating Abandoned Mine Shafts.* Department of the Environment, London.

Anon. 1977. *Ground Subsidence.* Institution of Civil Engineers, London.

Anon. 1982. *Treatment of Disused Mine Shafts and Adits.* National Coal Board, London.

Anon. 1988. Engineering geophysics: Report by the Geological Society Engineering Group Working Party. *Quarterly Journal Engineering Geology*, **21**, 207–271.

Anon. 1999a. *Code of Practice for Site Investigations.* BS5930. British Standards Institution, London.

Anon. 1999b. *Standard Guide for Selecting Surface Geophysical Methods.* Designation D-6429. American Society for Testing and Materials, Philadelphia, PA.

Bell, F.G. 1986. Location of abandoned workings in coal seams. *Bulletin of the International Association of Engineering Geology*, **30**, 123–132.

Bell, F.G. 1988a. The history and techniques of coal mining and the associated effects and influence on construction. *Bulletin of the Association of Engineering Geologists*, **24**, 471–504.

Bell, F.G. 1988b. Land development: state-of-the-art in the search for old mine shafts. *Bulletin of the International Association of Engineering Geology*, **37**, 91–98.

Bell, F.G. 1992. Ground subsidence: a general review. *Proceedings of the Symposium on Construction over Mined Areas*, Pretoria, South African Institution of Civil Engineers, Yeoville, 1–20.

Bell, F.G., Cripps, J.C., Culshaw, M.G. and Stacey, T.R. 1989. Investigation and treatment of some abandoned mine workings. *Proceedings of the International Conference on Surface Crown Pillars and Abandoned Mine Workings*, Timmins, Ontario, CANMET, 155–167.

Bell, F.G., Culshaw, M.G., Moorlock, B.S.P. and Cripps, J.C. 1992. Subsidence and ground movements in chalk. *Bulletin of the International Association of Engineering Geology*, **45**, 75–82.

Bieniawski, Z.T. 1967. *An Analysis of the Results from Underground Tests Aimed at Determining the In Situ Strength of Coal Pillars.* Report MEGY569. Council for Scientific and Industrial Research, Pretoria, South Africa.

Bieniawski, Z.T. 1970. *In situ* large scale testing of coal. *Proceedings Conference In Situ Testing of Rocks and Soil*, British Geotechnical Society, London, 67–74.

Bishop, I., Styles, P., Emsley, S.J. and Ferguson, N.S. 1997. The detection of cavities using the microgravity technique: case histories from mining and karstic environments. In: *Modern Geophysics in Engineering Geology*, Engineering Geology Special Publication No. 12, McCann, D.M, Eddleston, M., Fenning, P.J. and Reeves, G.M. (eds), Geological Society, London, 153–166.

Bock, O. and Thom, C. 2001. Sub-CM subsidence measurements with the wide-angle airborne laser ranging system. *Surveys in Geophysics*, **22**, 537–548.

Braithwaite, P.A. and Cole, K.W. 1986. Subsurface investigations of abandoned limestone workings in the West Midlands of England by use of remote sensors. *Proceedings of the Conference on Rock Engineering in an Urban Environment*, Hong Kong, Institution of Mining and Metallurgy, London, 27–39.

Briggs, H. 1929. *Mining Subsidence.* Arnold, London.

Browne, M.A.E., Forsyth, I.H. and Macmillan, A.A. 1986. Glasgow: a case study in urban geology. *Journal of Geological Society*, **143**, 509–520.

Bruhn, R.W., Magnuson, M.O. and Gray, R.E. 1981. Subsidence over abandoned mines in the Pittsburg Coalbed. In: *Proceedings of the Second International Conference on Ground Movements and Structures*, Cardiff, Geddes, J.D. (ed.), Pentech Press, London, 142–156.

Bryan, A., Bryan, J.G. and Fouche, J. 1964. Some problems of strata control and support in pillar workings. *Mining Engineer*, **123**, 238–266.

Bullock, S.E.T. and Bell, F.G. 1997. Some problems associated with past mining in the Witbank Coalfield, South Africa. *Environmental Geology*, **33**, 61–71.

Cameron-Clarke, L.S., Barrett, A.J. and Blight, G.E. 1992. Support of areas undermined by coal workings for road corridors. *Proceedings of the Symposium on Construction over Mined Areas*, Pretoria, South African Institution Civil Engineers, Yeoville, 193–198.

Carnec, C. and Delacourt, C. 2000. Three years of mining subsidence monitored by SAR interferometry, near Gardanne, France. *Journal of Applied Geophysics*, **43**, 43–54.

Carter, P.G. 1985. Case histories which break the rules. In: *Mineworkings '84, Proceedings of the International Conference on Construction in Areas of Abandoned Mineworkings*, Edinburgh, Forde, M.C., Topping, B.H.V. and Whittington, H.W. (eds), Engineering Technics Press, Edinburgh, 20–29.

Carter, P., Jarman, D. and Sneddon, M. 1981. Mining subsidence in Bathgate, a town study. In: *Proceedings of the Second International Conference on Ground Movements and Structures*, Cardiff, Geddes, J.D. (ed.), Pentech Press, London, 101–124.

Coates, D.F. 1965. *Pillar Load: Part I, Literature Survey and Hypothesis.* Department of Mines and Technical Surveying, Research Report RI 68, Department of Mines and Technical Surveying, Ottawa, Canada.

Corin, L., Couchard, B., Dethy, B., Halleux, L., Mojoie, A., Richter, T. and Wauters, J.P. 1997. Radar tomography applied to foundation design in a karstic environment. In: *Modern Geophysics in Engineering Geology*, Engineering Geology Special Publication No. 12, McCann, D.M, Eddleston, M., Fenning, P.J. and Reeves, G.M. (eds), Geological Society, London, 167–173.

Cripps, J.C., McCann, D.M., Culshaw, M.G. and Bell, F.G. 1988. The use of geophysical methods as an aid to the detection of abandoned shallow mine workings. In: *Minescape '88, Proceedings of the Symposium on Mineral Extraction,*

Utilisation and Surface Environment, Harrogate, Institution Mining Engineers, Doncaster, 281–289.

Dean, J.W. 1967. Old mine shafts and their hazard. *Mining Engineer*, **127**, 368–377.

Denkaus, H.O. 1962. Critical review of the present state of scientific knowledge related to the strength of mine pillars. *Journal of the South African Institute of Mining and Metallurgy*, **63**, 59–75.

Donnelly, L.J. 1998. *Mining Subsidence at Carrickfergus Salt Mines Northern Ireland*. British Geological Survey and Geological Survey of Northern Ireland, Report PN/98/7, Keyworth, Nottinghamshire.

Donnelly, L.J. and McCann, D.M. 2000. The location of abandoned mine workings using thermal techniques. *Engineering Geology*, **57**, 39–52.

Edmonds, C.N., Green, C.P. and Higginbottom, I.E. 1987. Subsidence hazard prediction for limestone terrains, as applied to the English Cretaceous Chalk. In: *Planning and Engineering Geology*, Engineering Geology Special Publication No. 4, Culshaw, M.G., Bell, F.G., Cripps, J.C. and O'Hara, M. (eds), the Geological Society, London, 283–294.

Forster, A. 1988. The geology of Castlefields Mine, Dudley, and its effect on the stability of the mine roof. In: *Engineering Geology of Underground Movements*, Engineering Geology Special Publication No. 5, Bell, F.G., Culshaw, M.G., Cripps, J.C. and Lovell, M.A. (eds), the Geological Society, London, 287–291.

Garrard, G.E.G. and Taylor, R.K. 1988. Collapse mechanisms of shallow coal mine workings from field measurements. In: *Engineering Geology of Underground Movements*, Engineering Geology Special Publication No. 5, Bell, F.G., Culshaw, M.G., Cripps, J.C. and Lovell, M.A. (eds), the Geological Society, London, 181–192.

Goodman, R.E., Korbay, S. and Buchignani, A. 1980. Evaluation of collapse potential over abandoned room and pillar mines. *Bulletin of the Association of Engineering Geologists*, **17**, 27–37.

Gowan, S.W. and Trader, S.M. 2000. Mine failure associated with a pressurized brine horizon: Retsof Salt Mine, western New York. *Environmental and Engineering Geoscience*, **6**, 57–70.

Gray, R.E. and Meyers, J.F. 1970. Mine subsidence support methods in the Pittsburg area. *Proceedings American Society Civil Engineers, Journal of the Soil Mechanics and Foundations Division*, **96**, 1267–1287.

Greenwald, H.P., Howarth, H.C. and Hartman, I. 1941. *Experiments on Strength of Small Pillars of Coal in the Pittsburg Bed*. United States Bureau of Mines, Technical Paper 605, U.S. Government Printing Office, Washington, DC.

Gregory, O. 1982. Defining the problem of disused coal mine shafts. *Chartered Land Surveyor/Chartered Mine Surveyor*, **4**, No. 2, 4–15.

Gunn, D.A., Ager, G.J., Marsh, S.H., Raines, M.G., Waters, C.N., McManus, K.B., Forster, A., Jackson, P.D. and Lowe, D.R. 2004. *The Development of Thermal Imaging Techniques to Detect Mineshafts*. British Geological Survey, Internal Report IR/01/24, Keyworth, Nottinghamshire.

Hasan, S. 1996. Subsidence hazard from limestone mining in an urban setting. *Environmental and Engineering Geoscience*, **2**, 497–505.

Healy, P.R. and Head, J.M. 1984. *Construction over Abandoned Mine Workings*. Construction Industry Research and Information Association (CIRIA), Special Publication 32, London.

Hoek, E. and Brown, E.T. 1980. *Underground Excavations in Rock*. Institution of Mining and Metallurgy, London.

Holland, C.T. 1964. The strength of coal in mine pillars. *Proceedings of the Symposium on Rock Mechanics*, University of Missouri, Rolla, Missouri, Pergamon Press, New York, 450–466.

Jackson, P.D. and McCann, D.M. 1997. Cross-hole seismic tomography for engineering site investigation. In: *Modern Geophysics in Engineering Geology*, Engineering Geology Special Publication No. 12, McCann, D.M, Eddleston, M., Fenning, P.J. and Reeves, G.M. (eds), Geological Society, London, 247–264.

Karfakis, M.G. 1993. Residual subsidence over abandoned coal mines. In: *Comprehensive Rock Engineering, Volume 5, Surface and Underground Case Histories*, Hoek, E. (ed.), Pergamon Press, Oxford, 451–476.

Koerner, R.M., Lord, A.E., Bowders, J.J. and Dougherty, W.W. 1982. C W microwave location of voids beneath paved areas. *Proceedings of the American Society Civil Engineers, Journal of the Geotechnical Engineering Division*, **108**, 133–144.

Leggo, P.J. and Leach, C. 1982. Subsurface investigations for shallow mine workings and cavities by the ground impulse radar technique. *Ground Engineering*, **15**, No. 1, 20–24.

Littlejohn, G.S. 1979. Consolidation of old coal workings. *Ground Engineering*, **12**, No. 3, 1–21.

Lloyd, B.N., Cripps, J.C. and Bell, F.G. 1995. The estimation of grout take for small scale developments in areas of shallow abandoned coal mining: some examples from the East Pennine Coalfield. In: *Engineering Geology and Construction*, Engineering Geology Special Publication No. 10, Culshaw, M.G., Cripps, J.C. and Walthall, S. (eds), Geological Society, London, 135–141.

McCann, D.M., Suddaby, D.L. and Hallam, J.R. 1982. *The Use of Geophysical Methods in the Detection of Natural Cavities, Mineshafts and Anomalous Ground Conditions.* Open File Report No. EG82/5, Institute of Geological Sciences, London.

McCann, D.M., Jackson, P.D. and Culshaw, M.G. 1987. The use of geophysical surveying methods in the detection of natural cavities and mineshafts. *Quarterly Journal Engineering Geology*, **20**, 59–73.

MacCourt, L., Madden, B.J. and Schuman, E.H.R. 1986. Case studies of surface subsidence of bord and pillar workings. *Proceedings of the Symposium on Effect of Underground Mining on Surface*, Johannesburg, South African National Group for Rock Mechanics (SANGORM), Johannesburg, 25–32.

McDowell, P.W., Barker, R.D., Butcher, A.P., Culshaw, M.G., Jackson, P.D., McCann, D.M., Skipp, B.O., Matthews, S.L. and Arthur, J.C.R. 2002. *Geophysics in Engineering Investigations.* Construction Industry Research and Information Association (CIRIA), Report C592, Engineering Geology Special Publication No. 19, Geological Society, London.

McMillan, A.A and Browne, M.A.E. 1987. The use and abuse of thematic information maps. In: *Planning and Engineering Geolgy*, Engineering Geology Special Publication No. 4, Culshaw, M.G., Bell, F.G., Cripps, J.C. and O'Hara, M. (eds), the Geological Society, London, 237–246.

Madden, B.J. and Hardman, D.R. 1992. Long-term stability of bord and pillar workings. *Proceedings of the Symposium on Construction over Mined Areas*, Pretoria, South African Institution Civil Engineers, Yeoville, 37–51.

Marino, G.G. and Choi, S. 1999. Softening effects on bearing capacity of mine floors. *Proceedings of the American Society Civil Engineers, Journal of Geotechnical and Geoenvironmental Engineering*, **125**, 1078–1089.

Marino, G.G. and Gamble, W. 1986. Mine subsidence damage from room and pillar mining in Illinois. *International Journal of Mining and Geological Engineering*, **4**, 129–150.

Merad, M.M., Verdal, T., Roy, B. and Kouniali, S. 2004. Use of multi-criteria decision-aids for risk zoning and management of large area subjected to mining induced hazards. *Tunnelling and Underground Space Technology*, **19**, 125–138.

Michalski, S.R. and Gray, R.E. 2001. Ash disposal – mine fires – environment: an Indian dilemma. *Bulletin of Engineering Geology and the Environment*, **60**, 23–29.

Miller, A., Richards, J.A. and McCann, D.M. 1988. Microseismic monitoring of the infill trial at Castlefields Mine, Dudley. In: *Engineering Geology of Underground Movements*, Engineering Geology Special Publication No. 5, Bell, F.G., Culshaw, M.G., Cripps, J.C. and Lovell, M.A. (eds), the Geological Society, London, 319–324.

Mooijman, O.P.M., Van der Kruk, J. and Roest, J.P.A. 1998. The detection of abandoned mine shafts in the Netherlands. *Environmental and Engineering Geoscience*, **4**, 307–316.

Orchard, R.J. 1964. Partial extraction and subsidence. *Mining Engineer*, **123**, 417–427.

Palchik, V. 2000. Prediction of hollows in abandoned underground workings at shallow depth. *Geotechnical and Geological Engineering*, **18**, 39–51.

Piggott, R.J. and Eynon, P. 1978. Ground movements arising from the presence of shallow abandoned mine workings. *Proceedings of the First International Conference on Large Ground Movements and Structures*, Cardiff, Geddes, J.D. (ed.), Pentech Press, London, 749–780.

Price, D.G. 1971. Engineering geology in the urban environment. *Quarterly Journal of Engineering Geology*, **4**, 191–208.

Price, D.G., Malkin, A.B. and Knill, J.L. 1969. Foundations of multi-storey blocks on coal measures with special reference to old mine workings. *Quarterly Journal of Engineering Geology*, **1**, 271–322.

Russell, O.R., Amato, R.V. and Leshendok, T.V. 1979. Remote sensing and mine subsidence in Pennsylvania. *Proceedings of the American Society of Civil Engineers, Transportation Engineering Journal*, **195**, 185–199.

Salamon, M.O.D. 1967. A method of designing bord and pillar workings. *Journal of the South African Institution of Mining and Metallurgy*, **68**, 68–78.

Salamon, M.D.O. and Munro, A.H. 1967. A study of the strength of coal pillars. *Journal of the South African Institution of Mining and Metallurgy*, **68**, 53–67.

Sheorey, P.R., Loui, J.P., Singh, K.B. and Singh, S.K. 2000. Ground subsidence observations and a modified influence function method for complete subsidence prediction. *International Journal of Rock Mechanics and Mining Science*, **37**, 801–818.

Singh, K.B. and Dhar, B.B. 1997. Sinkhole subsidence due to mining. *Geotechnical and Geological Engineering*, **15**, 327–341.

Siriwardane, H.J., Kannan, R.S.S. and Ziemkiewicz, P.F. 2003. Use of waste materials for the control of acid mine drainage and subsidence. *Proceedings of the American Society Civil Engineers. Journal of Environmental Engineering*, **129**, 910–915.

Sizer, K.E. and Gill, M. 2000. Pillar failure in shallow coal mines – a recent case history. *Transactions Institution of Mining and Metallurgy*, Section A, Mining Technology, **109**, A146–A152.

Smith, G.J. and Rosenbaum, M.S. 1993. Abandoned shallow mine workings in chalk: a review of the geological aspects leading to their destabilisation. *Bulletin of the International Association of Engineering Geology*, **48**, 101–108.

Speck, R.C. and Bruhn, R.W. 1995. Non-uniform mine subsidence ground movement and resulting surface-structure damage. *Environmental and Engineering Geoscience*, **1**, 61–74.

Statham, I., Golightly, C. and Treharne, G. 1987. The thematic mapping of the abandoned mining hazard – a pilot study for the South Wales Coalfield. In: *Planning and Engineering Geology*, Engineering Geology Special Publication No. 4, Culshaw, M.G., Bell, F.G., Cripps, J.C. and O'Hara, M. (eds), the Geological Society, London, 255–268.

Steart, F.A. 1954. Strength and stability of pillars in coal mines. *Journal of the South African Society of Chemistry, Metallurgy and Mining*, **54**, 309–325.

Szechy, K. 1970. *The Art of Tunnelling*. Akademia Kiado, Budapest.

Taylor, R.K. 1968. Site investigation in coalfields. *Quarterly Journal of Engineering Geology*, **1**, 115–133.

Terzaghi, K. 1943. *Theoretical Soil Mechanics*. Wiley, New York.

Terzaghi, K. 1946. Introduction to tunnel geology. In: *Rock Tunnelling with Steel Supports*, Proctor, R.V. and White, T. (eds), Commercial Shearing and Stamping Company, Youngstown, Ohio, 17–99.

Thorburn, S. and Reid, W.H. 1978. Incipient failure and demolition of two-storey dwellings due to large ground movements. *Proceedings of the First International Conference on Large Ground Movements and Structures*, Cardiff, Geddes, J.D. (ed.), Pentech Press, London, 87–99.

Tincelin, E. 1958. *Pression et Deformations de Terrain dans les Mines de Fer de Lorraine*. Jouve Editeurs, Paris.

Van Besian, A.C. and Rockaway, J.D. 1988. Influence of overburden on subsidence development over room and pillar mines. In: *Engineering Geology of Underground Movements*, Engineering Geology Special Publication No. 5, Bell, F.G., Culshaw, M.G., Cripps, J.C. and Lovell, M.A. (eds), the Geological Society, London, 215–220.

Waite, R.H. and Allen, A.S. 1975. Pumped slurry backfilling of inaccessible mine workings for subsidence control. Information Circular 8667, United States Bureau of Mines, *United States Government Printing Office*, Washington, DC.

Walton, G. and Cobb, A.E. 1984. Mining subsidence. In: *Ground Movements and Their Effects on Structures*, Attewell, P.B. and Taylor, R.K. (eds), University College Press, London, 216–242.

Wardell, K. and Eynon, P. 1968. Structural concept of strata control and mine design. *Transactions Institution of Mining and Metallurgy*, **77**, Section A, A125–A150.

Wardell, K. and Wood, J.C. 1965. Ground instability problems arising from the presence of shallow old mine workings. *Proceedings of the Midland Society Soil Mechanics and Foundation Engineering*, 7, 7–30.

White, H. 1992. Accurate delineation of shallow subsurface structure using ground penetrating radar. *Proceedings of the Symposium on Construction over Mined Areas*, Pretoria, South African Institution of Civil Engineers, Yeoville, 23–25.

Wilson, A.H. 1972. A hypothesis concerning pillar stability. *Mining Engineer*, **131**, 409–417.

Yager, R.M., Miller, T.S. and Kappel, W.M. 2001. Simulated effects of 1994 salt mine collapse on groundwater flow and land subsidence in a glacial aquifer system, Livingstone County, New York. United States Geological Survey, Professional Paper 1611, *United States Government Printing Office*, Washington, DC, 1–80.

Chapter 3

Longwall mining and subsidence

With the coming of the Industrial Revolution the demand for coal in Britain increased. Longwall mining evolved more or less at the same time, probably originating first in the Shropshire Coalfield. However, longwall mining has been used to mine other stratiform deposits, although it primarily is used to mine coal.

The great majority of the coal mined in Britain in the twentieth century and at present, although now greatly reduced in amount, is by panel working, which is a development of the longwall system suited to mechanized extraction techniques. This method of mining involves the total extraction of a series of panels of coal that are separated by pillars whose width is small compared to overburden thickness. The coal is exposed at a face 30–300 m in width between two parallel roadways. The roof is supported temporarily at the working face, and near the roadways. After the coal has been won and loaded the face supports are advanced leaving the rock, in the areas where coal has been removed, to collapse (Whittaker and Reddish, 1989). Subsidence at the surface more or less follows the advance of the working face and may be regarded as immediate.

Total extraction of coal, although not the predominant method of mining in the United States, began in the latter part of the nineteenth century (Bruhn *et al.*, 1981; Yokel *et al.*, 1982). This involved the extraction of long narrow pillars that were left between the stalls formed during the initial phase of mining. Subsidence of the ground was contemporaneous with pillar extraction. Block systems of mining were introduced in the 1920s wherein square pillars, with sides varying from 18 to 30 m, were separated by narrow stalls and entries. Again pillars were removed subsequently. This represents a more successful system of total extraction than that involving long narrow pillars. In the latter method 20% or more coal could be lost as a result of roof falls, squeezes or other complications that meant that some pillars had to be left in place. Hence, the likelihood of subsidence occurring remains a possibility for years after mines have been abandoned.

Subsidence associated with longwall mining differs in two important respects from that associated with abandoned pillar and stall workings.

First, longwall mining can be predicted with a reasonable degree of accuracy. Second, longwall mining is more or less contemporaneous with extraction. These factors mean that subsidence associated with longwall mining can be dealt with by surface developers either by designing structures to accommodate the surface movements or by delaying development until subsidence has ceased (Bell, 1988). However, it must be admitted that some of the factors that influence subsidence due to longwall mining, such as the reactivation of faults, cannot be quantified with any precision.

3.1. Subsidence and longwall mining

Shadbolt (1978) provided a historical review of the various theories relating to mining subsidence due to total extraction of coal. According to Lehman (1919), the subsidence that occurs over a completely mined out area in a flat seam is trough-shaped and extends outwards beyond the limits of mining in all directions. In trough subsidence the resulting stratal and surface ground movements are regarded as largely contemporaneous with mining, producing more or less direct effects at the surface (Brauner, 1973a). Trough shaped subsidence profiles develop tilt between adjacent points that have subsided different amounts and curvature results from adjacent sections that are tilted by differing amounts. Maximum ground tilts are developed above the limits of the area of extraction and may be cumulative if more than one seam is worked up to a common boundary. Where movements occur, points at the surface subside downwards and are displaced horizontally inwards towards the axis of the excavation (Fig. 3.1). Differential horizontal displacements result in a zone of apparent extension on the convex part of the subsidence profile (over the edges of the excavation) whilst a zone of compression develops on the concave section over the excavation itself. Differential subsidence can cause substantial damage, the tensile strains thereby generated usually being the most effective in this respect (Fig. 3.2). Comparatively slight deviations in the subsidence profile are accompanied by appreciable variations in strain. In fact, ground movement is three-dimensional and movements of the vertical and two horizontal components may occur simultaneously.

As mentioned, removal of roof support in longwall mining is followed by collapse of those rocks that are immediately above the coal seam since they are subjected to bending and tensile stresses. These broken rocks offer partial support to the superincumbent roof layers. Nevertheless, stresses in the rock mass remaining in place are significantly increased, and the resultant fracture and associated dilation mean that the rock strength is reduced from a peak to a residual value with loss in load bearing capacity and the redistribution of stress (Farmer and Altounyan, 1981). In addition, laminated rock masses may suffer bed separation. The fracture zone defined by the extent of dilation extends at least half the face width above seam level.

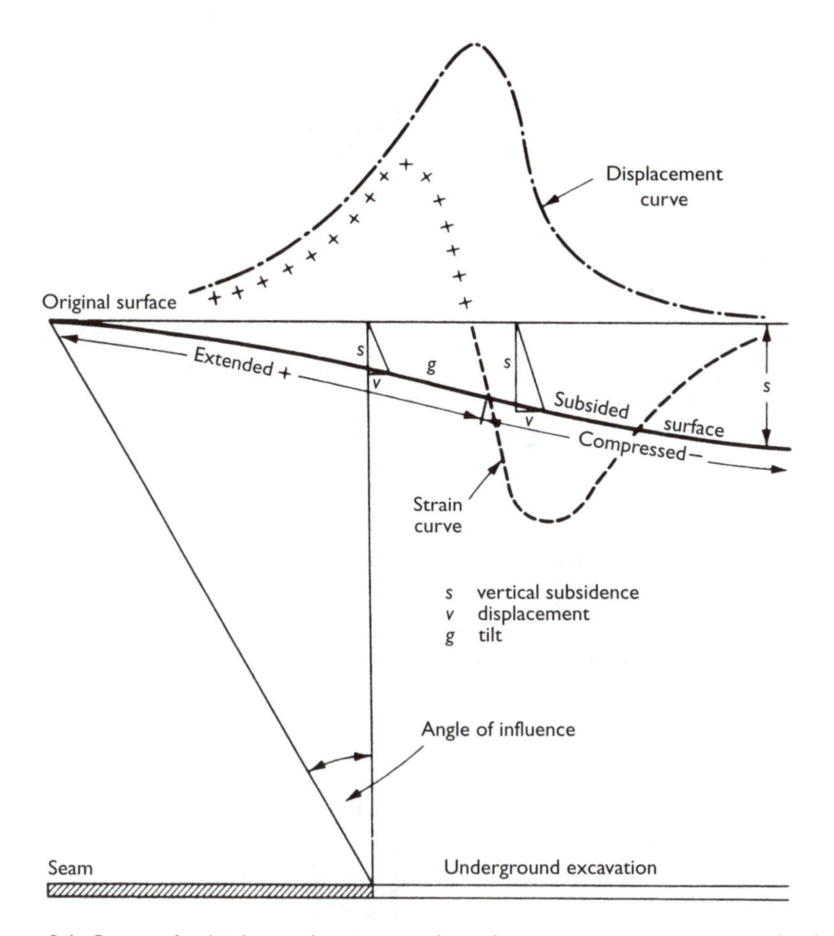

Figure 3.1 Curve of subsidence showing tensile and compressive strains, vertical subsidence and tilt, together with angle of influence or draw (not to scale).

The surface area affected by ground movement is greater than the area worked in the seam. The boundary of the surface area affected is defined by the limit angle of draw or angle of influence, which varies from 8° to 45°, depending on the coalfield (Fig. 3.1). Table 3.1 shows some quoted angles of draw for different countries. It would seem that the angle of draw may be influenced by depth, seam thickness and local geology, especially the location of self-supporting strata above the worked coal seam. For inclined seams, the surface subsidence trough is displaced towards the down-dip side. The angle of draw depends on the dip of the seam, it being least at the rise side of the goaf and increases towards the dip side (Fig. 3.3).

All other things being equal, the thicker the coal seam (assuming the full thickness is extracted), the larger is the surface subsidence. If more than one

Figure 3.2 Examples of subsidence damage due to longwall mining of coal, Elsecarr, South Yorkshire, England, showing extensive cracking, walls being shored, and doors and windows blocked up after frames being severely twisted.

Table 3.1 Some examples of angles of draw (From Brauner, 1973a; Singh and Singh, 1998; reproduced by kind permission of Elsevier)

Country	*Angle of draw*
Netherlands, Limburg coal field	35–45°
Germany	30–45°
Northern France	35°
Russia	30°
Britain	25–35°
United States: Eastern	15–27°
United States: Central	0–8.5°
United States: Western	12–16°
Austrian Tertiary Coal Basin	41–42°
Slovakia, Ostrava-Karvina coal field	25–39°
South Africa	11°
Australia, Newcastle coal field	35°
India, Pench and Kanhan coal fields	41°
Japan, Kuho (II) mine	40–50°

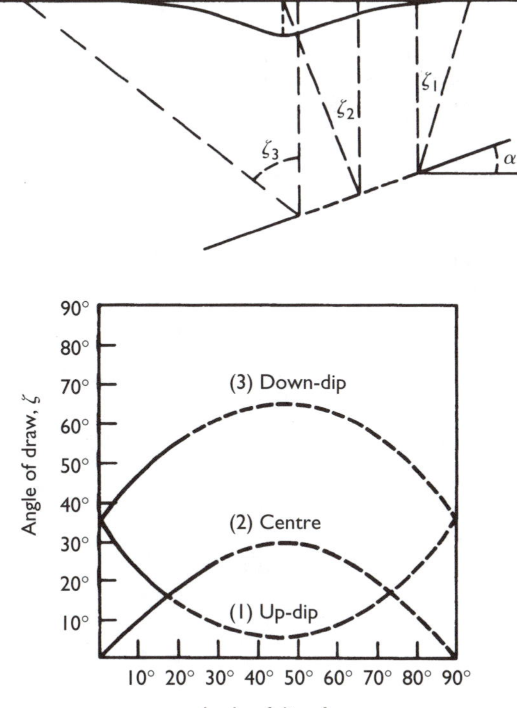

Figure 3.3 Effect of seam inclination on angle of draw. (After Anon., 1975; reproduced by kind permission of the Coal Authority.)

seam is worked simultaneously beneath the same area, then the subsidence effects are cumulative. The maximum possible subsidence, S_{max}, is

$$S_{max} = Ha \tag{3.1}$$

where H is seam thickness (or thickness of seam extracted) and a is a subsidence factor that ranges from 0.1 to 0.9 (Table 3.2). The subsidence factor generally is regarded as being independent of depth. However, it does depend on whether or not the mined out area has been filled or packed. Unfortunately, the character of any infill and the manner by which it is placed is so variable that only qualitative assessments of the subsidence factor can be made. In fact, it has been suggested that it is only in a single mining area where character and placement of infill are similar, that quantitative assessment of the effects of the subsidence factor can be made properly (Wardell and Webster, 1957).

Horizontal displacement along a subsidence trough is proportional to the slope of the subsidence profile. Tensile strain ($\varepsilon+$) and compressive strain ($\varepsilon-$) occur on both sides of the subsidence profile. The point of transition from

Table 3.2 Subsidence factor and ratio of maximum horizontal displacement to maximum possible subsidence (After Brauner, 1973a; reproduced by kind permission of the United States Bureau of Mines)

Coal field and method of packing	*Subsidence factor*	*Ratio*
British coal fields		
Solid stowing	0.45	0.16
Caving or strip packing	0.90	0.16
Ruhr coal field, Germany		
Pneumatic stowing	0.45	0.35–0.45
Other solid stowing	0.50	0.35–0.45
Caving	0.90	0.35–0.45
North and Pas de Calais coal field, France		
Hydraulic stowing	0.25–0.35	0.40
Pneumatic stowing	0.45–0.55	0.40
Caving	0.85–0.90	0.40
Upper Silesia, Poland		
Hydraulic stowing	0.12	—
Caving	0.70	—
Russia and Ukraine		
Donbass district	0.80	0.30
Lvov-Volyn district	0.80–0.90	0.34
Kizelov district	0.40–0.80	0.30
Donetz, Kuznetsk and Karaganda districts	0.75–0.85	0.30
Sub-Moscow and Cheliabinski districts	0.85–0.90	0.35
Pechora	0.65–0.90	0.30–0.50
United States		
Central	0.50–0.60	—
Western	0.33–0.65	—

compression to tension is referred to as the point of inflection and coincides with the point of half-maximum subsidence. Theoretically, the area under compressive strain is equal to that under tensile strain. In a strain profile, maximum tensile strain is located directly above or near but outside, the ribside (i.e. position of the gate roads). Maximum compressive strain is located either above the centre of the goaf or near there. The maximum compressive strain possible occurs in subcritical openings whereas maximum possible tensile strain is found in supercritical openings (see later). This means that in terms of a structure that is located above the centreline of an advancing longwall panel that it is subjected initially to tensile ground strains and later to compressive strains in a direction parallel to the centreline of the panel (Speck and Bruhn, 1995). At the same time, the structure is subjected to compressive strains of changing magnitude in a direction normal to the centreline of the panel. On the other hand, a structure that is located between the centreline of a panel and the panel boundary is subjected to tensile strains of varying magnitude that fan out from the centre of the panel, as well as to compressive strains of varying magnitude that are orientated perpendicular to the tensile strains.

Tilt of a subsidence profile is found by dividing the difference in subsidence by the distance between two points or surveying stations. The maximum tilt in each subsidence profile occurs at the point of inflection and decreases towards the centre and edges of the profile where the slope is reduced to zero. The maximum possible tilt occurs at 2.75 S_{max}/d or 33.35S/d, where $S_{max}$is maximum subsidence, S is subsidence (see later) and d is depth of seam. Slope profiles vary with width–depth ratios of excavations. For example, the greatest maximum slope occurs when the width–depth ratio of an excavation is 0.45 (Fig. 3.4).

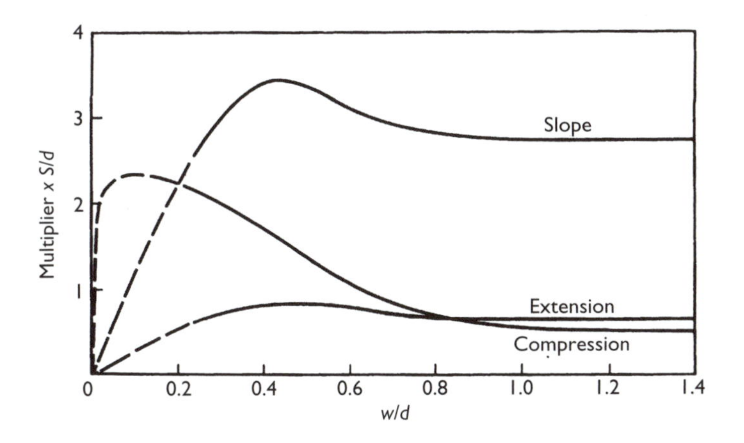

Figure 3.4 Graph for predicting maximum slope and strain for various width–depth ratios of a panel. S = subsidence, d = depth, w = width. (Reproduced by kind permission of the Coal Authority.)

Curvature can be expressed by the difference in slope between two stations or radius of curvature, p:

$$p = l_b/d^2S \qquad (3.2)$$
$$= l_b/\theta$$

where l_b is bay length (i.e. distance between two stations), d^2S is the second derivative of the subsidence profile and θ is the differential slope between the two points. Strain (ε) is, according to Orchard and Allen (1965), proportional to the ratio between the differential slope, θ, and the distance between observation points, that is, the bay length, on average $\varepsilon = (0.08/l_b)^{1/2}$. As with strain, curvature, tilt and vertical and horizontal displacements also change in magnitude and direction as a subsidence profile progresses.

Usually, there is an appreciable difference between the volume of mineral extracted and the amount of subsidence at the surface. For example, Orchard and Allen (1970) showed that where maximum subsidence of 90% of seam thickness occurred, the volume of the subsided ground was only 70% of the coal extracted. This is mainly attributable to bulking.

In a level seam the greatest amount of subsidence occurs over the centre of the working, diminishing to zero at approximately 0.7 of the depth outside the boundaries of the panel. However, as far as noticeable movement and damage are concerned a distance of 0.5 times the depth is more appropriate.

One of the most important factors influencing the amount of subsidence is the width–depth relationship of the panel removed. In fact, Orchard (1954) maintained that the maximum subsidence in British coalfields generally began at a width–depth ratio of 1.4:1. This is the critical condition above and below which maximum subsidence is and is not achieved respectively (Fig. 3.5). The concept of the area of influence or critical area of extraction was developed by Wardell (1954). He indicated that for a given point P on the surface, the area of influence in a level seam is the circle at the base of an imaginary cone with its axis passing upwards through the overlying strata to P at its apex. The diameter of the area equals 1.4 × seam depth. Any workings outside this area do not affect point P whereas all workings within the area do. However, all the coal from within this critical area must be extracted before point P undergoes maximum subsidence. If only part of the critical area is worked, then point P only suffers partial subsidence. Hence, the development of subsidence attributable to a given working is influenced by its width in relation to its depth. Three stages were distinguished, namely, subcritical (width less than 1.4 × depth), critical (width equal to 1.4 × depth) and supercritical (width greater than 1.4 × depth).

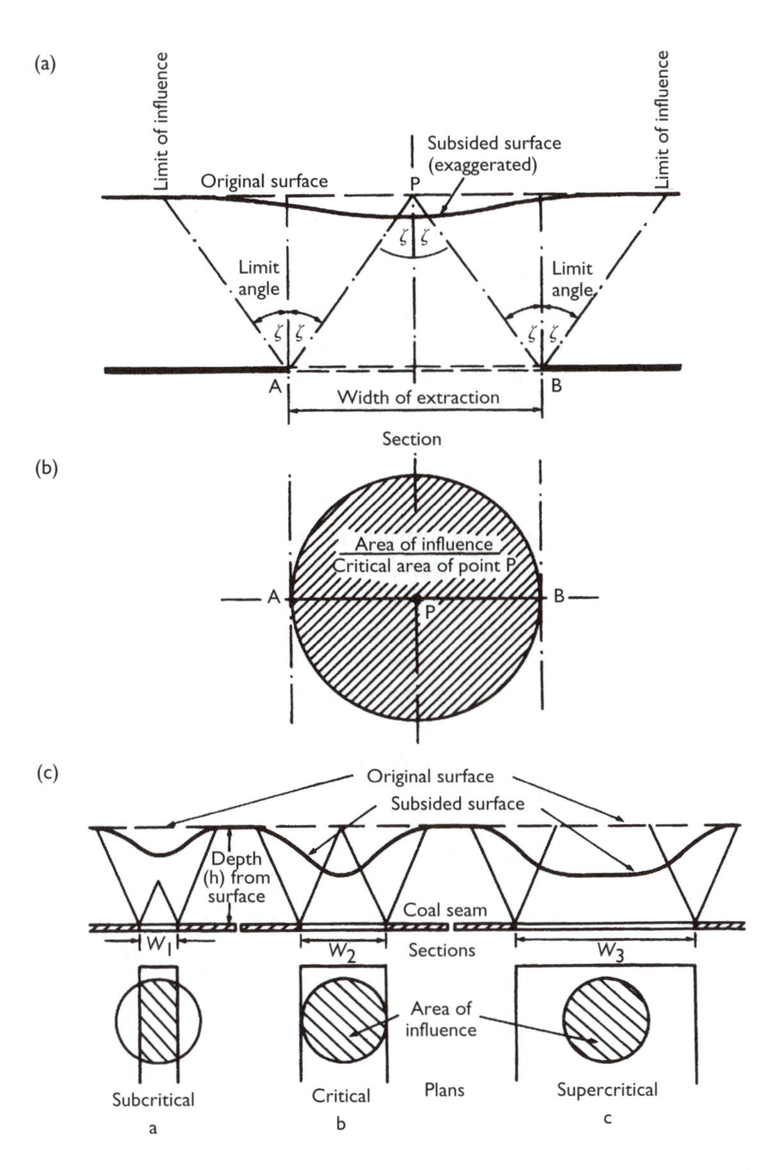

Figure 3.5 Subcritical, critical and supercritical conditions according to the depth–width ratio. Depth–width ratios (a) less than 1.4 (b) 1.4 (c) greater than 1.4. (After Anon., 1975; reproduced by kind permission of the Coal Authority.)

However, Orchard and Allen (1970) subsequently pointed out that the large width–depth ratio necessary to cause 90% maximum subsidence usually can only be achieved in shallow workings because with deeper workings the critical area of extraction is made up of a number of panels often with narrow pillars of coal left *in situ* to protect one or other of the roadways. These pillars reduce the subsidence and some coalfields in Britain, as a result, had no experience of subsidence in excess of 75% to 80%. They further pointed out that with shallow workings the supported or partially supported zones in a goaf, such as roadways and gateside packs, compose a much greater proportion of the width than in deeper workings. Therefore, deep panels can cause more subsidence than shallow ones with the same width–depth ratio.

Previously, Orchard (1957) had found that ground movements over shallow workings revealed many trends that were not so apparent over deep workings, mainly because of the masking effect of thick overburdens. For instance, for a given maximum subsidence, the curvature of the ground surface is more marked over shallow workings than over deep workings owing to the smaller distance over which the subsidence curve is spread. The horizontal strains are proportional to subsidence and inversely proportional to the depth of workings. The maximum slope in the ground in a subsidence trough also is proportional to the subsidence at the bottom of the trough and inversely proportional to the depth of the workings. It follows that maximum slope is proportional to the maximum horizontal strain.

As long as mining is carried out within the area of influence of a surface point, that point continues to subside. However, once coal has been removed from the whole area all the subsidence at the surface point will have taken place.

Wardell (1954) demonstrated that the time taken for subsidence at a surface point to be completed is more or less inversely proportional to the rate of forward advance of the workings (Fig. 3.6). The transition of movement to the surface is almost instantaneous, commencing when the strata at seam horizon begin to relax. Although the precise beginning and end of subsidence are difficult to determine, measurable subsidence occurs when the face is within a distance of $0.75d$ (d = depth to coal seam) and reaches approximately 15% of maximum when the face is directly below a given point, P, on the surface. For all practical purposes, it is complete when the face has advanced $0.8d$ beyond this given point P. Residual subsidence then occurs, those points that are subsiding fastest experiencing the most residual subsidence. The time factor is minimal under normal conditions.

Subsequently, Brauner (1973b) stated that in addition to the rate of advance and size of critical area, the duration of surface subsidence depends on geological conditions, depth of extraction, type of packing and previous extraction. For example, it lasts longer for thick bedded or stronger overburden and for complete caving. Depth, in particular, influences the rate of subsidence. First, this is because the diameter of the area of influence and therefore the time taken for a working, with a given rate of advance, to transverse it, increases with depth and, second, because at greater depths several workings may be necessary before the area of influence is

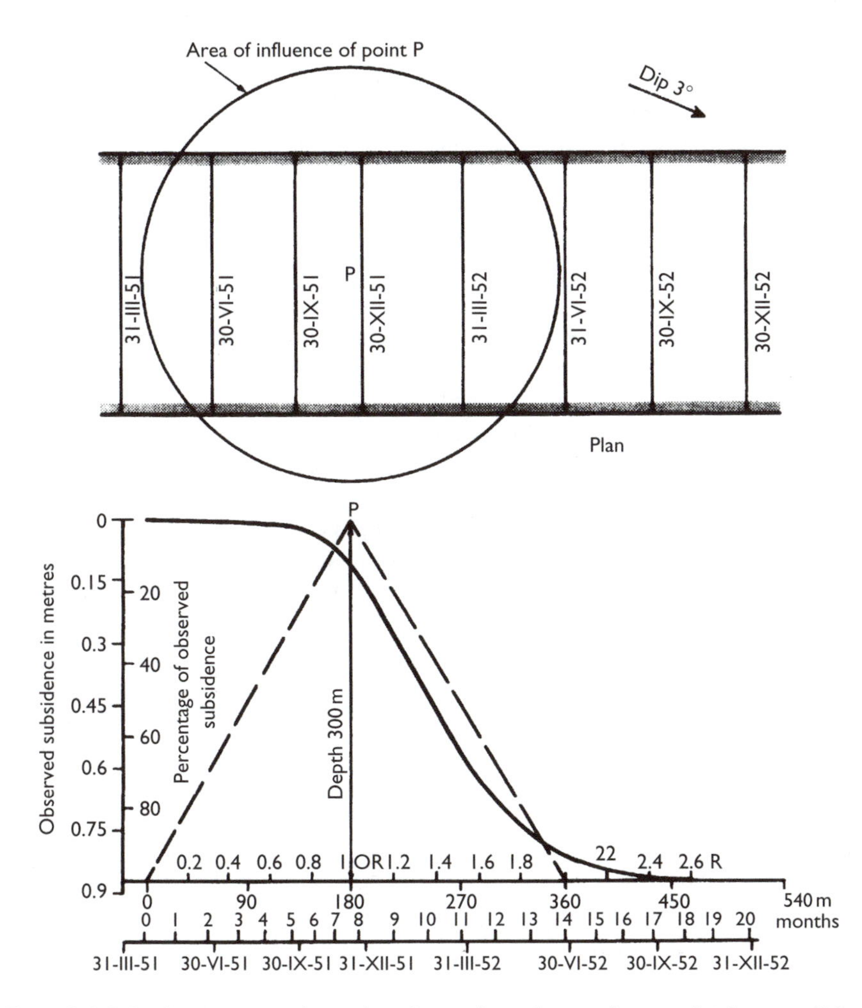

Figure 3.6 Subsidence–time relationship. R = radius of critical area of influence. (After Wardell, 1954; reproduced by kind permission of the Coal Authority.)

completely worked out. Consequently, the time that elapses before subsidence is complete varies according to circumstances.

Drent (1957) confirmed the connection between the moment of maximum effect and depth of working. He suggested that at the moment of maximum rate of subsidence in the Limburg Coalfield in the Netherlands, the surface point affected had undergone some 50% of the final subsidence to which it could be subjected. The thickness of non-Carboniferous strata above the Coal Measures appeared to influence the amount of maximum rate of subsidence more so than did the Coal Measures (Carboniferous) strata.

More specifically, Pottgens (1979) found that in the Limburg Coalfield the maximum rate of subsidence was reached 45 days after the working face had passed a given point at the surface when seams are overlain by thick overburden. In fact, the maximum rate of subsidence appeared to be inversely proportional to the thickness of the overburden. It was found that in the Netherlands 90% of subsidence took place upon completion of mining in the critical zone. In the example quoted by Pottgens, this was about six months after the working face had passed beneath the observation point. Generally, in the Netherlands ground movement has been observed to stop about one year after the cessation of mining operations.

Residual subsidence takes place at the same time as instantaneous subsidence and may continue after the latter for periods normally up to two years. The magnitude of residual subsidence is proportional to the rate of subsidence of the surface and is related to the mechanical properties of the rocks above the coal seam concerned. For instance, strong rocks produce more residual subsidence than weaker ones. Residual subsidence rarely exceeds 10% of total subsidence if the face is stopped within the critical width, but falls to 2–3% if the face has passed the critical width. Very occasionally values greater than 10% have been recorded. For instance, after a study of five coal mines in Britain, Ferrari (1997) found that residual subsidence ranged from 8% to 45% of total subsidence and continued for up to 11 years after mining operations ceased. Singh and Singh (1998) noted that residual subsidence in the Kamptee coal field, India, varied between 7.4% and 22.4% of total subsidence and took place in less than two years. According to Yao and Reddish (1994) the maximum residual subsidence in a level seam occurs at the half-subsidence point along a longitudinal line to the workings while along a transverse line it occurs at the centre of the workings. The maximum residual subsidence in an inclined seam occurs at the ribside point. Residual subsidence also may be influenced by mine water rebound and the reactivation of faults.

According to Shadbolt (1978) basic ground movements are fairly consistent in typical Coal Measures rocks. Such movements decrease with increasing depth and generally can be predicted at the surface to within ±10% in exposed coalfields. The situation can be complicated if seams have been worked previously, especially by partial extraction methods. Interaction effects tend to enhance the basic movements and may give rise to erratic subsidence. If a seam has been worked previously by total extraction methods, this frequently causes the effective movements to increase by about 10% when another seam is worked subsequently above or below. This is because the first seam extracted will have disturbed the ground.

The dip of a coal seam influences the direction in which longwall mining takes place in that when the dip exceeds 30° working commonly takes place along the strike. Whittaker and Reddish (1989) noted that the principal method of working such seams is by horizon mining, which involves driving several roadways through the strata in which the coal seams occur.

Subsidence attributable to horizon mining layouts tends to be concentrated with most subsidence troughs having their major axes parallel to the strike of the seam. However, Whittaker and Reddish did point out that it is possible to phase longwall extractions so that the development of ground strain at the surface is more widespread. As pointed out above, caving of the roof behind the longwall in dipping seams tends to displace the maximum subsidence from over the centre line of the face towards the dip side, giving rise to an asymmetrical development of ground movement. Maximum subsidence occurs at the point normal to the centre of the goaf and the angle of draw depends on the dip of the seam, it being least at the rise side and increases towards the dip side. Hence, the area of influence at the dip side is broader, which means that the area of tensile strain is wider (Fig. 3.3). Degirmenci *et al.* (1988), for instance, showed experimentally that the dip side experiences appreciably more lateral displacement for equal vertical subsidence on each side of the longwall panel and hence correspondingly increased tensile strain. If the seam is at shallow depth, then the subsidence at the rise side of the face has a more marked effect at the surface than at the dip side, especially with the ground strain being concentrated. Steeply dipping coal seams (e.g. over 75°) can generate large strains on the sides of a longwall working that, in turn, can lead to discontinuities being opened at the surface, along with the development of stepping. Diez and Alvarez (2000) maintained that when very steep coal seams are mined flexuring and breakage of strata occur normal to bedding whilst slippage and shearing take place parallel to the bedding. This can complicate subsidence, giving rise to multiple subsidence profiles.

Ground movement is asymmetrical about the centre of an extraction panel in areas of high relief. Frequently, tensile strains produced by subsidence associated with longwall mining, especially of dipping seams, can give rise to significant fissuring on steep hillsides that can be responsible for stability problems (Whittaker and Reddish, 1993). For example, Donnelly *et al.* (2001) investigated the occurrence of landslides induced by subsidence in four regions of the central Andes in Colombia. On the other hand, Holla (1997) found that lateral movement on both sides of a valley in New South Wales caused large compressive strains. The strains were 3 to 4 times those that normally would be expected in areas of flat or very gentle topography. Maximum horizontal movement was as much as 40% of observed vertical movement. Beyond the goaf the horizontal movements generally exceeded the vertical movements but the former appeared to be rigid body movements with small deformations that may produce insignificant surface impacts.

3.2. Geological factors and subsidence due to longwall mining

Ground movements induced at the surface by mining activities are influenced by variations in the ground conditions, especially by the near surface rocks and superficial deposits. However, the reactions of surface deposits to ground movements are usually unpredictable. Indeed, some 25% of all cases of mining subsidence undergo some measure of abnormal ground movement that, at least in part, is attributable to the near surface strata.

3.2.1. Surface soils and rocks

Many sand, silt and clay soils may deform uniformly in a subsiding area. Nonetheless, Speck and Bruhn (1995) noted some examples of non-uniform soil behaviour. These included soils on slopes that might shear along one or more surfaces (e.g. a soil-rock interface), resulting in soil creep or slippage thereby distorting the distribution of ground strain. Soils may be loosened or consolidate as a consequence of ground movements. The latter is likely to occur if the water table is lowered by fractures developing due to subsidence, which bring about drainage of near surface sediments. In addition, soil may be washed into such fractures (see later). Speck and Bruhn indicated that the influence of soil on ground strain depended on its thickness, as well as its properties.

Superficial deposits are often sufficiently flexible to obscure the effects of movements at rockhead. In particular, thick deposits of till tend to obscure tensile effects. On the other hand, superficial deposits may allow movements to affect larger areas than otherwise.

Rigid inclusions in stratal sequences above coal seams being mined, such as sandstone lenses in mudrocks, often present problems when subjected to compressive ground strains. These lenses tend to be forced upwards by the compressive forces with obvious consequences to foundations placed on or over them (Donnelly and Melton, 1995).

It has been suggested by Peng and Geng (1982) that the subsidence factor decreases with increasing proportion of strong rocks in the overburden above a coal seam. They also maintained that stiffer rocks, such as strong sandstones, which may behave as cantilever beams, take longer to react to longwall mining. As a consequence, ground movements at the surface take place more slowly and take more time to develop maximum values. Non-uniform subsidence therefore may occur above a panel when abnormally thick beds of sandstone are present in the overburden or when facies changes give rise to irregular displacements in other rock units. In addition, Whittaker and Reddish (1989) pointed out that bed separation can occur between competent and weak strata, and that such separation is more permanent over the ribside of a panel.

The necessary readjustment in weak strata to subsidence effects usually can be accommodated by small movements along joints. However, as the strength of the surface rock and the joint spacing increases so the movement tends to become concentrated at fewer points so that in massive limestones and sandstones movements may be restricted to master joints. Hence, well developed joints or fissures in such rocks concentrate differential displacement. Tensile and compressive strains many times the basic value have been observed at such discontinuities (Bell and Fox, 1988). For example, joints may gape anything up to a metre in width at the surface (Fig. 3.7). According to Shadbolt (1978) it is quite common for the total lateral movement caused by a given working to concentrate in such a manner. In such instances, no strain is measurable on either side of the discontinuity concerned.

Although some authorities have been of the opinion that fissures are formed in surface strata by longwall mining and that their orientation is aligned in relation to the geometry of the workings (one set of fissures running parallel to the coal face and the other parallel to the ribside), this is most unlikely except perhaps in the case of near surface workings. For instance, the ground strains at the surface generated by extraction of a seam at 500 m depth are small and could not give rise to major fissures in massive rock. On the other hand, they could open existing joints. Hence, the relationship between mine geometry and surface fissuring is probably

Figure 3.7 Joint gaping at the surface as a result of subsidence due to longwall mining of coal.

coincidental. Some fissures may be associated with the reactivation or dilation of faults (Donnelly and Rees, 2001).

In concealed coalfields the strata overlying the coal bearing measures often influence the basic movements developed by subsidence. In fact, abnormal subsidence behaviour and inconsistent movements are much more common in concealed than exposed coalfields. The occurrence of abnormally thick beds of sandstone can modify stratal movement due to mining, especially when width–depth ratios are small. Such beds may resist deflection, in which case stratal separation occurs and the effective movements at the surface are appreciably less than otherwise would be expected. The differences in behaviour disappear when the extraction becomes wide enough for the sandstone to collapse, then subsidence behaviour reverts to normal. Whittaker and Breeds (1977) noted a predominant jointing pattern in the Permo-Triassic strata overlying the Coal Measures in Nottinghamshire, England. They showed that in both the Sherwood Sandstone (Trias) and the Magnesian Limestone (Permian) there was a tendency to produce less subsidence than that predicted for width–depth ratios exceeding 1.0. They suggested that this was due to the two formations behaving as block jointed media in which the individual blocks did not return exactly to their former positions after compressional ground movement had ceased. Obviously, the friction between the blocks plays an important role. On the other hand, rock type did not appear to have a marked influence when the width–depth ratio was less than 1.0. Whittaker and Breeds also found that where the surface rocks consisted of Sherwood Sandstone, then surface strain was noticeably more irregular and tensile strain was higher than when Magnesian Limestone occurred at the surface. By contrast, there was no recognizable difference in the maximum compressional strains recorded. The type of rock did not appear to have a notable influence on the maximum ground slope resulting from subsidence. The relationship between subsidence over the ribside and width–depth ratio in the Sherwood Sandstone again suggested that block behaviour occurred at the surface. Furthermore, the maximum tensile strain appeared to be displaced further away from the extraction area, that is, towards the area of less constraint on the line of surface blocks. Hence, there is a greater likelihood of discontinuities opening further from the compressional zone rather than nearer to it. This did not appear valid for width–depth ratios less than 0.4, which suggests that magnitudes of strain need to be sufficiently large to promote surface block behaviour. However, in this situation it appears that the discontinuities in the Magnesian Limestone do not necessarily influence subsidence behaviour as much as those in the Sherwood Sandstone and that the limestone tends to behave normally. Such block movement is controlled primarily by the nature of the discontinuities in the sandstone.

There are many areas where joints in rockhead strata, beneath overlying superficial deposits, have been opened and the gape been enlarged by mining subsidence. Where the superficial deposits are granular and permeable, there is a tendency for subsurface water to wash them into these discontinuities, causing localized surface subsidence. The exact locations of joints that may open cannot be predicted but can adversely affect residential, commercial or industrial developments. The jointing pattern, however, can be exposed on site by digging trenches through the overburden. When joints are found, their condition (e.g. open or choked) is recorded. In this way the position, orientation, elevation and width of all joints encountered in trenches can be recorded on a site plan.

A thick compacted fill immediately overlying a jointed rock mass, provided it is well graded and possesses some cohesion, can form an arch between the adjacent surfaces of a joint that is opened by subsidence. On the other hand, reliance on a certain thickness of soil cover to provide self-arching is difficult to justify where soil conditions are unknown and the degree of support provided by the bedrock cannot be quantified. What is more, the capacity of a soil arch to withstand overlying foundation loads is more difficult to assess than the span distance or arch thickness. Small isolated foundations are probably the most vulnerable, and where the thickness of the arch is limited their failure probably occurs by punching shear. Larger more extensive and rigid foundations do have a greater ability to span open joints and to redistribute load.

Clay that has been strengthened by the addition of cement or lime can be used to fill large open joints in rock masses, being either compacted into place or used to help compact surface soil more effectively. The soil–cement arch method of foundation support is probably most suitable in those situations where the depth of superficial cover exceeds 3 m. The degree to which the properties of a clay soil can be enhanced by the addition of cement or lime depends not only on the quantity added but also upon the type and proportion of clay minerals present. In general, however, an increase in unconfined compressive strength of several hundred per cent can be obtained by the addition of a few per cent, by weight, of either cement or lime to a clay soil (a rough rule of thumb is 1.0% for every 10% of clay minerals present; rarely do clay soils contain more than 80% clay minerals).

Open joints can be washed out and then filled with a grout material, although the grout mix does not need to be of an equivalent strength to that of the host rock mass. Cement, fly-ash and sand mixes can be used to fill joints. For example, gaping joints may be occupied with grouts consisting of 1 part cement, 3–5 parts fly-ash and 10–15 parts sand. A small quantity of bentonite may be added to minimize segregation of the grout. Gravel may be added to provide extra bulk when grout is used to fill widely gaping joints. Rock paste also may be used to fill joints. It essentially consists of colliery spoil mixed with water.

Bell and Fox (1988) described a site for residential development where joints opened by mining subsidence occurred beneath 1 or 2 m of superficial deposits. Where the site investigation revealed that an open joint occurred beneath a position in which a foundation structure was to be constructed, it was first exposed for a distance beyond the perimeter of the foundation equal to the depth of the backfill or to where it terminated, whichever was the least distance. Then, the open joint was filled with a bentonite grout and structural geowebbing was laid over the filled fissure (to accommodate any movement), after the initial set of the grout had taken place. An impermeable membrane was laid over the geowebbing and turned up the face of the excavation for a distance of 350 mm. The trench then was backfilled to foundation formation level with sand. The latter was compacted in layers (Fig. 3.8). Such treatment allowed strip footings to be adopted for both single and two-storey dwellings.

3.2.2. Faults and their reactivation

Faults also tend to be locations where subsidence movement is concentrated, which may lead to reactivation along fault planes thereby causing deformation of the surface (Fig. 3.9). Reactivation may result in the formation of a fault scarp or step at the surface that may be accompanied by fissures or compression ridges. Dykes can have similar effects on subsidence. Whilst subsidence damage to structures located close to or on the surface outcrop of a fault or dyke can be very severe, in any particular instance the areal extent of such damage is limited, often being confined to within a few metres of the outcrop (Fig. 3.10). Also, many faults have not reacted adversely when subjected to subsidence (Hellewell, 1988). On the other hand, several phases of fault reactivation are possible during multi-seam mining operations, these being separated by periods of relative stability. The extent to which faults influence and modify subsidence movements cannot be quantified accurately, and unfortunately the exact location of faults at the surface is not always easy to determine.

In fact, faults tend to act as boundaries controlling the extent of the subsidence trough (Fig. 3.10). If a fault is encountered during seam extraction and its throw is large, then if the workings terminate against the fault permanent strains are induced at the surface, which probably are accompanied by severe differential subsidence in the zone of influence of the fault. Indeed, a subsidence step may occur at the outcrop of such faults. Buildings located above such faults may be so severely affected that they have to be demolished. This happened in the village of Elsecarr, near Barnsley in South Yorkshire, England (Fig. 3.11). When workings terminate against a fault plane that has an angle of hade larger than the angle of draw, then the subsidence profile extends uninterrupted to the surface. Conversely, when the hade of the fault is less than the angle of draw, the fault determines the extent of the subsidence trough, which in this case is less than that normally expected. Faults are most likely to react adversely when their hade is less

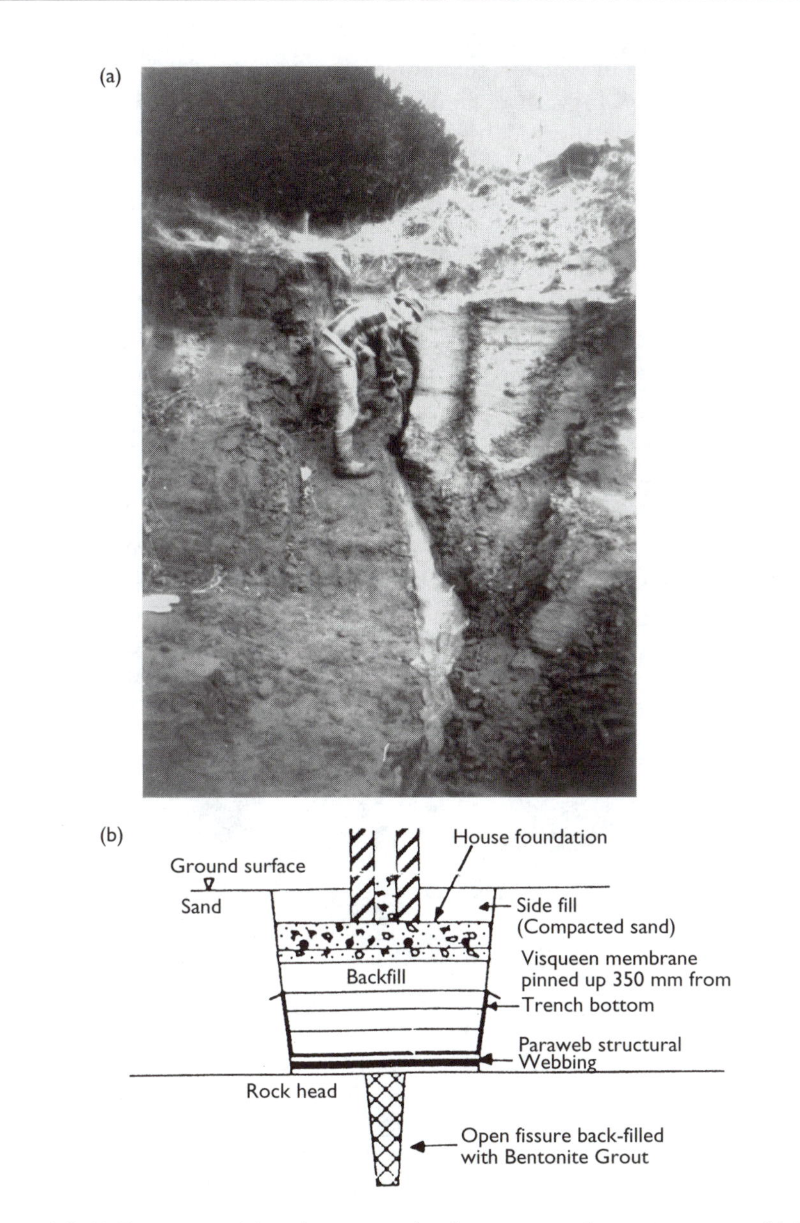

Figure 3.8 (a) Treatment of foundation trench when an open fissure is present. (b) More detailed treatment. (After Bell and Fox, 1988.)

than 30°, when they have simple form and the material occupying the fault zone does not offer high frictional resistance. For example, fault steps tend to occur when faults represent single sharp stratal breaks (Lee, 1966). By contrast, if a fault consists of a relatively wide shatter zone, then the surface

Figure 3.9 Examples of structural damage caused by fault reactivation. (a) Houses subjected to tensional stress in Nottinghamshire, England. (Reproduced by kind permission of the Coal Authority.) (b) Compressive stress affecting a wall in Eastwood, West Midlands, England. (Reproduced by kind permission of Elsevier and Dr David Reddish.)

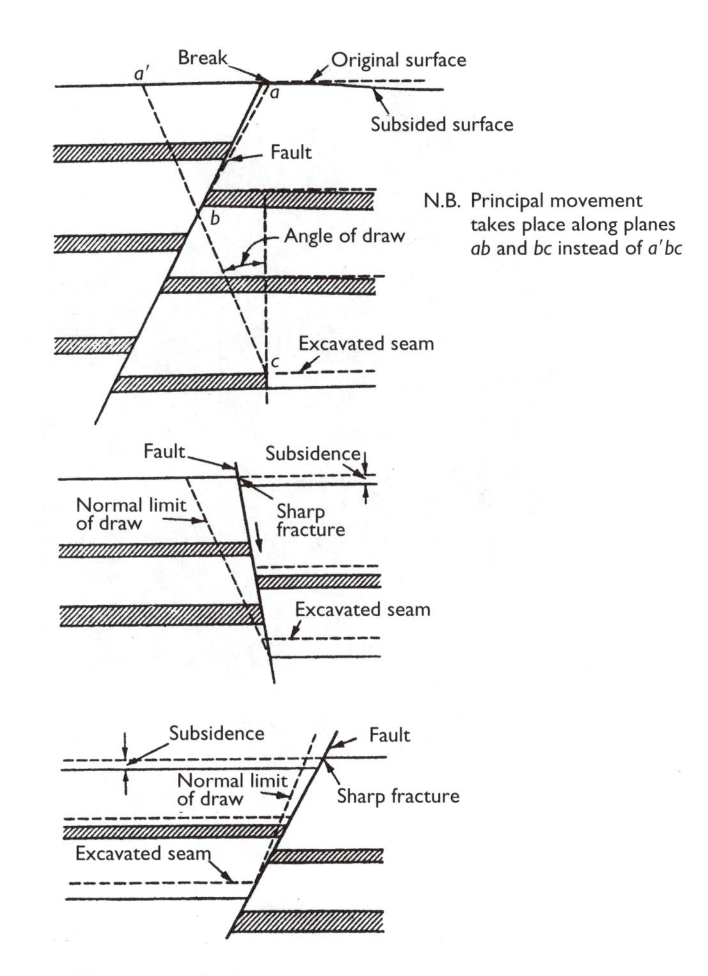

Figure 3.10 Influence of faults on subsidence.

subsidence effects usually are less pronounced but more predictable in terms of location and amount.

Phillips and Hellewell (1994) proposed that the abnormal subsidence associated with faults is influenced by the type of surface rocks. This explains why stepping is not consistent along a fault affected by subsidence. In addition, stronger surface rocks such as some sandstones tend to fracture and form blocks that cause more widespread damage because of the cantilevering effect of the fractured blocks. This, they suggested, explains the reverse stepping that occurs at times. Also, ground movements associated with faults tend to continue for longer periods than normally would be expected. The thickness and nature of the unconsolidated deposits above bedrock can influence the magnitude of fault steps.

Figure 3.11 Subsidence damage in the neighbourhood of a fault step (windows fell out, door frames twisted and masonry severely cracked) Elsecarr, South Yorkshire, England. Fault step corresponds with dip in road.

The type and extent of subsidence associated with faults also depends on the methods and extent of mining in relation to the fault plane. For example, where the width of the extraction is large enough, the surface subsidence at a point vertically above the intersection of the fault and the workings may be approximately one-half that of the maximum subsidence for that working. The most notable steps occur when the coal is worked beneath the hade of a fault because the strain relieving process encourages increased movement along the fault plane, faces in other positions being much less likely to cause differential movement (Fig. 3.10). Workings on the up thrown side of a fault are less likely to cause stepping than similar workings on the downthrown side. Steps are usually down towards the goaf but if old workings exist, then steps may occasionally occur away from the face. The vertical displacement can vary along a fault step and may be accompanied by horizontal displacement. Accordingly to Lee (1966), the size of fault steps is on average one-third of the maximum subsidence that takes place but this value varies appreciably. However, the size of step often appears to be consistent where the underground conditions are uniform. The extent of a step is very much limited to the area worked. Furthermore, a single working of small width–depth ratio approaching a fault at right angles is less likely to cause a step than a large width–depth working parallel to a fault. A fault step is much more likely to develop when the fault has been affected by previous workings in shallow seams than it is for a single working in a virgin area. Once differential movement has occurred further movement in the area can cause renewed movement along a fault, which at times may be out of proportion to the thickness and extent of the coal extracted. As successive workings increase the effect on a fault plane, so the chances of steps developing are increased.

Obviously, building over faults in areas likely to be subjected to subsidence should be avoided wherever possible because of the relatively uncertain nature of the surface ground movements and the fact that structures normally cannot be designed to withstand highly localized and attenuated differential subsidence. Indeed, Anon. (1959) recommended that structures be set back at least 16 m from the line of surface outcrop of a fault.

In a concealed coalfield where a fault passes through a block jointed formation that outcrops at the surface, then severe fissuring or fracturing can occur at the surface. The fissures generally run parallel to the line of the fault but can occur up to 300 m away from it.

Fault reactivation occurs simultaneously with mining subsidence via the processes of aseismic creep over periods ranging from weeks to years and may continue after normal subsidence has ceased, but the precise duration is impossible to determine. Moreover, it occurs in phases separated by periods of inactivity. Since instantaneous stick-slip fault mechanisms do not occur, there is an absence of seismicity during fault reactivation. The likelihood of fault reactivation increases with increasing number of seams worked and is associated more frequently with shallow workings. The cumulated effect of reactivation following multi-seam extraction may result in the generation of large fault scarps in excess of 2 m in height.

The mechanisms involved in mining induced fault reactivation have been reviewed by Donnelly (2000). Such reactivation during longwall mining basically is controlled by several inter-relating geological and mining factors (Table 3.3). The principal geological factors that influence fault reactivation include the prevailing and pre-existing stress field; the geological history of the fault; the geotechnical properties of the fault (friction, cohesion and pore fluid pressures); the proximity of the fault to the ground surface; the local hydrological regime; and the density and orientation of discontinuities in adjacent rock masses. In general, it is the main faults that have the greatest tendency to undergo reactivation when they are subjected to mining

Table 3.3 Geological and mining factors that influence the reactivation of faults (After Donnelly and Rees, 2001)

Geological factors	*Influence on fault reactivation*	*Mining factors*	*Influence on fault reactivation*
1 Pre-existing and prevailing stress fields	Faults oriented perpendicular to stress field may be relatively stable. Faults oriented parallel to stress field are least stable	1 Depth of extraction	At greater depth of mining ground movements are spread over a larger region and reduce strain accumulations on faults
2 Geotechnical properties of fault zones	Shear strength, friction, cohesion and pore fluid pressures will vary depending upon rock type, weathering, ground-water, mineralogy, type of superficial cover and width of fault zone	2 Fault dip and mining orientation	Faults that dip between 70° and vertical are more likely to reactivate than low angled faults
3 Geological history and age of the fault	Several generic categories of faults exist in the Coal Measures. First order (master) faults define crustal blocks; second order (main) occur in trends of one or two evenly spaced across the blocks; third order (minor) are limited to a few horizons. Master faults are most likely to reactivate during mining	3 Distance from fault	The closer the mining to the fault, the greater the influence on reactivation, the thicker the seam the greater the influence on reactivation
4 The proximity of the fault to the ground surface	Faults that crop out are more likely to create ground disturbances, than faults that in-crop	4 Location of mining	Fault reactivation is more likely when workings are located below the fault plane (footwall), than when situated above (hangingwall)
5 Hydrogeology	Groundwater flow into and within faults will affect the geotechnical	5 Rate of longwall advance	The greater the rate of mining the less time is available to dissipate the ground

Table 3.3 Continued

Geological factors	*Influence on fault reactivation*	*Mining factors*	*Influence on fault reactivation*
	properties of the fault and influence stability		movements and the greater the likelihood of reactivation
6 Rock mass discontinuities	Joints, bedding planes and fissures concentrate mining subsidence ground strains. The greater the density of discontinuities to dissipate ground movements, the lower the likelihood that reactivation will manifest on a single fault plane	6 Dip of seam	Extractions in horizontal coal seams and coal seams that dip towards the fault will have a positive influence on reactivation. Seams dipping away from faults will reduce the likelihood of reactivation
		7 Mining history	Multi-seam mining operations may cause several phases of fault reactivation separated by periods of stability

subsidence. The principal mining factors that influence fault reactivation are the depth of extraction; the mine and fault geometry; the horizontal distance of the workings; extraction thickness; the rate of mining; seam dip; and the history and intensity of mining.

The surface expression of reactivated faults, that is, their morphology, height and persistence vary considerably and are controlled by the geotechnical properties of the surface materials and the geomorphology (Fig. 3.12) They range from subtle topographic deflections and flexures merely recognizable across agricultural land or road side verges, to distinct high-angled fault scarps 3–4 m high and at least 4 km long where thick sandstones crop out (Donnelly *et al.*, 2000; Fig. 3.13). More commonly, they are less than a metre high, less than a metre wide and a few hundreds of metres long. In addition, where a thin superficial cover is overlain by brittle material such as concrete surfaces or roads, then scarps tend to be distinctive, and may extend over hundreds of metres in length and be up to 2 m in height. However, the ground deformation on either side of the fault is limited to a few metres. On the other hand, where weaker ground exists such as thick (10 m +) tills, sands and gravels or clay, then these may not support the formation of a fault scarp or may allow temporary scarps to develop before erosion destroys them. In such instances a flexure develops, with a reduced height, but affects

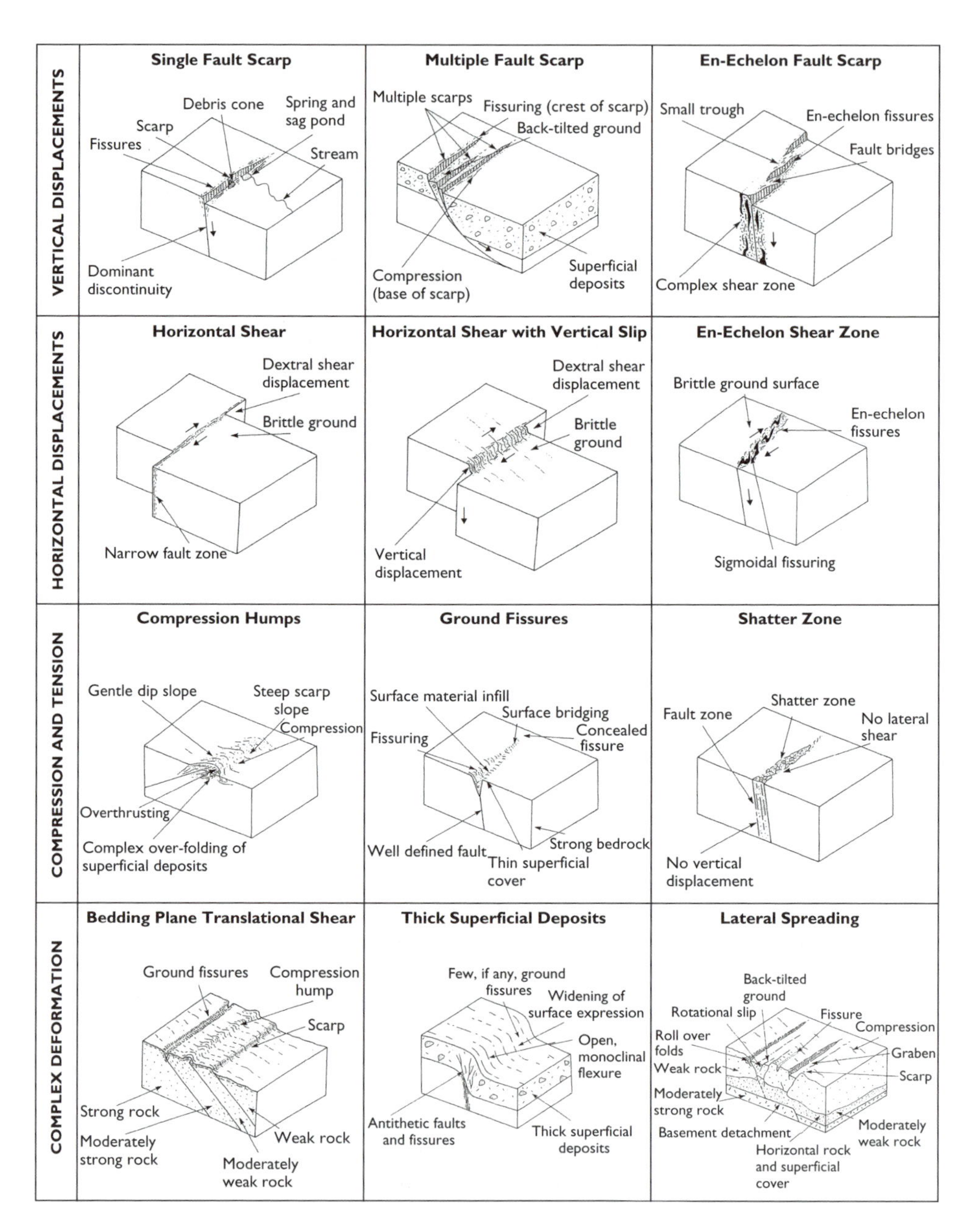

Figure 3.12 Schematic diagram (not to scale) illustrating the types of surface expression of mining induced fault scarps that may be formed at the ground surface. (After Donnelly and Rees, 2001.)

(a)

(b)

Figure 3.13 The 3–4 m high and 4 km long Tableland Fault Scarp, South Wales.

a much broader area on either side of the fault. What is more, in weak or unconsolidated superficial sediments the dominant fault plane may become detached or splay into multiple fractures. Some faults have been observed in trenches to generate listric slip surfaces and intricate deformation structures. Layer parallel detachment may occur when the superficial deposits are stratified. This may lead to the generation of secondary faults that may branch out into a multitude of fractures or die out before reaching the ground surface.

The formation of graben structures are also a common surface expression of fault reactivation, especially on road or concrete surfaces or where lateral spreading has occurred (Donnelly *et al.*, 2001).

The majority of the faults in areas of former mining subsidence become stable after mining has been abandoned. However, in recent years reactivation, along some faults in British and European coalfields have been observed (Bekendam and Pottgens, 1995; Donnelly, 1998). These ground movements have occurred after mining has ceased, some faults undergoing renewed reactivation several years after the mining has been completed. In certain circumstances reactivation may be due to groundwater discharge along faults or to mine water rebound following the cessation of mine water pumping when a mine is closed. Mine water rebound may increase the pore fluid pressures within faults, causing a reduction in shear strength along the fault planes. This is capable of counteracting part of the normal stress acting across the faults and therefore may result in reactivation. In this way fault reactivation may occur several years after mining has ended. Hence, faults that have undergone induced reactivation should be considered as potentially hazardous since they can be associated with ground movements, and with groundwater discharge and/or gas emissions. Therefore, it is possible that renewed fault reactivation may occur.

Ground fissures may be present at fault outcrop positions, in areas where mining subsidence has occurred. But it may be difficult to distinguish between those fissures that have originated due to mining subsidence and those that have been formed by other processes, such as mass wasting and cambering. As illustration, subsidence induced ground fissures in County Durham, England, have caused widespread disruption to houses and land (Donnelly, 1998). In particular, fissure zones up to 500 mm wide and over 2 km long have been observed at Easington, Quarrington Hill, Witch Hill and Cassop, these being associated with abandoned workings at Cassop, Thornley, Sherborne and East Hetton collieries (Young and Culshaw, 2001). The fissures have presented difficulties for the development of new housing, civil engineering construction and landfill. Similar features also have been documented in South Wales and the West Midlands of England (Donnelly and Rees, 2001). In some instances the fissures have developed into fault breccia pipe collapses and in extreme cases sinkholes develop. The depths of these fissures are not known but in County Durham they probably extend through the Magnesian Limestone (Permian), which reaches to 70 m below ground level. Fissures in the Magnesian Limestone that have undergone dilation aid percolating water to dissolve the limestone, further opening the fissures until the fissures walls ultimately collapse. In the West Midlands of England, ground fissures in the Sherwood Sandstone (Trias), up to 1.5 m wide and 3 km long, have proved a particular problem in the Downes Bank area, where high pressure gas mains have been damaged due to fissuring of the ground caused by mining subsidence.

Compression ridges on the ground surface can be similar in appearance to reactivated fault scarps. Like reactivated faults, they cause widespread damage to land, buildings and structures in areas that have been subjected to mining subsidence. Compression ridges typically form in the centre of longwall subsidence troughs in the area of maximum compression, as well as on the flanks of fissures. Compression ridges have been observed from a few centimetres high, to distinct linear or sinusoidal ridges up 0.5 m high, and several hundreds of metres in length. These are often short-lived features, being destroyed by necessary repairs to roads or ploughing of agricultural land (Donnelly and Melton, 1995). Where excessive compressive ground strains occur, then overthrusting may be observed. Compression ridges may form at relatively low ground strains. Generally, they are formed at lithological contacts via the process of bedding plane translational shear, where weaker mudstones may undergo swelling and heave. Moreover, compression ridges tend to coincide with the outcrop of lithological contacts of contrasting geotechnical properties, for instance, where a weak seat earth or shale lies adjacent to a strong sandstone in the bottom of a subsidence trough. Compression ridges usually exist in conjunction with fissures.

Lateral shear usually only forms a component of ground movement at the outcrop of a fault. However, in some instances lateral shear displacements can dominate, with little or no vertical slip. Lateral shear tends to occur along faults when there is a significant change in strength of rock mass across the fault, as for instance, where a fault displaces sandstone against mudstone. It also is controlled by the geometrical relationships between mining and a fault (Donnelly and Reddish, 1994).

3.3. A review of prediction methods of subsidence due to longwall mining

An important feature of subsidence due to longwall mining is its high degree of predictability. Usually, movements parallel and perpendicular to the direction of face advance are predicted. Although adequate for many purposes such methodology does not consider the three-dimensional nature of ground movement. For instance, observations have shown that individual points move on approximately helical paths, that the pitch and radii change from point to point, and that the direction of rotation is different on opposite sides of a subsidence basin.

Methods of subsidence prediction can be grouped into three fundamental categories. First, empirical methods attempt to fit subsidence functions to field measurements. Second, analytical or theoretical methods of prediction derive subsidence functions from elastic theory and rock mechanics, and are based entirely on theory. Third, semi-empirical methods develop subsidence functions based on theory but that are related to field data by the use of constants and correlation coefficients.

3.3.1. Empirical methods

Empirical methods of subsidence prediction such as those that were developed by the National Coal Board (NCB) were refined by continuous study and analysis of survey data from British coalfields (Anon., 1975). Unfortunately, however, empirical methods tend not to take topography, the nature of the strata and geological structure involved, and how the rock masses are likely to deform into account. Consequently, Voight and Pariseau (1970) emphasized that such empirical relationships can be applied only under conditions similar to those of the original observations. Nevertheless, these methods are being improved continuously so that they can yield more accurate results. For example, the prediction methods developed by the NCB allow the amount of subsidence due to longwall mining in Britain to be predicted usually within ± 10%.

The subsidence prediction methods developed by the NCB are based initially on the width–depth ratio of the working face, all other factors being related to this ratio (Fig. 3.14). When a panel is not worked far enough to cause maximum subsidence, in other words when the critical condition is not attained, the subsidence predicted from Figure 3.14 must be adjusted by reference to Figure 3.15, that is, by considering the distance travelled by the face as a fraction of the depth and applying the appropriate partial subsidence curve.

Observance of the shapes of subsidence troughs has revealed that they conform to a general pattern from which standard subsidence profiles can

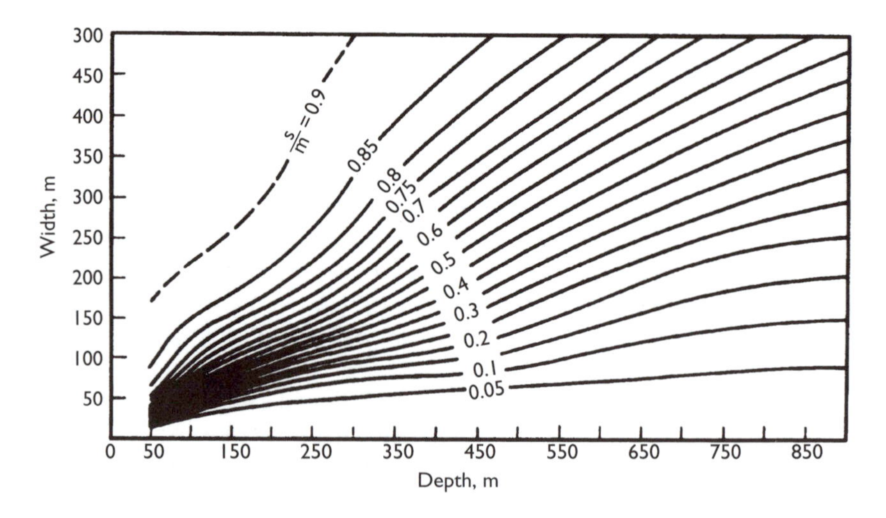

Figure 3.14 Relationship of subsidence to depth and width; prediction of maximum subsidence (*S*); *m* = thickness of seam extracted. (After Anon., 1975; reproduced by kind permission of the Coal Authority.)

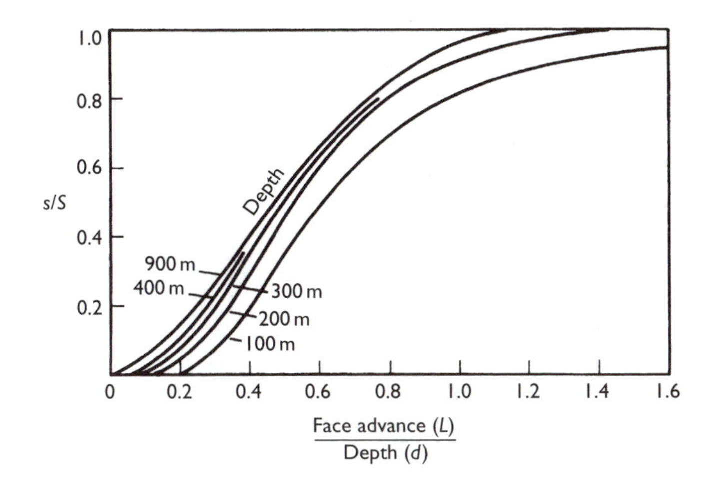

Figure 3.15 Correction graph for limited face advance to be used with Figure 3.14; s = subsidence; S = maximum subsidence. (After Anon., 1975; reproduced by kind permission of the Coal Authority.)

be produced. By itself the value of maximum subsidence over a particular extracted panel is of limited importance, the whole profile of subsidence having to be ascertained in order to study the effects of mining. The shape of a subsidence profile varies with the width–depth ratio of the extraction and for any value a curve can be plotted relating partial subsidence at various points to distance from the centre of the panel as a proportion of depth. Figure 3.16 represents a series of profiles from which the subsidence value for any profile can be determined.

The zone of maximum extension coincides with the position of the ribside where the width–depth ratio exceeds 1.35:1, lying outside the rib when the ratio is smaller. The maximum compression occurs in the centre of a panel in narrow panels (width less than 0.42 × depth), whilst at higher ratios the profile develops two compression zones (Fig. 3.17). Figure 3.18 is a graph for predicting strain profiles and the proportion of extension to compression for a given width–depth ratio can be derived from Figure 3.4, which shows that with narrow panels the intensity of compression greatly exceeds that of extension whereas that of extension is approximately 25% greater than compression when the critical condition is reached.

Because all principal factors are related to the width–depth ratio this means that to be absolutely reliable this ratio has to remain constant throughout the production life of a working face. For various geological and operational reasons, for example, inclined seams or changes in the width of working face, this is not the case and consequently the above mentioned prediction methods must be modified.

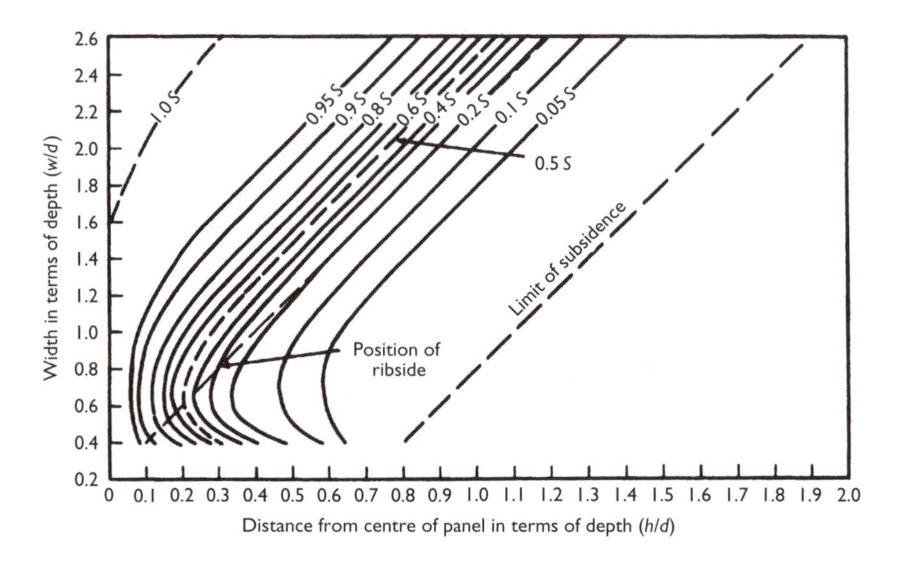

Figure 3.16 Graph for predicting subsidence profiles. w = width, d = depth, h = horizontal distance. (After Anon., 1975; reproduced by kind permission of the Coal Authority.)

Burton (1978) developed an empirical three-dimensional model of subsidence prediction that can be represented graphically. In this way a series of average subsidence and strain profile shapes that are related to width–depth ratios, can be portrayed as a single graph, which Burton referred to as a static set. The model consists of two basic patterns, namely, the static loop and the dynamic loop. A static set can be represented by a loop that is related to static values of displacements of lines of points across a subsidence trough. The pattern that is produced indicates the relationship between an increase in vertical and horizontal displacement with an increase in the amount of extraction. The depth of the static loop is equivalent to that of the maximum subsidence and it closes where the surface point concerned is not influenced by subsidence effects. The direction of the slope of the loop shows the general location of tension and compression, the latter occurring towards the centre of the line. Dynamic loops describe the position of a given point at a certain moment in time. They can be developed either in relation to vertical or horizontal displacement. The vertical pattern of loops becomes smaller on moving from the centre of the trough. Hence, a three-dimensional picture of the subsidence basin can be established for any point in time.

Burton (1978) showed that surface points move inwards across a panel and also move backwards and forwards in the line of advance. A series of differential movements is obtained by fixing one point and then calculating

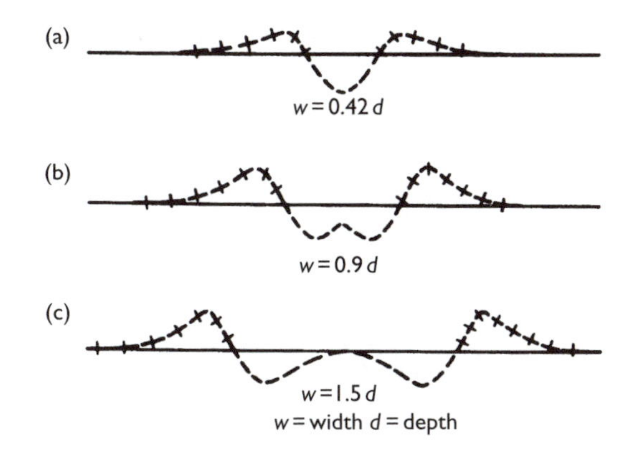

Figure 3.17 Types of strain profile. (a) Maximum compression occurs at the centre of a panel in the narrower panels, that is, when $w = 0.42d$ the single compression zone has a greater intensity than the extension. (b) A hump occurs in the compression curve when $w = 0.9d$. (c) Two separate compression zones occur when $w = 1.5d$. (After Anon., 1975; reproduced by kind permission of the Coal Authority.)

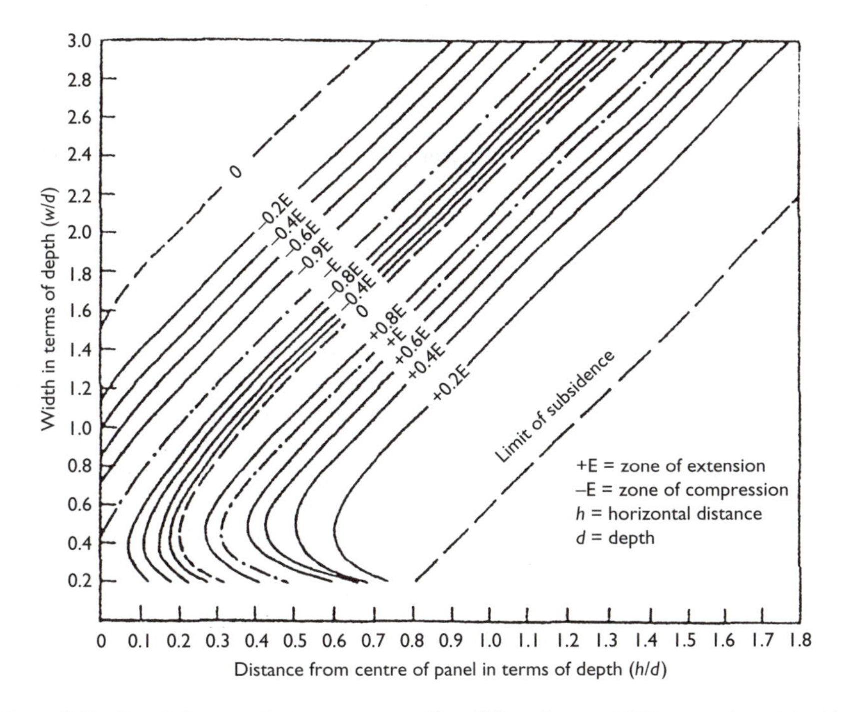

Figure 3.18 Graph for predicting strain profiles. (After Anon., 1975; reproduced by kind permission of the Coal Authority.)

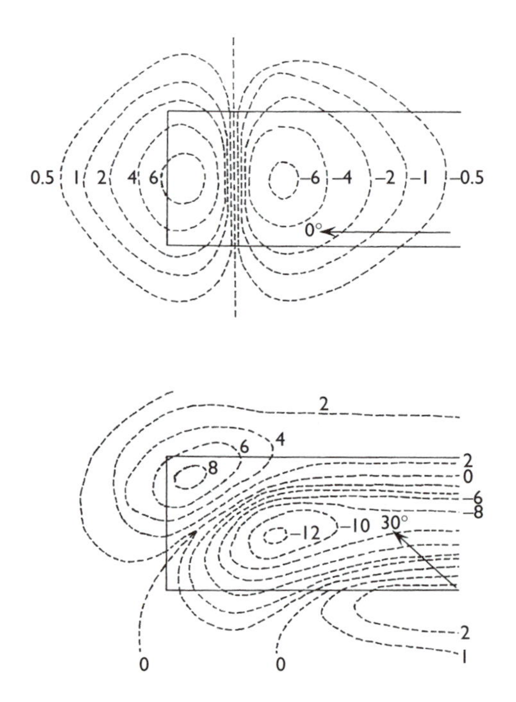

Figure 3.19 Strain contour maps produced by the same extracted panel indicating the difference in arrangement according to the line of direction chosen. (Modified after Burton, 1978.)

the displacements of another point in relation to this. Accordingly, strains can be calculated along the line between the two points under consideration. In other words, strain can be calculated in any direction with respect to face advance. This is important because the value of strain, or change in surface length, is direction related, that is, a zone of tension or compression, which is related to a line normal to the ribside has no meaning for a line running in another direction. Hence, a strain contour map must be related to a surface direction (Fig. 3.19).

In 1981 Burton produced a mathematical model, based on his previous work, which formed the basis of a computer program. This, in addition to permitting subsidence problems to be looked at in three-dimensions, also can produce subsidence profiles and maps rapidly. The advantage of the mathematical model is that it permits feedback control in that it can be amended by future input of data and in this way accuracy of subsidence prediction is continuously being improved. Because of the feedback control system, Burton maintained that the method could be applied to any coalfield and provided accurate predictions as long as the feedback of data was available. Burton (1985) subsequently presented a program that can predict displacements, stresses and strains at any time at any horizon above the worked seam rather than just at the surface, whether the face of the panel is static or moving.

3.3.2. Theoretical and numerical methods

The analytical or theoretical methods of subsidence prediction assume that stratal displacement behaves according to one of the constitutive equations of continuum mechanics over most of its range. The continuum theories have been developed from the analysis of a displacement continuity produced by a slit in an infinite elastic half-space. Analytical procedures subsequently were developed for three types of subsurface excavations based on elastic ground conditions, that is, non-closure, partial closure and complete closure. Further work extended the closed form solution to transversely isotropic ground conditions in both two- and three-dimensions.

Berry (1978) reviewed various theoretical methods that have been developed for predicting mining subsidence. Deformation of the ground is brought about by several different mechanisms depending, in part, on the nature of the strata concerned. For instance, a fireclay beneath a coal seam will flow into roadways, whilst a sandstone roof typically breaks at a forward angle over the face. In Britain gross fracturing in the roof rocks is not believed to extend upwards much more than three times the seam thickness, although minor fractures and bed separation may continue some distance further. As a consequence, most of the ground that undergoes deformation as a result of longwall extraction has been regarded as a continuous medium. Indeed, Whetton and King (1958) and then Hackett (1959) suggested that in some cases of deep mining, most of the deforming ground behaved in a more or less elastic manner. Although Berry disagreed and maintained that the ground is not sufficiently homogeneous to be treated by the methods proposed by these authors, he nevertheless concluded that elastic models simulate most ground deformations with reasonable accuracy. The disadvantage is that for any particular problem estimation of the various elastic constants for massive layered rock masses is difficult.

Previously, Salamon (1964) had developed a means of calculating subsidence, assuming elastic behaviour, based on the concept that stresses and displacements due to mining a seam can be predicted from the convergence distribution of the seam, regardless of the complexity of the mining layout. The mined out area was divided into a network of small areas called face elements. First, the elementary stresses and displacements induced by closure and ride of a single face element are determined, and then using the principle of superposition, the total displacements and stresses can be obtained by summation of the effects of all the face elements in the excavation.

Numerical models permit quantitative analysis of subsidence problems and are not subject to the same restrictive assumptions required for the closed form analytical solutions. Finite element modelling frequently has been applied to subsidence problems since it can accommodate non-homogeneous media, non-linear material behaviour and complicated mine geometries.

The finite element method has been applied to problems of ground movement, for example, by Zienkiewicz *et al.* (1966) and Stacey (1972). Dahl (1972) analysed mining subsidence by using two-and three-dimensional finite element models assuming both elastic and elastoplastic rock behaviour. Data from the National Coal Board and from surveys in the United States served as a basis for comparison. Isotropic elastic properties were assumed for the models in the pre-failure state and the Coulomb criterion was used for the extension of the analysis into the post-failure state. Dahl found that the models of relatively low strength approximated more closely to the field subsidence results. Berry (1978) indicated that the finite element method has the advantage of allowing the elastic constants to differ for each element; hence the appropriate constants can be used for each layer. Nonetheless, he felt that the use of the finite element technique for estimation of subsidence, because of the involvement of considerable depths below the surface and either side of the excavation zone, was of questionable value.

Alternatively, finite difference models can be used for the large strain, non-linear phenomena associated with subsidence development. Other elastic approaches employing numerical techniques include boundary element methods.

Alejano *et al.* (1999) referred to a finite difference method of modelling subsidence using the FLAC computer model. They showed that for British data the results tended to fit empirical observations for horizontal and gently inclined coal seams.

Coulthard and Dutton (1988) described two, two-dimensional numerical methods of subsidence prediction, one being a non-linear finite difference method, the other a distinct element method. The rock mass was modelled as a non-linear medium with horizontal joints along which slip could occur in the finite difference method. The distinct element method models the rock mass as an assemblage of discrete deformable blocks, which may rotate, slide against each other and separate from each other. Coulthard and Dutton considered that realistic non-linear representation of a rock mass was an essential part of any method of subsidence prediction.

Tsur-lavie and Denekamp (1981), and Tsur-lavie *et al.* (1985) presented a boundary element method for the evaluation of subsidence. Their method was based on a fundamental solution of stresses and deformation around a single rectangular indentation at the boundary of an infinite elastic half plane. The solution can be used for both compressible and incompressible material. The model represents a state of uniform vertical displacement of the roof of the mined area, equal to the width of the mined coal seam causing total closure of the area. The model was used for an analysis of ground subsidence as a function of the span and height of the longwall opening, assuming various values of Poisson's ratio. It was found that in the case of deep mines the use of small values of Poisson's ratio gave close agreement with actual measurements of subsidence whereas for shallow mines use of large values yield the best agreement. These results suggest that subsidence in shallow mines is associated with a state of failure extending to the ground surface. In deep mines failure is confined to a relatively limited zone above the subsiding roof.

After reviewing the literature, Aston *et al.* (1987) concluded that no analytical method currently existed that could predict surface subsidence reliably. In a comparison with results derived from empirical and semi-empirical methods, they found that analytical methods yielded smaller vertical displacements by almost an order of magnitude. Most analytical models assume elastic bending, whereas additional mechanisms are involved in subsidence. It therefore is necessary to identify and understand each of the mechanisms involved, as well as their role in the overall process, in order to improve subsidence prediction.

3.3.3. Semi-empirical methods

Most methods of subsidence prediction fall into the category of semi-empirical methods since fitting field data to theory often results in good correlations between predicted and actual subsidence. There are two principal methods of semi-empirical prediction, namely, the profile functions and the influence functions methods (Brauner, 1973b). The profile function method basically consists of deriving a function that describes a subsidence trough. The equation produced is normally for one-half of the subsidence profile and is expressed in terms of maximum subsidence and the location of the points of the profile. In supercritical extraction the central position of the curve is S_{max} and changes to zero subsidence at the edges of the critical area. For subcritical extraction the profile is determined from the critical profile produced from an empirical/mathematical relationship.

The general relationships of the method are shown in Figure 3.20. The profile is described by a function $S(x)$ normally derived from field measurement or model investigations. Generally, the profile is asymmetrical about the point of inflection, with half full subsidence occurring at this point. The distance, d, from the inflection point to the ribside must be either determined from empirical relationships or measured. Normally, the relationship is given as a proportion of the depth of extraction. The depth, and therefore critical radius, influence the slope of the profile. Thus, the subsidence of a point at position x is a function of the maximum subsidence, S_{max}, the distance, d, and the critical radius, B, that is:

$$s = f(xS_{max}dB) \tag{3.3}$$

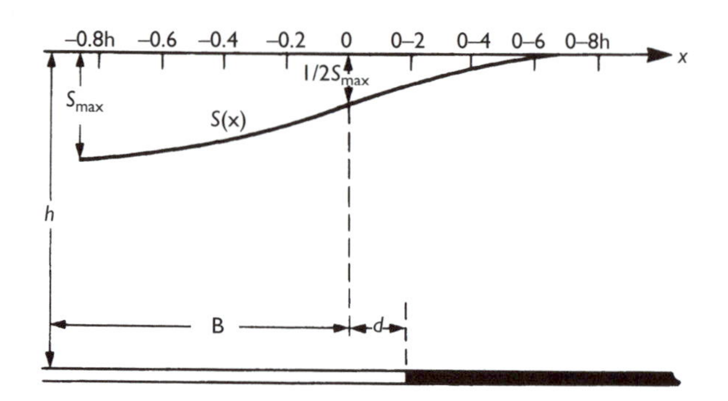

Figure 3.20 Subsidence profile functions.

In subcritical extraction a further parameter, the width–depth ratio, may be included. Subcritical extraction also means that the field determination of *B* is difficult since the slope at the centre of the trough vanishes over a very small distance. Diez and Alvarez (2000) recently developed a semi-empirical profile function method to determine subsidence caused by mining very steep or near vertical coal seams.

Provided the input data are available, the profile function method provides an accurate technique for the prediction of subsidence where the mine geometry is relatively simple as in longwall mining. On the other hand, the influence function approach, according to Hood *et al.* (1983), is better suited to subsidence prediction above irregularly shaped panels. The influence function approach to prediction of subsidence is based upon the principle of superposition. It uses several infinitesimal parts of the extraction and assumes that the subsidence of a point is influenced to a greater extent by these elements. The subsidence trough is regarded as a combination of many infinitesimal troughs formed by a number of infinitesimal extraction elements (Fig. 3.21). The contributions of a single extraction element on a surface point can be derived as the product of its area, dA, and a value p indicating the degree of the influence of dA on the surface point P (Fig. 3.22). The influence of p depends on the horizontal distance, r, between the point and element, that is, $p = f(r)$. The influence value refers to the surface point and therefore the extraction element directly beneath the point has the greatest influence. Accordingly, the function $p(r)$ reaches a maximum value when $r = 0$. If P is a point at the centre of the critical area, then it is influenced by all the extraction elements and therefore undergoes full subsidence, S_{max}. Thus, S_{max} is represented by the complete area under the curve $p(r)$. If P is at the edge (i.e. the ribside) of the extraction, then it is influenced by only half the sum of the possible influences and so only experiences half the full subsidence. This is also the point of inflection of the profile and so the distance between the point of full subsidence and the

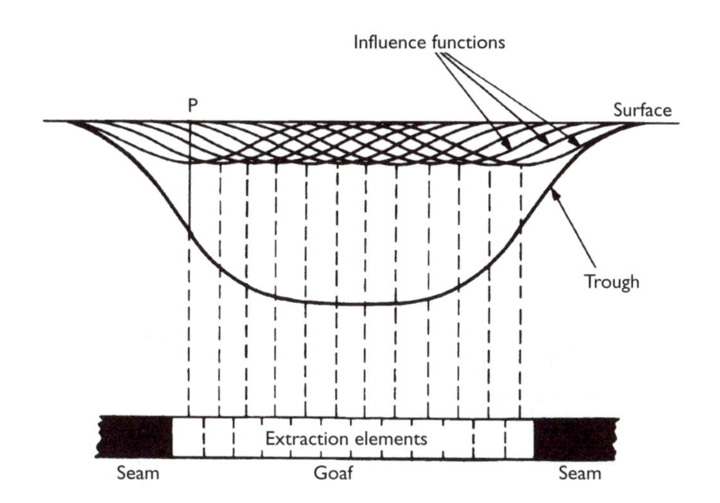

Figure 3.21 Superposition of infinitesimal influences.

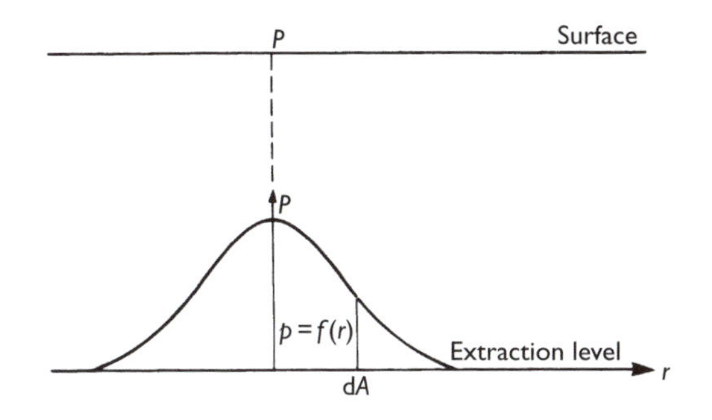

Figure 3.22 Influence function.

inflection point is equal to half the width of the critical area, namely, B. In three dimensions the full subsidence can be referred to as the volume of the solid revolution of $p(r)$ around the z axis, the radius of the solid being B. Consequently, the full subsidence can be expressed as:

$$S_{max} = 2\pi \int_0^B rp(r)\, dr \tag{3.4a}$$

or

$$S_{max} = 2\pi \int_0^\infty rp(r)\, dr \tag{3.4b}$$

depending on whether $p = 0$ at $r = B$, hence Eq. (3.4a); or p is asymptotic to zero, hence Eq. (3.4b).

There are two problems associated with the use of influence functions. First, the subsidence profile obtained is asymmetrical about the inflection points of the sides of the trough and symmetrical about the centre of the panel. This often is not the case in an actual subsidence profile. Second, the method assumes that the point of inflection is situated immediately above the edge of the panel (the ribside). Special measures have to be taken to relocate the predicted curve when this is not the case and if the distance involved is large, then inaccuracies arise in the subsidence predicted.

Donnelly *et al.* (2001) referred to the use of the SWIFT program (Subsidence With Influence Function Technique), initially developed by Ren *et al.* (1987) that is based on subsidence observations in British coal fields, to predict the magnitude of subsidence at a mine in Colombia. Although the program overestimated the maximum subsidence, the shape of the subsidence profile, area of influence and location of maximum subsidence were similar to those derived by precise surveying. The overestimation of predicted subsidence was attributed to the presence of strong rhyolite sills in the overburden, which acted as beams during subsidence. This emphasizes the fact that subsidence prediction methods based on particular geological settings cannot necessarily be applied to other settings where the geological characteristics differ. Bello *et al.* (1996) proposed a method of subsidence prediction based on influence functions that considered the three-dimensional nature of the problem.

The fundamental concept of complementary influence functions is that the separate influence functions describing the response of mined and unmined zones act together to produce subsidence (Sutherland and Munson, 1984). Each influence function is defined by the response of a limit element, that is, an unmined element for the coal left in place and a mined element for the void created by extraction. The amount of subsidence is predicted by appropriately summing these elements over the entire seam, it being the sum of the influence of both the mined and unmined response. Hence, complementary influence functions can be used to compute subsidence above mines with complex geometry (i.e. for room and pillar as well as longwall workings). This method, however, tends to overestimate the subsidence directly over the ribside.

Marr (1975) described the zone area or circle method of subsidence prediction used in many coalfields of western Europe. Basically, surface subsidence is estimated by constructing a number of concentric zones around a surface point, the radius of the outer zone being equal to the radius of the area of influence (Fig. 3.23). The subsidence at the surface point is obtained from the summation of the proportions of coal that were extracted in each zone, multiplied by its particular subsidence factor. In practice it has been found that 3, 5 or 7 zones are suitable for most estimations. The method permits estimation of subsidence that develops when panels of irregular shape are mined. Marr used seven zones, the width of

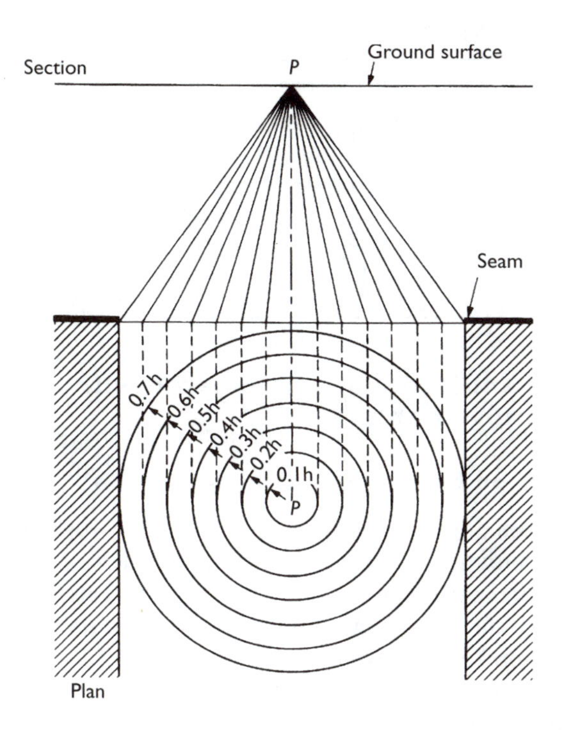

Figure 3.23 Zone area or circle method of subsidence prediction. (After Marr, 1975; reproduced by kind permission of the Institute of Materials, Minerals and Mining.)

each one being equal to one-tenth of the depth of the workings. The subsidence, S, at the surface point is derived from the following expression:

$$S = A^n a + B^n b + C^n c + D^n d + E^n e + F^n f + G^n g \tag{3.5}$$

where A, B, C, D, E, F and G are the proportions of the zone area extracted; a, b, c, d, e, f and g are the zone factors; and n is a constant that is 2.3 for the Coal Measures in Britain. Marr calculated zone factors for maximum subsidence at the centre of the critical area of extraction and for subsidence over the ribside (here subsidence is 0.183 of the thickness of the seam extracted compared with maximum subsidence that is 0.9 times extracted seam thickness). The zone factors are shown in Table 3.4. In order to obtain more realistic values of subsidence allowances can be made for the effect of pillars of coal left between mined out areas and for inclined seams. Ren *et al.* (1987) compared the results obtained by using the zone area method of prediction with those derived by the NCB method (Anon., 1975) and found that they were similar for both subsidence and displacement.

Table 3.4 Zone factors (After Marr, 1975)

Zone	*Maximum*	*Ribside*	*Mean*
a	0.019	0.007	0.013
b	0.063	0.029	0.045
c	0.116	0.101	0.109
d	0.217	0.220	0.219
e	0.235	0.265	0.259
f	0.187	0.211	0.199
g	0.045	0.067	0.056

3.4. Damage and problems associated with ground movements

3.4.1. Influence on building and structures

The different types of ground movement associated with mining subsidence affect different buildings and structures in different ways. For instance, vertical subsidence may seriously affect drainage systems, and tilt may cause serious concern as far as railways and tall structures, such as chimneys, are concerned. Damage to buildings generally is caused by differential horizontal movements and the concavity and convexity of the subsidence profile that give rise to compression and extension in the structure itself, the latter generally being the more serious. Usually, however, it is not just a simple matter of examining the reaction of a structure to a particular value of tensile or compressive strain. For example, it is quite common for a structure to be subjected to compressive strains in one direction and tensile strains in another direction. It also may be subjected to alternative phases of tensile and compressive ground movements so there is a dynamic effect to consider. Consequently, any acceptable design for a structure situated in an area of active longwall mining must have regard for the nature, degree and periodicity of the ground movements likely to be caused by mining.

Ground movement that adversely affects the safety or function of a building or structure is unacceptable. However, the appearance of many buildings is also of concern and therefore significant cracking of architectural features is unacceptable. Hence, an estimation of the amount of subsidence that will adversely affect structural members and/or architectural features is required. This is influenced by many factors, including the type and size of the building or structure, and the properties of the materials of which it is constructed, as well as the rate and nature of the subsidence. Because of the complexities involved, critical movements have not been determined analytically. Instead, almost all criteria for tolerable subsidence or settlement have been established empirically on the basis of observations of ground movement and damage in existing buildings (Boone, 1996).

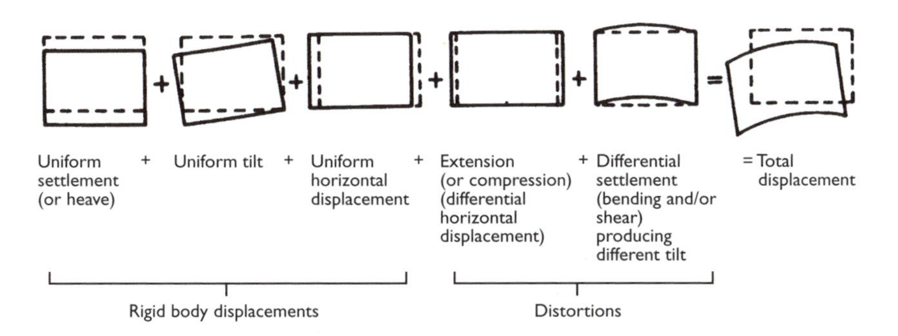

Figure 3.24 Components of ground movement.

Geddes (1984) indicated that total displacement of a structure may be regarded as consisting of five components, three of which are rigid body displacements, the remaining two being distortional components (Fig. 3.24). The latter components are consequent on divergences within the structure from uniform movements and represent the main causes of damage.

As far as the effects of subsidence on buildings or structures are concerned, it should not be assumed that ground movements produced in the presence of buildings or structures are the same as those when they are absent. Moreover, although in many cases the contact between the ground and building or structure is maintained during the period of ground movement, in some situations gaps develop between the ground and building or structure, and slip type displacements may take place. Friction and adhesion at the interface between the ground and foundation structure during subsidence generate structural deformation equal to at least a fraction of the horizontal displacement. The strength of structural components may be high enough to withstand damage when the ground strains are compressive but this may not be the case when the ground undergoes tensile strain. According to Geddes (1981), the horizontal displacements of a foundation structure and the ground located in the zone of tension are compatible as long as the shear stresses along the interface between the two are less than the shear strength of the interface. If the latter is exceeded, then slippage occurs along the underside of the foundation structure. The ground–structure interaction also is influenced by the fact that buildings or structures are not perfectly flexible, hence the magnitude of vertical support reaction may vary during the passage of a subsidence wave. As noted earlier, discontinuities may concentrate ground strain and so a building or structure founded above such a feature may be damaged even though the horizontal strain over the length of the building or structure was rather modest.

Two parameters commonly have been used for developing correlations between damage and differential settlement, namely, angular distortion and

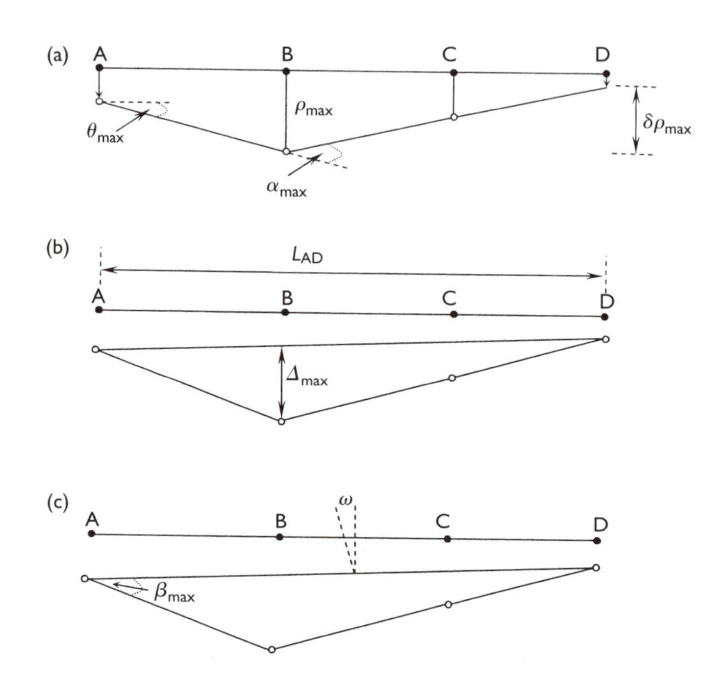

Figure 3.25 Definitions of foundation movement. (a) Definition of settlement, ρ, relative settlement, $\delta\rho$, rotation, θ and angular strain, α. (b) Definition of relative deflection, Δ and deflection ratio, Δ/L. Definition of tilt, ω and relative rotation (angular distortion) β. (Modified after Burland *et al.*, 1977.)

deflection ratio (Fig. 3.25). Angular distortion, δ/l, is the differential movement between two points divided by the distance separating them. When related to building damage, angular distortion commonly is modified by subtracting the rigid body tilt, ω, from the measured displacement. In this way the modified value is more representative of the deformed shape of the building. The deflection ratio, Δ/L, is defined as the maximum displacement, Δ, relative to a straight line between two points divided by the distance, L, separating the points.

Skempton and MacDonald (1956) selected angular distortion (relative rotation) as the critical index of ground movement (Table 3.5). They concluded that cracking of load bearing walls or panel walls in frame structures is likely when δ/l exceeds 1/150. Skempton and MacDonald also suggested that a value of $\delta/l = 1/500$ could be used as a design criterion that provides some factor of safety against cracking. Bjerrum (1963) also suggested limits for damage criteria based angular distortion (Table 3.6). Grant *et al.* (1974) supported the conclusions of Skempton and MacDonald but noted that cracking should be anticipated when δ/l exceeds 1/300. Both these groups of authors regarded tilting as rigid body rotation that does not

Table 3.5 Limitations of ground movement (After Skempton and MacDonald, 1956; reproduced by kind permission of Thomas Telford Publishing)

Criterion	*Independent footings*		*Rafts*
Angular distortion (δ/l)	1/300		1/300
Greatest differential movement	Sands	30 mm	30 mm
	Clays	45 mm	45 mm
Maximum movement	Sands	50 mm	75 mm
	Clays	75 mm	75–125 mm

Table 3.6 Limiting angular distortion (After Bjerrum, 1963; reproduced by kind permission of Elsevier)

Category of potential damage	δ/l
Danger to frames with diagonals	1/600
Safe limit for no cracking of buildings[a]	1/500
First cracking of panel walls	1/300
Tilting of high rigid buildings becomes visible	1/250
Considerable cracking of panel and brick walls	1/150
Danger of structural damage to general buildings	1/150
Safe limit for flexible brick walls, L/H> 4[a]	

Note
a Safe limits include a factor of safety.

contribute to the distortion of the structure. Hence, both removed the differential movement due to tilting from the computed values of angular distortion. Leonards (1975), however, maintained that in the case of framed structures supported on isolated spread footings, the validity of this assumption was questionable. He maintained that in such a case, tilting contributes to the stress and strain in the frame unless each footing tilts or rotates through the same angle as the overall structure. Because this is unlikely to occur, he suggested that the effects of tilt should be included in the differential movement criteria.

The use of angular distortion as a criterion of structural damage has been criticized as an oversimplification, it being argued that the effect of ground curvature and type of structure was not fully appreciated. This is particularly the case in relation to the behaviour of load bearing brick walls undergoing hogging or sagging. What is more, angular distortion implies that damage is due to shear distortion within the structure, which is not necessarily the case.

Polshin and Tokar (1957) defined allowable displacement in terms of deflection ratio, Δ/L. Their displacement criteria are given in Tables 3.7 and 3.8. There are a number of differences between these criteria and those presented

Table 3.7 Allowable settlement criteria (After Polshin and Tokar, 1957; reproduced by kind permission of Elsevier)

Type of structure	*Sand and hard clay*	*Plastic clay*
$\beta = (\delta/l)$		
Civil and industrial building column foundations:		
For steel and reinforced concrete structures	1/500	1/500
For end rows of columns with brick cladding	1/145	1/1000
For structures where auxiliary strain does not arise during non-uniform settlement of foundations	1/200	1/200
Tilt of smoke stacks, towers, silos, etc.	1/250	1/250
(Δ/L)		
Plain brick walls:		
For multi-storey dwellings and civil buildings		
at $L/H < 3$	0.0003	0.0004
at $L/H > 5$	0.0005	0.0007
For one-storey mills	0.0010	0.0010

Table 3.8 Allowable average settlement for different building types (After Polshin and Tokar, 1957; reproduced by kind permission of Elsevier)

Kind of building	*Allowable average settlement (mm)*
Building with plain brick walls	
$L/H > 2.5$	80
$L/H < 1.5$	100
Building with brick walls, reinforced concrete or reinforced brick	150
Framed building	100
Solid reinforced concrete foundations of smoke stacks, silos, towers, etc.	300

by Skempton and MacDonald (1956). For instance, frame structures and load bearing walls are treated separately. The allowable displacement for frames is expressed in terms of the slope or the differential displacement between adjacent columns. This is very similar to the angular distortion of Skempton and MacDonald without correction for tilt. The limiting values quoted by Polshin and Tokar (1957) vary between 1/500 and 1/200. The maximum allowable deflection ratio was assumed to be related to the development of a critical level of tensile strain in a wall. For brick walls, the critical tensile strain was taken as 0.05%. Polshin and Tokar adopted more stringent limits for differential movement of load bearing brick walls than did Skempton and MacDonald. The deflection ratio at which cracking occurs in brick walls was related to the length to height ratio, L/H, of the wall.

Burland and Wroth (1975) and Burland *et al.* (1977) also used the deflection ratio, Δ/L, at which the critical tensile strain, 0.075%, is reached as a criterion for allowable ground movement. They proposed that limiting deflection criteria should be developed for at least three different cases. Diagonal strain is critical in the case of framed structures, which are relatively flexible in shear, and for reinforced load bearing walls, which are relatively stiff in direct tension. Bending strain is critical for unreinforced masonry walls and structures that have relatively low tensile resistance. Hence, unreinforced load bearing walls, particularly when subjected to hogging, are more susceptible to damage than frame buildings. For unreinforced load bearing walls in the sagging mode Burland and Wroth gave values of 0.4×10^{-3} (1/2500) for an *L/H* ratio of 1, and 0.8×10^{-3} (1/1250) for a ratio of 5. However, they pointed out that cracking in the hogging mode occurs at half these values of deflection ratio. For frame buildings with *L/H* less than 3, the Burland and Wroth, and Skempton and MacDonald (1956) criteria provide reasonable limits for the deflection ratio. When *L/H* exceeds 3, the Skempton and MacDonald criterion is the more conservative of the two. In the case of load bearing walls in the sagging mode, the Skempton and MacDonald criterion appears to be unconservative while the Polshin and Tokar (1957) limits appear reasonable. However, load bearing walls in the hogging mode may crack at deflection ratios that are much smaller than the Polshin and Tokar limits.

Wahls (1982) summarized the findings of the aforementioned authors. He concluded that for frame buildings with panel walls angular distortions of 1/300 are likely to produce some cracking of architectural features, while angular distortions of 1/150 probably will cause structural damage. The tolerable differential movement for load bearing walls is smaller than that which can be withstood by framed structures and is influenced by the *L/H* ratio of the wall.

Charles and Skinner (2004) pointed out that when differential ground movement occurs, a building or structure may undergo both distortion and tilt. In the case of tall structures, the large height to length ratio generally gives rise to a predominantly rigid body rotation, although some distortion may occur. In fact, tilt may lead to some tall structures collapsing. On the other hand, when low-rise buildings are subjected to differential subsidence, then distortion normally is of more concern than tilt. When low-rise buildings in areas of mining subsidence have been constructed on rafts with adequate stiffness to resist horizontal tensile forces, then differential subsidence causes buildings to tilt as a rigid body, and in this way prevents distortion of the building and cracking of walls. Nonetheless, there comes a point when tilt becomes unacceptable in terms of aesthetics, serviceability (e.g. doors swinging open, drainage falls becoming insufficient etc.) or stability. The tolerability of tilt depends on the type of building and the purpose it serves. According to Charles and Skinner, a design limit value for tilt can be regarded as 1/400. Tilt of walls and floors of low-rise buildings becomes noticeable between 1/250 and 1/200. Problems associated with serviceability are unlikely until the tilt is appreciably greater and structural distress does not begin until tilt reaches 1/50 (Table 3.9).

According to Brauner (1973a,b), it is generally accepted in the Ukraine that most surface structures are safe if ground disturbances do not exceed values of 4×10^{-3} for slope, 5000 m for radius of curvature and 2×10^{-3} for strain. Table 3.10 provides a classification of allowable disturbances for various types of structure found in the Donetz Coalfield, Ukraine. The coefficient of safety is a dimensionless factor used to multiply seam thickness in order to determine the depth below which mining does not give rise to deformations greater than allowable for the category concerned (α refers to the angle of dip of the seam). Buildings are classified according to height in Table 3.11.

Investigations carried out by the National Coal Board (Anon., 1975) have revealed that typical mining subsidence damage starts to appear in conventional structures when they are subjected to effective strains of 0.5–1.0 mm m^{-1} and damage can be classified as negligible, slight, appreciable, severe and very severe (Fig. 3.26 and Table 3.12). However, this relationship between damage and change in length of a building or structure is only valid when the average ground strain produced by mining subsidence

Table 3.9 Indicative values for tilt of low-rise buildings (After Charles and Skinner, 2004; reproduced by kind permission of Thomas Telford Publishing)

Classification	*Tilt*	*Comment*
Design limit value	1/400	The maximum differential subsidence across a building is related to the design limit value for tilt. If a building is likely to tilt more than this limit value, then ground treatment of deep foundations may be required
Noticeability	1/250	The point at which the tilt of a building becomes noticeable depends on the type and purpose of the building, and the powers of observation and perception of the occupants. Tilt of low-rise buildings typically is noticed between 1/250 and 1/200
Monitoring	1/250	When tilt is noticed it is advisable to make some measurements to confirm that the building has tilted. If the measured tilt is greater than 1/250, then monitoring should be carried out to determine whether the tilt is increasing
Remedial action	1/100	Where tilts of this magnitude are measured or the measured rate of increase of tilt indicates that this degree of tilt will be exceeded, some remedial action should be taken. This is likely to include re-levelling the building, perhaps by grouting or underpinning and jacking
Ultimate limit	1/50	If tilt reaches this level, then the building may be regarded as in a dangerous condition and remedial action either to re-level the building, or to demolish it will be required urgently

Table 3.10 Categories of protection, Donetz District, Ukraine (After Brauner, 1973b; reproduced by kind permission of the United States Bureau of Mines)

Category	*Allowable tilt* ($\times 10^{-3}$)	*Allowable radius of curvature (m)*	*Allowable strain* ($\times 10^{-3}$)	*Coefficient of safety*	
				$\alpha < 45°$	$\alpha => 45°$
I	4.0	20 000	2.0	400	500
II	4.5	18 000	2.5	350	400
III	5.0	12 000	3.5	250	300
IV	8.0	5500	6.0	150	200
V	10.0	3000	7.5	100	150
VI	25.0	1000	14.0	50	75

Table 3.11 Category according to height, Donetz District, Ukraine (After Brauner, 1973b; reproduced by kind permission of the United States Bureau of Mines)

Height (m)	*Number of floors*	*Category*
Over 15	5 or more	II
10–15	3–4	III
5–10	1–2	IV or V
Less than 5	1	VI

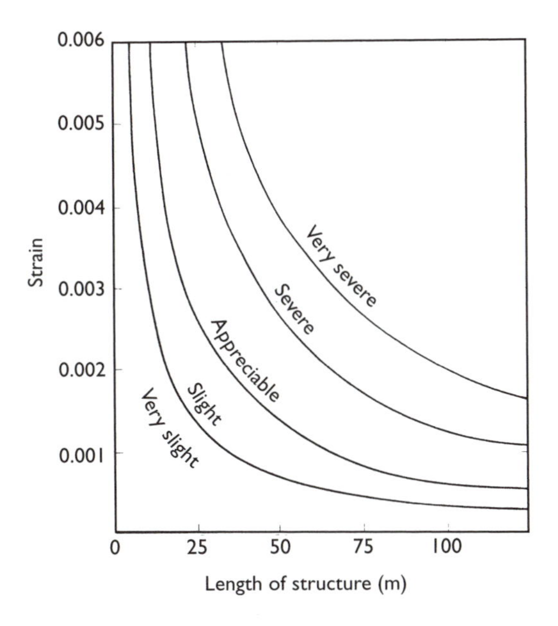

Figure 3.26 Relationship of subsidence damage to length of structure and horizontal strain. (After Anon., 1975; reproduced by kind permission of the Coal Authority.)

Table 3.12 National Coal Board classification of subsidence damage (After Anon., 1975; reproduced by kind permission of the Coal Authority)

Change in length of structure (mm)	*Class of damage*	*Description of typical damage*
Up to 30	1 Very slight or negligible	Hairline cracks in plaster, perhaps isolated slight fracture in the building, not visible from the outside
30–60	2 Slight	Several slight fractures showing inside the building. Doors and windows may stick slightly. Repairs to decoration probably necessary. Not visible from the outside
60–120	3 Appreciable	Slight fractures showing on outside building (or one main fracture). Doors and windows sticking. Service pipes may fracture
120–180	4 Severe	Service pipes disrupted. Open fractures requiring rebonding and allowing weather into the structure. Window and door frames distorted. Floors sloping noticeably. Some loss of bearing in beams. If compressive damage, overlapping of roof joints and lifting of brickwork with open horizontal fractures
>180	5 Very severe	As above but worse and requiring partial or complete rebuilding. Roof and floor beams loose or non-bearing and in need of shoring up. Windows broken with distortion. Severe slopes on floors. If compressive damage, severe buckling and bulging of the roof and walls

is equalled by the average strain in the building or structure. In fact, this commonly is not the case, strain in the structure being less than it is in the ground (Geddes, 1984). Accordingly, Figure 3.26 only serves as a guide to the likely damaging effects of ground strains induced by mining subsidence. In addition, it takes no account of the design of a building or structure, or of construction materials. Nevertheless, it indicates that the larger a building or structure, the more susceptible it is to differential vertical and horizontal ground movement.

More recent criteria for subsidence damage to buildings have been proposed by Bhattacharya and Singh (1985) who recognized three classes of damage, namely, architectural that was characterized by small-scale cracking of plaster and doors and windows sticking; functional damage that was characterized by instability of some structural elements, jammed doors and windows, broken window panes and restricted building services; and structural damage in which primary structural members were impaired,

Table 3.13 Recommended damage criteria for buildings (After Bhattacharya and Singh 1985)

Category	*Damage level*	*Angular distortion (mm/m)*	*Horizontal strain (mm/m)*	*Radius of curvature (km)*
1	Architectural	1.0	0.5	—
	Functional	2.5–3.0	1.5–2.0	20
	Structural	7.0	3.0	—
2	Architectural	1.3	—	—
	Functional	3.3	—	—
	Structural	—	—	—
3	Architectural	1.5	1.0	—
	Functional	3.3–5.0	—	—
	Structural	—	—	—

there was a possibility of collapse of members and complete or large-scale rebuilding was necessary. Their conclusions are summarized in Table 3.13. They observed that basements were the most sensitive parts of houses with regard to subsidence damage, and therefore that basements usually suffered more damage than the rest of a building.

3.4.2. *Influence on groundwater*

Longwall mining of coal gives rise to significant changes in the hydraulic properties and groundwater levels in overlying aquifers because of fracturing and bed separation associated with subsidence that, in turn, influence recharge, well yield and possible pollution. For instance, Singh and Singh (1998) recorded an average fall in water level in wells in an unconfined aquifer of sand in the Kamptee Coalfield, India. This they attributed to the development of tensile strains of about 4.5 mm m^{-1} caused by longwall mining of three panels of coal leading to an increase in the void space of the sand. Unfortunately, this resulted in an acute shortage of water for irrigation and domestic use during the dry season. Booth *et al.* (2000) indicated that the permeability of a sandstone aquifer in Jefferson County, Illinois, which is some 24 m thick, increased by 1–2 orders of magnitude and its storativity by one order, due to subsidence. Well yields increased but the quality of the groundwater deteriorated, becoming more saline with a higher sulphate content. The potentiometric levels in the moderately transmissive sandstone declined rapidly during the tensional phase of subsidence but then partially recovered during the compressional phase. The levels made a full recovery over several subsequent years. By contrast, a sandstone of low transmissivity in Saline County experienced only slight increases in permeability due to fractures associated with subsidence. The potentiometric levels in the sandstone declined rapidly and no significant recovery occurred. Booth *et al.* concluded that the variations in hydrogeological properties, continuity and geometry of the hydrogeological units on a scale as local as a panel strongly affect the initial potentiometric response and critically control recovery. What also should be borne in mind, is that fractures opened by mining subsidence could act as pathways for contaminants that could affect aquifers at shallow depth. The numerous impacts of fractures associated with subsidence on domestic water supplies from wells and springs in Virginia have been considered by Zipper *et al.* (1997). Carpenter (1997) described how such fractures, which may extend to the water table, could be mapped by resistivity and electromagnetic (EM) methods.

Dumpleton (2002) referred to drillholes being sunk in the Sherwood Sandstone (Trias) prior to two panels being worked in the Barnsley coal seam, which is 2.5 m thick and occurs at a depth of 550–600 m in that part of the Selby Coalfield, North Yorkshire, England. Pumping tests were carried out in the drillholes over a period of two years in order to determine the effects of subsidence on the hydrogeological properties of the sandstone. The results indicated increases in the post-mining transmissivity of up to 234% directly over one of the panels and of up to 149% around the margins of the panel. However, post-mining storativity remained largely unchanged. More strikingly, the greatest effects were noticed during the closest approach of the second panel, which caused some additional subsidence over the first. This gave rise to an increase in maximum transmissivity of 1979%, with increases in storativity of up to 625%. Dumpleton attributed the anomalous intra-cycle recovery-drawdown events that occurred during the latter phase as due to rapid dilation and compression of fractures in the aquifer associated with mining. Although different behaviour patterns do occur in aquifers subjected to mining subsidence, the results obtained bear some similarities to those found in the United States as mentioned in the previous paragraph. However, the Barnsley seam is located at greater depth and so the investigation shows the subsidence effects still can influence the properties of shallow aquifers.

3.5. Measures to reduce or avoid subsidence effects

The contemporaneous nature of subsidence associated with longwall mining sometimes affords the opportunity to planners to phase long-term surface development in relation to the cessation of subsidence (Bell, 1987). However, the relationship between future programming of surface development and that of subsurface working may be difficult to coordinate because of the differences that may arise between the programmed intention and the performance achieved. Usually, the relationship between the two programmes cannot be established closely for more than a few months ahead.

Damage attributable to longwall mining subsidence can be controlled and influenced by precautionary measures incorporated into new structures in mining areas, by preventative works applied to existing structures, and by mine design involving special underground layouts or any combination thereof (Anon., 1975; Anon., 1977). Several factors have to be considered when designing buildings for areas of active mining. First, where high ground strains are anticipated, the cost of providing effective rigid foundations may be prohibitive. Second, experience suggests that buildings with deep foundations, on which thrust can be exerted, suffer more damage than those in which the foundations are more or less isolated from the ground. Third, because of the relationship between ground strain and size of structure, very long buildings should be avoided unless their long axes can be orientated normal to the direction of principal ground strain. Finally, although tall buildings may be more susceptible to tilt rather than to the effects of horizontal ground strain, tilt can be corrected by using jacking devices. Another important aspect, as far as planning and development at the surface are concerned, is that this type of subsidence is predictable.

The most common method of mitigating subsidence damage is by the introduction of flexibility into a structure (Bell, 1978). In flexible design, structural elements deflect according to the subsidence profile. The foundation therefore remains in contact with the ground as subsidence proceeds. One of the most notable examples of a flexible form of construction in Britain is the CLASP (Consortium of Local Authorities Special Programme) system, which incorporates a structural framework that is flexible enough to accommodate differential subsidence by being able to deflect sufficiently to ride the approaching subsidence wave without cantilevering over it (Anon., 1977). In other words, the building is constructed on a jointed floor slab, with a superstructure of lightweight steel, jointed with pins and braced with a spring loaded system. The joints preferably should be located between the stanchions. In addition, the slab is reinforced to accommodate strains as the ground moves beneath it and is laid on a compacted bed of sand some 250 mm in thickness. A membrane separates the sand and slab to reduce friction between them. The frame is clad so that movement can occur without distortion and special attention is paid to the flexibility of window openings, stairs and services.

If a large building is required, it is desirable to separate it into small units and to provide a gap of at least 50 mm between each pair of units, the space extending to foundation level.

Flexibility also can be achieved by using specially designed rafts. Raft foundations should be as shallow as possible, preferably above ground, so that compressive strains can take place beneath them instead of transmitting direct compressive forces to their edges. They should be constructed on a membrane so that they will slide as ground movements occur beneath them. For instance, reinforced concrete rafts, laid on granular material

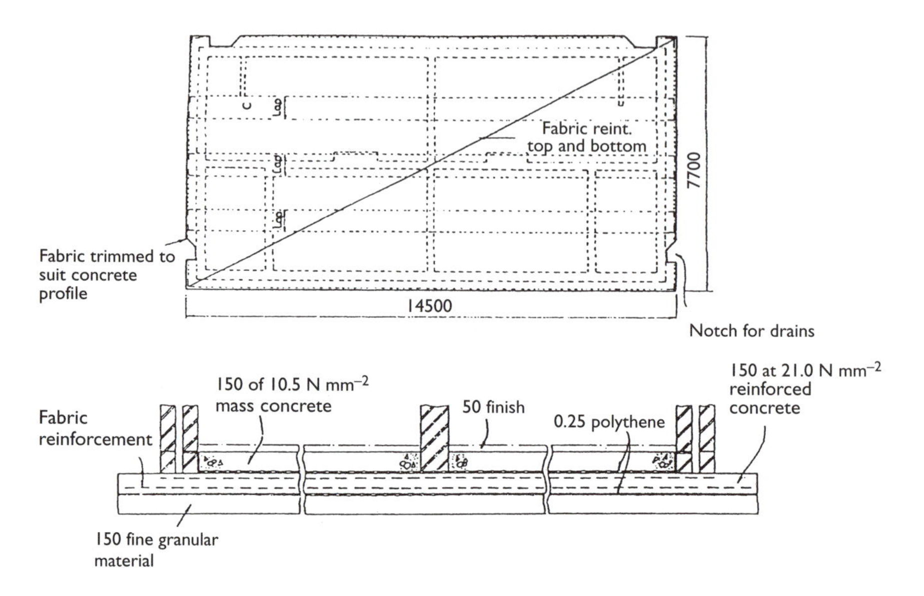

Figure 3.27 A two-layer (sandwich) raft foundation (dimensions in mm).

reduce friction between the ground and the structure. Where relatively small buildings (up to about 30 m in length) are concerned, they can be erected on a sandwich raft foundation (Fig. 3.27). Cellular rafts have been used for multi-storey buildings (Fig. 2.20).

The use of piled foundations in areas of mining subsidence presents its own problems. The lateral and vertical components of ground movement that occur as mining progresses mean that the pile caps tend to move in a spiral fashion, and that each cap moves at a rate and in a different direction according to its position relative to the mining subsidence. Such differential movements and rotations normally would be transmitted to the structure with a corresponding readjustment of the loadings on the pile cap. In order to minimize the disturbing influence of these rotational and differential movements it often is necessary to allow the structure to move independently of the piles by the provision of a pin joint or roller bearing at the top of each pile cap. It may be necessary to include some provision for jacking the superstructures where severe dislevelment is likely to occur.

Preventative techniques can frequently be used to reduce the effects of movements on existing structures. An engineering procedure, for example, was developed by Peng *et al.* (1996) to predict the potential damage to houses due to critical curvature and strain associated with longwall mining. Again, one of the principal objects is to introduce greater flexibility. In the case of buildings longer than 18 m, damage can be reduced by cutting them into smaller, structurally independent units (Ji-xian, 1985). The space produced should be large enough to accommodate deflection. In particular, such items as chimneys, lift shafts, machine beds, etc, can be made independent of the other, generally lighter, parts of a building and separated from the main structure by joints through foundations, walls, roof and floor that allow freedom of movement. Extensions and outbuildings similarly should be separated from the main structure by such joints.

The excavation of trenches around buildings subjected to compressive ground strains has reduced the damaging effects of ground movement significantly. The trench is about 1 m from the perimeter wall of a structure and extends down to foundation level, which effectively breaks the continuity of the surface and hence the foundations are isolated from side-thrust. The trenches are backfilled with compressible material and covered with concrete slabs.

Buildings that are weak in tensile strength can be afforded support by strapping or tie bolting together where they are likely to undergo extension. However, some distortion can occur at the points where the ties are fixed.

Damage to surface buildings and structures can be reduced by adopting a specially planned layout of underground workings that takes account of the fact that surface damage to structures is caused primarily by ground strains. Thus, to minimize the risk of damage underground extraction must be planned so that surface strain is reduced or eliminated (Brauner, 1973b). Harmonious mining, as used in parts of Europe, involves mining three or more seams simultaneously but with careful selection of mining dimensions and rates of advance so that the area in the centre of the subsidence basin undergoes uniform vertical subsidence, and the resultant strains from each panel tend to cancel each other out (Fig. 3.28(a)). In Britain, however, mining conditions seldom lent themselves to simultaneous extraction and the stepped-face layout was used (Fig. 3.28(b)). This method allows for the effects of travelling movements that accompany an advancing face, as well as transverse movements. The degree of cancellation is governed by the distance between the two faces and if the distance is equivalent to 0.45R (R = the radius of the area of influence about a surface point), then the tension from the leading face generally will not develop beyond 50% of its maximum. Unfortunately, it is seldom possible to arrange the direction and location of underground workings to suit the need to minimize surface movement in the neighbourhood of a particular surface structure.

Pillars of coal can be left in place to protect surface structures above them, as was the case beneath Selby Abbey in North Yorkshire, England. The size of the pillars can be determined by the horizontal distance from a surface structure at which the advancing face must stop so that the total strain at the surface is less than the allowable values (Brauner, 1973a). In panel and pillar mining, pillars are left between relatively long but narrow panels. Adjacent panels are designed with a pillar of sufficient width in between so

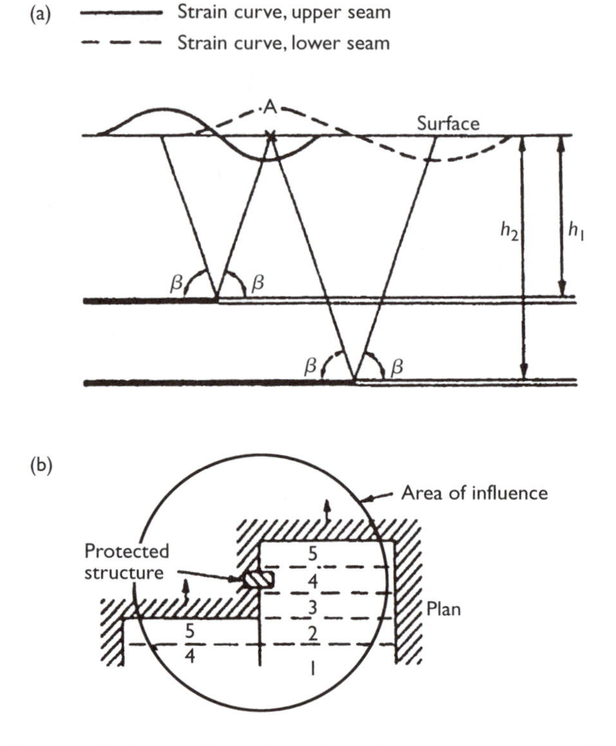

Figure 3.28 (a) Strain reducing configuration of workings in two seams. (b) Extraction with staggered faces.

that the interaction of the ground movements results in flat subsidence profiles at the surface with low ground strains. The resultant surface subsidence ranges from 3% to 20% of the thickness of the seam. Subsidence, however, increases with depth of mining due to the greater loading carried by the pillars. Hence, pillar widths have to be increased with increasing depth. Narrow shortwall (e.g. panels 40 m wide with intervening pillars about 50 m wide) or single entry panel methods of extraction can be used that are narrow enough to allow strata to bridge the goaf, so resulting in little collapse and therefore reduced ground movement at the surface.

Subsidence of the ground surface can be reduced by lowering the width to depth extraction ratio. This was practised beneath the town of Mansfield, England, in the 1980s, where longwall panels were lessened by at least 50% to reduce the expected subsidence when working in the vicinity of the Sheepbridge Lane Fault, which outcropped in a suburban area to the south of Mansfield.

Maximum subsidence can be reduced by packing the goaf. The method is influenced by the particular packing method used. This subsidence factor, because it varies with method, needs to be determined from actual observations of individual cases in each mining district. Variations also are due to different overburden pressures at different depths, different types of packing materials, and the differences in the quality and speed of pack construction. According to Anon. (1975), pneumatic stowing in British coalfields reduced subsidence by up to 50%. Furthermore, Whittaker and Reddish (1993) maintained that experience with stowing behind longwall faces in British coalfields showed that the width–depth ratio usually should exceed 0.6 for stowing to achieve a significant reduction of surface subsidence. Subsidence investigations were conducted by Singh and Singh (1998) over three superimposed longwall panels in the Kamptee Coalfield in central India that were hydraulically filled with sand after the coal was extracted. They found that the maximum subsidence over the panels was 8.75%, 9.0% and 2.7% of the extracted thickness. Sanzotti and Bise (1996) examined the potential for controlling longwall subsidence in south-west Pennsylvania by pneumatically backstowing colliery waste into the mined out area by scheduling the process as part of the mining operation.

The most common types of subsidence damage to highways are undulations, distortions and cracking of the carriageway caused by compressive and tensile stresses in the subgrade. Malkin and Wood (1972) indicated that it was necessary to consider the anticipated effects of subsidence during and after construction of a highway at the design stage, and to make final assessment during construction. In areas of current mining, the mine operators will be able to provide plans of their workings and data concerning future workings. Predictions may be able to be made from the data received regarding the area affected or to be affected, the amount of subsidence likely to occur and the resulting ground strains. Hence, an assessment of the compatibility of the mining proposals, in particular, with the design specification for a route should be able to be made. In some instances it may be necessary to obtain agreement for a modification or restriction of mining in order to avoid damaging ground movements. Also, close liason with mine operators may indicate that re-phasing mine working or changes in mine layout can minimize or eliminate subsidence damage. Faults and dykes in mining areas can concentrate the effects of mining subsidence giving rise to surface cracking or the development of steps, which can lead to severe surface disruption of a highway. For example, Bell (1987) referred to up to 2 m of subsidence occurring along a fault, consequent on longwall working of coal, which affected a motorway in Lancashire, England (see Section 3.6, Case history 1). Such movement entails local resurfacing of the motorway. Treatment of faults and dykes involves locating their outcrops. Some idea of their locations, at times, may be obtained from geological maps or mine plans. Their actual location during a site investigation frequently involves trenching or boring large diameter auger holes. The surface effects of movements along fault or dyke planes may be reduced by

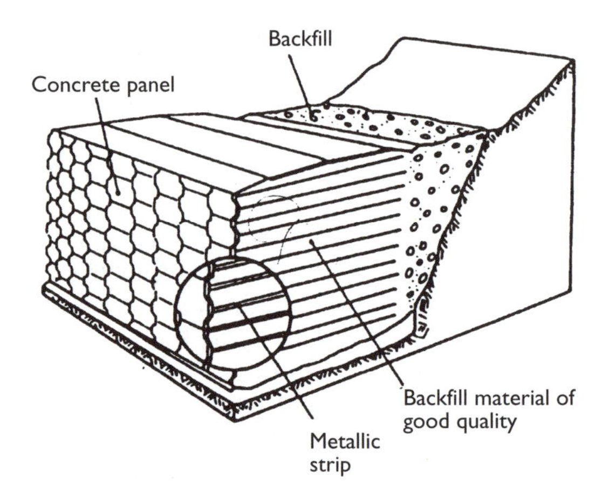

Figure 3.29 Reinforced earth system.

excavating along the zones of potential movement and replacing with cushions of granular material. Block jointed rock masses can give rise to irregular movements when affected by mining subsidence with joints gaping by anything up to a metre or so. Because such movements are unpredictable, Malkin and Wood suggested that the rock masses should be broken down prior to road construction in order to reduce the amount of ground movement. A rigid form of road pavement is not recommended for high-speed roads in areas where mining subsidence has occurred or may occur, the pavement preferably being of the flexible type with bituminous base and surfacing.

Reinforced earth is a composite material consisting of soil in which occur reinforcing elements that generally consist of strips of galvanized steel or plastic geogrids. It also is necessary to provide some form of barrier to contain the soil at the edge of a reinforced earth structure (Fig. 3.29). This facing can be either flexible or stiff but it must be strong enough to retain the soil and to allow the reinforcement to be fixed to it. Reinforced earth structures frequently are used to carry roads. As reinforced earth is flexible and the structural components are built at the same time as backfill is placed, it is particularly suited for use in areas where differential subsidence may occur during or soon after construction.

Subsidence movements can cause relative displacements in all directions and so subject a bridge to tensile and compressive stresses. Although a bridge can have a rigid design to resist such ground movements, it usually is more economical to articulate it thereby reducing the effects of subsidence (Fig. 3.30). Bearings and expansion joints must be designed to accommodate the movements. In the case of multi-span bridges, the piers

Figure 3.30 An articulated bridge to reduce the effects of longwall mining subsidence.

should be hinged at the top and bottom to allow for tilting or change in length, rocker bearings being incorporated at each pier. Jacking sockets can be used to maintain the level of the deck.

Flexible joints can be inserted into pipelines to combat the effects of subsidence. Thick walled plastics that are more able to withstand ground movements have been used for service pipes. Where pipes already are laid they can be exposed and flexible joints inserted or be freed from contact with the surrounding ground. The latter reduces the ground strains transmitted to the pipeline. The pipe trench can be backfilled with, for example, pea gravel so that the friction between the pipe and the surrounding soil is lowered. If failure is likely to cause severe problems, then services can be relocated or duplicated.

One of the consequences of subsidence due to mining, especially longwall working of coal in low-lying areas alongside rivers, is flooding. Peng *et al.* (1996) investigated ponding phenomena along a number of streams caused by longwall mining subsidence. They found that two important factors were the change in stream gradient and the angle of stream flow. The data collected, together with subsidence data, was used to model the formation characteristics of stream ponding and accumulated water volume. One of the most notable regions where coal mining has taken place is the Ruhr Basin in Germany, where the maximum subsidence recorded is 24 m. By the end of the nineteenth century subsidence had caused the reversal of natural drainage in extensive areas, and this gave rise to problems with sanitation and associated outbreaks of typhus and cholera. In fact, flooding was characteristic of this area before mining of coal began and consequently has been exacerbated by subsidence, giving rise to a situation where much of the River Emscher area is now a ‘polderland’ (Bell and Genske, 2001). Accordingly, areas now have to be drained by a large number of pumping stations to protect them from flooding. Similar conditions exist in the eastern lowlands of the River Lippe so that a belt affected by subsidence also extends from the Rhine along the northern Ruhr district to Hamm. By 1989 the ‘polderland’ along the River Emscher totalled approximately 340 km^2, while along the River Lippe there were around 243 km^2 of ‘polderland’.

The development of subsidence basins that extend below the water table lead to surface areas being inundated, resulting in the formation of ponds and lakes. In densely populated areas these can have amenity value in that they can be developed for recreational purposes and as nature reserves. On the other hand, the development of such lakes may mean that roads and railways have to be re-routed, that existing buildings have to be protected and that agriculture is adversely affected. Some lakes in the aforementioned areas have been filled and rivers have had to be realigned and channelled.

3.6. Case history 1

Warrington New Town in Lancashire, England, which is an extension of Warrington, was designated by the government in 1968 and its population was planned to increase from 120 000 to some 190 000 by 1991. This meant that new residential, commercial and industrial areas had to be developed. For example, some 600 to 1000 houses were planned to be built per year over the period of development. However, about 30% of the area involved would be subjected to subsidence due to longwall mining of coal at some time (Fig. 3.31). This placed a constraint on the timing of development in those areas undergoing active subsidence, and on the type and form of construction in areas to be undermined in future. Obviously, the aim of the Development Corporation was to minimize subsidence damage at acceptable cost. Hence, the influence of mining had to be considered at all stages, from planning, through financial appraisal and design, to construction. Over the time scale involved in full development of the New Town, it was considered that mining would not restrict the completion of the development programme (Wilde and Crook, 1985).

The Coal Measures outcrop to the north of Warrington and beneath the town are covered by 700–1000 m of Triassic sandstone. The latter is overlain by Pleistocene deposits that vary from 0 m up to a maximum of 50 m thick.

Two collieries, Bold and Parkside, worked coal by longwall mining from beneath Warrington New Town. Up to six seams were worked at Bold Colliery and five seams were considered economically recoverable at the more modern Parkside Colliery. Subsidence within the designated area due to workings at the former colliery was up to 1 m with calculated resultant strains amounting to 6 mm m^{-1} and in places gave rise to very severe damage. At the time, maximum subsidence from future workings at Bold Colliery was estimated to be up to 2 m with resultant maximum strains up to 3 mm m^{-1}. Maximum subsidence in the designated area above Parkside Colliery was estimated to be as much as 1 m with associated strains of up to 3 mm m^{-1}.

Future programming of surface development and subsurface working may be difficult to coordinate because of the differences that may arise

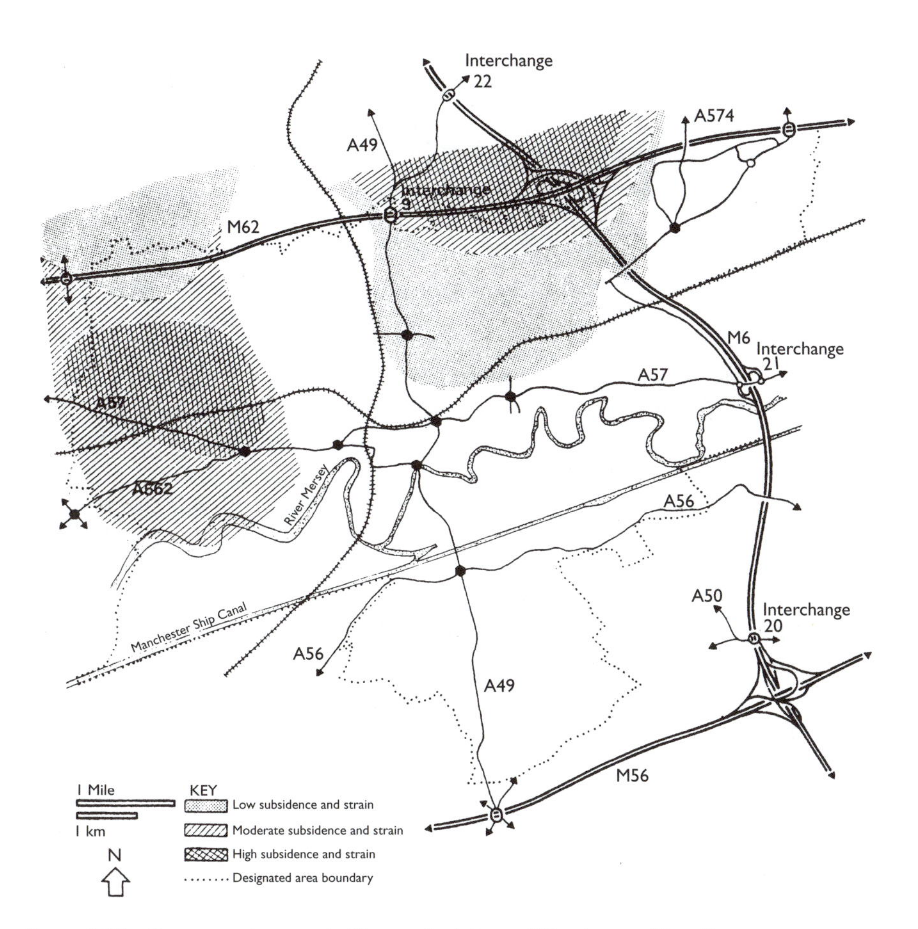

Figure 3.31a Extent and general effects of mining at Warrington New Town.

between the programmed intention and the performance achieved (Bell, 1987). Hence, the relationship between the two programmes cannot be established closely for more than a few months ahead. However, many surface developments at Warrington required a three to five year preparation time. Accordingly, the Development Corporation established a system for gathering and retrieval of information to allow realistic assessments to be made of the mining situation. This system facilitated decisions relating to whether development in a particular area went ahead or what amount of investment in subsidence precautions was justified.

The progress of working faces, tunnel drives and other underground developments were recorded by the National Coal Board and then British Coal on plans at a scale of 1:2500. The Development Corporation also possessed plans for the coal seams within the designated areas that were

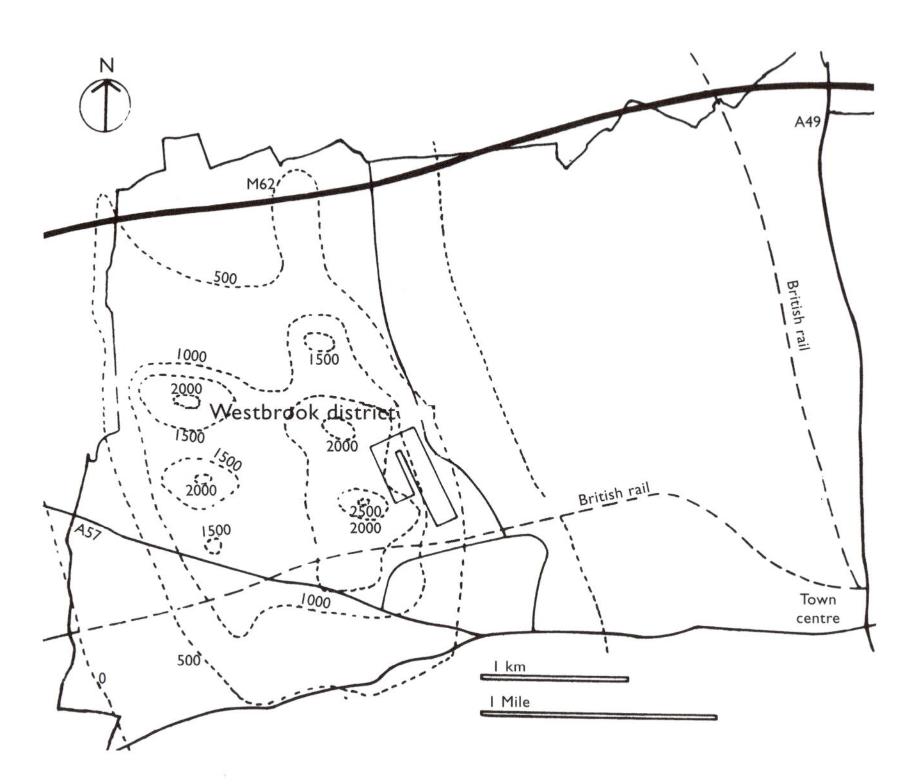

Figure 3.31b Anticipated subsidence at completion of a proposed mining programme in the Westbrook District (subsidence contours are in millimeters).

updated every two or three weeks. These plans provided the first indications of changes in layout, level changes, faulting and other underground factors, which could require the Development Corporation to alter its response to the mining situation. The positions of working coal faces were drawn on plans of the New Town area and the ground movements for these areas were predicted, thereby providing an indication of likely subsidence damage.

A network to monitor subsidence movement was established at the beginning of the development. Permanent levelling stations were established to monitor subsidence and the results were stored in a database, so that immediate responses could be made to requests for subsidence information. Levels were taken at least every three months in areas of active subsidence. Buildings and road surfaces in these areas also were periodically surveyed for subsidence damage such as tension and compression cracks. Of particular importance were differential movements across faults. The Twenty Acre Fault, which formed the eastern boundary of Bold Colliery, represents a good example. It gave rise to severe surface disruption, notably where the fault intersected the M62 motorway. Up to 2 m of subsidence occurred in the affected area, with a differential movement across the fault of 1 m. Strain measurements in excess of 50 mm m^{-1} were obtained from successive linear traverses between adjacent surface levelling stations across the fault. Inclinometers installed across the fault recorded a maximum lateral movement of 400 mm. Piezometers installed across the fault not only showed excess pore water pressure in the clastic material occupying the fault but that, as undermining proceeded, pulses of very high pore water pressure occurred that were related to spasmodic subsidence movements. On the other hand, because the fault accommodated most of the subsidence effects, the areas to the east of it were affected less than would normally have been the case. In this instance, very large former aircraft hangars, which were located only 150 m east of the fault, suffered relatively little from the effects of undermining.

As the northern area of Warrington New Town was likely to be affected to varying degrees by mining subsidence, the mining programme within this area was assessed and, where necessary, development was phased. In other words, development was not undertaken until the majority of the subsidence and strain effects had taken place. If this was not possible, then underground and surface developments were not programmed to run concurrently. Much of the development in the south of the New Town was completed prior to the occurrence of mining, but it was predicted that some of the area could be subjected to subsidence of up to 1 m and strains up to 2 mm m^{-1}. Consequently, appropriate precautionary measures were incorporated into the foundation design of buildings.

The design features that were incorporated into the proposed structures to mitigate the effects of subsidence included three types of protective measures. Standard foundations with a restriction on block lengths were used in areas where only negligible strain (less than 1 mm m^{-1}) was likely to occur. Steps in the foundation were avoided. Plane reinforced concrete raft foundations, thickened under the external walls and resting on a layer of hardcore (compacted crush-rock), were employed in those areas likely to experience low to intermediate strains between 1 and 2 mm m^{-1}. Any unsuitable soils beneath the thickened edges of a raft were removed and replaced with pulverised fly ash (PFA). A polythene membrane on a thin layer of sand separated the hardcore from the raft. A plane reinforced concrete raft, with reinforced concrete ground beams under all load-bearing walls, was used in areas likely to undergo higher strains. The sides of the beams were surrounded by expanded polystyrene to allow for lateral movements. All foundations in the subsidence areas were kept as high in the ground as possible, after bearing capacity, frost susceptibility, and other relevant factors were taken into account. Precautionary measures also were taken in road and drainage works. For example, flexible surfacing materials were specified in roadworks and in drainage works, gradients were kept to a maximum and short pipe lengths with flexible joints were used.

As mentioned earlier, subsidence associated with faults can cause significant damage due to localized high strain. If mining occurs after an area has been developed, it is difficult to afford protection against possible future problems if the location of a fault trace is not known. Regular monitoring by the subsidence levelling system allowed problem areas to be identified as quickly as possible thereby allowing the interested parties to be informed that some surface disruption could be experienced. Ideally, if early warnings were obtained, the coal operator would restrict or modify the underground workings in the area concerned in order to prevent further damage, which could lead to expensive remediation and adverse publicity. Unfortunately, there always could be instances when structures could be severely damaged and therefore would have to be demolished. If a known fault was likely to form a mining boundary then, where possible, its surface position had to be located. For example, the surface outcrop of the Winwick Fault in a particular area was located by trenching and boring. Because development was concurrent with mining in this area, development of the area cut by the fault could be deferred. When mining ceased temporarily, the area was developed, but no structures were constructed over the fault. Where development occurred after initial mining in an area, the zones of movement associated with faulting were identified by the subsidence levelling system. The data available permitted surface development to proceed by designating these zones for open space or landscaping for adjoining housing or industrial development.

3.7. Case history 2

St Wilfrid's Church, Hickleton, South Yorkshire, England, dates from the mid-twelfth century. The church stands on the Lower Magnesian Limestone (Permian), which varies in thickness from 10.2 to 12.3 m in this locality. It is overlain by a metre or so of clay and gravel on top of which rests a metre of topsoil and fill. The Lower Magnesian Limestone rests unconformably upon the Middle Coal Measures, the uppermost unit of which is a sandstone, which is between 5.5 and 6.0 m thick. Mudrocks occur below the sandstone.

Six coal seams were worked beneath the church, five by the longwall method of extraction and the sixth by the room and pillar method (Roscoe, 1988). However, only one seam was extracted from beneath the church, this being worked in 1911, whereas when the others were worked coal was left in place beneath the church. Nevertheless, the church still fell within the area of influence of the workings and therefore experienced some subsidence (Table 3.14). Subsidence damage both before and after the Second World War had necessitated extensive repairs (Bell, 1990). In December 1981, movement with associated cracking occurred, the cracks mostly developing in a narrow zone crossing the church from north-west to south-east, some gaping up to 100 mm in width (Fig. 3.32). In addition, the wall of the South

Table 3.14 Subsidence elements in relation to St Wilfrid's Church (After Roscoe, 1988; reproduced by kind permission of the Geological Society of London)

Seam, date of mining	*Least distance from church to ribside (m)*	*Extracted thickness (m)*	*Assumed panel width (m)*	*Subsidence at panel centre (mm)*	*Subsidence at church (mm)*	*Extension/unit at church ($mm\,m^{-1}$)*
Shafton (1949)	170	1.0 (1.52)	500	900	3	+0.05
Low main (1962)	180	1.22	200	512	22	+0.15
New Hill (1954)	160	1.22	200	451	32	+0.25
Barnsley (c.1911)	N/A	2.40	500	1700	1700	Unknown
Dunsil (1952)	200	1.75	180	315	60	+0.3
Dunsil (1958)	200	1.75	180	315	60	+0.3
Parkgate (1934)	290	1.57	500	863	35	+0.13
Parkgate (1947)	200	1.57	500	863	51	+0.22

Chancel Aisle was lowered through 285 mm and voids were found when floor slabs were lifted in the immediate vicinity of the cracked zone. The calculated horizontal movements, however, were significantly less than the width of the cracks. This, together with the fact that mining had ceased some 20 years earlier, meant that some other factor than subsidence must have been responsible for the movements.

The cracked zone more or less corresponded with the location of a small fault (which has a throw of approximately 5 m to the south-west) running beneath the church. Ground movements generated by mining subsidence tend to be concentrated along faults. It could well have been that the limestone along the fault was broken and gaped due to previous ground subsidence. Collapse of the uppermost limestone and overlying material into the void zone could have taken place along the fault and so initiated the movements concerned.

It was assumed that further differential ground movements associated with the fault could occur. Unfortunately, there was no way that the amount or character of any future movements could be predicted. Consequently, it was decided that the church would have to be underpinned with new foundations that would ensure that it would remain unaffected by any future ground movements.

A prestressed concrete substructure was constructed to underpin the old foundations and transfer the weight of the church by means of three

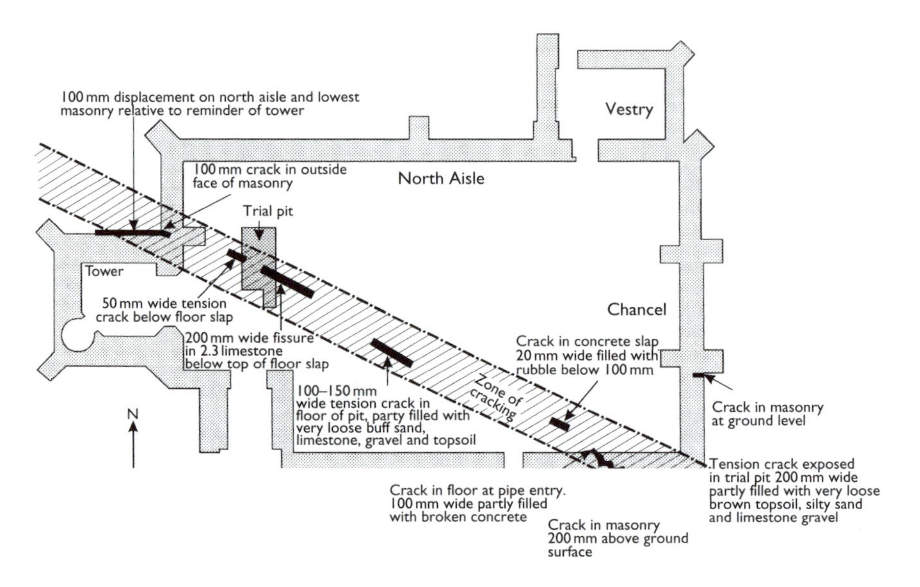

Figure 3.32 Plan of Hickleton Church showing crack zone.

structural bearings to the new foundations (Fig. 3.33). Three pad foundations were located outside the fault zone and provision was made for levelling the church on these foundations by jacking in the event of further subsidence. The pad beneath the tower is 8 m by 4 m and carries a load of approximately 20 MN. It is founded 13 m below ground level and the longer side runs in the same direction as the long axis of the church. A column, 2 m in diameter, transmits the load to the pad. The north and south pads were founded on limestone at 5.3 m and sandstone at 4.3 m below the surface respectively. They each carry a load of 9 MN that is slightly less than 300 kPa (i.e. half the load of the tower pad).

The first operation was to cast reinforced concrete beams beneath the bases of the columns. These extended beyond the column bases and contained projecting reinforcement for incorporation into the crossbeams. The crossbeams connect up with the main east-west running beams, which are 2 m wide and 3 m in depth (Fig. 3.34). The latter run alongside the underpinning to the walls, which consist of concrete blocks, 1 m deep by 1.25 m wide, which are positioned at 2.4 m centres. These were placed before the main beams and again were reinforced with projecting steel bars that connected them to the main beams. Reinforced concrete beams spanned between the blocks to help underpin the walls.

The north-south beam runs between the bearings and is 2 m wide by 3 m in depth. Finally, the beam beneath the tower is connected with the two main

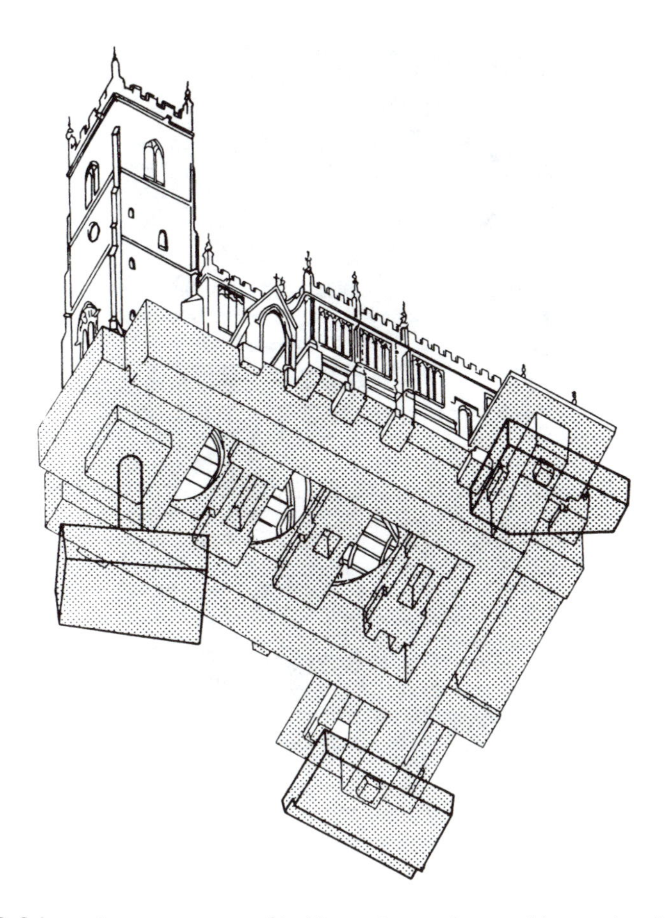

Figure 3.33 Schematic arrangement of jacking points and ground beams for the restoration of Hickleton Church.

east-west beams. This beam was cast in nine operations, each corner section being cast first so that the weight of the tower could be underpinned. The foundation beneath the tower was constructed before the underpinning of the church walls was completed and a shaft was sunk from inside the tower by using precast segmental concrete rings that were grouted. This provided the formwork for the column above the pad foundation.

Parts of the church at the north-west and south-east corners, as well as the whole vestry, were dismantled, and each stone was marked and stored. They were rebuilt subsequently in the same manner, damaged areas being replaced. These elaborate and expensive remedial works have proved successful, the church having suffered no further distress.

Figure 3.34 View of one of the main east-west beams showing steel reinforcement being fixed at Hickleton Church.

3.8. Case history 3

Barlaston lies about 6 km south of Stoke-on-Trent, England. The area has been undermined extensively and Barlaston has been affected by workings at Hem Heath (previously Trentham) and Florence collieries (Donnelly and Rees, 2001). Since the 1960s over 50 km of fault scarps have been formed in the Barlaston area that can be directly related to mining induced fault reactivation. These scarps may reach 2.0 m in height and may be traced across the ground surface for distances of several hundreds of metres. The affects of fault reactivation can be severe in terms of structural damage, the sterilization of land and financial loss. For instance, where the faults intersect roads, buildings, residential property, structures and utilities, these may be damaged severely (Fig. 3.35). Houses frequently have been rendered unsafe for habitation and therefore demolished.

The Carboniferous Coal Measures rocks that outcrop in and around Barlaston range from moderately strong, fine and coarse sandstones with

Figure 3.35 Damage to Barlaston Church caused by fault reactivation at a sandstone-mudstone contact.

mudstone and siltstone interbeds, to weak to moderately strong mudstones and siltstones with shaley clay. The superficial cover of glacial till is thin or absent and the topography is almost flat, apart from a few slopes on dipping sandstone beds and valley sides. From east to west, the faults that occur within the outcrop are the Newcastle Fault, Hollybush Fault, Bedcroft Fault and Crowcrofts Fault (Fig. 3.36).

The reactivation of the Newcastle Fault is most clearly observed in the Trentham area where it was seen as a scarp or break-line that can be traced for a distance of over 2 km through a housing estate, farmland and across a major road. The fault tends to traverse through gardens and between houses. Nevertheless, some houses have had to be demolished and others have required underpinning and structural repairs. The scarp is distinct on brittle pavements and road surfaces and shows extensive compression and en-echelon fracturing, though on grass verges and across gardens the trace of the fault is marked only by a gentle downwarp and subtle flexuring.

The Hollybush Fault was reactivated in Newstead during two separate phases of the working of the Yard and Great Row seams in the early 1990s. The style of reactivation was notable in that dextral shear occurred with little vertical step displacement (Donnelly and Reddish, 1994). The fault resulted in severe damage to several premises on the Newstead industrial

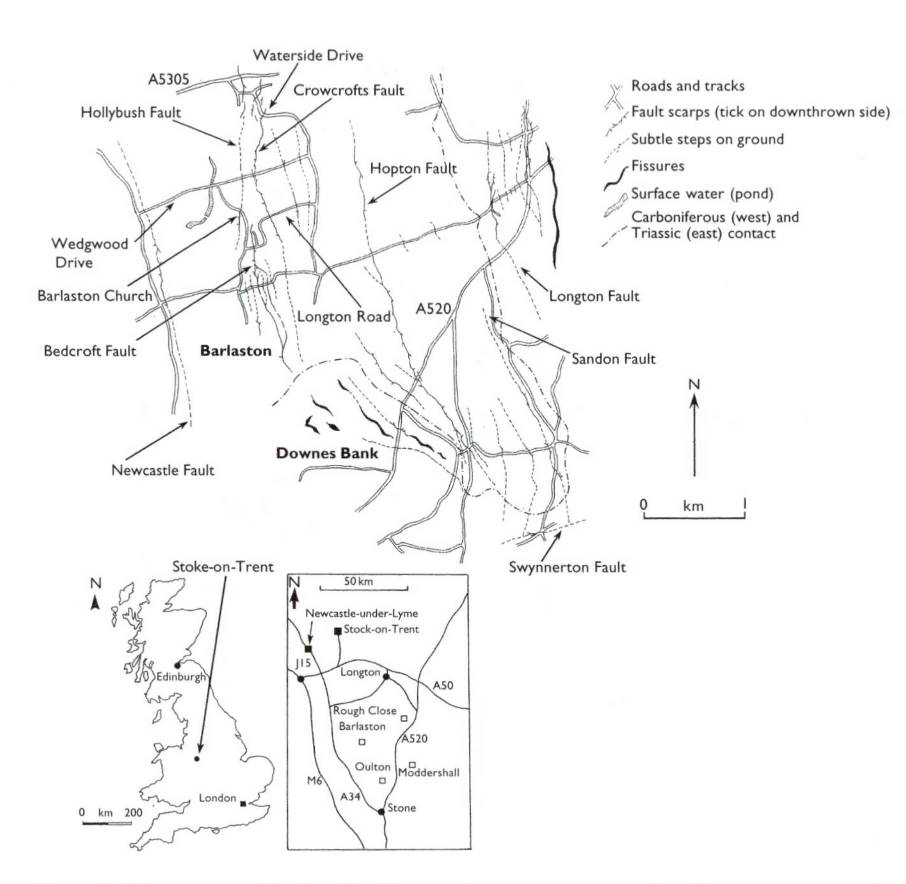

Figure 3.36 Location of faults that have undergone mining induced reactivation in the vicinity of Barlaston, Staffordshire, England. (After Donnelly and Rees, 2001.)

estate, from which it could be readily traced southwards for a distance of 400 m (Fig. 3.37). By 1991, the reactivation of the Hollybush Fault along a lithological contact caused severe damage to Barlaston Hall and Barlaston Church (Fig. 3.35). The two structures were rendered unsafe, but could not be demolished since they are listed as sites of national heritage, the church has since been repaired and the hall underpinned. In the period from 1995 to 2000, further damage was observed to industrial premises located above the fault in the absence of any active mining.

A splay of the Crowcrofts Fault, named the Bedcroft Fault, was reactivated during mining subsidence in the late 1980s to early 1990s. This caused extensive and severe damage to several houses near Longton Road, which were subsequently demolished. The fault scarp was visible as a broad open flexure with up to 1.8 m of vertical slip, 0.5 m of lateral shear, en-echelon shearing and severe compression for a distance of at least 500 m. Its reactivation caused severe damage to numerous houses that subsequently were

Figure 3.37 The reactivation of the Hollybush Fault, Newstead, Staffordshire, showing later 0.1–0.3 m lateral shear of the kerbstones, pavements and buildings, with no observable vertical displacement.

demolished. This fault was traced southwards where it formed distinct shatter zones in the centre of Barlaston village.

The Crowcrofts Fault underwent several phases of reactivation separated by periods of relative stability, from the 1960s to the 1990s, caused by the longwall working of multiple coal seams (Fig. 3.38(a)). A fault step, varying from a subtle flexure to a distinct steep-sided fault scarp over 1 m high, could be traced for a distance of over 3.5 km. This caused widespread damage to houses on Waterside Drive in Newstead and Longton Road in Barlaston, many were subsequently demolished (Fig 3.38(b)). Agricultural land was damaged due to alteration of the gradients and flooding. Road and bridge repairs were frequent along Wedgwood Drive and several other roads in this area, due to fault reactivation. Furthermore, a gas main was exposed in a trench, to the south of Wedgwood Drive, where it crossed the fault and it was monitored for subsidence and strain accumulations to prevent disruption (Fig. 3.39). Occasionally, groundwaters were observed to discharge from the scarp (Fig. 3.39(a), (b)). Some discharges were under a hydrostatic head of pressure that was sufficient to create an artesian effect.

Faults that have been reactivated on the outcrop of Triassic sandstones to the south and east of Barlaston are different from those developed within the Carboniferous rocks. This can be illustrated by the Triassic rocks of Downes Bank, which rise to almost 50 m in height above the Carboniferous

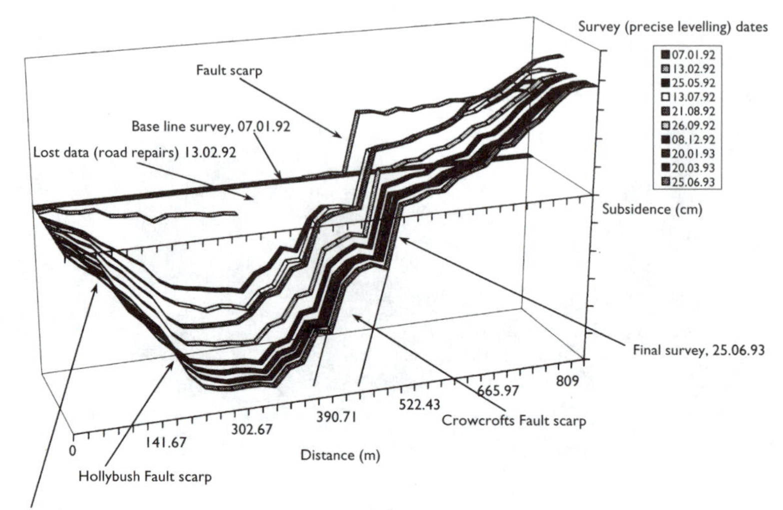

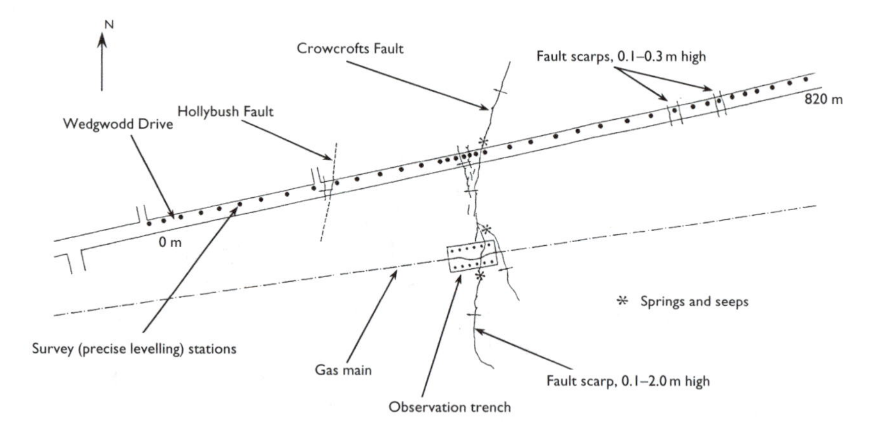

Figure 3.38a Schematic section showing the third phase of reactivation of the Crowscroft Fault along Wedgwood Drive, Barlaston, during the period 1992–93. The outcrop position of the fault was observed as multiple scarps, fissures and temporary springs. The subsidence monitoring data recorded the development of the subsidence trough and the reactivation of the fault. (After Donnelly and Rees, 2001.)

rocks of the area and are generally free of Quaternary deposits. There several faults have been subject to recent movement. This causes the dilation of the fault plane with, or without, a component of vertical slip, and resulted in the generation of ground fissures. Many fault fissures occur for a distance

Figure 3.38b Reactivation of the Crowscroft Fault, Longton Road, Barlaston, causing damage to a house, the fault scarp running immediately in front of the house.

Figure 3.39a The Crowscroft Fault scarp exposed in the walls of a gas main trench south of Wedgwood Drive, Barlaston, showing groundwater plumes and debris deposited along the fault scarp.

Figure 3.39b Road repair being undertaken to Wedgwood Drive following the reactivation of Crowscroft Fault in 1992. The fault scarps are visible in the grassed verge and fields to the left of the road.

Figure 3.40 Reactivated fault exposed in a trench on Downes Bank. The trace of the fault is marked by a ground fissure about 0.1 m wide. Remediation of the fault was to be made using a geotextile mat and granular fill.

of over 1 km, their depths are unknown. These represent the dilation of faults and joint sets during mining subsidence. Where observed in cross-section, in 10 m deep trenches, the faults consist of distinct parallel and sub-parallel dislocations in a loose sandy gouge. In the early 1990s some faults were remediated by excavation of trenches and reinstatement with gravel wrapped in a geotextile membrane (Fig. 3.40). This reduced the surface effects of subsequent movement in 1994 but did not completely remediate the problem.

The reactivation of faults occurs over relatively long periods, from weeks to years via the processes of aseismic creep. This is evidenced in the Barlaston area by the relatively slow and episodic movement of faults. Although mining induced seismicity was relatively common in the region (Lovell *et al.*, 1996), this is unlikely to have been caused by instantaneous shear failure (stick-slip mechanisms) on faults. The recorded seismicity was likely to have been generated by rock movements or failures resulting from changes in the state-of-stress in the rock mass in the vicinity of the mining excavation. These include the collapse of the strong sandstone roof strata into the goaf, bed separation or bed collapse, pillar failure in abandoned room and pillar mines, and gas outbursts. In the East Midlands of England there is evidence linking mining induced fissuring in Triassic sandstones and seismic activity (Bishop *et al.*, 1993).

References

Alejano, L.R., Ramirez-Oyanguren, P. and Taboada, J. 1999. FDM predictive methodology for subsidence due to flat and inclined coal seam mining. *International Journal of Rock Mechanics and Mining Science*, **36**, 475–491.

Anon. 1959. *Report on Mining Subsidence*. Institution of Civil Engineers, London.

Anon. 1975. *Subsidence Engineers Handbook*. National Coal Board, London.

Anon. 1977. *Ground Subsidence*. Institution of Civil Engineers, London.

Aston, T.R.C., Tammenagi, H.Y. and Poon, A.W. 1987. A review and evaluation of empirical and analytical subsidence prediction techniques. *Mining Science and Technology*, **5**, 59–69.

Bekendam, R.F. and Pottgens, J.J. 1995. Ground movements over the coal mines of southern Limburg, the Netherlands and their relation to rising mine waters. *Proceedings of the Fifth International Symposium on Land Subsidence*, The Hague, International Association Hydrological Sciences, 3–12.

Bell, F.G. 1987. The influence of subsidence due to present day coal mining on surface development. In: *Planning and Engineering Geology*, Engineering Geology Special Publication No. 4, Culshaw, M.G., Bell, F.G., Cripps, J.C. and O'Hara, M. (eds), the Geological Society, London, 359–368.

Bell, F.G. 1988. the history and techniques of coal mining and the associated effects and influence on construction. *Bulletin of the Association of Engineering Geologists*, **24**, 471–504.

Bell, F.G. 1990. Preservation and restoration of historic buildings: two case histories. *Proceedings of the Conference on Techniques of Restoration and Methods of Evaluation of Safety*, Bangkok, A.A. Balkema, Rotterdam, 54–64.

Bell, F.G. and Fox, R.M. 1988. Ground treatment and foundations above discontinuous rock masses affected by mining subsidence. *Mining Engineer*, **148**, 278–283.

Bell, F.G. and Genske, D.D. 2001. The influence of subsidence attributable to coal mining on the environment, development and restoration: some examples from western Europe and South Africa. *Environmental and Engineering Geoscience*, **7**, 81–99.

Bell, S.E. 1978. Successful design for mining subsidence. *Proceedings of the First International Conference on Large Ground Movements and Structures*, Cardiff, Geddes, J.D. (ed.), Pentech Press, London, 562–578.

Bello, G.A., Menendez, D.A., Ordieres, M.J.B. and Gonzalez, N.C. 1996. Generalization of the influence function method in mining subsidence. *International Journal of Surface Mining, Reclamation and Environment*, **10**, 195–202.

Berry, D.S. 1978. Progress in the analysis of ground movements due to mining. *Proceedings of the First International Conference on Large Ground Movements and Structures*, Cardiff, Geddes, J.D. (ed.), Pentech Press, London, 781–811.

Bhattacharya, S. and Singh, M.M. 1985. *Development of Subsidence Damage Criteria*. Office of Surface Mining, United States Department of the Interior, Contract No. J5120129, Engineers International Inc., Washington, DC.

Bishop, I., Styles, P. and Allen, M. 1993. Mining induced seismicity in the Nottinghamshire Coalfield. *Quarterly Journal of Engineering Geology*, **26**, 253–279.

Bjerrum, L. 1963. Discussion, Section IV. Proceedings of the European Conference on *Soil Mechanics and Foundation Engineering*, Wiesbaden, **2**, 135–137.

Boone, S.J. 1996. Ground movement related building damage. *Proceedings of the American Society Civil Engineers, Journal of Geotechnical Engineering*, **122**, 886–896.

Booth, C.J., Curtiss, A.M., Demaris, P.J. and Bauer, R.A. 2000. Site-specific variation in the potentiometric response to subsidence above active longwall mining. *Environmental and Engineering Geoscience*, **6**, 383–394.

Brauner, G. 1973a. *Subsidence due to Underground Mining. Part I: Theory and Practice in Predicting Surface Deformation*. Bureau of Mines, Department of the Interior, United States Government Printing Office, Washington, DC.

Brauner, G. 1973b. *Subsidence due to Underground Mining. Part II: Ground Movements and Mining Damage*. Bureau of Mines, Department of the Interior, United States Government Printing Office, Washington, DC.

Bruhn, R.W., Magnuson, M.O. and Gray, R.E. 1981. Subsidence over abandoned mines in the Pittsburg Coalbed. *Proceedings of the Second International Conference on Ground Movements and Structures*, Cardiff, Geddes, J.D. (ed.), Pentech Press, London, 142–156.

Burland, J.B. and Wroth, C.P. 1975. Allowable and differential settlement of structures including damage and soil–structure interaction. In: *Settlement of Structures*, Pentech Press, London, 611–654.

Burland, J.B., Broms, B.B. and De Mello, V.F.B. 1977. Behaviour of foundations and structures. *Proceedings of the Ninth International Conference on Soil Mechanics and Foundation Engineering*, Tokyo, **2**, 495–547.

Burton, D.A. 1978. A three dimensional system for the prediction of surface movements due to mining. *Proceedings of the First International Conference on Large Ground Movements and Structures*, Cardiff, Geddes, J.D. (ed.), Pentech Press, London, 209–228.

Burton, D.A. 1981. The introduction of mathematical models for the purpose of predicting surface movements due to mining. *Proceedings of the Second International Conference on Ground Movements and Structures*, Cardiff, Geddes, J.D. (ed.), Pentech Press, London, 50–64.

Burton, D.A. 1985. Program in BASIC for the analysis and prediction of ground movement above longwall panels. *Proceedings of the Third International Conference on Ground Movements and Structures*, Cardiff, Geddes, J.D. (ed.), Pentech Press, London, 338–353.

Carpenter, P.J. 1997. Use of resistivity and EM techniques to map subsidence fractures in glacial drift. *Environmental and Engineering Geoscience*, **3**, 523–536.

Charles, J.A. and Skinner, H.D. 2004. Settlement and tilt of low-rise buildings. *Proceedings of the Institution of Civil Engineers, Geotechnical Engineering*, **157**, 65–75.

Coulthard, M.A. and Dutton, A.J. 1988. Numerical modelling of subsidence induced by underground coal mining. *Proceedings of the Twenty Ninth United States Rock Mechanics Symposium, Key Questions in Rock Mechanics*, Rolla, Missouri, 529–536.

Dahl, H.D. 1972. Two and three-dimensional elastic-elastoplastic analysis of mine subsidence. *Proceedings of the Fifth Conference on Strata Control*, London, 1–5.

Degirmenci, N., Reddish, D.J. and Whittaker, B.N. 1988. A study of surface subsidence behaviour arising from longwall mining of steeply pitching coal seams. *Proceedings of the Sixth International Congress on Coal*, Zonguldak, Turkey, 27 p.

Diez, R.R. and Alvarez, J.T. 2000. Hypothesis of multiple subsidence trough related to very steep and vertical coal seams and its prediction through profile functions. *Geotechnical and Geological Engineering*, **18**, 289–311.

Donnelly, L.J. 1998. *Fault Reactivation and Ground Deformation Investigation, Easington Colliery, County Durham*. Report No. WN/98/9, British Geological Survey, Nottingham.

Donnelly, L.J. 2000. The reactivation of geological faults during mining subsidence from 1859 to 2000 and beyond. *Transactions of the Institution of Mining and Metallurgy*, Section A, Mining Technology, **109**, A179–A190.

Donnelly, L.J. and Melton, N.D. 1995. Compression ridges in the subsidence trough. *Geotechnique*, **45**, 555–560.

Donnelly, L.J. and Reddish, D.J. 1994. The development of surface steps during mining subsidence not due to fault reactivation. *Engineering Geology*, **36**, 243–255.

Donnelly, L.J. and Rees, L. 2001. Tectonic and mining induced fault reactivation around Barlaston on the Midlands Microcraton. *Quarterly Journal of Engineering Geology and Hydrogeology*, **34**, 195–214.

Donnelly, L.J., Northmore, K.J. and Jermy, C.A. 2000. Fault reactivation in the vivinity of landslides in the South Wales Coalfield. *Proceedings of the Eighth Symposium on Landslides, Landslides in Research, Theory and Practice*, Bromhead, E.N., Dixon, N. and Ibsen, M.L. (eds), 481–486.

Donnelly, L.J., De La Cruz, H., Asmar, I., Zapata, O. and Perez, J.D. 2001. The monitoring and prediction of mining subsidence in the Amaga, Angelopolis, Venecia and Bolombolo Regions, Antioquia, Columbia. *Engineering Geology*, **59**, 103–114.

Drent, S. 1957. Time curves and thickness of overlying strata. *Colliery Engineer*, **34**, 271–278.

Dumpleton, S. 2002. Effects of longwall mining in the Selby Coalfield on the piezometry and aquifer properties of the overlying Sherwood Sandstone. In: *Mine Water Hydrogeology and Geochemistry*, Special Publication 198, Younger, P.L. and Robins, N.S. (eds), the Geological Society, London, 75–88.

Farmer, I.W. and Altounyan, P.F.R. 1981. The mechanics of ground deformation above a caving longwall face. *Proceedings of the Second International Conference on Ground Movements and Structures*, Cardiff, Geddes, J.D. (ed.), Pentech Press, London, 75–91.

Ferrari, C.R. 1997. Residual coal mining subsidence – some facts. *Mining Technology*, **79**, 177–183.

Geddes, J.D. 1981. Subgrade restraint and shearing force effects due to moving ground. *Proceedings of the Second International Conference on Ground Movements and Structures*, Cardiff, Geddes, J.D. (ed.), Pentech Press, London, 288–306.

Geddes, J.D. 1984. Structural design and ground movements. In: *Ground Movements and their Effects on Structures*, Attewell, P.B. and Taylor, R.K. (eds), Surrey University Press, London, 243–267.

Grant, R., Christian, J.T. and Vanmarke, E.H. 1974. Differential settlement of buildings. *Proceedings of the American Society of Civil Engineers, Journal of Geotechnical Engineering Division*, **100**, 973–991.

Hackett, P. 1959. An elastic analysis of rock movements caused by mining. *Transactions Institution of Mining Engineers*, **118**, 421–433.

Hellewell, F.G. 1988. The influence of faulting on ground movement due to coal mining. The UK and European experience. *Mining Engineer*, **147**, 334–337.

Holla, L. 1997. Ground movement due to longwall mining in high relief areas in New South Wales. *International Journal of Rock Mechanics and Mining Science*, **34**, 775–787.

Hood, M., Ewy, R.T. and Riddle, R.L. 1983. Empirical methods of subsidence prediction. A case study from Illinois. *International Journal of Rock Mechanics and Mining Science and Geomechanical Abstracts*, **20**, 153–170.

Ji-xian, C., 1985. The effects of mining on buildings and structural precautions adopted. *Proceedings of the Third International Conference on Ground Movements and Structures*, Cardiff, Geddes, J.D. (ed.), Pentech Press, London, 402–422.

Lee, A.J. 1966. The effect of faulting on mining subsidence. *Mining Engineer*, **125**, 417–427.

Lehmann, K. 1919. Movements in open diggings and troughs. *Gluckauf*, Springer-Verlag, Berlin, 249–256.

Leonards, G.A. 1975. Discussion of differential settlement of buildings. *Proceedings of the American Society Civil Engineers, Journal of Geotechnical Engineering Division*, **101**, 700–702.

Lovell, J.H., Ford, G.D., Henni, P.H.O., Baker, C., Simpson, I. and Pettitt, W. 1996. *Recent Seismicity in the Stoke-on-Trent Area, Staffordshire*. British Geological Survey (NERC), Technical Report WL/96/20, Keyworth, Nottingham.

Malkin, A.B. and Wood, J.C. 1972. Subsidence problems in route design and construction. *Quarterly Journal of Engineering Geology*, **5**, 179–194.

Marr, J.E. 1975. The application of the zone area system to the prediction of mining subsidence. *Mining Engineer*, **135**, 53–62.

Orchard, R.J. 1954. Recent developments in predicting the amplitude of subsidence. *Journal of the Royal Institution of Chartered Surveyors*, **86**, 864–876.

Orchard, R.J. 1957. Prediction of the magnitude of surface movements. *Colliery Engineer*, **34**, 455–462.

Orchard, R.J. 1964. Partial extraction and subsidence. *Mining Engineer*, **123**, 417–427.

Orchard, R.J. and Allen, W.S. 1965. Ground curvature due to coal mining. *The Chartered Surveyor*, **34**, 86–93.

Orchard, R.J. and Allen, W.S. 1970. Longwall partial extraction systems. *Mining Engineer*, **129**, 523–535.

Peng, F.F., Sun, V.Z. and Peng, S.S. 1996. Modeling the effects of stream ponding associated with longwall mining. *Mining Engineering*, **48**, 59–64.

Peng, S.S. and Geng, D.Y. 1982. Methods of predicting the subsidence factor, angle of draw and angle of critical deformation. In: *State-of-the-Art of Ground Control in Longwall Mining and Mining Subsidence*, American Institute Mining Engineers, New York, 211–215.

Peng, S.S., Luo, Y. and Dutta, D. 1996. An engineering approach to ground surface subsidence damage due to longwall mining. *Mining Technology*, **78**, 227–231.

Phillips, K.A.S. and Hellewell, F.G. 1994. Three-dimensional ground movements in the vicinity of a mining activated geological fault. *Quarterly Journal of Engineering Geology*, **27**, 7–14.

Polshin, D.E. and Tokar, R.A. 1957. Maximum allowable non-uniform settlement of structures. *Proceedings of the Fourth International Conference on Soil Mechanics and Foundation Engineering*, London, **1**, 402–406.

Pottgens, J.J.E. 1979. Ground movements by coal mining in the Netherlands. *Proceedings of the Speciality Conference of American Society Civil Engineers, Evaluation and Prediction of Subsidence*, Gainesville, FL, Saxena, S.K. (ed.), 267–282.

Ren, G., Reddish, D.J. and Whittaker, B.N. 1987. Mining subsidence and displacement prediction using influence function methods. *Mining Science and Technology*, **5**, 89–104.

Roscoe, G.H. 1988. Saint Wilfrid's Church, Hickleton: mining subsidence and remedial works. In: *Engineering Geology of Underground Movements*, Engineering Geology Special Publication No. 5, Bell, F.G., Culshaw, M.G., Cripps, J.C. and Lovell, M.A. (eds), Geological Society, London, 257–264.

Salamon, M.D.G. 1964. Elastic analysis of displacements and stresses induced by mining of seam and reef deposits. Part I: Fundamental principles and basic solutions as derived from idealized models. *Journal of the South African Institute of Mining and Metallurgy*, **64**, 128–149.

Sanzotti, M.J. and Bise, C.J. 1996. Concomitant backstowing: potential for alleviating concernes associated with high-production longwall mining. *Mining Engineering*, **48**, 54–58.

Shadbolt, C.H. 1978. Mining subsidence. *Proceedings of the First International Conference on Large Ground Movements and Structures*, Cardiff, Geddes, J.D. (ed.), Pentech Press, London, 705–748.

Singh, K.B. and Singh, T.N. 1998. Ground movements over longwall workings in the Kamptee Coalfield, India. *Engineering Geology*, **50**, 125–139.

Skempton, A.W. and MacDonald, D.H. 1956. Allowable settlement of buildings. *Proceedings of the Institution of Civil Engineers*, **5**, Part III, 727–768.

Speck, R.C. and Bruhn, R.W. 1995. Non-uniform mine subsidence ground movement and resulting surface-structure damage. *Environmental and Engineering Geoscience*, **1**, 6–74.

Stacey, T.R. 1972. Three dimensional finite element stress analysis applied to two problems in rock mechanics. *Journal of the South African Institute of Mining and Metallurgy*, **72**, 251–256.

Sutherland, H.J. and Munson, D.E. 1984. Prediction of subsidence using complementary influence functions. *International Journal of Rock Mechanics and Mining Science and Geomechanical Abstracts*, **21**, 195–202.

Tsur-lavie, Y. and Denekamp, S. 1981. A boundary element method for the analysis of subsidence associated with longwall mining. *Proceedings of the Second International Conference on Ground Movements and Structure*, Cardiff, Geddes, J.D. (ed.), Pentech Press, London, 65–74.

Tsur-lavie, Y., Denekamp, S. and Fainstein, G. 1985. Geometry of subsidence associated with longwall mining. *Proceedings of the Third International Conference on Ground Movements and Structures*, Cardiff, Geddes, J.D. (ed.), Pentech Press, London, 324–337.

Voight, B. and Pariseau, W. 1970. State of predictive art of subsidence engineering. *Proceedings of the American Society of Civil Engineers, Journal of the Soil Mechanics and Foundations Division*, **96**, 721–750.

Wahls, H.E. 1982. Tolerable settlements of buildings. *Proceedings of the American Society of Civil Engineers, Journal of the Geotechnical Engineering Division*, **107**, 1489–1503.

Wardell, K. 1954. Some observations on the relationship between time and mining subsidence. *Transactions of the Institution of Mining Engineers*, **113**, 471–483 and 799–814.

Wardell, K. and Webster, N.E. 1957. Surface observations and strata movement underground. *Colliery Engineer*, **34**, 329–336.

Whetton, J.T. and King, H.J. 1958. Mechanics of mine subsidence. *Colliery Engineer*, **35**, 247–252 and 285–288.

Whittaker, B.N. and Breeds, C.D. 1977. The influence of surface geology on the character of mining subsidence. *Proceedings of the Conference on Geotechnics of Structurally Complex Formations*, Capri, Associazone Geotechnica Italiana, **1**, 459–468.

Whittaker, B.N. and Reddish, D.J. 1989. *Subsidence: Occurrence, Prediction and Control*. Elsevier, Amsterdam.

Whittaker, B.N. and Reddish, D.J. 1993. Subsidence behaviour of rock structures. In: *Comprehensive Rock Engineering: Principles, Practice and Projects, Volume 4, Excavation, Support and Monitoring*, Brown, E.T, Fairhurst, C. and Hoek, E. (eds), Pergamon Press, Oxford, 751–780.

Wilde, P.M. and Crook, J.M. 1985. The significance of ground movements due to deep coal mining and their effects on large surface developments at Warrington New Town. *Proceedings of the Third International Conference on Ground Movements and Structures*, Cardiff, Geddes, J.D. (ed.), Pentech Press, London, 240–247.

Yao, X.L. and Reddish, D.J. 1994. Analysis of residual subsidence movements in U.K. coalfields. *Quarterly Journal of Engineering Geology*, **27**, 15–23.

Yokel, F.Y., Salomone, L.A. and Gray, R.E. 1982. Housing construction in areas of mine subsidence. *Proceedings of the American Society of Civil Engineers, Journal of Geotechnical Engineering Division*, **108**, 1133–1149.

Young, B. and Culshaw, M.G. 2001. *Fissuring and related Ground Movements in the Magnesian Limestone and Coal Measures of the Houghton-le-Spring area, City of Sunderland*. Report No. WA/01/04, British Geological Survey, Nottingham.

Zienkiewicz, O.C., Cheung, V.K. and Stagg, K.O. 1966. Stresses in anisotropic media with particular reference to problems in rock mechanics. *Journal of Strain Analysis*, **1**, 172–182.

Zipper, C., Balfour, W., Roth, R. and Randolph, J. 1997. Domestic water supply impacts by underground coal mining in Virginia, USA. *Environmental Geology*, **29**, 84–93.

Chapter 4

Metalliferous mining and subsidence

4.1. Mining methods

The formation of orebodies is highly varied and complex with several processes being involved in their generation (Evans, 1993). Consequently, there are several underground mining methods used for the extraction of metalliferous deposits. These may be classified according to the type of support used. Supported mining methods may be classified further depending on whether rock or artificial supports are used. In general the main methods are summarized in Figure 4.1.

In partial extraction methods, solid pillars are left unmined to provide support to the underground workings (Fig. 4.2). For example, sill pillars and shaft pillars are designed to protect important underground areas and barrier pillars provide regional support for the safety of underground operations. Pillars may be variable in shape and size, their geometry being determined largely by the shape and variation in the grade of orebody. For instance, in stope and pillar mining openings are driven in the mineral deposit in a regular or random pattern to form pillars for ground support. The irregular shape and size of pillars, together with their random location, is so as to locate them in low grade ore or waste. If the deposits are less than 10 m thick, then the room may be advanced in a single operation, consisting of drilling, blasting, loading and transportation, this being termed full face advance. In thicker deposits the extraction cycle may require the development of a series of benches to extract the orebody.

Pillars are designed in modern practice for the purpose they have to serve (Hedley, 1978). This design takes account of the strength of the pillars, the strength of the floor and roof rocks, the capacity of the rock to span between the pillars, and the stresses acting on the pillars. For example, for flat-lying deposits the stresses are induced by the overburden whilst in steeply dipping deposits they are a consequence of the *in situ* horizontal stress field. For intermediate dips, the active stresses are a combination of the two and the pillars may be subjected to significant shear stresses. It therefore is possible to design against the occurrence of surface subsidence.

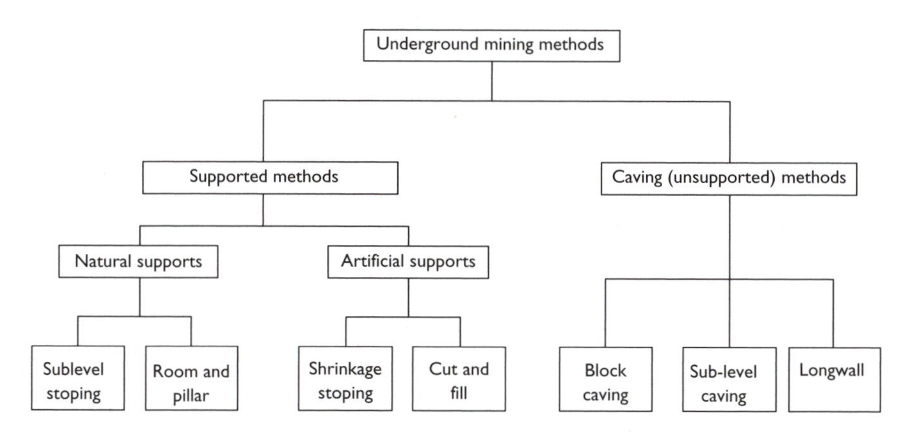

Figure 4.1 Classification of metalliferous mining support methods.

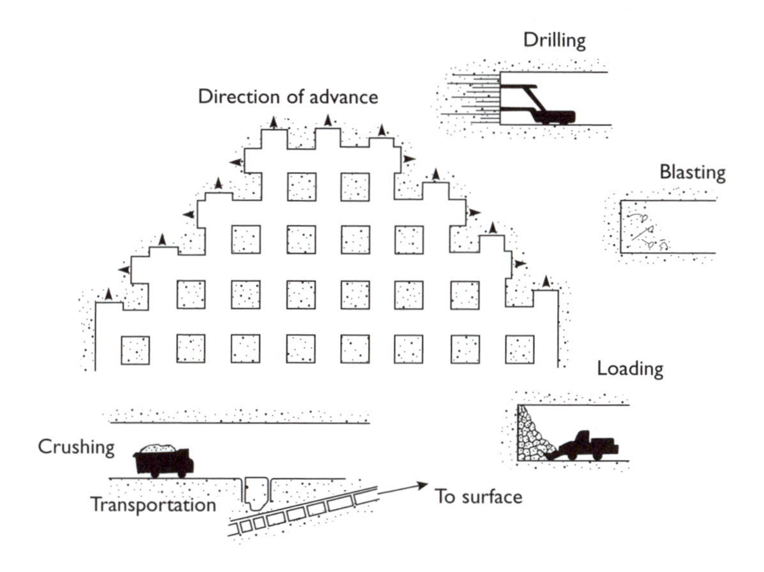

Figure 4.2 Partial extraction mining leaving pillars for support.

However, this procedure was not practised in earlier times. In such situations subsidence of the surface is a possibility. The collapse of the roof or hangingwall of the workings between pillars in mines at shallow depth can result in localized surface subsidence. The subsidence profile may be very severe, with large differential subsidence, tilts and horizontal strains. Collapse of a pillar or pillars can give rise to localized, or substantial areas of subsidence respectively. In the case of regular pillar systems, failure of a single pillar may lead to overstressing of surrounding pillars causing some of them to collapse and so give rise to trough like subsidence at the surface. If variable sized pillars have been used, or if larger barrier pillars are provided at regular spacings, then the area of subsidence is likely to be restricted. The resulting subsidence profile generally is irregular, with large differential subsidence, tilts and horizontal strains at the perimeter of the subsidence area. The presence of faults can give rise to fault steps. Failure of roof and/or floor leads to a similar type of subsidence behaviour. When mining inclined orebodies, the steeper the dip, the more localized the subsidence is likely to be. As in other pillar mining, pillar stability is time dependent in that in open or waterlogged abandoned workings, the pillars, roof and floor rocks may deteriorate over time leading ultimately to failure and associated subsidence.

In many cases, since pillars often represent reserves of ore, they are extracted or reduced in size towards the end of the life of a mine. For instance, Szwedzicki (1989) mentioned that a large proportion of the copper ore reserves of Miriam Mine in Zimbabwe were held in pillars and that these were worked systematically, indeed 60% of production came from pillar recovery and caved pillars. Because pillar robbing can affect pillar stability adversely, an attempt was made to retain the stress levels in the pillars at a minimum by developing an appropriate sequence of recovery. In other words, pillar recovery progressed from the uppermost to the lowest orebodies and from the lowest extraction level of each orebody upwards. In addition, backfilling was undertaken to ensure safe working conditions. Szwedzicki noted that backfilling with sand provided strata control but that when the backfill consisted of leached tailings it did not fulfil its support role because of its high compressibility.

Sub-level stoping is a vertical stoping method whereby longhole drilling and blasting are carried out from sub-levels to break the ore (Fig. 4.3). The ore flows through the stope by gravity and is drawn off at the haulage level. This method normally is applied to massive or steeply dipping orebodies with irregular geometries. The rock surrounding the mineralized zone must be strong.

Cut and fill stoping is a mining method in which horizontal slices of ore are excavated in the stope and replaced by waste as fill, the fill being placed in the mined out areas to allow mining to continue (Fig. 4.4). Hence, the filling operation forms part of the mining operation. This method originally was developed to mine steeply dipping mineral veins, although it also has been used to extract massive deposits. The orebody is divided into sections, at intervals of approximately 30–100 m, by a series of main haulage levels. These often are driven into the footwall. Cross-cuts are driven to intersect the orebody at 15–30 m intervals and so the orebody is divided into a series of blocks. As well as providing support to pillars, and hangingwall and footwall rocks, the fill also provides a working platform from which the

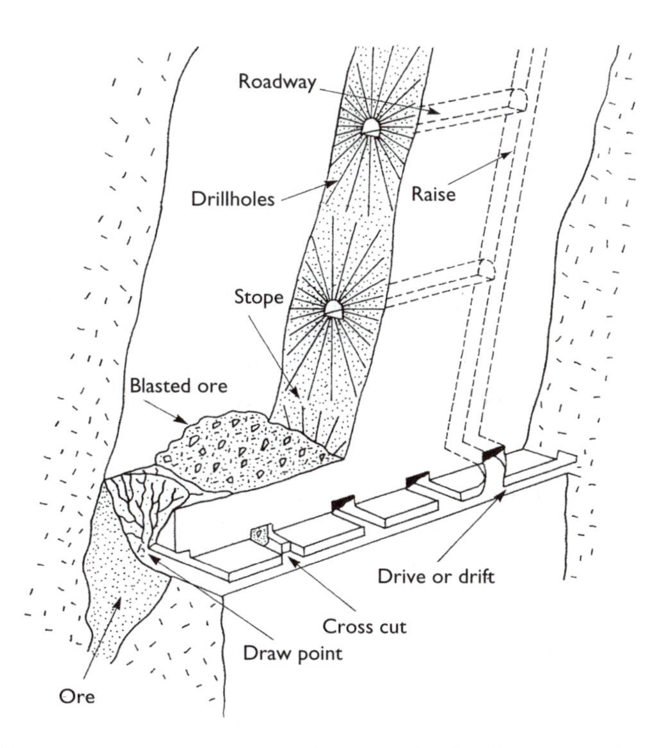

Figure 4.3 Diagrammatic representation of a stope that has been partially extracted by sub-level stoping.

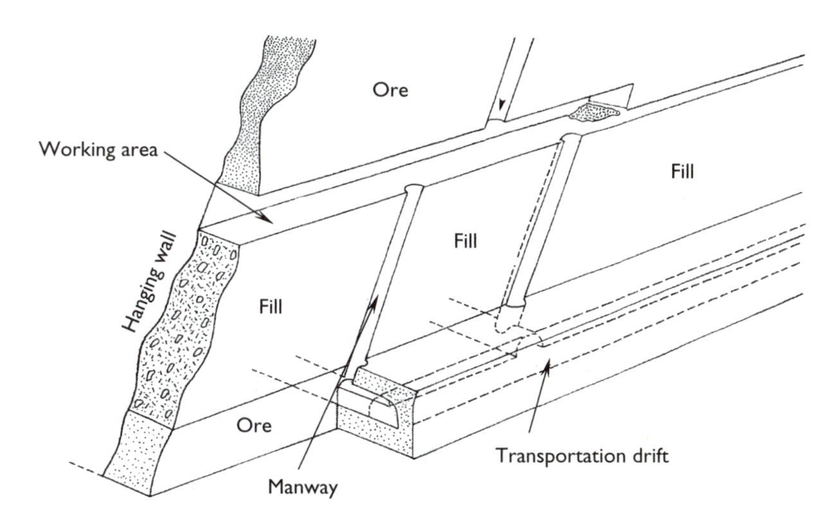

Figure 4.4 Diagrammatic representation of cut and fill stoping.

next slice of ore is drilled and blasted, and onto which the ore falls. Fill generally is placed hydraulically but occasionally pneumatic placement is used. Hydraulic filling requires special placement and drainage techniques as the slurry contains 30–40% water. Hence, all access openings into the stope from below have to be equipped with bulkheads to prevent flooding the haulage drift during filling. Dams or barricades are erected in the stope to control the placement of the fill. Drainage of the fill while it sets is provided by drains installed along the stope sill with water being disposed of by the drainage system in the level below. The tops of passageways and ore chutes are extended above the fill in order for them to remain open to provide access and ore passes, and may be lined with timber, steel arches or concrete. The timing of the placement of the fill is crucial for the success of the method since it must be in place in time to take up some or the entire overburden load on the ore in the stope. Cut and fill stopes usually are bounded by pillars for major ground support. By providing overall support to the workings, pillars subsequently can be removed either partially or totally from within the fill. In the United States stope height tends to range between 45 and 90 m and stope width from 2 to 30 m, depending on rock mass behaviour and placement of fill (Hartman, 1987). The length of a stope depends on the requirements of mechanization, varying from 60 to 600 m. Because the mine openings are backfilled substantially, the potential for collapse of the workings is limited. Although some surface subsidence may still occur, catastrophic collapse is unlikely.

The shrinkage stoping method is similar to cut and fill (Fig. 4.5). The main difference is that in shrinkage stoping the broken rock is left in place to support the sidewalls and to form a working floor as the stope progresses

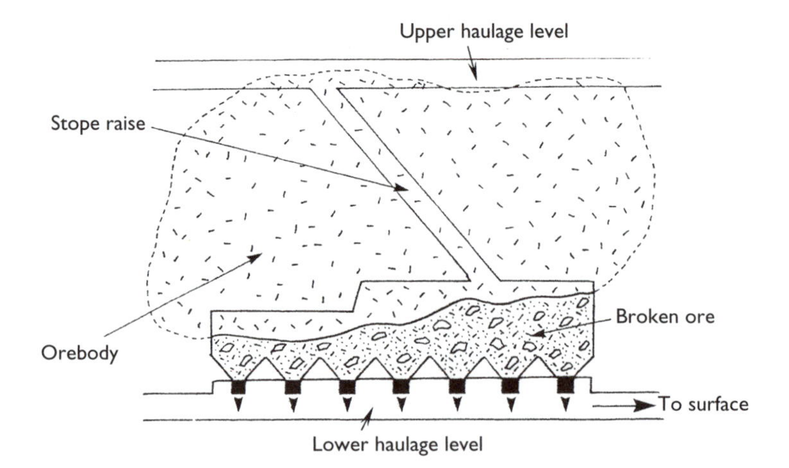

Figure 4.5 Diagrammatic representation of shrinkage stoping.

upwards. Access ways still must be maintained for personnel. Following blasting, there is an increase in volume of broken rock. Fragmented ore then is extracted to maintain a working space between the rock of the stope and the floor. When the stope reaches the upper haulage level, then the remaining fragmented ore is extracted. The disadvantages of this method include up to 60% of the ore remaining until the stope is mined out; weaker wall rocks may dilute the orebody due to slabbing of the sidewalls; and large open voids are left behind that can cause subsidence and affect the stability of adjacent stopes. These may require filling with waste.

Many metalliferous deposits occur in large disseminated orebodies and often the only way in which such orebodies can be mined economically is by means of very high volume production. When the rock mass is sufficiently competent, large open stopes can be excavated. The sizes of open stopes are the subject of design, and some form of pillar usually separates adjacent open stopes. The method therefore involves partial extraction, but open stopes may have spans of 50–100 m. Such open stopes will remain as stable underground openings with proper design, so that usually no surface subsidence will occur. However, mining usually aims to maximize extraction so that the long-term stability of the surface above an open stope is open to question.

As in longwall coal mining, total extraction implies extensive extraction over large areas, although complete extraction is not achievable. For instance, the hematite deposits of Cumbria, England, were worked by total extraction methods that gave rise to subsidence (Fig. 4.6). In particular, subsidence due to extraction of ore from the Hodbarrow Mine, Millom,

Figure 4.6 Subsidence, up to 20 m in places, above Hodbarrow Mine, Millom, Cumbria. (Reproduced by kind permission of the British Geological Survey. © NERC. All rights reserved. IPR/61–05C.)

was responsible by 1922 for the formation of four large depressions each about 400 m in width and 15 m deep. This type of mining using longwalls or tabular stopes with substantial spans, is practised in the gold and platinum mines of southern Africa. Extraction of one, two, or more reefs has taken place, each typically with stoping widths of 1 to 2 m. When this type of mining occurs at considerable depth, subsidence often occurs at the surface. The subsidence profile, however, is smooth and subsidence is not noticed readily. For instance, observations at the surface made by Ortlepp and Nicoll (1964) above a mined area at a depth of some 1300 m, with an effective span of 1000 m, showed a total subsidence of nearly 0.3 m and a surface tilt of 0.05%. By contrast, when this type of mining takes place at shallow depth it frequently generates detrimental subsidence (Hill, 1981). The effects are dependent on the dip of the orebody, which in South Africa varies from as little as 10° to vertical, although adverse subsidence generally is not observed when the mining depth exceeds 250 m. The surface profile commonly is very irregular and tension cracks often have been observed. Closure of stopes is more likely for flatter dips and hence surface movements are more likely to occur under such conditions. Steeply dipping stopes tend to remain open and present a longer term hazard. Most of the detrimental subsidence that has occurred has been associated with geological features such as dyke contacts and faults (Stacey and Rauch, 1981). Collapse of workings may take place after any artificial support in them in the form of fill has been washed away.

In caving methods, the orebody and surrounding rock from above is allowed to cave into the opening created by the mined extraction. This is in direct contrast to supported methods of mining where there is a requirement to prevent the collapse of the ground. Caving relies on the ability to control the failure of the rock mass under load and cannot be used where subsidence needs to be controlled, where surface waters and aquifers are present, or to extract minerals susceptible to spontaneous combustion such as iron sulphides.

Sub-level caving methods are applicable to near-vertical ore deposits. In sub-level caving, mining progresses downward while the ore between sub-levels is broken overhead (Fig. 4.7). The overlying waste rock (hangingwall) caves into the void created as the ore is drawn off. Mining is conducted on sub-levels from development drifts and cross-cuts, connected to the main haulage level below by ramps, ore passes and raises. Each area divided by the cross-cuts is referred to as a stope and is extracted separately. Since only the waste is caved, the ore must be drilled and blasted, usually with fan-hole rounds. Because the hangingwall eventually caves to the surface, all main and secondary developments are located in the footwall. In vertical cross section, the sub-level drifts and cross-cuts are staggered so that those on adjacent sub-levels are not directly above one another. Hence, drillholes driven from one sub-level penetrate vertically to the second level above.

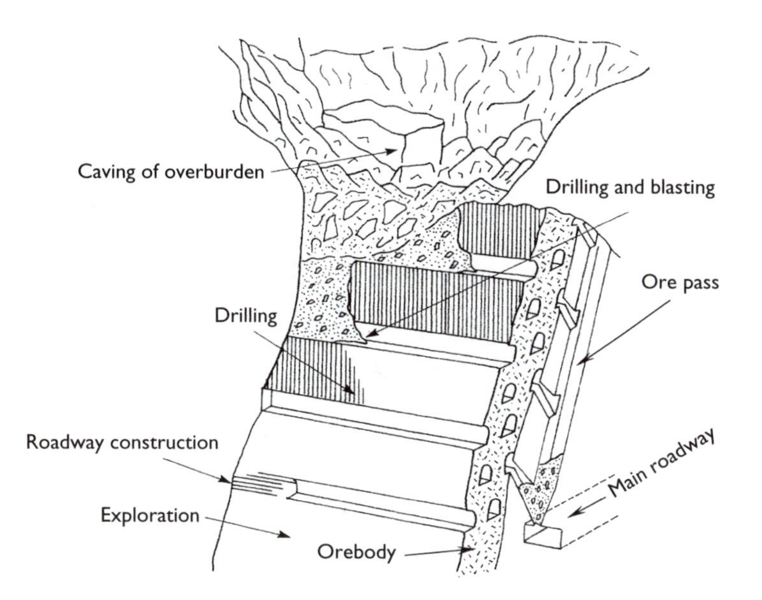

Figure 4.7 Diagrammatic section of sub-level caving.

Mining is planned so that sub-levels are engaged in sequential operations. The optimum interval between sub-levels varies from 9.1 to 13.7 m, the centre spacing of cross-cuts between 7.6 and 10.7 m, and cross-cuts tend to be around 1.8 to 2.7 m^2.

Block caving is used to extract massive, tabular or steeply dipping orebodies. In block caving, a rectangular block of the orebody is undercut, by drilling and blasting, to induce the block to cave, permitting the broken rock to be drawn off below (Fig. 4.8(a)). If the deposit is overlain by capping or bounded by a hangingwall, it caves into the void created by drawing the ore. Unlike sub-level caving, both the ore and waste are caved in block caving. The area and volume of the ore removed at the bottom of the block during undercutting must be sufficiently large to induce caving in the mass above, which then continues progressively on its own. Steady drawing of the caved ore from the underside of the block provides space for more broken ore to accumulate and causes the caving action to continue upward until all the ore in the original block has caved and been drawn by gravity or mechanical loading. The horizontal dimensions commonly employed in block caving vary from 30×30 m to 90×90 m, and for panel caving the width ranges from 30 to 90 m, with pillars between the panels of 3 to 9 m. Main haulageways often are paralleled by laterals and interconnected by cross-cuts. To provide ore drawing facilities, chutes, draw-points or trenches are prepared in the orebody beneath the block to

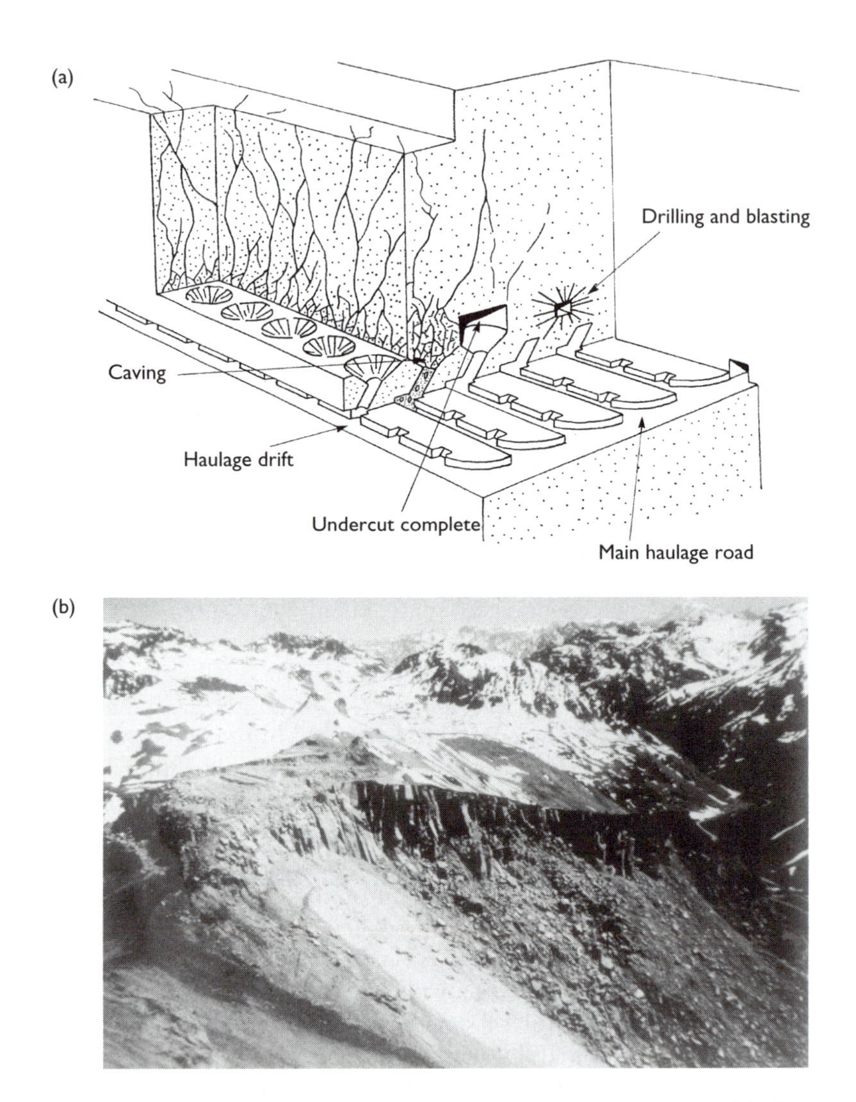

Figure 4.8 (a) Diagrammatic representation through a block caving layout. (b) Cratering due to block caving at a copper mine in central Chile.

be caved. They lie adjacent to the haulage drifts and cross-cuts on the main level. Finger raises to serve as ore passes then are driven to the sub-levels above, if any, and to the undercut sub-level, where they are belled out, other finger raises serve as passageways. Undercutting to initiate caving is undertaken by the carefully controlled removal of pillars. Extraction takes place by drawing the ore from the base without further blasting. Since caving occurs above the orebody, it can progress through to the surface to form a subsidence crater (Fig. 4.8(b)). This crater usually has several scarps around its perimeter, and contains material that has subsided *en block* overlying caved and rotated material, and loosely consolidated ravelled material. Owing to the large volumes that are extracted during mining, the extent and depth of the subsidence crater can be very large, for instance, craters can exceed 1000 m in diameter with their base being at 100 m or more below the ground surface. Hence, caving methods tend to be restrictive in application because of the resultant subsidence.

4.2. Collapse of crown pillars

Discontinuous subsidence over hard rock orebodies may pose a hazard at the surface through the formation of crown holes or craters. Most crown holes (sometimes referred to as sinkholes) form following the collapse of surface crown pillars above mined out stopes. Crown pillars refer to rock masses of variable geometry that are located above the uppermost underground workings of a mine, which serve to ensure both the stability of the workings and the surface. Hence, crown pillars are critical support elements.

Potentially unstable crown pillars may extend substantial depths below the ground surface depending upon the mining method and existing stress field; the pillar dimensions; the strength and deformability of the rock masses; the presence of any weak zones within the rock masses; the weatherability of the rocks forming the crown pillar; the presence, thickness and character of any soil horizon overlying the pillar; the volume and stability of the underlying stope concerned; the proximity of adjacent mining voids; and the presence and nature of any backfill (Hutchinson *et al.*, 2002). Failure of a crown pillar can prove catastrophic in an active mine, leading to air-blast damage underground and inflow of water, or sometimes tailings, from the surface (Stacey and Swart, 2001). As noted, the collapse of a crown pillar that gives rise to a crown hole or crater can present a hazard at the surface and so, if possible, must be prevented. Unfortunately, the determination of whether remedial action is necessary to ensure the long-term stability of crown pillars over active or abandoned near surface stopes is difficult because of the differing nature of the material of which pillars are composed. In particular, the upper part of a pillar may be weathered to varying degrees, and may be covered with a soil horizon of varying thickness. Other factors in addition to rock mass characteristics that are required to help assess the stability of crown pillars include pillar and stope configuration, geological and groundwater conditions, and the presence and type of any backfill. A summary of some of the data required for an assessment of the stability of a crown pillar is given in Table 4.1. Access to the underground workings may not be available at some mines so that direct visual observation of the rock masses of crown pillars and stopes is not possible. Data collection in such situations is obtained from drillholes and by geophysical techniques (see Chapter 2).

Table 4.1 Crown pillar analysis data requirements (*) (Modified after Hutchinson *et al.*, 2002; reproduced by kind permission of Springer)

Data required	*Information Source*										
	Mine plans and sections	*Mine reports*	*Conversation with mine personnel*	*Review of existing core*	*Drilling holes*	*Core logging and soil sampling*	*Testing core and soil samples*	*Borehole observations*[a]	*Structural mapping*	*Observation of rock mass stability*[b]	*Monitoring with instruments*[c]
Crown pillar dimensions	*	*			*			*		*	*
Stope geometry	*	*	*		*			*		*	*
Stope mining and stability history	*	*	*					*		*	*
Backfill: type, competence and stability	*	*	*		*		*	*			
Artificial support of crown pillars or stope	*	*	*								
Rock mass strength and classification		*		*	*	*	*		*	*	*
Structural geology: average rock mass, faults and shears	*	*	*	*	*	*		*	*		
Stage of stress: induced		*			*			*			*
Overburden: stratigraphy, profile location of water table	*	*		*	*	*	*	*			

Notes

a Including data from borehole camera and downhole laser distance meter.

b Underground access via mine development excavations or by holes from surface. Observations may be made visually and/or by cavity laser distance meter.

c Instrumentation may include extensometers, time domain reflectometry (TDR) and stress cells.

Carter and Miller (1995) maintained that the assignment of a single factor of safety to the unique set of dimensions and properties of a particular crown pillar generally is not valid. They indicated that arching may develop in good quality rock masses that are not adversely affected by discontinuities and thereby promote stability. By contrast, failure of a crown pillar usually involves a number of mechanisms. In blocky rock masses failures of crown pillars most frequently develop where several adversely orientated discontinuities intersect or where a certain system of major discontinuities afford a means of release for gravity collapse. If crown pillars are characterized by poor quality rock, small block sizes and unfavourable geometry, then failure tends to occur by ravelling from the underside of the crown zone. Failure also can take place due to shearing through the rock mass at the boundaries of the pillar or disruption of the competence of the pillar by sagging. After a survey of crown holes in Western Australia and Northern Territory, Szwedzicki (1999) concluded that most crown holes formed where the surface crown pillars consisted of oxidized rock material, where the orebodies were steeply dipping, and where the underground openings had a large open span. He found that the thickness of failed crown pillars varied between 25 and 60 m and the estimated unsupported span beneath a crown pillar ranged from 70 to 270 m. The crown holes were circular or elliptical in plan with equivalent diameters between 20 and 65 m.

Hutchinson *et al.* (2002) provided a review of the various methods that have been used to assess the stability of crown pillars, which can be grouped into empirical methods, mechanistic methods (i.e. analysis according to definable failure modes such as sliding failure, lamination bending failure and chimney failure) and numerical modelling methods. However, as Carter and Miller (1995) had pointed out, the value of the results obtained from these methods depends on the quality of the data relating to many of the factors already mentioned, especially those concerning the rock mass characteristics that are required to determine rock mass quality (e.g. the rock mass rating (RMR) of Bieniawski, 1989: or the *Q* value of Barton *et al.*, 1975). If the data obtained is limited, then there is a limitation placed on the validity of the results. Because of this, Carter and Miller suggested that risk-based decision making procedures could be used to assess crown pillar stability. An assessment of crown pillar stability is complicated further by the fact that the geotechnical conditions of the mine workings are likely to deteriorate with time. Szwedzicki (1999) found that in those cases that he studied in which oxidized hard rock material formed the crown pillars, then the time between the formation of underground openings and pillar collapse ranged from 10 to 63 years.

Where crown pillar stability analysis suggests that the rock structure is likely to be unstable at some time in the future, then some form of rehabilitation may have to be carried out, depending on the degree of risk to property or person. Rehabilitation techniques depend on whether the pillar is left in place or is removed and commonly include some form of backfilling. The use of backfill provides support to stopes and pillars and so increases their stability, helps constrain loose rock thereby reducing deformation, and reduces or eliminates the risk of caving and associated subsidence. Backfill consists of crushed waste or tailings from which the fines may have been removed. It is important that the gain in strength of backfill occurs as quickly as possible. A small percentage of cement can be added to backfill to increase its strength when set. Alternatively, Hutchinson *et al.* (2002) mentioned that a crown pillar can be removed by blasting and the void then backfilled from the surface. They also indicated that the void could be allowed to flood after removal of the crown pillar. Brose (1996) provided a review of the various factors such as site investigation, environmental compliance and reclamation, that should be taken into account when assessing the need for rehabilitation of old mines in areas where urban development is likely to take place.

4.3. Subsidence prediction in mining massive deposits

When mining is practised in massive deposits, particularly with caving methods, it is almost certain that subsidence will occur, or that the risk of subsidence will be high. Such mining therefore is unlikely to be permitted where it will pose a threat to public safety or would be located beneath sensitive or historical structures. The stability of mine structures such as shafts, underground accesses and haulages, surface plant and facilities are usually the main concern. In such circumstances the prediction of the magnitude of subsidence is unlikely to be required. Rather prediction will be required as to whether subsidence will affect the mine structures and adjacent areas or not. This is a common problem in mining, which generally arises from poor initial planning or from the fact that mining operations often expand beyond the initially planned limits, as a result of increased ore prices and discovery of additional or extended orebodies. Although surface plant can be moved relatively easily, waste dumps, tailings dams and shafts, and sometimes residential areas, may be located too close to the limits of the orebody to be confident of their safety. Owing to the critical importance of such facilities to a mine, and the prohibitive cost of replacing or moving them, it is important to determine the potential influence on them of any extended mining. This assessment can be carried out using a theoretical approach or by extrapolation of appropriate field data, if available, or by interpretation from similar cases to the problem case.

Theoretical methods of subsidence prediction also make use of some of the numerical methods referred to in Chapter 3. For example, boundary element and finite element methods are applicable for calculating the deformation around massive openings. Finite element analysis may be preferable if the geology is substantially non-homogeneous and non-linear failure behaviour of the mass requires modelling. However, the openings created in massive mining are rarely sufficiently two-dimensional in nature for a valid two-dimensional analysis to be carried out. It therefore generally is necessary to make use of a three-dimensional stress analysis method. Although both finite and boundary element methods are available, the magnitude of the problem of using the former method for complicated three-dimensional modelling makes the latter method far more attractive. A drawback of the boundary element method is its limited capacity to model non-homogeneity. However, from a practical point of view, this is not considered to be a significant drawback, and considerable success has been achieved using a combination of three-dimensional boundary element analyses to model the full geometrical situation, and two-dimensional finite element analyses, based on these results, to model the detailed geological situation (Diering and Laubscher, 1987). In such a tandem approach, the three-dimensional results are used as a calibration for the subsequent two-dimensional analyses.

Owing to the limited knowledge available regarding the deformation characteristics of rock masses, and the field stresses, the assumption of linear elasticity for the three-dimensional analyses is considered to be the most suitable choice for practical applications. To predict failure and subsidence, it is necessary to apply a suitable failure criterion to the results. The choice of the criterion should be based on local conditions. The validity of predictions will be enhanced substantially if it is possible to back-analyse a failure, and so establish a credible failure criterion. This approach has proved to be successful in underground situations (Diering and Laubscher, 1987). The use of non-linear two-dimensional analyses for more detailed local analysis of failure, in which boundary conditions for the two-dimensional model are derived from the three-dimensional results, is also practical.

Hoek (1974) developed a method for predicting the extent of caving resulting from mining of an inclined tabular deposit. This approach is based directly on limit equilibrium analysis of the stability of rock slopes, and allows the failure plane angle and an angle of break to be determined. These define the position on the surface of the furthest influence of mining. Curves from which these values can be obtained are given in Figure 4.9.

Some of the empirical methods of prediction referred to in Chapter 3 may be applicable, or at least sufficiently so for preliminary estimates to be made, in cases in which the horizontal extent of mining is sufficient to result in collapse of the workings. Besides these established methods, there are no empirical techniques for prediction of subsidence in massive mining situations. There are, however, some published records of monitored subsidence behaviour, which can be used for comparative purposes (Brumleve and Maier, 1981; Panek, 1981; Touseull and Rich, 1981). These records show that the subsidence behaviour, described by the angle of draw, is very site specific.

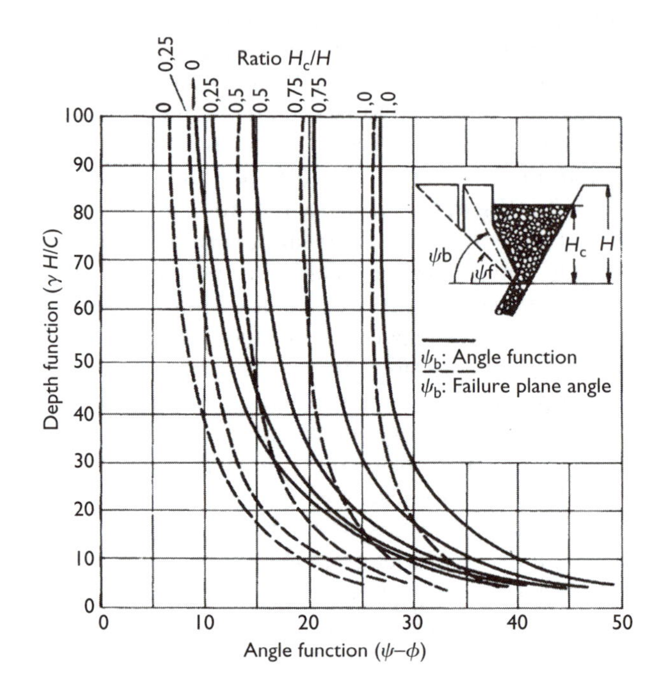

Figure 4.9 Caving break back angles. (Modified after Hoek, 1974; reproduced by kind permission of the Institute of Materials, Minerals and Mining.)

In many cases the angle is near vertical while in others it approaches values typical of coal mining. For example, Panek (1981) noted that the angles of draw varied from 33° to 55° at San Manuel Mine in the United States. The area of surface subsidence therefore may be affected by the ratio between the extracted span and the depth of overburden; the extent of bulking that takes place during the caving process; the structural and deformational characteristics of the overburden strata; the surface topography; the climate; and any major geological structural features. In summary, whilst empirical data may provide background comparative information, each case must be evaluated individually.

4.4. Gold mining in the Johannesburg area, South Africa

Taking the Johannesburg area as an example, gold deposits were extensively worked within the central area where three reefs were mined. The gold bearing deposits of the central Witwatersrand are found in thin conglomeratic reefs that are interbedded with quartzites. At outcrop the dip of the reefs varies between 20° and vertical. Mining commenced in 1886 and has taken place predominantly along three reefs, namely, the South Reef, the Main Reef and the Main Reef Leader. In the early years of mining attention was focused on the Main Reef and the Main Reef Leader. In the South Reef and Main Reef Leader, which have widths of some 1 m and 1.5 m respectively, development began with reef drives at successively greater depths, which were connected by raises. Stoping commenced from the raises and support was installed as mining progressed. Once the stoping span exceeded around 30 m, the hangingwall strata tended to sag and settle on the supports. This sagging gave rise to bedding plane separation and the formation of a dome of disturbed or fractured rock above the stoped out area. The size of the fracture zone largely depends on the dip of the reefs. In a near vertical reef only a small zone exists in the sidewalls but as the dip decreases the fracture zone above the stope increases in size.

The middling (distance between adjacent reefs, normal to the plane of the reef) between the reefs varies from zero up to 30 m. This has meant that although reefs normally have been worked separately, in some areas where two reefs were very close together, then they were mined as one. In such instances, extensive narrow tabular stopes were formed that were supported by occasional pillars, timber props and waste packs. Moreover, stopes at times were tip-filled from the surface with waste, which provided some support. Nonetheless, the timber props and waste packs tended to deteriorate with time leading to stope closure, which was manifested at the surface in the form of subsidence. One of the problems is that such stope closure and associated subsidence cannot be predicted. Hill (1981) noted that sudden collapse of surface material into old mine workings has not been uncommon in the Johannesburg area (Fig. 4.10(a) and (b)). There appears to be no time limit on the occurrence of such surface collapses.

A further problem is associated with the early days of mining, that is, that only poor records were kept and those mine plans that are available frequently are inaccurate. Often when a mine was nearing the end of its working life pillars were robbed and this went unrecorded. Consequently, inspection of mine plans does not necessarily aid the determination of ground stability conditions. Also, records of subsidence were not kept strictly so that frequently there is no certainty as to whether subsidence has occurred or whether the full subsidence potential still remains.

4.4.1. The subsidence problem

Several factors influence the character of the subsidence that develops. These are summarized in Table 4.2. Both continuous and discontinuous subsidence have taken place above shallow gold mines. Stacey (1986) described typical mechanisms of subsidence (Fig. 4.11). For instance, continuous subsidence can result from regular closure of underground workings in which the rocks in the hangingwall are competent. A number

Table 4.2 Factors affecting mining (After Stacey and Bell, 1999)

Factors	
Dip of reefs	Closure of stopes more likely for flatter dips. Hence, more likely that vertical surface movements will occur over dipping stopes. These are more likely to occur soon after mining. Steeply dipping stopes are more likely to remain open and thus present a restricted, but longer term hazard. If local collapse from the hangingwall occurs, the collapsed material is likely to ravel down steeply dipping stopes. In stopes dipping at angles shallower than approximately 35°, the collapsed rock will not move down the stope, but is likely to bulk, ultimately providing support between hangingwall and footwall. Such stopes are therefore likely to be self-stabilizing
Stoping width	The greater the stoping width, the greater the potential magnitude of closure, and hence the greater the potential movements at the surface. Increase in the stoping width has an adverse effect on stability of pillars, and on the effectiveness of artificial support. Differential stoping widths may be associated with potential movement differentials
Extent of mining	The greater the areal extent of mining without effective solid support, the greater the potential for subsidence at the surface. Systematic pillar support not common in areas of shallow undermining. If large areas have been mined, it is likely that closure will have taken place soon after mining is completed, and that any subsidence that may occur is likely to be uniform. Whether this has occurred or not, the greater the area of mining, the less likely it is that differential subsidence will occur. The potential danger area in this regard is close to the mining abutment
Number of reefs mined	As for stope widths above, the greater the number of reefs mined, the greater the potential for larger closures, and hence the greater the potential for surface movements
Separation of reef	The closer that reefs together, the more likely that they will interact, and hence the greater the potential for surface movements. The greater the separation between reef horizons the greater the motivation for considering the stopes independently of one another
Competence of rocks	Weak hangingwall or footwall rocks may allow punching of pillars that renders the solid support ineffective. Weak hangingwall rock will allow easier collapse of hangingwall between supports, with greater consequent potential surface effects. This has the advantage that it is more likely to have occurred soon after mining has been completed. In a multi-reef situation, weak strata will result in greater collapse potential
Type of support	Solid pillar support is most effective support for preventing subsidence at surface, provided that the spans between pillars not excessive. The only types of installed support that can be considered to provide some control on surface behaviour are sand filling and waste pack support, provided that their existence can be proved and their permanence assured. These supports will not prevent subsidence, but will reduce hangingwall collapse and promote even subsidence. Their effectiveness depends on the area that they cover, and the spacing between adjacent areas of such support will determine the potential for local collapse of the hangingwalls
Time elapsed since mining	Results of surface levelling have shown that the rate of mining-induced surface movement reduces with time after mining ceased. Most records have shown that the rate of these movements reduces rapidly, and effectively ceases within 20 years of completion of mining. Records also show that reactivation of mining, for example, as in reclamation mining when solid pillars are removed, reactivates movement at surface. In partial extraction systems, collapses of pillars can occur many years after cessation of mining, leading to subsidence of surface

Figure 4.10 (a) Undermining of a building in Johannesburg, the old workings were revealed after heavy rainfall. (b) Old stope workings revealed during remedial treatment of the ground at a site in Johannesburg.

of situations can lead to discontinuous subsidence. First, discontinuous subsidence can develop due to cantilevering of a wedge of the hangingwall defined by a shallow dipping stope and the ground surface. Such behaviour normally is confined to a zone adjacent to the outcrop of a stope (i.e. generally within some 30 m of the outcrop). This may lead to the formation of a subsidence step at the ground surface, or a number of steps, which may be accompanied by open cracks. Hill (1981) recorded that the occurrence of surface cracks was particularly prevalent in the pre-1920 period of mining when mining occurred predominantly at shallow depth. Nevertheless, he mentioned more recent occurrences of surface fracturing associated with mining. Second, discontinuous subsidence may result from block movement, in which the subsided block of ground is defined by a major fault or dyke contact, the stope and the ground surface. Again this may give rise to stepping and open cracks occurring at the surface, with associated differential subsidence (Fig. 4.12). Such subsidence may not be confined to the outcrop area and may occur over a wide range of mining depths. Third, discontinuous subsidence, in the form of crown hole development can occur as a result of ravelling or washing away of fill from a stope outcrop area. Also, the walls of stopes that outcrop can deteriorate and collapse into the old workings below leaving a cavity at the surface (Fig. 4.13). Vertical reef outcrops are especially suspect, particularly if the footwall consists of shale

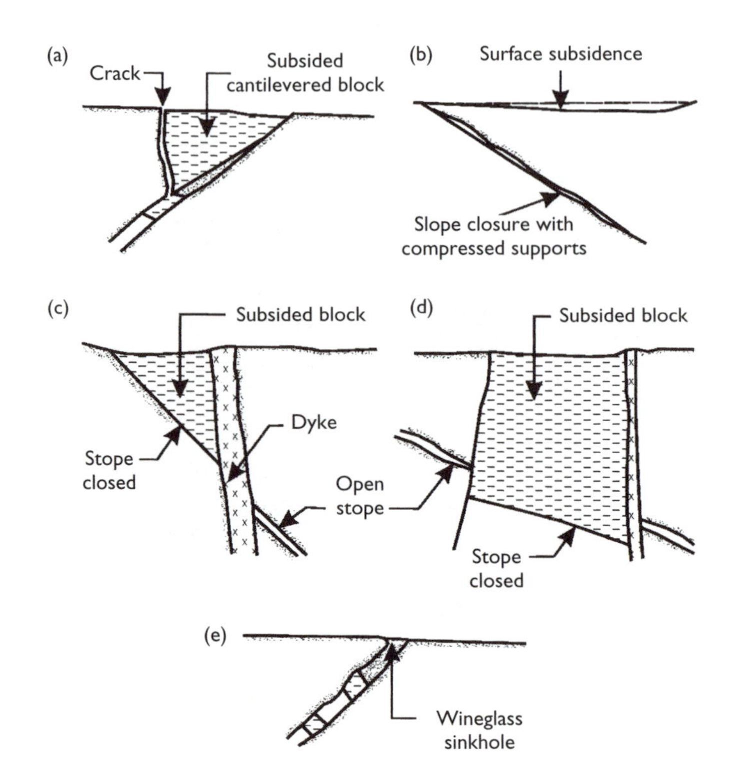

Figure 4.11 Mechanisms of subsidence.

Figure 4.12 Subsidence on a faulted dyke contact leading to cracking of ground.

Figure 4.13 Collapse of surface outcrop of a reef into stope.

that weathers more easily and is much weaker than the quartzites of the hangingwall. These outcrop areas therefore always are suspect.

Stacey and Meintjes (1994) mentioned that from 1903 until 1927 the Government Mining Engineer recorded numerous subsidences in the annual reports. All the cases involved mining operations in the Central Rand area where mining was taking place from the outcrops to a depth of approximately 250 m below the surface. Unfortunately, however, there was no sustained systematic monitoring of surface subsidence during this period. After 1927 several surface effects attributable to mining in the Central Rand area were reported to the Inspector of Mines. As a result of these, where mining took place at depths exceeding the 240 m depth contour, the latter was adopted as the depth beyond which restrictions on mining in relation to surface development normally were not imposed. Subsequently, the depth factor concept has been embodied in the standard building restrictions that have been developed by the Government Mining Engineer (Hills, 1981). These limit the permissible heights of proposed buildings in relation to the depth of mining below the site. As can be seen from Table 4.3, these restrictions are relaxed progressively up to a mining depth of 240 m, beyond which they do not apply unless the mining situation is unusual. However, due to the development of new methods of construction and a better understanding of the behaviour of undermined ground, consideration is

Table 4.3 Summary of existing building restriction guidelines (After Stacey and Bell, 1999)

Depth of reef (m)	*Number of storeys*	*Height of walls (m)*
0–90	Nil	—
90–120	One, with one basement	5.0
120–150	Two, with one basement	8.5
150–180	Three, with one basement	12.0
180–210	Four, with one basement	15.5
210–240	Five, with one basement	19.0
Over 240	No restrictions unless mining circumstances are unusual	

now given to increasing the allocated heights of buildings when the depths of the reefs are from 90 to 240 m. In such instances, the stability of the ground has to be proven or methods whereby it will be stabilized provided. In fact, Stacey and Bakker (1992) have suggested new guidelines for the development of undermined ground that involve the recognition of what can be regarded as hazard zones. They identified four types of zone, the outcrop zone, the shallow zone, the deep zone and the special zone. These take into account three main factors, namely, that surface movements, if any, are gradual; that total surface movements are a small fraction of stoping widths; and that changes in surface movements after completion of mining are very small. By using the depth of mining for definition of development zones, the dip of the reef and the distance from the outcrop can automatically be taken into account. Account can be taken of the extent of mining by relating the potential surface movement to the total stoping width.

The outcrop zone extends from 3 m on the footwall side of the stope (reef) to positions at the surface where the depth of mining measured vertically is 25 m beneath. When more than one reef has been mined, depending on the middlings between the stopes, the one extreme is for each stope to occur separately in its own outcrop zone, and the other is for all stopes to occur in a single outcrop zone. This means that there may be more than one outcrop zone on a single site. No development is permitted unless the outcrops are 'safe' to the satisfaction of the Regional Director of Mines. Developments are designed to accommodate potential subsidence effects without endangering the safety of human lives. As a guide, a localized differential subsidence of 100 $mm\,m^{-1}$ of the total stoping width beneath the outcrop zone, and associated tilt, and a horizontal strain of 35 $mm\,m^{-1}$, is suggested. The potential differential subsidence is assumed to occur across a base length of 5 m. No buildings greater than two storeys in height, and no basements, are permitted in this zone. Special consideration is given in the case where appropriate stabilization measures are carried out below the surface in the old stopes. In the outcrop zone, no buildings should be permitted to straddle significant dyke and fault contacts. In fact, no building foundations should be located closer than 3 m to a surface contact. This restriction may be reviewed if it can be proved that specific professional design consideration has been given, by the designers, to the possibility of differential subsidence at the contact. This relaxation also applies to the development of other structures across contacts.

The shallow zone extends from the outcrop zone to positions at the surface where the mining depth, measured vertically, is 200 m. Developments should be designed to accommodate the effects of subsidence without endangering the safety of human lives. As a guide, the following deformations should be considered, namely, a localized differential subsidence, assumed to occur across a base length of 5 m, calculated using the following relationship:

$$S = 11 - 0.03D \tag{4.1}$$

where S is the value of differential subsidence expressed as a percentage of the total stoping width beneath the shallow zone and D is the vertical depth to the shallowest stope. For example, at a depth of 90 m, S is calculated to be 8.3%. If the total stoping width beneath the site is 2.5 m, then the localized differential subsidence that must be accommodated without distress to a structure is 0.21 m, or 42 mm m^{-1} across a 5 m base. Structures must be designed to accommodate a horizontal strain of $(S/3)$% without distress. No limits should be placed on the heights of buildings in this zone. Also, no limits are placed on the number of basements, provided that the closest distance from the basement to the shallowest hangingwall is at least 25 m. In the shallow zone, significant fault and dyke contacts should be located. Buildings should not be permitted to straddle the contact if the angle between the contact and the stope on the mined out side, is uncertain or is less than 95°. Other structures such as roads and bridges should be permitted to cross the contact, and the restriction on building development should be reviewed, provided that specific professional design consideration has been given to the possibility of differential subsidence at the contact. If the angle noted above is greater than 95°, then buildings and other structures should be permitted to cross the contact, provided that particular attention is given by the designers to the fact that differential subsidence could take place at the contact.

The deep zone should include all areas where the mining depth, measured vertically, is greater than 200 m. No restrictions are placed on development within this zone.

All ground within 5 m of the perimeters of shafts and winces, and within 5 m of significant dyke or fault contacts should be included in special zones. Shafts should be located and stabilized to the satisfaction of the Regional

Director of Mines. Founding on stabilized shaft locations should be permitted provided that specific professional design consideration has been given by the designers, to the possibility of differential movements at the location.

4.4.2. *Stabilization measures and specialized construction*

As can be inferred from earlier discussions, much of the Johannesburg area where the reefs outcrop is zoned for commercial or light industrial development. As such, it is suitable for the erection of low-rise, relatively low cost warehouses and factories with limited office accommodation. Such development cannot justify expensive stabilization measures, nonetheless remedial measures have to provide local support that will prevent collapses. The design of a flexible building structure has to accommodate any minor surface movements that may occur subsequently. Accordingly, the aim of shallow stabilization methods is to produce a competent surface zone to promote arching across the stopes.

Dynamic compaction has been used in areas where the ground at the surface consisted of very loose sandy gravel and imported ash fill (see later). Residual quartzite generally occurs at a depth of approximately 2.5 m. In addition, some mine workings have been filled with loose rock fill and sand, which were probably tipped into the open workings from the surface. In such instances, dynamic compaction can improve the characteristics of the soil down to bedrock and close any cavities that exist near the ground surface as a result of mine workings (Fig. 4.14). Dynamic compaction also can form an *in situ* ground arch that can span the reef outcrops.

A far more common method of providing the necessary local support is to plug the reef outcrops or the stopes and then place compacted backfill above the plug to provide a stabilizing ground arch (Fig. 4.15). This approach has the added advantage that stopes are exposed for inspection during the construction phase, ensuring that construction can be varied to suit local variations in stope conditions. The procedure involves excavating along the strike of the outcrop until the narrow throat of the stope is exposed with competent hangingwall and footwall surfaces. The plug, usually concrete, then is formed in the throat. The dimensions of the plug must be sufficient to span the throat of the stope in a stable manner and to provide a suitable base for the backfill. The concrete plug also seals the stope against ingress of water.

The 'porcupine' technique represents a novel method of plugging stopes and consists of drilling holes from the surface, approximately 100–120 mm in diameter, to intersect the workings more or less at right angles at some 5 m beneath the top horizon of fresh rock and to insert the 'porcupines' into the holes (Parry-Davies, 1992). The porcupines are fabricated lengths of tube, 50 mm in diameter, with holes drilled along the length into which

Figure 4.14 Dynamic compaction used for stabilization of shallow old mine workings.

Figure 4.15 (a) Plugging reef outcrops with concrete. (b) Stabilization of shallow outcrop of three adjacent stopes by a concrete plug.

high tensile steel wire is inserted. The wires extend to give a diameter of 500 mm but when the tube is inserted into the drillhole, the wires fold against the tube. They spring outwards when they enter the mined out stope. The porcupines bridge the stope and are socketed into the footwall. An intermesh of wires accordingly is formed within the stope. Holes then are drilled from the surface to just above the porcupines (Fig. 4.16). Oversanded concrete containing hemp fibres is pumped down these drillholes. The fibres become entangled in the wire mesh and so prevent the concrete passing through the mesh. A concrete plug is formed thereby, which allows the zone above to be filled with cheap grout.

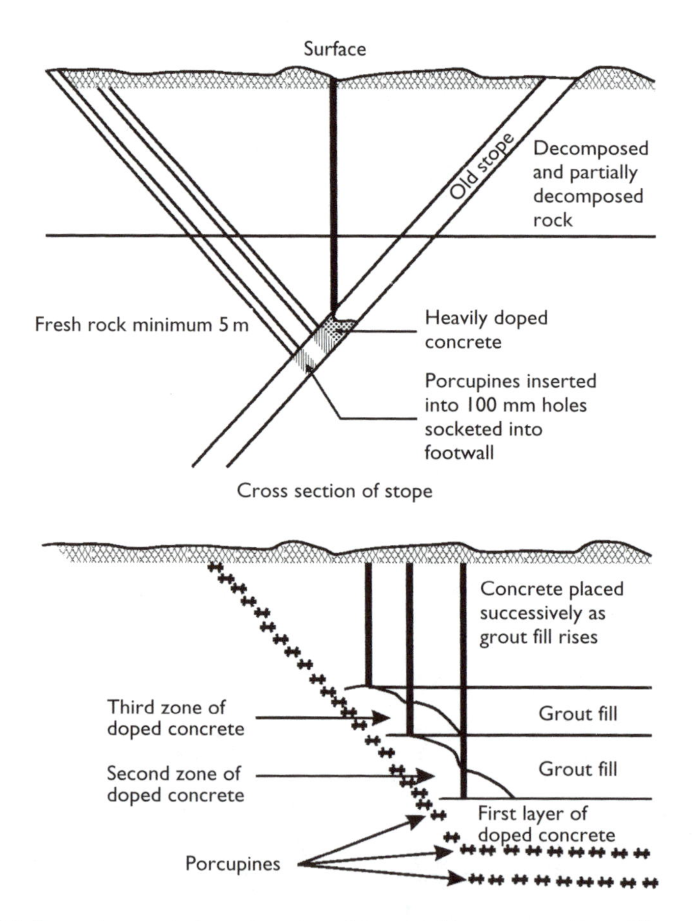

Figure 4.16 Porcupine technique for consolidating old mine workings. This technique was used because the Government Mining Engineer refused man-access to the workings and a portion of the Main Reef Road between Johannesburg and the West Rand had collapsed. (After Parry-Davies, 1992; reproduced by kind permission of the South African Institution of Civil Engineers.)

For major developments a different approach must be adopted since the consequences of any subsidence require that the risk be minimized. Owing to the higher cost of such development, greater expenditure on remedial works can be accommodated and usually can be offset against the lower cost of the 'restricted' land. The shortage of land suitable for development close to the central business district of Johannesburg has resulted in these 'restricted' areas becoming very attractive. In such cases shallow outcrop stabilization can be provided by rigid support between footwall and hangingwall surfaces such that in the event of collapse or instability at greater depth, the near-surface rigid zone forms an arch. The installation of pillars offers in-stope support. These can take the form of dip pillars, concrete strike pillars, concrete distributed 'blob' pillars and distributed 'blob' pillars formed by grouting of backfill (Stacey and Bell, 1999).

Where there is a possibility of a crown hole suddenly appearing in a road, leading to potentially serious consequences, a cable mat can be used as protection (Herman *et al.*, 1992). The cable mat is formed of steel tendons covered with plastic sheath. These bridge the possible crown hole position and are anchored well outside the likely crown hole zone (Fig. 4.17). An anchor block therefore occurs at each end of the cable mat with intermediate shear keys. Concrete, cast *in situ*, is used as a spacer and protects the tendons against corrosion. In order to avoid local failure of the tendons at cracks in the concrete, debonding or partial bonding of the tendons to the concrete is necessary over specific lengths. Lateral ties in the form of transverse rebars are required to secure the concrete in position, notably if large deformation occurs due to a crown hole forming beneath a road. The surface depression arising from such an occurrence offers evidence of a cavity beneath the road and appropriate repairs then are carried out. Catenary nets of welded mesh or geogrids also can be incorporated into a road in a similar fashion, especially when the possible occurrence of a small crown hole is suspected.

The design and construction of structures should minimize potential damage should any movements occur. What has to be borne in mind is that the

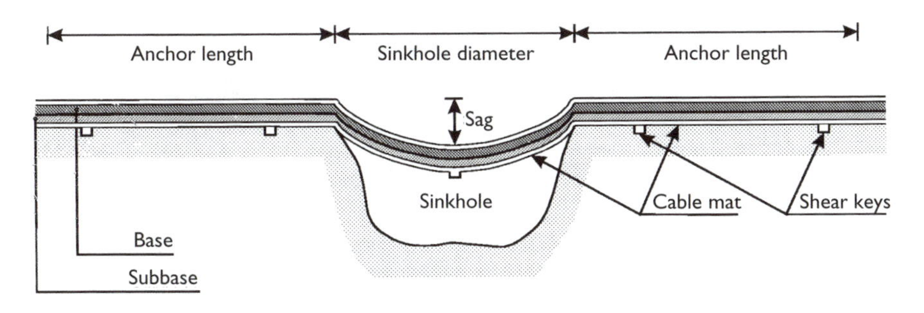

Figure 4.17 Use of cable mat to bridge a sinkhole (crown hole).

Table 4.4 Suggested limits of ground movements for buildings, roads, railway tracks and pipelines (After Stacey and Bell, 1999)

Building category[a]	*Damage level*	*Horizontal strain* ($\times 10^{-3}$)	*Tilt* ($\times 10^{-3}$)
1	Superficial	0.5	1.0
1	Hazardous	2.0	3.0
2	Superficial	2.0	1.5
2	Hazardous	4.0	3.5
3	Superficial	1.0	3.5
3	Hazardous	3.0	1.5
Road category[b]	*Damage level*	*Horizontal strain* ($\times 10^{-3}$)	*Tilt* ($\times 10^{-3}$)
1	Minor	2.0	10.0
1	Severe	5.0	15.0
2	Minor	1.0	5.0
2	severe	2.0	10.0
Railway line category[c,e]		*Horizontal strain* ($\times 10^{-3}$)	*Tilt* ($\times 10^{-3}$)
1		3.0	10.0
2		2.0	5.0
Pipeline category[d,e]		*Horizontal strain* ($\times 10^{-3}$)	*Tilt* ($\times 10^{-3}$)
1		2.0	4.0
2		1.0	2.0

Notes

a Category 1 – brick and masonry structures with brick load-bearing walls; Category 2 – steel and reinforced concrete frame structures; Category 3 – massive structures of considerable rigidity, including those of central core design.

b Category 1 – low volume gravel roads and paved roads; Category 2 – high volume paved roads. These would have kerbs and comprehensive stormwater drainage facilities.

c Category 1 – goods lines and lines carrying low volume passenger traffic, on which speed limits can easily be imposed; Category 2 – main lines and lines carrying high volume passenger traffic. On such lines it normally will not be practical to impose permanent speed limits from a system operating point of view.

d Category 1 – secondary sewers, water supply pipelines and stormwater drains, and all steel pipelines carrying non-hazardous materials; Category 2 – main sewers, water supply pipelines, stormwater drains and pipes carrying hazardous fluids.

e A damage level is not appropriate for railway lines and pipes, especially the former. Any movements are corrected easily by reballasting which, in any event, is done on a regular basis.

susceptibility of a structure to damage does not depend solely on the dimensions and nature of the ground movements, it also depends on the type of structure, its design, the method and quality of construction, and the types of materials used. In South Africa four principal types of structure have been recognized by Stacey and Bakker (1992) in relation to subsidence effects caused by shallow gold mining, namely, buildings, roads, railway tracks and pipelines. Over and above this, a number of categories within these structural types have been identified. The suggested limits for ground movements for these different structures are given in Table 4.4. The deformation limits for the four types of structures suggested by Stacey and Bakker (1992) can be extended by special features built into the structure.

4.5. Case history 1

Force Crag Mine is situated at the head of Coledale, west of Keswick, in the English Lake District. It has had a long history as a producer of lead (galena, PbS) and zinc (sphalerite, ZnS) ores, and barite ($BaSO_4$). As is common with the majority of metalliferous mines in Britain, periods of active mining have alternated with periods of abandonment. The date of the earliest mining at Force Crag is not known, although here is evidence of past mining in the region dating from the late sixteenth century or even earlier (Dumpleton *et al.*, 1996). The present mine was opened in 1838, and was abandoned in 1991.

The Force Crag epigenetic vein occupies a roughly east-west trending fault that displaces mudstones, siltstones and sandstones of the Skiddaw Group (Ordovician). The vein may be traced for at least 4.8 km from the foot of Gasgale Gill in the west to Coledale Beck in the east. The regional dip of the strata in the vicinity of Force Crag is to the south, the Force Crag vein cutting obliquely through this sequence at about 70°. The Skiddaw Group rocks around the mine are disturbed by complex slump folds, which are commonly recumbent, faulted and sheared along bedding planes. The age of the mineralization is possibly Carboniferous and may be associated with the Crummock Water Aureole, an elongate metamorphic and metasomatic aureole associated with a buried intrusion and a major basement fracture (Stanley and Vaughan, 1982).

Mineralization is not present continuously along the Force Crag Vein, but is abundant in lodes, rakes and scrins, reaching 1.5 m in width, but 6.1 m have been encountered locally. Unlike many orefields, notably the near by Pennine Orefield, the Lake District fields exhibit little or no lateral zonation of constituent minerals but large vertical zonation is apparent. Force Crag provides one of the best examples of this. In the higher levels of the mine, barite is the main constituent of the vein, accompanied by manganese, iron oxides, galena and spaherite. Traced downwards, the middle sections of the vein are barren, but payable mineralization occurs in the lower parts of the vein, near the foot of Force Crag. Here, galena and spaherite become important constituents of the vein with barite usually present in subordinate amounts.

At the ground surface of the site there are a variety of mine building and plant in various stages of dereliction, which have a degree of archaeological value. Around and above the buildings are extensive tips of mine waste and debris fans of natural scree derived from the hill side above the site.

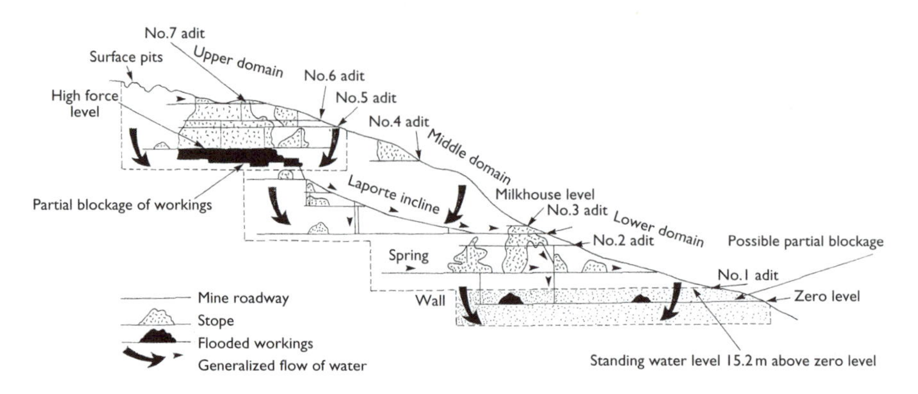

Figure 4.18 Schematic cross section showing mine workings and hydrogeological domains, Force Crag Mine, Cumbria, England.

A tailings lagoon is situated adjacent to the Coledale Beck. The present day state of the mine roadways, inclines and open stopes varies considerably. Some are directly accessible and still in a comparatively safe condition, whilst other parts of the mine have collapsed and are choked with fallen rock. Yet other parts of the mine are accessible but in a dangerously unstable condition.

Levels (adits) were driven into the hillside with a slight gradient to allow water to flow out. There are seven main levels that have surface entrances, some are accessible, others are blocked by the collapse of the roof materials at or just inside the level entrances (Fig. 4.18). The levels mostly were driven for considerable distances along the strike or adjacent to the vein. At intervals cross-cuts of similar dimension to the main levels or adits were driven more or less at right angles to the strike of the vein. The majority of the levels and roadways within the mine were driven in solid rock and are self-supporting, apart from where friable clay or fault gouge is intersected, or where unconsolidated superficial scree and mine waste lie close to or at the roof of the levels. Here, the levels are supported by timber and some steel props, and archways. In common with many other metalliferous mines, the various levels were connected by vertical shafts. These were driven for access, for ventilation and as ore passes.

The minerals were mined at Force Crag from a series of roadways, shafts, rises and stopes. The stopes were extracted by blasting, resulting in large open cavities, partially filled with broken ore. These may, or may not be supported, depending upon the conditions of the wall rocks and the presence of shale or fault gouge. Platforms, built systematically by miners, enabled the upper parts of the stope to be reached.

Parts of the mine are partially or completely flooded. The walls, roof and floor of the mine are, in places coated with ochre, in some parts of the mine clay-rich ochre reaches 1.5 m thick, delineating a high water mark above head height and occasionally reaching the roof. In other parts of the mine, the flooded drivages contain a blue-green discoloured water, caused by dissolved copper mineral salts. Ochreous mine water issues from some of the adit entrances. The discharge from No. 1 Level is particularly ochreous, orange precipitates occurring on the base of its channel, which also affect the Coledale Beck for some distance downstream of the confluence point. The collapse of some of the mine entrances has resulted in the build-up of mine water and there is concern that this may develop a sufficient head of pressure to breach the collapsed zones. Should this occur, it may further pollute Coledale Beck.

There are several environmental considerations that need to be addressed, although these are typical of small-scale abandoned metalliferous mines, their impact on the local environment is potentially damaging. As referred to, these primarily include discharges and accumulations of mine water that may cause pollution of Coledale Beck; risks to livestock and the public arising from the instability of both underground and surface workings; and stability problems associated with the potential failure of spoil heaps.

To address the environmental concerns the methodology adopted involved a desk study including an inspection of geological maps and mine abandonment plans. Full underground surveys of the mine workings were undertaken to ascertain their condition, as were surveys of all spoil heaps, scree and the tailings lagoon. Furthermore, surveys of all water courses that may flow into or out of the mine workings via adits, open stopes, or seepages beneath tips and scree were carried out. Water samples were taken and analysed to establish the chemistry of the inflowing and outflowing water courses, ponds and tailings lagoon in order to assess the degree of pollution in the vicinity of the mine. Hydrogeological assessments of the mine recharge-discharge system facilitated the determination of the quality and quantity of water entering and leaving the mine, as well as the routes and pathways involved. An assessment of the mine atmosphere included the identification of any hazardous gases particularly carbon dioxide, hydrogen sulphide and radon. Finally the stability of mine entries, roadways and stopes was determined, and a record of cases of subsidence and formation of crown holes was made.

When examined the timber and steel roof supports in the mine showed variable signs of deformation and bowing. Severe roof failures have occurred throughout the mine complex. Roof falls and slabbing occur adjacent to the mineralized zone, in open stopes, where the vein is barren, in heavily fractured rock and in weak fault gouge. Hence, evidence for subsidence, ground movements and the complete collapse of mine workings, resulting in crown holes, is abundant on the hill slopes throughout the mine area (Fig. 4.19(a)). The largest crown hole measures 6.0 m in diameter. Large active subsidence depressions, accompanied by ground compression and fissures are present above some of the level or adit entrances (Fig. 4.19(b)). The subsidence depressions are circular in plan, but asymmetrical in section, the flank of the upslope being the steeper. Further failures have occurred caused by the rotational slip of the thick overburden forming the sides of the crown holes and so extending their diameter. Some of the crown holes have been fenced off to prevent accidental access, being located close to a footpath, within a popular area of the Lake District National Park. There is the possibility for new crown holes to develop, at

Figure 4.19 (a) Force Crag Mine showing a chain of subsidence depressions above the underground position of the stope. (b) Crown holes associated with Force Crag Mine.

or close to public footpaths, and therefore regular inspection was recommended to observe signs of precursory ground movements. Almost vertical gullies occur along the outcrop of the Force Crag Vein. These have not been formed by the collapse of underground workings, but represent areas of historical surface extractions, and weathering and erosion along the faults.

The spoil heap material around the mine site consists of angular rock fragments of varying size, composed of the Skiddaw Group slates, siltstones and sandstones, with variable amounts of ore. Instability has occurred in the form of sliding of large boulders, surface sheet wash at times of prolonged heavy rain, gravitational rotational sliding, and fissuring at the heads of recent slip scars. Potentially polluted groundwater accumulations occurred at the toes of the spoil heaps.

Water samples were taken from surface streams and mine water discharges. In general, water from the higher streams was acid. The mine water discharges indicated high zinc and elevated concentrations of manganese, lead, copper and cadmium, with other metals present in low concentrations. It is likely that the precipitation of metals has occurred within the mine ochre sludges. Zinc concentrations exceed the environmental quality standards throughout the length of Coledale Beck, the mine water discharges exacerbating the pollution significantly. Sudden surges of excessive water throughout the mine complex during times of wet weather and snow melt could cause additional pollution of water courses.

4.6. Case history 2

The Standard Bank Administration Building in Johannesburg, South Africa, is a six-storey building and the site on which it stands is underlain by the Upper Witwatersrand Quartzites, which are gold bearing and in which a dolerite (diabase) dyke has been intruded, trending from east to west (Fig. 4.20). The quartzite at the site generally is coarse grained and heavily jointed. It dips southward at about 80°. The tabular conglomerates of the Main Reef Leader and the South Reef also run east to west across the site and the quartzite between the two is appreciably weaker and more friable than either the hangingwall or footwall quartzites. The dolerite is highly weathered.

Mining in the area reached a peak in the 1890s, and was initially from the surface down dip, with crown pillars at surface, and other pillars underground below each development level to ensure stability. The Main Reef Leader and the South Reef were worked in the area and average stoping widths in these two reefs were respectively 1.3 m and 1.4 m. The former had been extensively mined before the commencement of the twentieth century.

Since the shallow mining was completed mainly before the turn of the twentieth century, no detailed records of the extent of the mining were available. Many of the old stopes had been filled with waste rock, rubble and refuse from the surface. This fill, however, often has been subject to ravelling,

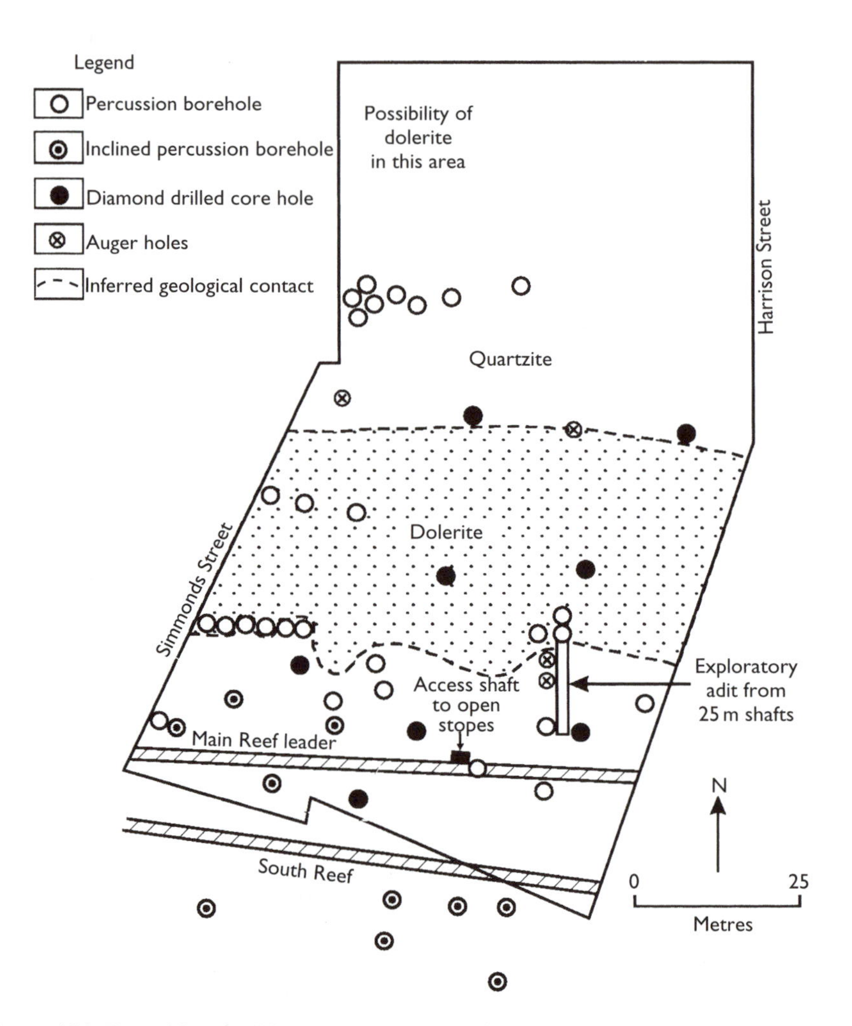

Figure 4.20 Plan of Standard Bank site showing surface geology.

frequently aided by run-off from rainfall, so that fill has moved to greater depths within the stopes to leave open cavities in the stopes at shallow depth.

It was decided to investigate the condition of the Main Reef Leader stope to a depth of 50 m and the condition of the footwall surface. Accordingly, a shaft was sunk to gain access to the workings in the Main Reef Leader stope. In fact, after sinking the shaft a short distance it was found to correspond with what was possibly an old exploratory shaft. A strong updraft of air in the old shaft indicated that it was in connection with workings at greater depth. Indeed, at the bottom of the old shaft it opened into the Main Reef Leader. This reef had been mined out almost to the surface, leaving only a 3 m crown pillar of solid rock. The stope was almost vertical, and had been tip-filled with quartzite rubble and other material from a crown hole close to the eastern boundary of the site. Exploration of the old workings was carried out to a depth of 46 m below surface (Fig. 4.21(a) and (b)). Access to the footwall rocks to examine the gouge-filled joints was effected

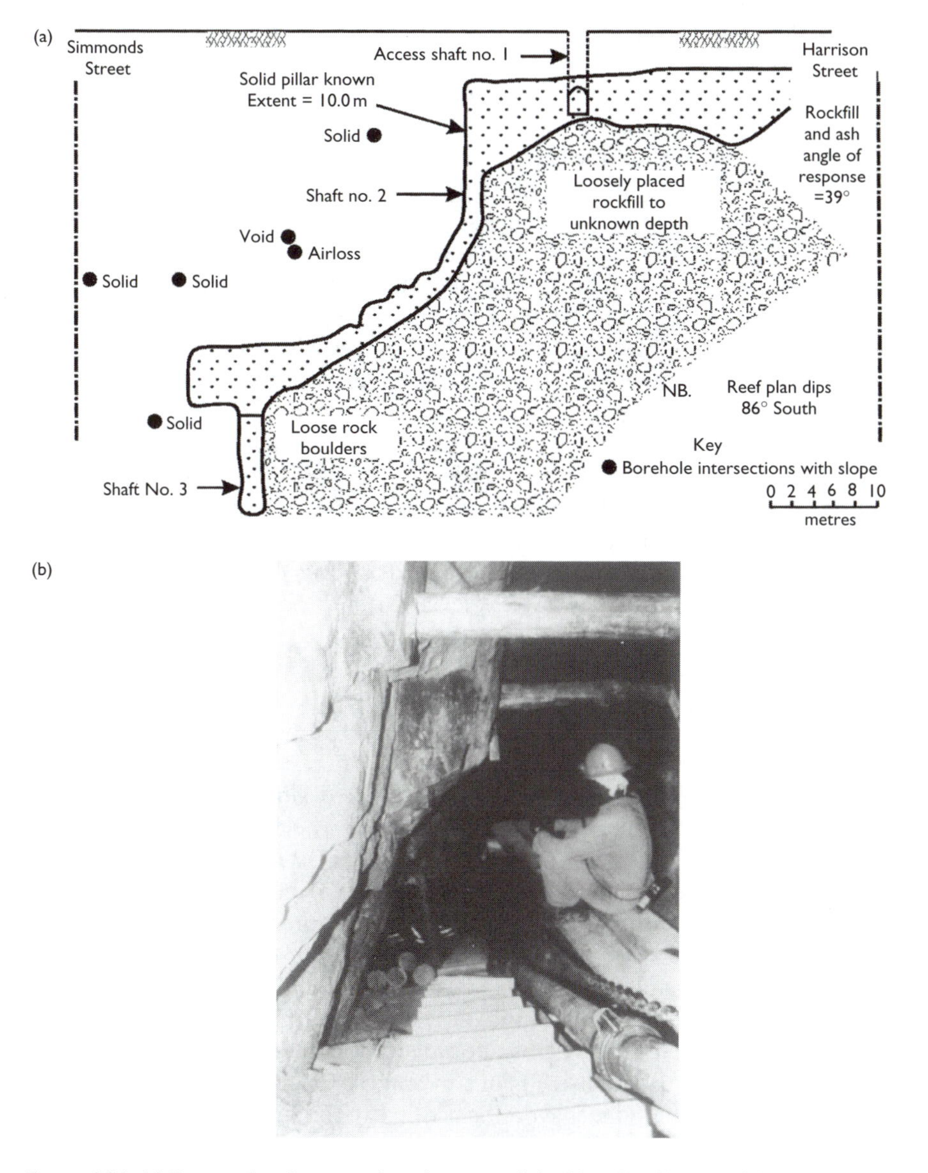

Figure 4.21 (a) Route of underground exploration of the Main Reef Leader. (b) Investigation of stopes via adit, Standard Bank site.

by augering a 25 m deep shaft in the weathered dolerite dyke and driving an adit back towards the stope. This also provided an opportunity to study the *in situ* condition of the quartzite–dolerite contact. This zone was very blocky and the joints exhibited appreciable evidence of slickensiding. The adit followed one of the major joints and it was occupied by clay gouge. The joints were very wavy and had an average dip of 5°, with a maximum of 25°. Because the dip was shallow this meant that large scale instability in the footwall was unlikely. This, together with the presence of the solid pillar, suggested that a remedial solution could be designed that would permit development of the site.

It was decided that measures were needed to provide rigid support between the footwall and hangingwall so that if some collapse or instability occurred in the mine workings at greater depth, this near-surface rigid zone would form an arch. Concrete dip pillars were to provide the rigid in-stope support. Two-dimensional displacement discontinuity stress analyses were used for most design calculations. However, three-dimensional mining simulation techniques with displacement discontinuity elements were used to determine the layout and size of the concrete pillars that it had been decided to use (Diering, 1980). The final design, as illustrated in Fig. 4.22, was developed from a number of analyses of a single stope. Simulation of regularly spaced pillars sunk to a depth of 60 m beneath the surface was followed by the removal of the rock crown pillar. This indicated that the stress in the boundary pillar, together with that in the lower third of the other pillars, could develop potentially unacceptably high levels (Fig. 4.22(a)). The calculated stress levels in the upper section of the inner pillars were very low. Accordingly, the model was modified, as shown in Figure 4.22(b), the intermediate pillars being shortened so that they extended to a depth of 35 m and a horizontal strike pillar linked the bases of the deeper (60 m) pillars. This led to much more acceptable levels of stress in the pillars.

Construction of the remedial works involved sinking winces (inclined shafts) 2 m wide and up to 8 m apart, along the line of the old stopes (Fig. 4.23). The stopes were loosely filled with quartzite rubble. Fortunately, however, small solid reef pillars had been left just above the second level (60 m) that corresponded to the required depth of the deeper pillars. The stopes were open along the full strike length just below this level. Some precariously supported rubble backfill formed the bases of these cavities that, after choking holes where some of the rubble had disappeared into the stope below, could support the construction of strike pillars by pouring in concrete grout in stages. Once the strike pillars had been completed, the winces were backfilled with concrete. The immediate footwall and hangingwall surfaces at the pillar locations were checked to depths of 3 m for open joints and fractures. If found, these were pressure grouted to ensure the integrity of the rock at the pillar locations.

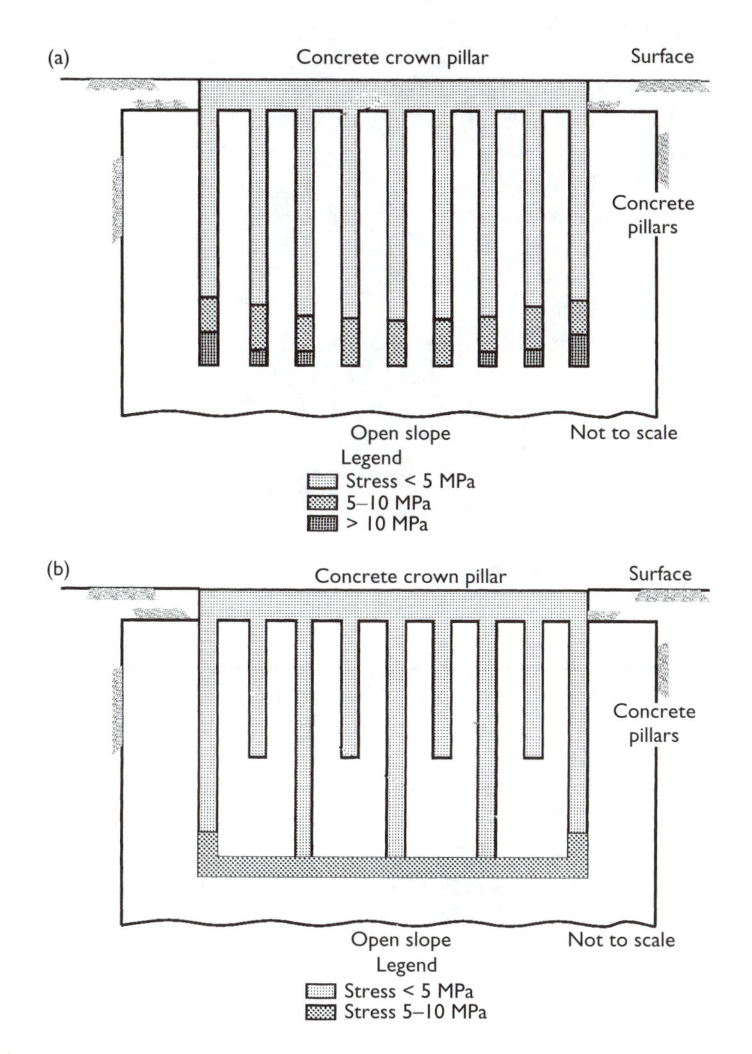

Figure 4.22 (a) Initial simulation of concrete dip pillars. (b) Simulation of final configuration of concrete pillar support.

4.7. Case history 3

When displacement occurred along the Johannesburg–Germiston railway line not far from the city boundary, it was essential to determine the cause of the movement as soon as possible. It was known that mining had taken place beneath the area affected and consequently subsidence was suspected (Fig. 4.24). In fact, mining had ceased in 1948 but reclamation mining, that is, extraction of remnants of solid reef, subsequently occurred.

Figure 4.23 Sinking of winces on stopes at Standard Bank site.

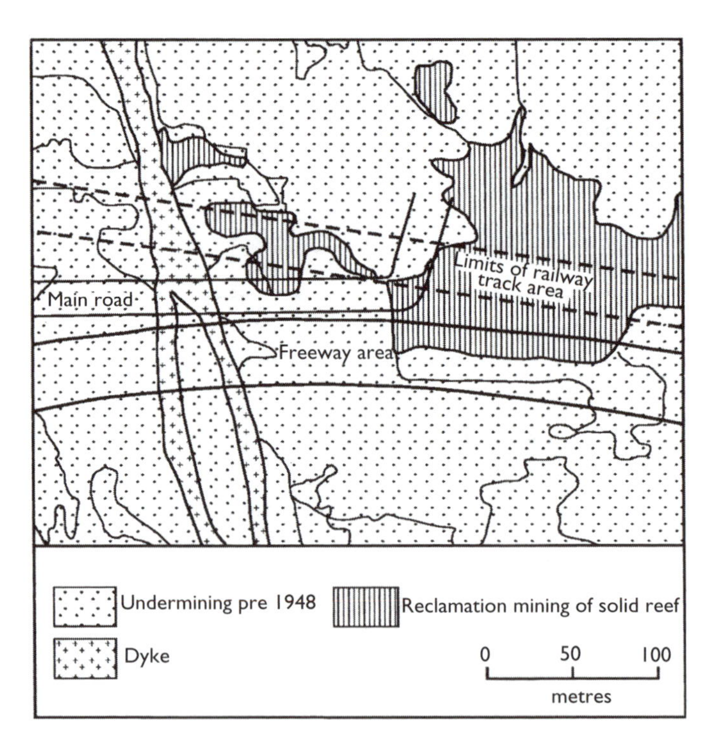

Figure 4.24 Plan of area in which subsidence occurred.

Hand-packed waste stone walls had been placed to provide support in those areas where reclamation mining was undertaken. Indeed, such stone walls have to be installed over at least 50% of an area of reclamation mining. In this case the area covered by stone walls appeared to exceed 75%. Nonetheless, such stone walls represent a passive form of support so that some degree of convergence of stopes takes place before reaction is developed. In some instances, because of the poor quality of the stone packing, up to 50% convergence could occur in order to develop the full load. Accordingly, this means that strata in the hangingwall can have significant freedom of movement.

Inspection of the mine plans for the area revealed that the South Reef, Main Reef Leader and Main Reef had been worked. These reefs are inclined at an angle of about 35° to the south. The South Reef is the shallowest, occurring some 200 m beneath the railway. The Main Reef Leader occurs approximately 4.5 m below the South Reef and, in turn, overlies the Main Reef by some 3 m. The heights of the stopes in the three reefs are 0.85, 0.6 and 0.75 m, respectively. In addition, the mine plans showed that a dyke, about 20 m in width with a north-south strike, occurred in the area where the displacement had occurred. The dip of the dyke is almost vertical and slightly to the east. Furthermore, the dyke would appear to have been

Figure 4.25 Cracks developed on the north bank of the M2 motorway due to subsidence.

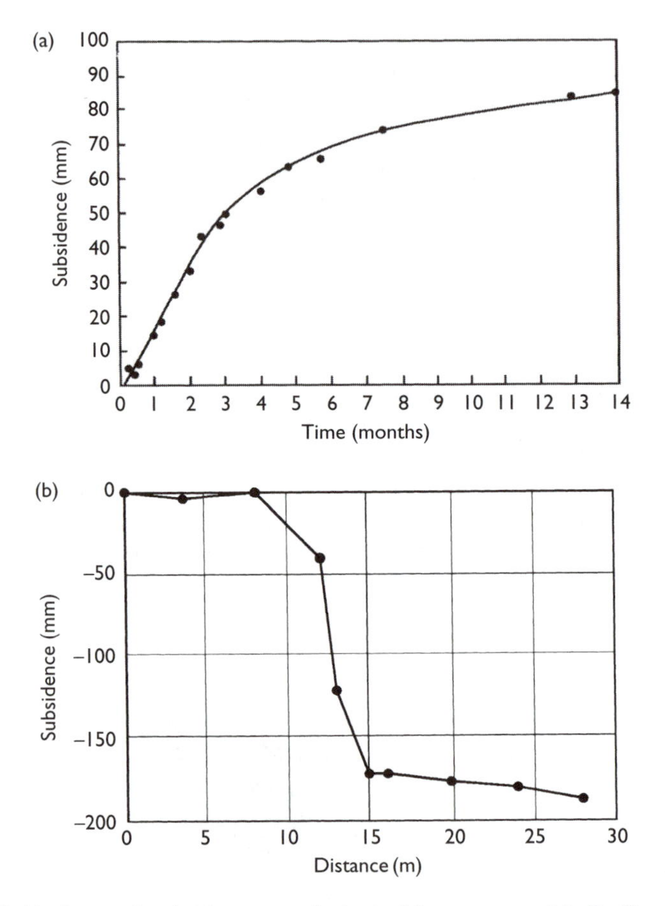

Figure 4.26 (a) Rate of subsidence on faulted dyke contact. (b) Profile of surface subsidence across faulted dyke contact.

intruded along a fault as the outcrops of the reefs have been displaced on either side of it. Hence, the dyke contact beneath the railway was examined in the stopes on the lower two reefs. Examination revealed a sharp contact with only a small amount of gouge.

A site investigation, involving trenching and sinking large diameter auger holes, was carried out in order to locate the dyke and assess its condition near the surface, as well as to determine whether there was any evidence of depressions that may be associated with past mining. The dyke was located and the area where movement had taken place seemed to correspond with its eastern contact. No surface depressions were found. However, cracks were found in an adjoining road and in a drain on the northern side of the

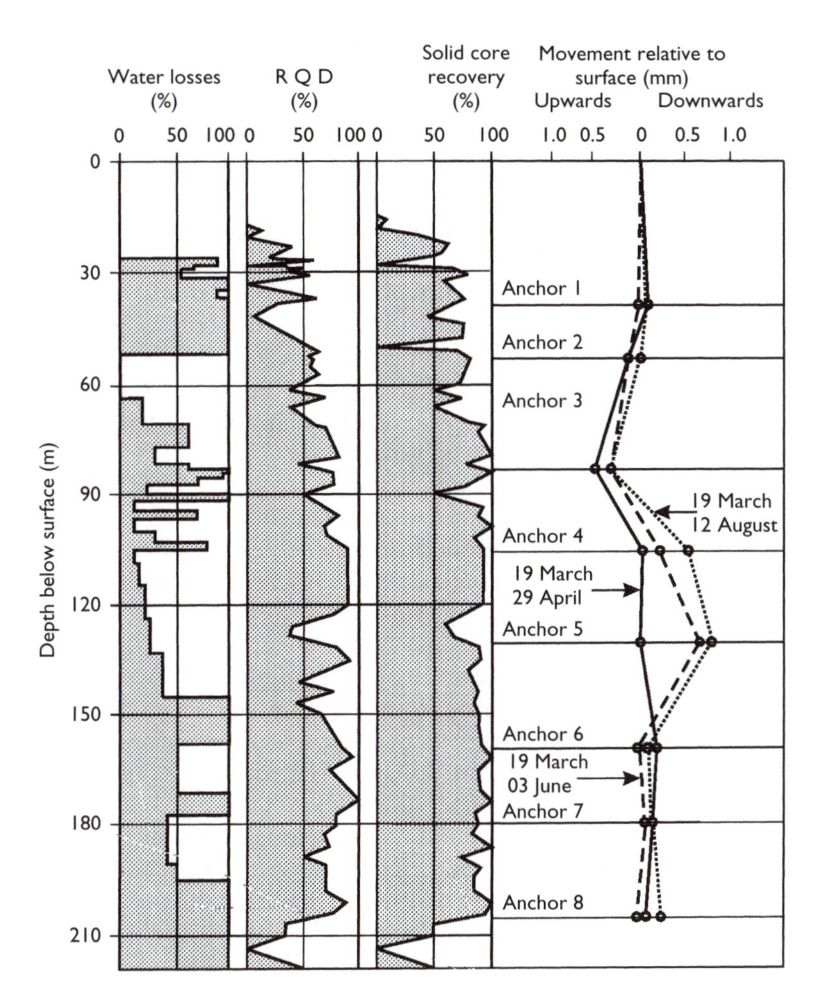

Figure 4.27 Results of drilling and extensometer monitoring.

Table 4.5 Tilting in relation to dyke contact along Johannesburg–Germiston railway (After Stacey and Bell, 1999)

Distance from dyke (m)	*Tilt ($\times 10^{-3}$)*
0–5	0
5–10	27
10–15	31.1
15–20	11
20–25	11

rail tracks. In addition, there were numerous cracks in the concrete footings of the old overhead gantries that supplied electricity to the trains.

Subsequently, surface cracking developed and could be traced some 250 m north of the railway reserve to a nearby motorway (Fig. 4.25). Surface monitoring by simple levelling showed that the dyke itself had not moved but that movement had taken place east of the dyke. Initially, the rate of subsidence was around 0.5 mm per day but this declined to around 0.1 mm per day (Fig. 4.26(a)). It was estimated that between 50 and 100 mm of subsidence had occurred, within a short period, prior to monitoring so that the total subsidence over a year was something like 150–200 mm. The subsidence profile in Figure 4.26(b) shows the abrupt nature of the movement. Such a step tends to be characteristic of subsidence associated with faults, and this dyke, as mentioned, was intruded along a fault.

Due to the depth at which mining took place, together with the lack of evidence of movement in the two lower stopes, it was proposed that as a consequence of collapse in the South Reef stope cavities may have developed in the rocks above. Although core drilling revealed fracture zones at shallow depth, no cavities of consequence occurred above 210 m. Two cavities, 0.35 m and 1 m in extent, were found beneath this level. An extensometer was installed in a drillhole to monitor the movements in the hangingwall. It showed that strata in the hangingwall were not moving *en mass* but that relative movements were taking place, giving rise to bed separation (Fig. 4.27). It was concluded that movement was not likely to occur over large areas of the dyke contact at any particular time but that further localized step subsidence could continue for a long period as there still are cavities at depth. Subsequent monitoring by conventional levelling revealed that tilting had taken place near the dyke contact as shown in Table 4.5. The tilt in the 10–15 m range was directly related to the dyke contact and its value exceeded the severe level both for roads and for railways in category 2 (Table 4.4). This entailed local resurfacing of the motorway and reballasting of the railway track to correct levels.

References

Barton, N., Lien, R. and Lunde, J. 1975. *Engineering Classification of Rock Masses for the Design of Tunnel Support*. Publication 106, Norwegian Geotechnical Institute, Oslo.

Bieniawski, Z.T. 1989. *Engineering Rock Mass Classifications*. Wiley-Interscience, New York.

Brose, R.J. 1996. Subsurface mine reclamation for urban construction. *Environmental and Engineering Geoscience*, 2, 73–83.

Brumleve, C.B. and Maier, M.M. 1981. Applied investigations of rock mass response to panel caving, Henderson Mine, Colorado, USA. In: *Design and Operation of Caving and Sub-level Stoping Mines*, America Institute of Mining Engineers, New York, 223–249.

Carter, T.G. and Miller, R.I. 1995. Crown pillar rish assessment – planning aid for cost effective mine closure remediation. *Transactions Institution Mining and Metallurgy*, 104, A41–A57.

Diering, J.A.C. 1980. An improved method for the determination, by a MINSIM type analysis, of stresses and displacements around tabular excavations. *Journal of the South African Institution of Mining and Metallurgy*, **80**, 425–430.

Diering, J.A.C. and Laubscher, D.H. 1987. Practical approach to numerical stress analysis of mass mining operations. *Transactions of the Institution of Mining and Metallurgy*, **96**, Section A, Mining Industry, A179–A188.

Dumpleton, S., Donnelly, L.J. and Young, B. 1996. *A Survey of Force Crag Mine, near Keswick, Cumbria*. Report No. WE/96/34, British Geological Survey, Keyworth, Nottingham.

Evans, A.M. 1993. *An Introduction to Ore Geology and Industrial Minerals*. Third Edition, Blackwell Scientific Publications, Oxford.

Hartman, H.L. 1987. *Introductory Mining Engineering*. Wiley, New York.

Hedley, D.F. 1978. Design guidelines for the multi-seam mining at Elliott Lake. *Canadian Center for Mineral and Energy Technology*, CANMET, Report No. 78–9, Ottawa.

Herman, S., Rautenbach, M.J. and Stuart, A.J. 1992. The use of a cable mat over shallow undermining on route K90, Boksburg. *Proceedings of the Symposium on Construction over Mined Areas*, Pretoria, South African Institution of Civil Engineers, 173–182.

Hill, F.G. 1981. The stability of the strata underlying the mined-out areas of the central Witwatersrand. *South African Institution of Mining and Metallurgy*, **81**, 145–160.

Hoek, E. 1974. Progressive caving induced by mining an inclined orebody. *Transactions of the Institution of Mining and Metallurgy*, **83**, Section A, Mining Industry, A133–A139.

Hutchinson, D.J., Phillips, C. and Cacante, G. 2002. Risk considerations for crown pillar stability assessment for mine closure planning. *Geotechnical and Geological Engineering*, **20**, 41–63.

Ortlepp, W.D. and Nicoll, A. 1964. A case history of subsidence resulting from mining at considerable depth. *Transactions Institution Mining Metallurgy*, **65**, 214–235.

Panek, L.A. 1981. Ground movements near a caving stope. In: *Design and Operation of Caving and Sub-level Stoping Mines*, America Institute of Mining Engineers, New York, 329–354.

Parry-Davies, R. 1992. Consolidation of old mine workings. *Proceedings of the Symposium on Construction over Mined Areas*, Pretoria, South African Institution of Civil Engineers, 223–227.

Stacey, T.R. 1986. Interaction of underground mining and surface development in a central city environment. In: *Rock Engineering and Excavation in an Urban Environment*, Institution of Mining and Metallurgy, London, 397–404.

Stacey, T.R. and Bakker, D. 1992. The erection or construction of buildings and other structures on undermined ground. *Proceedings of the Symposium on Construction over Mined Areas*, Pretoria, South African Institution of Civil Engineers, 282–289.

Stacey, T.R. and Bell, F.G. 1999. The influence of subsidence on planning and development in Johannesburg, South Africa. *Environmental and Engineering Geosciences*, **5**, 373–388.

Stacey, T.R. and Meintjes, H.A.C. 1994. Application of numerical modelling to assess potential movements resulting from shallow undermining. *Proceedings of the Symposium on Numeric Modelling in Geotechnical Engineering*, SANGORM, Pretoria, 115–119.

Stacey, T.R and Rauch, H.P. 1981. A case history of subsidence resulting from mining at considerable depth. *Transactions of South African Institution Civil Engineers*, **23**, 55–58.

Stacey, T.R. and Swart, A.H. 2001. *Practical Rock Engineering Practice for Shallow and Opencast Mines*. The Safety in Mines Research Advisory Committee (SIMRAC), Johannesburg.

Stanley, C.J. and Vaughan, D.J. 1982. Copper, lead and cobalt mineralization in the English Lake District: classification, conditions of formation and genesis. *Journal of the Geological Society*, **44**, 257–260.

Szwedzicki, T. 1989. Pillar recovery at Mhangura Copper Mines, Zimbabwe. *Transactions of the Institution of Mining and Metallurgy*, **98**, Section A, Mining Industry, A127–A136.

Szwedzicki, T. 1999. Sinkhole formation over hard rock mining areas and its risk implications. *Transactions of the Institution of Mining and Metallurgy*, **108**, Section A, Mining Industry, A27–A36.

Touseull, J. and Rich, C. 1981. *Documentation and Analysis of a Massive Rock Failure at Bautsch Mine, Galena, Illinois*. United States Bureau of Mines, Report No. 8453, Washington, DC.

Chapter 5

Abstraction of fluids and subsidence

Subsidence associated with the abstraction of fluids has occurred in many parts of the world, for example, according to Zhou *et al.* (2003) at present it affects over 150 areas. They went on to mention that such subsidence ranges up to 10 m or more and that the areas affected vary in size from several square kilometres to over 10 000 km^2. In many developing countries where the demand for groundwater is increasing continuously this presents a serious geoenvironmental problem because of the conflict between economic development and environmental protection. Land subsidence due to groundwater lowering represents a very complicated system consisting of various kinds of spatially distributed information such as geology, terrain, land use, precipitation, evapotranspiration, large aquifer systems, hydrological parameter distribution, groundwater abstraction, groundwater flow and groundwater recharge. Frequently, such subsidence has taken time to recognize and to do something about. Consequently, millions of pounds worth of damage may have occurred during the time elapsed. Subsidence most frequently is associated with groundwater withdrawal for supply for domestic, industrial or agricultural purposes but it may be caused by groundwater lowering in relation to mining, quarrying operations, land drainage or in relation to the production of geothermal energy. However, subsidence also is associated with the abstraction of oil, gas and brines.

A brief review of the history of the aquifer-drainage model of land subsidence attributable to the withdrawal of fluids, as developed in the United States, has been provided by Holzer (1998). The model maintained that subsidence due to the abstraction of fluids from aquifer systems is the result of non-recoverable consolidation of slowly draining fine grained layers, that is, aquitards. Subsequently, it was recognized that the same effect could be produced in sands with a resultant reduction in groundwater level. In fact, the concept was initially proposed by Pratt and Johnson (1926) to explain the subsidence associated with the Goose Creek oil field in Texas. Two years later, Meinzer (1928) concluded that the concept of rigid aquifers was incompatible with field evidence. In other words, an aquifer becomes compressed when groundwater is withdrawn, this being due to a reduction in internal support consequent on a decline in pore water pressure. This was similar to the concepts being advanced by Terzaghi (1925) on consolidation and effective stress in relation to soil mechanics (see later). Hence, the conceptual development of the aquifer-drainage model is associated with subsidence that has taken place where fluids have been withdrawn from petroleum reservoirs, as well as aquifer systems. Subsidence in the Santa Clara Valley, California, a notable subsidence area, was recognized in the early 1930s. Tolman and Poland (1940) explained this subsidence in terms of the aquifer-drainage model. Tolman and Poland were aware of the work of Terzaghi. They also suggested that part of the consolidation was elastic and that this could lead to the partial recovery of subsidence. However, as far as the Santa Clara Valley is concerned most of the subsidence has been permanent. The development of deep well vertical extensometers in the 1950s located zones of consolidation in the Santa Clara and San Joaquin valleys, California, and were able to demonstrate the cause and effect relationship between decline in groundwater head and subsidence.

If groundwater is withdrawn at a faster rate than it is replenished by recharge, then the water table is lowered. This increases the load that has to be carried by the strata affected and so can lead to lowering of the ground surface (Chilingarian *et al.*, 1995). As can be inferred from earlier discussion, the mechanism responsible for subsidence is more or less the same when a decline in head of oil or gas occurs as they are abstracted. However, when highly soluble materials such as rock salt are exploited by solution mining, any associated subsidence primarily is due to the removal of the dissolved rock material. The components of subsidence associated with declines in fluid pressures frequently are difficult to predict accurately whilst those associated with some forms of solution mining are virtually unpredictable.

5.1. Subsidence and groundwater abstraction

Groundwater accounts for over 90% of the world's fresh water and represents a fairly constant source that is not likely to dry up under natural conditions, as surface sources may do in some parts of the world. It also is frequently of high quality. In addition, unlike the cost of construction of a large reservoir and dam that require a large initial investment of capital, a groundwater abstraction scheme can be put into operation stage by stage relating yield to demand. Because of these advantages groundwater is a resource of great importance and in some regions of the world is the primary, if not the only, source available. Even in a country with a humid climate like England, over 30% of the total water supply is satisfied by groundwater and in the United States this figures rises to 50%. Consequently, if subsidence problems due to excessive groundwater withdrawal are to be controlled, then the exploitation of the groundwater resource must be managed and resource development must become an integrated part of community planning.

As noted earlier, subsidence of the ground surface occurs in areas where there is intensive abstraction of groundwater, that is, where abstraction exceeds natural recharge and the water table is lowered, the subsidence being attributed to the consolidation of the sedimentary deposits as a result of increasing effective stress (Bell, 1988). The total overburden pressure in partially saturated or saturated deposits is borne by their granular structure and the pore water. When groundwater abstraction leads to a reduction in pore water pressure by draining water from the pores, this means that there is a gradual transfer of stress from the pore water to the granular structure. For instance, if the groundwater level is lowered by 1 m, then this gives rise to a corresponding increase in average effective overburden pressure of about 10 kPa. As a result of having to carry this increased load the fabric of the deposits affected may deform in order to adjust to the new stress conditions. In particular, the void ratio of the deposits concerned undergoes a reduction in volume, the surface manifestation of which is subsidence. The significance of the forces transmitted through the grain structure of a deposit was recognized by Terzaghi (1925, 1943), who advanced the classic relationship:

$$\sigma' = \sigma - u \tag{5.1}$$

where σ is the total stress, σ' is the effective stress and u is the pore water pressure. Figure 5.1 illustrates the effect of lowering the water table on the effective and pore water pressures. Scott (1979) pointed out that surface subsidence does not occur simultaneously with the abstraction of fluid from an underground reservoir, rather it occurs over a larger period of time than that taken for abstraction.

The amount of subsidence that occurs is governed by the increase in effective pressure (i.e. by the magnitude of the decline in the water table), the thickness and compressibility of the deposits involved, the depth at which they occur, the length of time over which the increased loading is applied, and possibly the rate and type of stress applied (Lofgren, 1968). For instance, the water content of normally consolidated clays frequently exceeds 50–60% and their shear strength usually is low, between 20 and 30 kPa. Consolidation in such clay of appreciable thickness may represent 10–25% of the reduction in groundwater level. The rate at which consolidation occurs depends on the thickness of the beds concerned, as well as the rate at which pore water can drain from the system that, in turn, is governed by its permeability. Thick slow-draining fine grained beds may take years or decades to adjust to an increase in applied stress, whereas coarse grained deposits adjust rapidly. For example, in surface aquifers composed

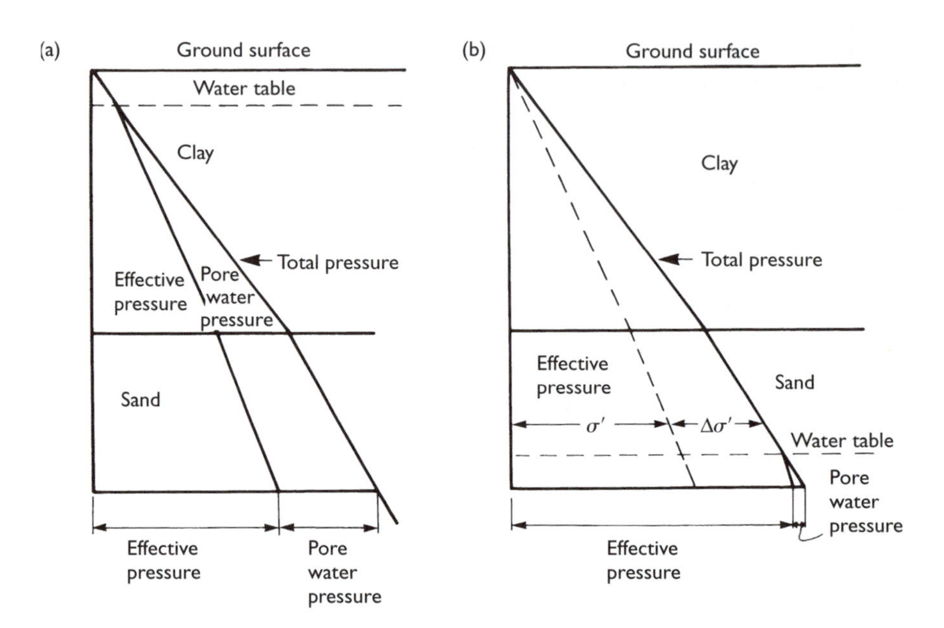

Figure 5.1 Pressures diagrams illustrating total, effective and pore water pressures. (a) Water table just below the surface. (b) Water table has been lowered into the sand and effective pressure is increased by an amount equal to the reduction in pore water pressure.

of sand and/or gravel the consolidation occurs rapidly as the increase in effective pressure is more of less immediate. However, the rate of consolidation of slow-draining aquitards reduces with time and is usually small after a few years of loading. In fact, the time required to reach a certain degree of consolidation of a clay layer normally increases with the square of the thickness of the layer.

Consolidation may be elastic or non-elastic depending on the character of the deposits involved and the range of stresses induced by a decline in the water level. In elastic deformation stress and strain are proportional, and consolidation is independent of time and recoverable. Non-elastic consolidation occurs when the grain structure of a deposit is rearranged to give a decrease in volume, that decrease being non-recoverable, that is, permanent. Generally, recoverable consolidation represents compression in the preconsolidation stress range, whilst non-recoverable consolidation represents compression due to stresses greater than the preconsolidation stress (Lofgren, 1979).

A formation is generally in equilibrium with the effective pressure prior to the water table being lowered by groundwater abstraction. As mentioned earlier, as the water table is lowered, the effective weight of the deposits in the dewatered zone increases since the buoyancy effect of the pore water is removed. This increase in load is transmitted to the deposits beneath the newly established level of the water table. If a confined aquifer is overlain by an unconfined aquifer, then the effective pressure is affected by changes in the level of the piezometric surface as well as in the position of the water table.

Subsidence is measured by periodic precision levelling of bench marks referenced to some stable datum outside the area affected. Use also can be made of marker points attached to structures and surface settlement plates. Photogrammetric techniques also have been used to monitor subsidence (Gubellini *et al.*, 1986).

Reference has been made in Chapter 2 to the use of laser and radar sensors on spaceborne platforms, such as the LIDAR system and InSAR, to produce high resolution (centimetre to metre) digital terrain models. Such systems send pulses from a spaceborne platform to the ground and measure the speed and intensity of the returning signal, and in this way changes in ground elevation can be mapped. The technique allows the production of subsidence maps by differentiating between images taken during two successive passes over the same area. Furthermore, maps showing the rates of subsidence, accurate to a few millimetres per year, can be drawn for periods currently up to a decade long. For example, Bell *et al.* (2002) reported using InSAR to investigate land subsidence in the Las Vegas area, Nevada. They were able to show that subsidence was located within four basins, each bounded by faults of Quaternary age, which controlled the spatial extent of the subsidence, along with the thickness of clay deposits. The maximum detected subsidence during the period April 1992–December 1997 was 190 mm. When Bell *et al.* compared the location of the subsidence basins with the distribution of pumping in the Las Vegas Valley they found that the subsidence was offset from the main areas where the groundwater abstraction was occurring. This they suggested may be due to heavy pumping up-gradient from compressible sediments in the subsidence areas. Bell *et al.* also were able to demonstrate from the InSAR data that the rate of subsidence had declined significantly since 1991 as a result of artificial recharge. In addition, data gathered by Global Positioning System (GPS) was used by Bell *et al.* in this survey. Sato *et al.* (2003) also described the use of GPS to determine land subsidence caused by compression of clay layers due to groundwater abstraction in Ojiya City, Japan. They were able to determine from their survey that the ground not only subsides but that some rebound also occurs during the winter months. Sato *et al.* also recognized that there was a good correlation between total strain (ratio of the height difference displacement to total thickness of clay layers) and the change in effective stress with change in groundwater level. Previously, Mes *et al.* (1998) had referred to the use of GPS to measure subsidence at offshore oil and gas platforms. The procedure is fully automated and provides reliable measurements of the magnitude and rate of platform subsidence. For instance, subsidence in the Ekofisk oil and gas field in the North Sea is now being monitored at platforms by GPS.

Borehole extensometers can be used to monitor the amount of consolidation suffered by subsurface deposits, the change in thickness of the deposits in the depth interval above some bench mark set in the formation being recorded. Figure 5.2 shows two types of borehole extensometers that have been used in California to measure movements in aquifers that occur as a result of groundwater pumping. Inclinometers can be used to observe horizontal ground movements.

According to Riley (1986) a number of requirements are necessary if borehole extensometers are to be operated successfully in relation to subsidence monitoring. First, the base of the extensometer system (i.e. the bottom hole anchor or subsurface benchmark) must be stable with respect to the base of the depth interval concerned. Second, the well casing must not act as a significant load bearing structure within the consolidating sedimentary column and so not disturb the system being measured. Third, the instrument platform, which forms the extensometer datum at the ground surface, must be stable with respect to the top of the interval of interest. Fourth, the extensometer pipe or cable must maintain a constant

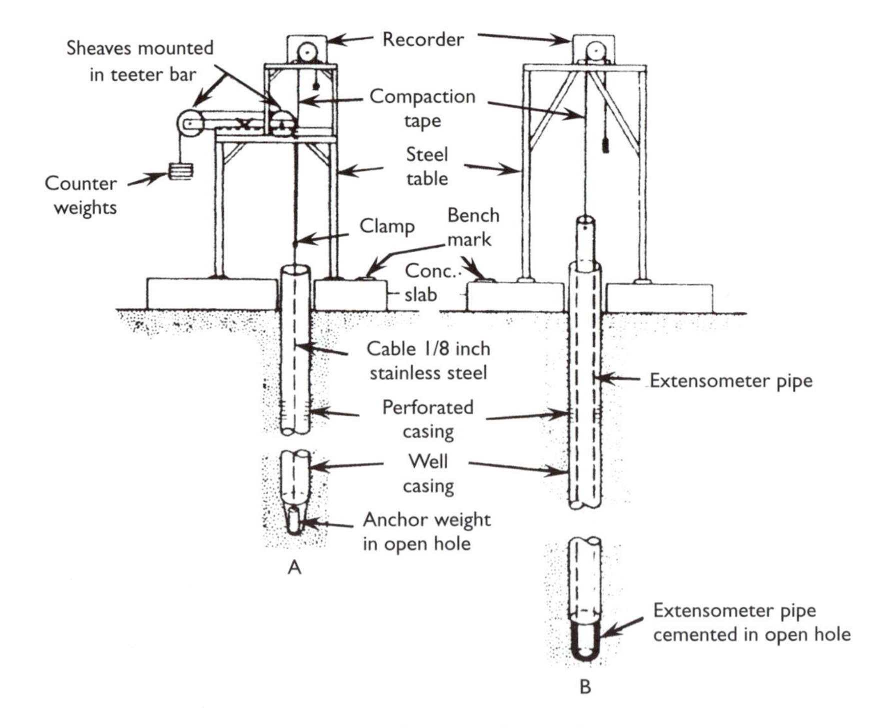

Figure 5.2 Borehole extensometer used to record ground movements.

length and the borehole must be as straight as possible in order to minimize downhole friction. Fifth, the means used above ground to support the extensometer pipe or cable (such as counterweights, levers or pulleys) must have negligible friction and must exert a constant uplift force on the pipe or cable irrespective of movement of the ground surface and instrument platform relative to the pipe or cable. Finally, the measurement and recording devices used to monitor displacement of the instrument datum (ground surface) relative to the extensometer pipe or cable must be linear, accurate, stable and sensitive, and must not impose significant frictional or spring loads on the extensometer. They also should provide sufficient redundancy to ensure that the record of cumulative displacement will not be lost during periods of routine maintenance, equipment failure or modification.

Long-term records from standard extensometers, in conjunction with those from piezometers, have made possible the determination of the properties of aquifer systems that control subsidence. Newer, extra high resolution extensometers permit definition of the compressibility and hydraulic conductivity of thin individual aquitards by means of short-term pumping tests. Downhole radioactive markers (i.e. weak radioactive bullets shot at regular intervals into the formation surrounding a newly drilled well) have been used to assess movements of reservoir consolidation by monitoring the relative displacement of the bullets.

Lofgren (1979) recorded that 20 years of precise field measurements in the San Joaquin and Santa Clara valleys of California had indicated a close correlation between hydraulic stresses induced by groundwater abstraction and consolidation of the water bearing deposits (Fig. 5.3). He went on to note that the stress-strain characteristics of the producing aquifer systems had been established and that the storage parameters of the systems had been determined from such data. What is more, this has provided a means by which the response of a groundwater system to future pumping stresses can be predicted. According to Lofgren, the storage characteristics of compressible formations change significantly during the first cycle of groundwater abstraction. Such abstraction is responsible for permanent consolidation of any fine grained interbedded formations, consolidation being brought about by the increase in effective pressure. The groundwater released by consolidation represents a one-time, and sometimes important, source of water to wells. For instance, Lofgren indicated that in subsidence areas of the San Joaquin Valley, the expelled water from consolidation ranged from a few percent to over 60% of groundwater abstracted. Over wide areas it averaged approximately one-third of total pumpage throughout the 50 years of groundwater overdraft. During a second cycle of prolonged pumping overdraft much less water is available to wells and the water table is lowered much more rapidly.

Methods that can be used to arrest or control subsidence caused by groundwater abstraction include reduction of pumping draft, artificial

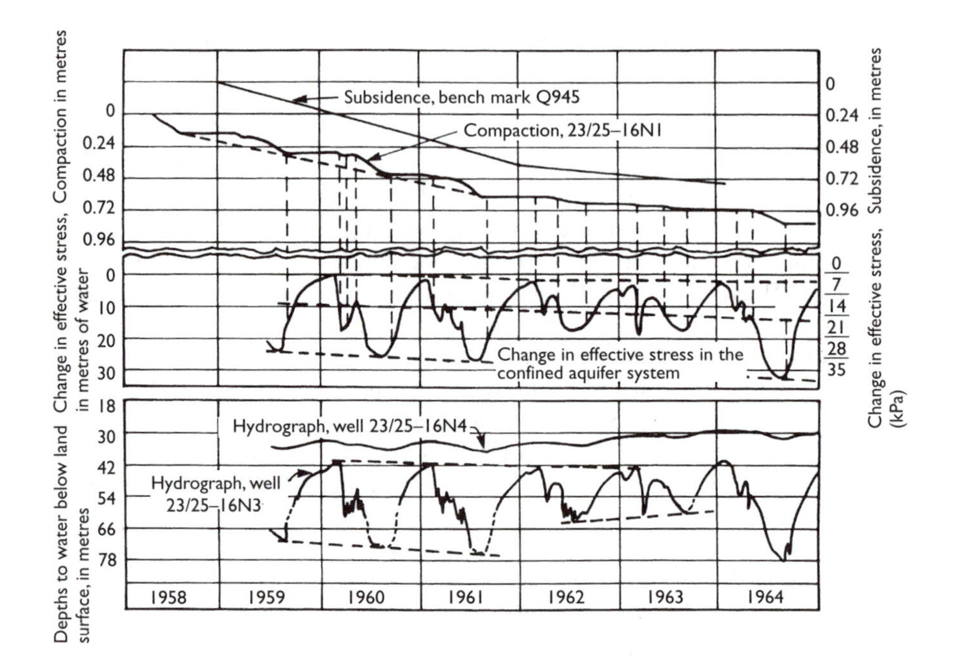

Figure 5.3 Land subsidence, consolidation, groundwater level fluctuation and changes in stress 4.8 km south of Pixley, California. (After Lofgren, 1968; reproduced by kind permission of the United States Geological Survey.)

recharge of aquifers from the ground surface and repressurizing the aquifer(s) involved via wells, or any combination thereof (Poland, 1972; Li and Helm, 2001). Reduction of pumping draft may be brought about by importing water from outside the area concerned, by conserving or reducing the use of water, by treating and re-using water, or by decreasing the demand for water. Some of these measures may require legal enforcement. The geological conditions determine whether or not the artificial recharge of aquifers from the ground surface is feasible. If confining aquitards or aquicludes inhibit the downward percolation of water, then such treatment is impractical. Artificial recharge can take place from trenches where the water table is at shallow depth. Although this is not likely to be satisfactory in relation to minimizing the effect of dewatering on water resources, it can be of use where dewatering is brought about by quarrying operations (Cliff and Smart, 1998). As far as artificial recharge from wells is concerned, the aim is to manage the rate and quantity of groundwater withdrawal so that its level in wells is either stabilized or raised somewhat so that the effective stress is not further increased. Repressurizing confined aquifer systems from wells may prove the only viable means of slowing down and eventually halting subsidence. Generally, the results of repressurizing aquifers from wells prove satisfactory when clean water is used; the problems that have occurred usually have involved clogging of well filters or aquifers. Some of the causes of clogging include air entrainment, micro-organisms and fine particles in the recharge water. An outline of the control measures taken to reduce subsidence due to groundwater abstraction in Taiwan has been provided by Pan (1997). These included the development of alternative surface water resources and recycling of cooling water. In addition, various water management measures were adopted such as issuing permission of the right to groundwater, the establishment of an inventory of existing pumping wells, prohibition of new wells being drilled, and setting up a groundwater monitoring network.

It is not only falling or low groundwater levels that cause problems, a rising or high water table can be equally troublesome. For example, since the mid-1960s the rate of abstraction from the chalk below London, England, has decreased significantly so that water levels in the 1980s were increasing by as much as 1 m $year^{-1}$ in places (Fig. 5.4; Marsh and Davies, 1983). The potential consequences of this include leakage into tunnels and deep basements, and a reduction in pile capacity. Similar problems exist in the Witton area of Birmingham, England, where factory basements have been flooded by rising groundwater in the Sherwood Sandstone aquifer. In central Liverpool, England, a similar problem affects railway tunnels by the water table rising, again in the Sherwood Sandstone. Similarly, in Louisville,

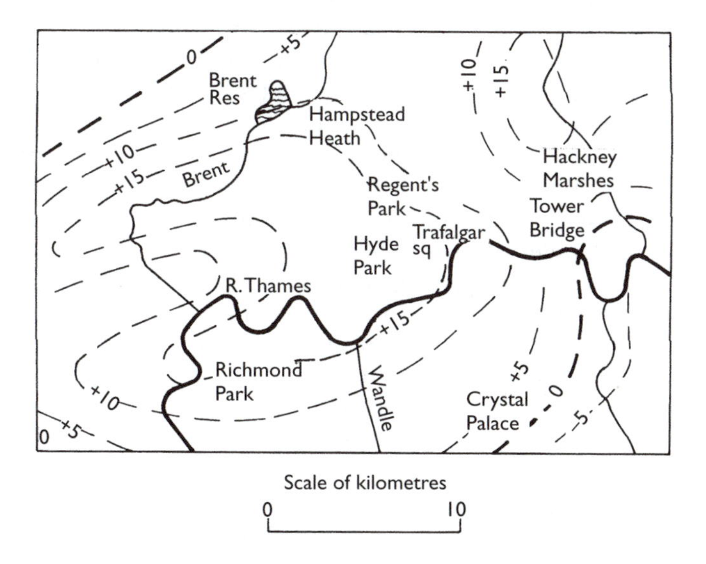

Figure 5.4 Changes in groundwater levels in the Chalk below London, 1960–1980 (contours in metres). (After Marsh and Davies, 1983; reproduced by kind permission of Thomas Telford Publishing.)

Kentucky, increased groundwater levels caused concern over the possibility of structural settlement, damage to basement floors and the disruption of utility conduits (Hagerty and Lippert, 1982).

The control of groundwater levels therefore has an importance that extends beyond water supply considerations. Clearly, if structures are built during a period when the water table is at an artificial or atypical level, then care must be taken to ensure that changing groundwater levels do not diminish the integrity of such structures. This obviously requires skilful long-term management of the groundwater resource to ensure that there are no large fluctuations in level. However, aquifers are sometimes deliberately over-pumped or over-mined during periods of water shortage. The usual assumption is that groundwater levels will recover during a following wet season and that no harm will be done. While this may be true generally, it is important that the risks involved in this type of operation are fully realized. Consequently, over-pumping should be carried out only infrequently, if at all, and only for short periods of time.

5.2. Some examples of subsidence due to the abstraction of groundwater

One of the classic areas where subsidence due to the withdrawal of groundwater has occurred is Mexico City. The Mexico City region consists of a series of flat plains formed on exceptionally porous deposits overlying a highly productive regional aquifer. Carillo (1948) revealed that in parts of Mexico City subsidence has occurred at a rate of 1 mm day^{-1}. This was due to the abstraction of water from several sand aquifers in very soft clay of volcanic, lacustrine origin. The aquifers extend under the city from an approximate depth of 50 m below ground to 500 m and groundwater has been abstracted since the mid-nineteenth century. Carillo noted that in 1933 the rate of withdrawal was 7 m^{3}s^{-1} from 2200 registered wells. The decline in head in the wells ranged from 0.4 to 2.05 m year^{-1}. Overall the piezometric level had fallen by some 30 m, corresponding to an increase in vertical effective stress of approximately 300 kPa. This was responsible for accelerating subsidence in the central area of the city. By 1959 most of the old city had undergone at least 4 m of subsidence, and in the north-east area as much as 7.5 m had been recorded (Fig. 5.5). This had serious consequences for both drainage and buildings (Figueroa Vega, 1976). The sewer system, for instance, which previously worked by gravity, now requires pumps. Buildings on end-bearing piles rose above the ground. What is more, the development of negative skin friction caused heavy overloading on piles that led to some sudden differential settlements. In 1953, however, a prohibition order was imposed whereby no more wells were to be sunk in the Valley of Mexico and subsequently there has been a slow reduction in the rate of groundwater withdrawal. However, according to

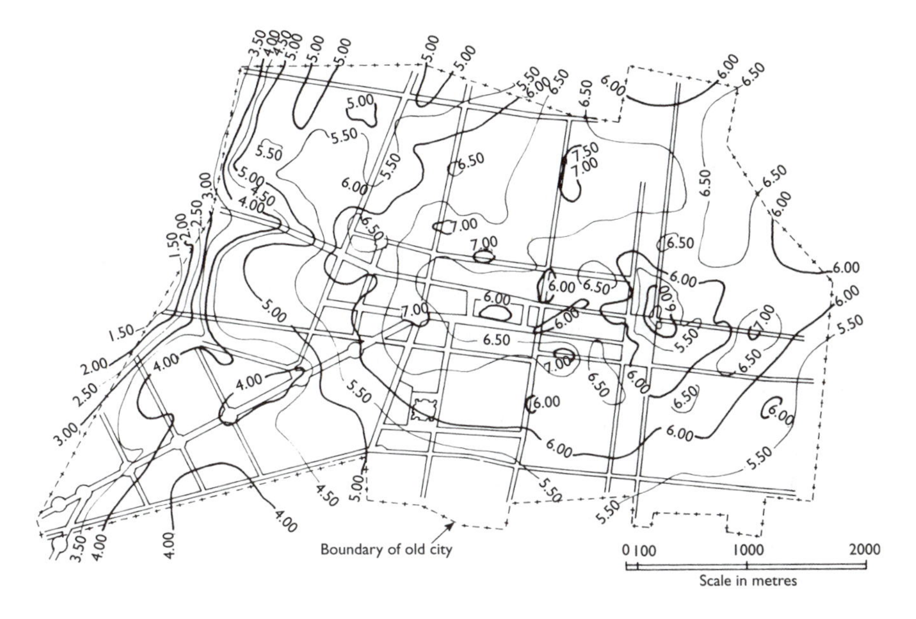

Figure 5.5 Subsidence due to the abstraction of groundwater in Mexico City, 1891–1959. Subsidence contours are given in half metre intervals.

Ortega-Guerrero *et al.* (1999), the restrictions on pumping led to increases in groundwater withdrawal in the outlying plains where communities have expanded rapidly. The Chalco Basin is one of these areas where pumping began in the 1950s and greatly increased in the 1980s. Furthermore, the lacustrine deposits in the Chalco Basin are significantly thicker than elsewhere in the Mexico City region, averaging 100 m and reaching a maximum of 300 m thick. Consequently, this area has the largest potential for land subsidence due to groundwater abstraction in the region. In fact, Ortega-Guerrero *et al.* mentioned that subsidence in the Chalco Basin had increased to 0.4 m annually and that by 1991 the maximum subsidence had reached 8 m. The subsidence means that farmland is flooded rapidly during the rainy season. If current pumping rates continue, then Ortega-Guerrero *et al.* estimated that by 2010 subsidence where the lacustrine deposits are thickest will amount to some 15 m. On the other hand, if the rate of groundwater abstraction is controlled so that there is no further decline in the water table, then by 2010 the maximum subsidence would be significantly less than 10 m.

Chi and Reilinger (1984) identified 48 localities in the United States that have experienced ground subsidence as a consequence of the withdrawal of groundwater. In fact, surface subsidence effects caused by groundwater withdrawal from unconsolidated sediments in the United States have affected approximately 22 000 km^2. Of these areas, the most notable examples have occurred in California and Texas. For instance, in the San Joaquin Valley of southern California large quantities of water have been abstracted from aquifers located from less than 30 m to over 200 m below the surface. Subsidence began in the San Joaquin Valley in the mid-1920s but the amount of subsidence remained small until after 1945. By 1970 over 13 500 km^2 of land had undergone subsidence, the maximum amount exceeding 8.9 m (Poland *et al.*, 1975). After 1970 the pressure on groundwater supplies was relieved as water became available from the Californian aqueduct. This allowed the water level to recover and by 1977 it had risen 73 m above the summer low of 1968. Unfortunately, the occurrence of notable droughts can mean that water supplies decline and so over-drafting of groundwater takes place (Anon., 1980). Because the water table had returned to its former level, it was believed that the water supply would be sufficient to last throughout a severe drought. In fact, the subsidence after one year of withdrawal associated with a severe drought equalled that which had taken place over the previous 40 years. Drawdown in the San Joaquin Valley takes place during a few months each year, for the rest of the time the water levels are recovering (Bull and Poland, 1975). Because of the cyclic nature of groundwater abstraction, the elastic component of storage change for a given net decline in groundwater level may be reduced and restored many times while the inelastic component is removed but once. A brief survey of subsidence in the San Joaquin Valley to 1995 has been provided by Swanson (1998).

Over 620 km^2 have been affected by subsidence due to the abstraction of groundwater in the Santa Clara Valley of California, where it reached a maximum in excess of 4 m. The cost of the associated damage amounted to more than $100 million (Anon., 1980). Fortunately, the over-drafting of groundwater in this area ceased in 1969 and there has been no further subsidence since then. Aquifers have been recharged from stored storm-water.

Thick unconsolidated lenticular deposits of sand and clay lie beneath the Houston-Galveston region in Texas. Beds of clay separate layers of sand retarding the vertical movement of water, and creating artesian conditions within the aquifers. The ratio of sand to clay, which is a major factor controlling consolidation, varies from place to place in the aquifer system (Gabrysch, 1976). Indeed, Delflache (1979) reported that the most notable subsidence in this region has occurred where the thickness of clay in the aquifer system is greatest, coupled with the largest declines in head. The use of groundwater in the Houston-Galveston region began in the 1890s. By 1974, abstraction of groundwater from the Chicot and Evangeline aquifers for municipal supply, industrial use and irrigation was about 23 m^3s^{-1}. Between 1943 and 1973 the water level in the Chicot aquifer was lowered by up to 61 m, whilst as much as 99 m of lowering occurred in the Evangeline aquifer. Pronounced regional subsidence was the result

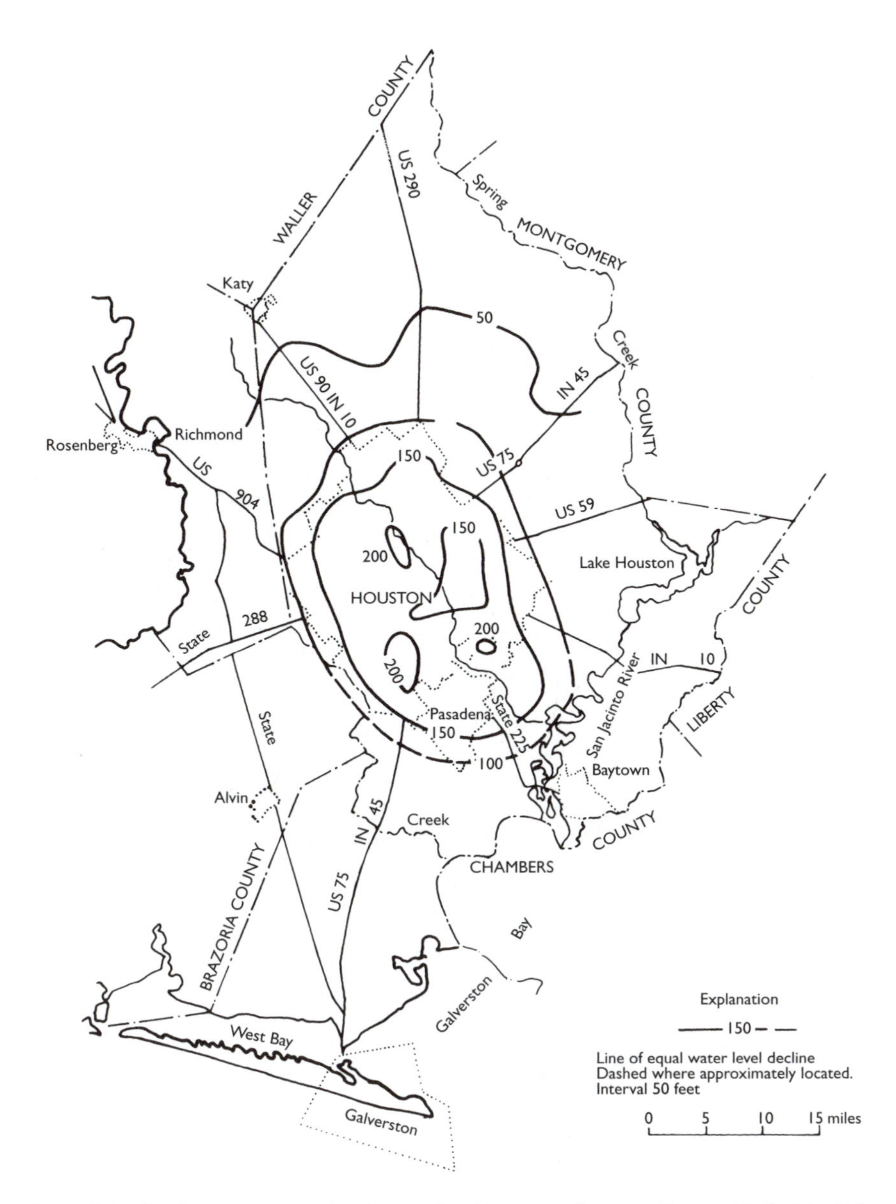

Figure 5.6a Decline in water levels in the Houston District, Texas, 1943–64. (After Gabrysch, 1969; reproduced by kind permission of the International Association of Hydrological Sciences.)

(Fig. 5.6(a), (b)). The area in which the subsidence was greater than 0.3 m increased from 906 km^2 in 1954 to about 6475 km^2 in 1973. By 1973 approximately 400 homes had been lost or damaged. Subsidence also has caused critical problems of inundation by normal tides and much of the

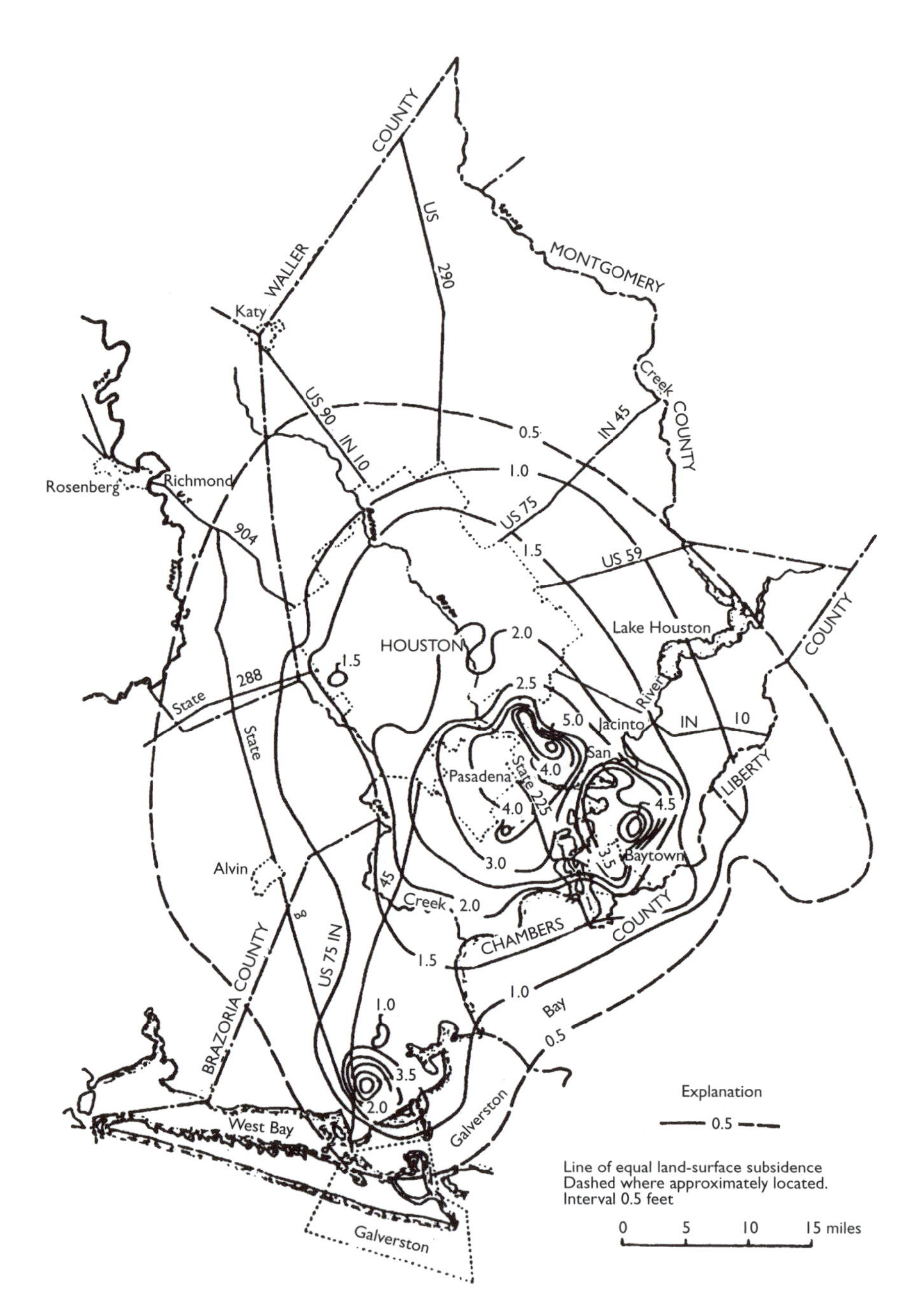

Figure 5.6b Subsidence of the land surface in the Houston District, Texas, 1943–64. (After Gabrysch, 1969; reproduced by kind permission of the International Association of Hydrological Sciences.)

region would be subject to catastrophic flooding by hurricane tides. In 1976 supplies of water became available from Lake Livingston. This has permitted less groundwater abstraction and hopefully the recovery of groundwater levels in the aquifers will lead to a substantial decrease in the rate of subsidence. Furthermore, the control of groundwater withdrawal was introduced in the Harris-Galveston region in 1975. A water management study undertaken in the Harris-Galveston Coastal Subsidence District in order to predict the effect of groundwater abstraction on surface subsidence has been described by Pollard and Johnson (1984). The first phase of study involved the compilation of data relating to subsidence, hydrogeology and water supply within the District. The second phase used the data to develop computer models to predict subsidence, the models being based on three possible scenarios of future groundwater abstraction. The study indicated that because of the restrictions placed upon groundwater withdrawal in the eastern part of the District, the rates of subsidence have been minimized. However, the models predict that the western part of the District will continue to subside by up to as much as 1.8 m in the centre of the area. Buckley *et al.* (2003) recently have claimed that subsidence east of Houston has ceased, however, west and north-west of Houston maximum subsidence rates are respectively in excess of 20 and 40 mm annually. This is consistent with current groundwater use patterns. Buckley *et al.* used radar interferometry, together with conventional geodetic measurements to determine these rates of subsidence.

Groundwater abstraction over the past 40 or so years has caused subsidence of part of the coastal plain of Thessalonoki, Greece, at a rate of up to 100 mm a year. Unfortunately, this has meant that the sea has invaded up to 2 km inland, with loss of some parts of the city of Thessaloniki while some industrial areas are at risk of flooding. According to Stiros (2001), the subsidence has been brought about not only by rising effective pressures causing consolidation of the deltaic sediments of the region but also is due to the oxidation of peat deposits in the vadose zone as groundwater levels are lowered. Stiros also suggested the load-induced consolidation of deeper sediments was contributing to subsidence.

Bangkok is gradually subsiding below sea level due to excessive abstraction of groundwater. Approximately four million people live within about 1.0 m of sea level, with subsidence on average reaching 1.0–1.5 m throughout most of the city. An estimated 15 000 wells abstract more than 1 000 000 m^3 of groundwater ever day from aquifers that receive relatively less recharge. The maximum rate of subsidence was 100 mm $year^{-1}$, averaging 10–40 mm $year^{-1}$, according to Rau and Nutalaya (1982). The area is underlain by highly compressive clays within the aquifer system, which comprise the source of the subsidence. The most serious problem resulting from groundwater lowering in Bangkok has been the reduction in hydraulic gradient for drainage and sewerage systems. The monsoon flooding and low gradients within the gravity operated system rendered many parts of the drainage system useless, and in places the gradient was reversed. Malfunctioning of the system caused many feedback problems, accentuated health problems, and helped to cause deterioration of foundations, masonry walls, footpaths, bridges and roads. Phien-wej *et al.* (1998) maintained that due to poor town planning and rapid economic growth of the city, the flow of groundwater within a safe limit was unlikely to be achieved within the near future. Hence, artificial recharge has been proposed as one way to restore the declining piezometric levels and mitigate subsidence. They described a field experiment that involved recharging the uppermost aquifer in the Bangkok system via a well, which demonstrated that rebound was possible. The evidence gathered from this trial will be applied elsewhere in the Bangkok area. An attempt at controlling the flooding due to land subsidence involved the construction of drainage tunnels many kilometres in length and 3.3 m in diameter to provide increased gradients for the system. Pumping stations were designed to lift the water 9.5 m and to be capable of discharging 16 m^3 of water per second into the river system.

According to Carminati and Martinelli (2002), the central-eastern area of the Po Plain in northern Italy is undergoing rapid subsidence, in other words, up to 70 mm annually. The maximum subsidence is occurring in the Po Delta and near Bologna, whilst the minimum is located above buried anticlines that probably are active tectonically. Intense groundwater withdrawal occurred during the second half of the twentieth century as the population expanded and industry grew. Unfortunately, the increasing subsidence has been associated with an increasing frequency of flooding. More specifically, subsidence has occurred in the Venice area and was due primarily to the abstraction of groundwater, from about 1920 onwards, from sand aquifers, interbedded with silts and clays, of Quaternary age, which underlie Venice to a depth of 1000 m. The withdrawal of natural gas in the industrial area of Ravenna enhanced the amount of subsidence that occurred (Carbognin *et al.*, 1976). For every metre of piezometric decline between 1952 and 1969 the subsidence in the industrial area and in Venice itself was 10 mm and 20 mm respectively. This is related to the greater thickness of clays beneath Venice. Consolidation of the deposits as a consequence of changes in hydraulic head took place over a relatively short period of time due to the fact that the aquitards are interrupted by thin sandy layers that facilitate drainage.

5.3. Subsidence and the withdrawal of geothermal fluid

The withdrawal of fluids from geothermal fields can result in land subsidence. Subsidence is caused by the change in volume of the reservoirs that have been depleted. In general, the morphology of the subsidence basin reflects the geometry of the underground reservoir, although faults can influence subsidence at the ground surface. As an example, Allis (2000) indicated that between 1950 and 1997 a maximum ground subsidence of 14 m occurred at the Wairakei geothermal field in New Zealand due to the abstraction of hot groundwater. This is predicted to increase to 20 $\pm$ 2 m by 2050. The cause of the subsidence is the result of consolidation of highly porous but low permeability lacustrine mudstone located at a depth of 100–200 m. These sediments continue to drain in spite of pore pressure stabilization in the underlying aquifers and principal reservoir during the early 1980s. Allis went on to note that the centre of the subsidence basin now is subsiding at 220 mm $year^{-1}$, after having reached a maximum of 480 mm annually during the 1970s. Most of the subsidence occurs in an area of less than 1 km^2, located 500 m from the production borefield. Smaller rates of subsidence (10–100 mm $year^{-1}$) have taken place over the rest of the area affected by groundwater withdrawal for the power plant, that is, an area of 30 km^2. In addition, Allis mentioned that horizontal ground movements ranging over 200 mm a year, along with extensional strain rates of 3×10^{-4} per year, were recorded around the flanks of the subsidence basin during the mid-1970s. As a consequence, fissures developed in the areas where the rate of extension was greatest. Any roads, pipelines, drains or transmission lines that crossed the area affected needed maintenance. The centre of the subsidence basin is now occupied by a pond. Other areas of subsidence relating to the abstraction of geothermal fluids are found at Larderello in Italy, at Cerro Prieto in Mexico and at the Geysers in California, and have been reviewed by Narasimham and Goyal (1984).

5.4. Subsidence and sinkholes

Many, if not most, sinkholes are induced by human activities, that is, they result from declines in groundwater level, especially those, due to excessive abstraction (Fig. 5.7). For example, Jammal (1986) recorded that 70 sinkholes appeared in Orange and Seminole counties in central Florida over the previous 20 years. Most developed in those months of the year when rainfall was least (i.e. April and May) and withdrawal of groundwater was high. Nonetheless, the appearance of a sinkhole at the surface represents a late stage expression of processes that may have been in operation for thousands of years.

Most collapses forming sinkholes result from roof failures of cavities in unconsolidated deposits. These cavities are created when the unconsolidated deposits move or are eroded downward into openings in the top of bedrock. Collapse of bedrock roofs, as compared with the migration of unconsolidated deposits into openings in the top of bedrock, is rare.

Some induced sinkholes develop within hours of the effects of human activity being imposed upon the geological and hydrogeological conditions.

Figure 5.7 A sinkhole formed in deposits overlying the Transvaal Dolomite, south of Pretoria, South Africa.

Several collapse mechanisms have been advanced and have included, first, the loss of buoyant support to roofs of cavities or caverns in bedrock previously filled with water and to residual clay or other unconsolidated deposits overlying openings in the top of bedrock. A second mechanism that has been suggested is an increase in the amplitude of groundwater level fluctuations. Yet other mechanisms include increases in the downward velocity of movement of groundwater, and movement of water from the ground surface to openings in underlying bedrock where most recharge previously had been rejected since the openings were occupied by water.

Areas underlain by highly cavernous limestones possess most sinkholes or dolines, hence doline density has proven a useful indicator of potential subsidence. As there is preferential development of solution voids along zones of high secondary permeability because these concentrate groundwater flow, data on fracture orientation and density, fracture intersection density and the total length of fractures have been used to model the presence of solution cavities in limestone. Therefore, the location of areas of high risk of cavity collapse has been estimated by using the intersection of lineaments formed by fracture traces and lineated depressions. Aerial photographs have proved particularly useful in this context.

Brook and Alison (1986) described the production of subsidence susceptibility maps of a covered karst terrain in Dougherty County, south-west

Georgia. These were developed using a geographical information system (GIS) that incorporated much of the data referred to in the previous paragraph. The county was partitioned into 885 cells each 1.18 km^2 in area. Five cell variables were used in modelling, namely, sinkhole density, sinkhole area, fracture density, fracture length and fracture intersection density. Broadly similar subsidence susceptibility models were developed from cell data by intersection and separately by linear combination. In the intersection technique cells having specified values for all variables were located and mapped. In the linear combination technique a map value, $MV = W_1r_1 + \cdots + W_nr_o$, where W is an assigned variable weight and r an assigned weight value, was calculated for each cell.

Throughout the south-east United States there are thousands of sinkholes of many different sizes and shapes. They may range from 1 or 2 m in diameter to hundreds of metres. Depths of a few metres are common, whilst the largest sinkholes exceed 40 m in depth. Whereas it takes thousands of years to create natural sinkholes, those induced by human activity largely have occurred since the early 1900s. More than 4000 sinkholes have been catalogued in Alabama as being caused by human activities with the great majority of these developing since 1950. Indeed, in Shelby County, Alabama, more than 1000 sinkholes developed between 1958 and 1973 in an area of 26 km^2. The largest was called the 'December Giant' because it suddenly developed in December 1972. It measures 102 m in diameter and is 46 m in depth. These sinkholes are particularly dangerous because they form more or less instantaneously by collapse and they often occur in significant numbers within a short time span. Sinkholes have produced costly damage to a variety of structures and serve as a major local source of groundwater pollution. They have been caused largely by continuous dewatering projects in carbonate rocks for wells, quarry and mining operations and drainage changes.

Spectacular sinkhole development due to groundwater abstraction also has occurred in the Hershey Valley, Pennsylvania. For example, in order to extend limestone quarry operations near Hershey the water table was lowered more than 50 m by pumping. Sinkholes started to form almost immediately in the surrounding area and numbered more than 100 in some of the nearby drainages. They ranged in size from 0.3 to 6 m in diameter with depths of 0.6–3 m.

5.5. Metalliferous mining, dewatering and sinkhole development: the Rand, South Africa

Dewatering associated with mining in the gold bearing reefs of the Rand, South Africa, which underlie dolomite and unconsolidated deposits, has led to the formation of sinkholes and subsidence depressions over large areas.

Sinkholes formed concurrently with the lowering of the water table in areas that formerly, in general, had been free from sinkholes. Normally, the cover of unconsolidated materials above the dolomite is less than 15 m, otherwise they choke the cavity on collapse due to bulking. Some of the first notable appearances of sinkholes were associated with the gold mining areas of the Far West Rand, catastrophic occurrences taking place in the dewatered compartments, with the disappearance of surface structures and dwellings, and the loss of life (De Bruyn and Bell, 2001). For example, sinkholes were noticed initially in 1959 and the seriousness of the situation was highlighted in December 1962 when a sinkhole appeared at the West Driefontein Mine and engulfed a three-storey crusher plant with the loss of 29 lives (see Case history 1, Section 5.12). Then, in August 1964 two houses and parts of two others disappeared into a sinkhole in Blyvooruitzicht Township with the loss of five lives (Fig. 5.8). Hence, certain areas became unsafe for occupation and so it became a matter of urgency that areas that were subject to the occurrence of sinkholes or subsidence depressions be delineated.

The dolomite bedrock frequently has a highly irregular pinnacled surface, with varying thicknesses of residual material overlying it (Fig. 5.9). In general profile, hard unweathered dolomite bedrock is overlain by slightly weathered rock and thereafter by low strength, insoluble residual material consisting mainly of manganese and iron oxides, and chert. Where this residual material consists chiefly of hydrated manganese oxides, it commonly is both readily erodible and compressible, and is referred to as wad. The horizon of this very low strength, porous and permeable material may be

Figure 5.8 A sinkhole developed at Blyvooruitzicht that claimed five lives.

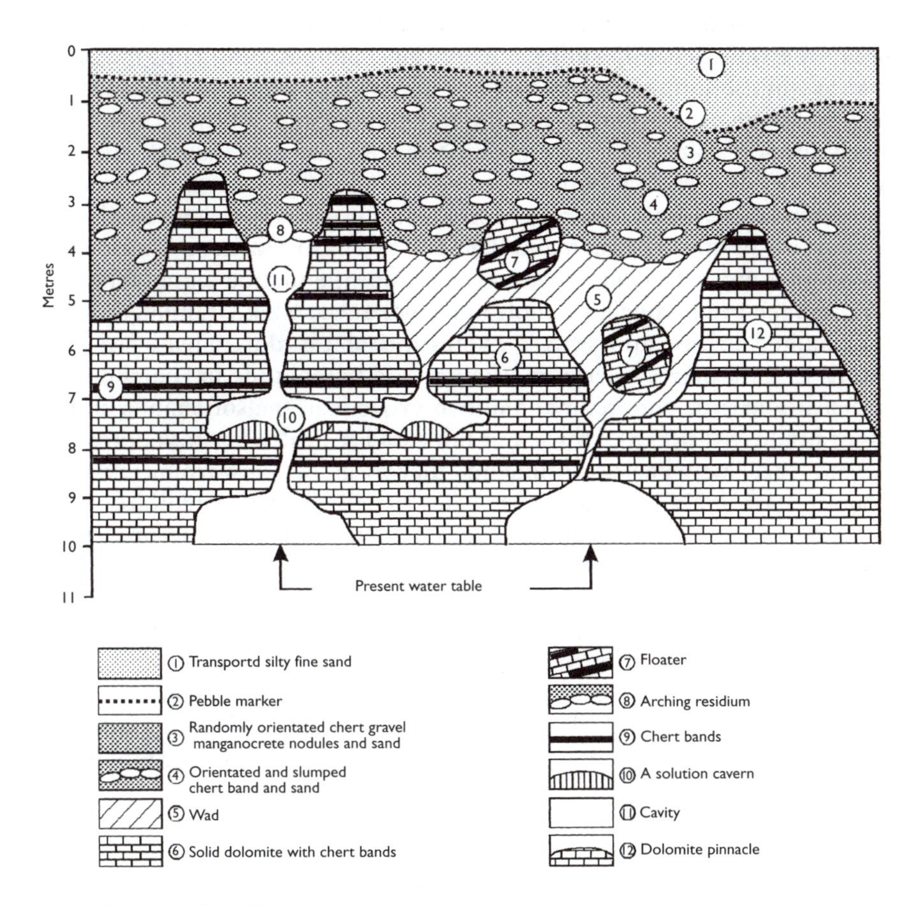

Figure 5.9 Typical profile in dolomite area.

up to several tens of metres thick, but is generally less than 10 m thick. Voids may be present in the wad. A high incidence of sinkhole development has been associated with formations that are chert rich, the chert content promoting differential leaching with associated development of cavities.

Sinkholes are steep-sided holes in the ground that, in these instances, occur at the surface rapidly (i.e. often within minutes). They commonly are of limited size, normally less than 50 m in diameter. Rapid subsidence can take place due to the collapse of caverns within dolomite that has been subjected to prolonged dissolution, this occurring when the roof rocks are no longer of sufficient strength to support themselves. However, as the solution of dolomite is a very slow process, contemporary solution therefore is very rarely the cause of collapse. Usually, in the gold mining areas of South Africa the downward movement of unconsolidated deposits into caverns within the dolomite or voids within the overburden occurs as a result of groundwater lowering and is responsible for the appearance of sinkholes at the surface (Van Schalkwyk, 1998). Subsidence depressions are large enclosed depressions, typically between 30 and 300 m in diameter, which develop slowly. They may form as a result of consolidation of compressible residual material derived from dolomite, again due to dewatering. Subsidence depressions also may form due to the premature termination of sinkhole formation. The geological factors that may influence the formation of sinkholes and subsidence depressions include the surface topography and drainage; the origin, thickness and character of the soil, be it transported or residual; the nature and surface expression of the dolomite, which frequently is pinnacled so that the frequency and amplitude of the pinnacles may be important; the presence and size of voids in the soil mantle and in the dolomite; the depth of the water table and its fluctuation; the presence of wad that may be susceptible to erosion by downward percolating groundwater; and the presence of dykes or sills or strata of Karoo age, which in the two latter instances occur above the dolomite and inhibit the development of sinkholes (De Bruyn and Bell, 2001).

Lowering the water table increases the downward seepage gradient, accelerates downward erosion, and removes buoyant support within the overburden. It also reduces capillary attraction and increases the instability of flow through narrow openings; and gives rise to shrinkage cracks in wad, which weaken the mass in dry weather and produce concentrated seepage during rains. Steeply sloping bedrock promotes subsurface instability in that groundwater flow is directed and concentrated along this surface, and therefore can develop higher flow velocities that lead to more effective erosion and dissolution. Pinnacles initially may offer arching support to the residual soil but subsequently sinkholes may form if the arches collapse. Sinkholes may develop along the trend of a major joint, the gape of which has been opened and enlarged by dissolution, or along the shear zones of faults. Palaeosinkholes, that is, ancient sinkholes that have been filled by the deposition of younger unconsolidated materials, may be reactivated under certain conditions.

As noted, the development of sinkholes is accelerated by artificial lowering of the water table and it may be that the incidence of development is related to the rate at which the water table is lowered. Dissolution of dolomites occurs along discontinuities, and it frequently is asserted that the most active dissolution occurs at and just below the water table (Trudgill and Viles, 1998). Hence, it is here, within the uppermost part of the phreatic zone, where caverns are developed. As such, they are often within 100 m of the ground surface. If the water table is lowered this can lead to the enlargement of caverns. When the water table is lowered rapidly by dewatering, then the equilibrium is upset and active subsurface erosion can occur as a result of the permeation of surface water. Grykes (narrow, deeply weathered joints in the dolomite bedrock) are flushed of residual material that can be transported downwards into the cavities or caverns in the bedrock. Cavities can grow and migrate upwards by progressive roof collapse, ultimately extending into the overlying residual material. Development of a cavity through the overlying material leads to the formation of a sinkhole once the cavity appears at the surface. The size of a sinkhole that develops depends not only on the depth of the water table but also on the thickness of the residuum. In other words, a large sinkhole does not develop in an area where the water table is high or the soil cover is thin.

Extensive faulting has occurred in the Far West Rand and these faults represent conduits that connect the groundwater in the dolomite with the gold bearing reefs below. Accordingly, many mines adopted dewatering techniques to dispose of the large amounts of groundwater that entered the mines from above. In addition, several dykes, from 6 to 60 m in width, occur in the Far West Rand (Fig. 5.10). These dykes represent barriers to groundwater movement, effectively compartmentalizing the groundwater regime. This allowed the gold mines to be dewatered, the water table being lowered within individual compartments. For example, active pumping from the Venterspost Mine began in 1935, this being discharged initially into the Wonderfontein Spruit. However, because pumping costs rose alarmingly, injection of cement grout was used in an attempt to control the flow of groundwater. By 1954, this was proving futile. Therefore, wells

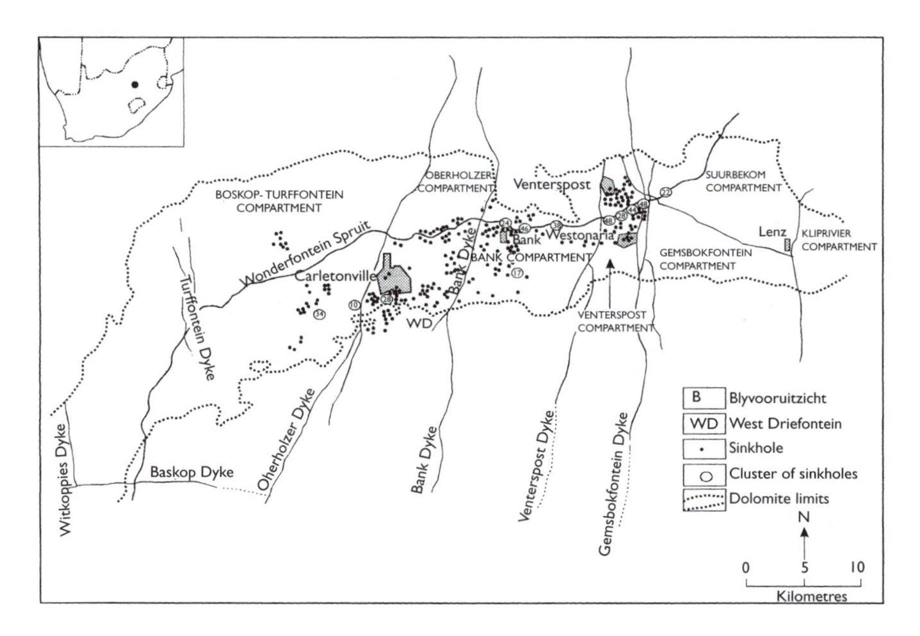

Figure 5.10 Compartments on the Far West Rand.

were sunk into the marshy bed of the Wonderfontein Spruit from which pumping took place, this being more efficient and much less costly. Unfortunately, however, within a few months a large sinkhole occurred at the surface (Wolmarans, 1996). Swart *et al.* (2003) reported that by 1987 there were 271 sinkholes in the 30 km stretch of streambed of the Lower Wonderfontein Spruit comprising an estimated total volume of 2 450 000 m^3. Palaeosinkholes have been reactivated. Also, in 1987 it was decided to rehabilitate the streambed and its banks by filling all the sinkholes within 100 m of the centre of the streambed and by constructing berms around any sinkholes located further away than 100 m. Most of the sinkholes concerned were backfilled by bulldozing soil obtained from borrow pits, the soil being compacted in layers. The soil then was grassed over. The remaining sinkholes were backfilled with waste rock and mine tailings, covered with soil and seeded. Refill and top-up operations of sinkholes occurred in 1995–96. Swart *et al.* admitted that the use of tailings, and to a lesser extent soil, can mean that shrinkage and cracking occurs as the material dries out, and that this can facilitate erosion by excess run-off during wet weather. The latter could lead to the collapse of backfill into a sinkhole. Nonetheless, these materials are cheap and available locally, and so in such circumstances can be used to top-up sinkholes when necessary.

Subsidence depressions tend to occur where the original water table was above bedrock and compressible material was located beneath the water table. The presence of such depressions frequently is indicated by sagging in the pebble marker (a pebble horizon often occurs beneath a cover of transported soil and, in turn, overlies residual soil) and chert horizons in the soil mantle, when exposed. Lowering of the water table by dewatering through the unconsolidated deposits, causes effective stress to increase and leads to consolidation within these deposits. In particular, where the thickness of wad is appreciable, the amount and rate of subsidence can be both significant and relatively rapid. The degree of subsidence that takes place tends to reflect the thickness and proportion of unconsolidated deposits that have consolidated. The thickness of these deposits varies laterally, thereby giving rise to differential subsidence that, in turn, causes large fissures to occur at the surface. In fact, the most prominent fissures frequently demarcate areas of subsidence. The total subsidence has varied from several centimetres to over 9 m. In addition, soil material can be transported due to the downward movement of the water table, via discontinuities, into the dolomite beneath. According to Wolmarans (1996), subsidence depressions have caused more disruption in townships on the Far West Rand than sinkholes. He quoted the example of Lupin Place in Carletonville where 24 houses had to be demolished, the area subsiding over 7 m in a four year period. Houses in the centre of the depression where the subsidence was more or less uniform were barely affected by the movements but those on the periphery were

Figure 5.11 Schutte's depression near Carletonville, Far West Rand.

subjected to notable differential subsidence. Nonetheless, the effective functioning of sewers and other services was adversely affected throughout the depression. The depression was filled with material from a nearby tailings dam. Schutte's depression near Carletonville eventually attained a depth of over 8 m and had a diameter of approximately 180 m (Fig. 5.11). Boreholes showed compressible material extending well below the depth of the original water table in both instances.

Bezuidenhout and Enslin (1969) used the gravimetric method in the Far West Rand in an attempt to try to delineate the potential locations of sinkholes, taking the zero gravity contour to coincide with the water table. Areas of positive gravity contours accordingly indicated areas where dolomite was above the water table, whilst negative gravity contours indicated where dolomite was below the water table. Hence, the former areas remain unaffected by groundwater lowering whereas the latter areas have been affected, that is, there is a greater potential for sinkhole and depression development because the water table has been lowered. Subsequently, Kleywegt and Enslin (1973) noted that as far as dewatering of compartments in the Far West Rand was concerned, most sinkholes developed in low lying areas where the original water table was up to 30 m beneath the ground surface. In fact, they found that most such sinkholes occurred in areas of gravity highs, in the transition areas between high and low gravity areas, especially where the gravity gradient was steep, and over narrow features. Few sinkholes developed where the original water table was between 30 and 60 m below the surface. Frequently, in the latter areas there is an extensive cover of rocks of Karoo age overlying the dolomite.

5.5.1. Classification of ground conditions and risk in dolomitic areas

Although other classifications of ground conditions in dolomitic areas in South Africa have been proposed, for example, by Wagener (1985) and by Van Rooy (1989), the most recently developed is the method of scenario supposition. These classifications although developed for use in the Rand, could be adapted for use elsewhere. The method of scenario supposition was described by Buttrick and Van Schalkwyk (1995), and Buttrick and Calitz (1995). It provides a means of evaluating the stability of the ground surface in dolomitic areas. The characterization of the potential stability of a site in a dolomitic area requires evaluation of the effects of human impact during the lifetime of a development. In the case of an undeveloped area the potential stability initially is reviewed in terms of the area being dewatered or not. A number of factors were chosen to help determine the likelihood of whether or not sinkholes or subsidence would occur. This allows an assessment to be made of the inherent risk of the hazard (i.e. occurrence of a sinkhole or subsidence depression) making its appearance. Such a risk can be described as low, medium or high. Over and above this, there is the financial risk related, for example, to the damage to property, and the risk of loss of life, which can be associated with the formation of a sinkhole or subsidence depression. The risk to property is either acceptable or unacceptable for a particular type of development.

The factors that Buttrick and Van Schalkwyk (1995) suggested should be used to evaluate the possible occurrence of sinkholes or subsidence depressions included the blanketing layer, the receptacles, the mobilizing agents and the maximum development space. The blanketing layer refers to the soil mantle occurring above the dolomite. Its thickness, character of the soil and position of the water table are important. The receptacles are voids either in the soil mantle or dolomite bedrock into which material from above can accumulate. Mobilizing agents are primarily water seeping into the ground from a surface source or the effect of groundwater lowering. Ground vibrations also may mobilize loosely unconsolidated soils. As far as sinkholes are concerned, the maximum development space refers to the maximum possible size of sinkhole that can form. It should be borne in mind that a sinkhole may grow with time as material sloughs from the sides. These evaluation factors are assessed in terms of a dewatering or non-dewatering scenario. Once the assessment of an area is made, it can be described in terms of the anticipated number of sinkholes or subsidence depressions that may occur, that is, the risk is low, medium or high. Low risk refers to no subsidence events occurring within a hectare in a 20 year period. If 0.07–0.7 events occur within a hectare within a 20 year period, then the area is characterized as of medium risk. Finally, an area is characterized as of high risk if more than 0.7 events occur in 20 years within a hectare. The characterization system can be used to zone an area in terms of risk. Such a zoning system then can be used in relation to the way an area is developed. In other words, do any restrictions need to be placed on development, do buildings or services have to include special design features or has the ground to be treated? The risk characterization and associated type of development, as suggested by Buttrick and Calitz (1995) and subsequently updated by Buttrick *et al.* (2001) are summarized in Tables 5.1 and 5.2.

Venter and Gregory (1987), after working on road construction over dolomite areas in the Far West Rand, suggested a multi-variable method of risk classification that considered the relative importance of the factors contributing to and against ground instability. The resistance to erosion and the competence of the overburden materials were considered the most important factors likely to prevent instability while the overall slope of the upper surface of the bedrock, the incidence of pinnacles and extent of cavities in the bedrock or residuum were regarded as the primary disturbing forces. They developed a rating system and low, medium and high categories of risk were assigned according to the sum total derived by the rating system.

The National Home Builders Registration Council in South Africa require that the risk of sinkhole and subsidence depression formation be established in areas where dolomites either outcrop at the surface or occur beneath a certain thickness of cover rocks. In the latter instance a cover of 60 m in thickness is considered safe where no dewatering has taken place and future dewatering is disallowed or strictly controlled. Where dewatering has occurred or where there is no control over groundwater levels, then a thickness of cover of up to 100 m is considered safe.

5.5.2. Investigation in and construction over dolomite areas

As in other areas where subsurface cavities exist, before a dolomitic area can be developed it needs to be investigated so that ultimately it can be zoned in terms of stability. On completion of the investigation and evaluation, the risks involved in developing an area must be specified, as must any precautionary measures needed to reduce the likelihood of sinkhole or subsidence depression formation, and any restrictions on land use, plot size, density and layout, services and the like.

In an investigation of the stability of a dolomitic area, it is important to investigate the bedrock topography, that is, to determine whether the bedrock is solid, fractured or pinnacled. Not only does the character of the

Table 5.1 Characterization: inherent risk of a specified size sinkhole or subsidence depression forming (After Buttrick *et al.*, 2001)

Inherent risk class sinkhole diameter	*Small sinkhole < 2 m*	*Medium sinkhole 2–5 m*	*Large sinkhole 5–15 m*	*Very large sinkhole > 15 m*	*Risk of subsidence depression formation*	*Recommended type of development in order to maintain acceptable development risk*
Class 1	Low	Low	Low	Low	Low NDS[b] or DS	Residential, light industrial and commercial development provided that appropriate water precautionary measures are applied. Other factors affecting economic viability such as excavatability, problem soils, etc. must be evaluated
Class 2	Medium	Low	Low	Low	Medium NDS[b]	Residential development with remedial water precautionary measures. No site and service schemes. May consider for commercial or light industrial development
Class 3	Medium	Medium	Low	Low	Medium NDS[b]	Selected residential development with exceptionally stringent precautionary measures and design criteria. No site and service schemes. May consider for commercial or light (dry) industrial development with appropriate precautionary measures
Class 4	Medium	Medium	Medium	Low	Medium NDS[b]	Selected residential development with exceptionally stringent precautionary measures and design criteria. No site and service schemes. May utilize for commercial or light (dry) industrial development with appropriate stringent precautionary measures

Class 5	High	Medium	Low	Low	High NDS[b]	These areas are usually not recommended for residential development but under certain circumstances selected residential development (including lower-density residential development, multi-storied complexes, etc.), may be considered, commercial and light industrial development The risk of sinkhole and subsidence depression formation is adjudged to be such that precautionary measures, in addition to those pertaining to the prevention of concentrated ingress of water into the ground are required to permit the construction of housing units
Class 6	High	High	Medium	Low	High NDS[b]	These areas are usually not recommended for residential development but under certain circumstances high-rise structures or gentleman's estates (stands 4000 m^2 with 500 m^2 proven suitable for placing a house) may be considered, commercial or light industrial development. Expensive foundation designs may be necessary. Sealing of surfaces, earth mattresses, water in sleeves or in ducts, etc
Class 7	High	High	High	Low	High NDS[b] DS	No residential development. Special types of commercial or light industrial (dry) development only (e.g. bus or trucking depots, coal-yards, parking areas). All surfaces sealed. Suitable for parkland
Class 8	High	High	High	High	Low-High NDS[a] or DS	No development, nature reserves or parkland

Notes
a Number of anticipated events per hectare over a period of 20 years with poor design and management.
b Non Dewatering Scenario and Dewatering Scenario.

Table 5.2 Recommended development types (After Buttrick and Calitz, 1995)

Development type	Description
1 Residential affordable	Stands 250–500 m^2 with a single dwelling plus possibly a second dwelling structure
2 Residential (medium to large stands)	Stands 600–1000 m^2 with a single dwelling, e.g. not exceeding 200 m^2 on a 600 m^2 stand and 350 m^2 on a 1000 m^2 stand
3 Residential group housing (single storey)	Maximum 30 units ha^{-1} with individual maximum floor area of units 100 m^2 (excluding garages or covered parking) and development designed to minimize pipe lengths for providing wet services to units (i.e. low density of of water bearing services)
4 Residential group housing (double storey)	Maximum cover of 30% with maximum floor: area ratio of 0.6 (excluding garages or covered parking). It is extremely important that the design should be such that wet services pipes should be shared and serve a number of units (i.e. low density of water bearing services)
5 High-rise residential	Very low percentage cover area (around 10% or less), but detailed foundation investigations and very high quality wet services to serve the buildings
6 Commercial	Low coverage by structures (30% or less) and large areas for parking and/or storage facilities. Limited ablution facilities
7 Light industrial[a]	As above
8 Light dry industrial[a]	As above, but very limited use of water
9 Heavy industrial[a]	Variable coverage of area, but limited ablution facilities. There may be large structures and the extensive controlled use of water is possible
10 Gentleman's estates	Stands 4000 m^2 or larger with single dwelling and minor outbuildings
11 Parklands	
12 Nature reserve	

Note

a Due consideration should be given not only to stability but also to environmental constraints, for example, groundwater pollution.

overlying soil mantle need to be determined but its continuity should be established. The particular dolomite formation(s) needs to be identified, as well as the occurrence of any dykes or sills, or overlying strata.

As far as direct exploration is concerned, percussion drilling is used most frequently in site investigations in dolomitic areas. Drillholes should be sunk at least 6 m into solid dolomite bedrock. Drill chippings are sampled for logging purposes and penetration rates are recorded, as is any loss of air flush. All holes must be sealed satisfactorily upon completion. Trenching of the soils above bedrock allows a continuous profile of the conditions in the trench to be recorded, as well as permitting freedom of sampling. Pits occasionally are dug. Auger drilling, using a 750 mm flight auger, is sometimes used for inspection of residual deposits. Augers, however, can meet with refusal in chert gravels in overburden, especially those containing cobbles or small boulders. Diamond drilling is used only infrequently in dolomitic areas and core recovery of residual soil containing chert is extremely difficult, if not impossible. Percussion drilled holes can be used for *in situ* testing and for monitoring purposes. In this context the position of the water table, the permeability of the ground and ground movements are important.

Most of the foundations structures and ground treatment methods mentioned below could be adapted for use in other areas where cavities exist in the ground. Wagener (1985) reviewed the types of foundation structures used in dolomitic areas. This subsequently was updated by Van den Berg (1996). Conventional foundation structures, that is, strip or pad footings may be used for houses and light steel-framed structures in dolomitic areas where the soil mantle consists predominantly of sandy chert gravel that is less than 3 m in thickness and the stability risk is acceptable. Where dolomite pinnacles or boulders approach the surface, these could present a problem of differential subsidence. If differential subsidence of between 10 and 25 mm is likely, then Van den Berg (1996) recommended the use of flexible construction.

Engineered soil mattresses have been used where the soil mantle is variable. The mattress limits both total and differential subsidence; reduces the foundation stress by spreading the load fairly uniformly beneath the mattress to an acceptable level; presents a relatively impermeable layer thereby limiting the ingress of water; and so reduces the risk of small sinkholes developing. A mattress also represents a relatively strong layer that may bridge a small cavity that develops beneath it, especially if the compacted soil is reinforced by geogrids. Wagener (1985) pointed out that the thickness of a mattress was related to the thickness and geotechnical properties of the soil on the one hand and the sensitivity of the structure to be erected to subsidence on the other. The method of construction depends on the site conditions and the fill material available. On sites where shallow pinnacles and boulders occur, and where chert gravel and waste rock are available as fill, Wagener suggested that material to a depth of 1 m below the tops of pinnacles and large boulders should be removed. In some instances it may be necessary to remove the tops of pinnacles that protrude above the general pinnacle level by blasting. Additional excavation may be required if pinnacles are spaced far apart or loose material is present. The base of the excavation first is levelled, and then lifts of fill are placed and compacted with a small vibratory roller. A layer of waste rock is placed about 200 mm above the tops of pinnacles and compacted with a vibratory roller. Uncontrolled ingress of water must be avoided during compaction. Ten passes of the roller are usually sufficient. Soilcrete or low strength mass concrete can be used as an alterative to waste rock. Then, the mattress is built up to the required level (above ground level to provide good drainage) using selected chert gravel, or other material (Fig. 5.12). Chert gravel normally is compacted to 95% modified AASHTO dry density at ± 2% optimum moisture content. Houses can be founded on a light raft with thickened edge beam on a mattress of lesser thickness. The treatment of a site with a thick cover over pinnacles and boulders is essentially the same. In this case mattresses consist of properly compacted chert gravel or waste rock, provided the latter is capped by a layer (greater than 1 m thick) of less permeable material to limit water ingress into the underlying dolomite. The foundation structure, for example, a reinforced concrete raft, should be placed in the mattress at shallow depth. A raft distributes the applied loads over a large area.

When numerous pinnacles occur near the ground surface, a layer of reinforced concrete can be used to span between pinnacles, a raft being placed above the concrete. In this way the foundation loading is transferred to the dolomite (Fig. 5.13). This type of solution normally does not find application

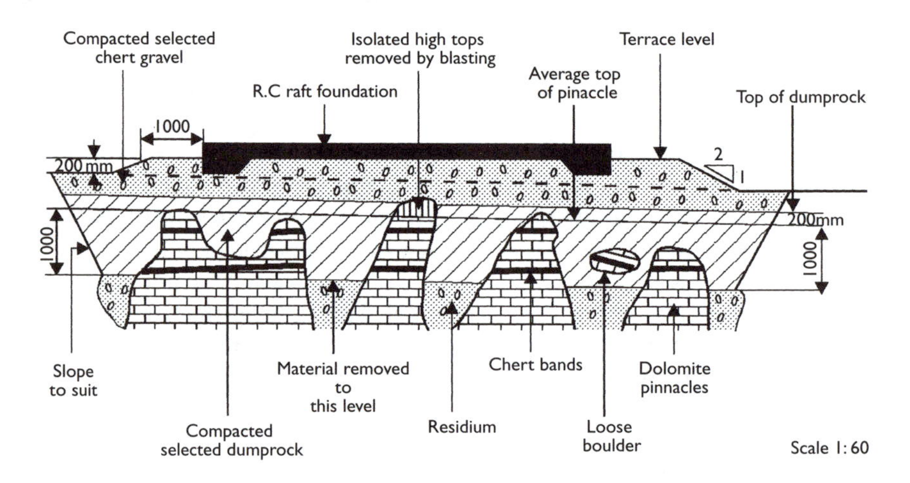

Figure 5.12 An engineering mattress with raft foundation. (After Wagener, 1985; reproduced by kind permission of the South African Institution of Civil Engineering.)

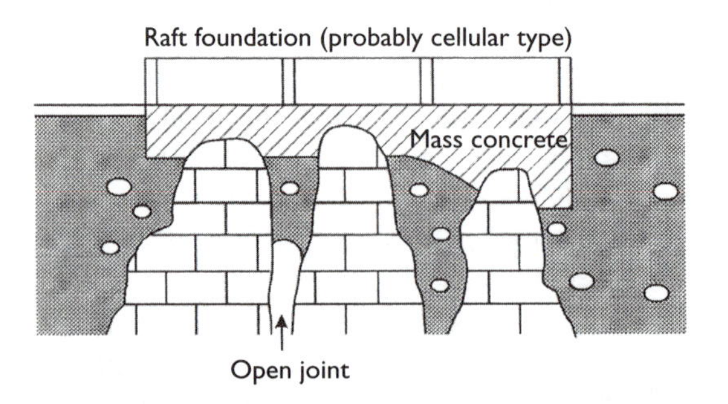

Figure 5.13 Raft foundation on concrete base spanning pinnacles of dolomite.

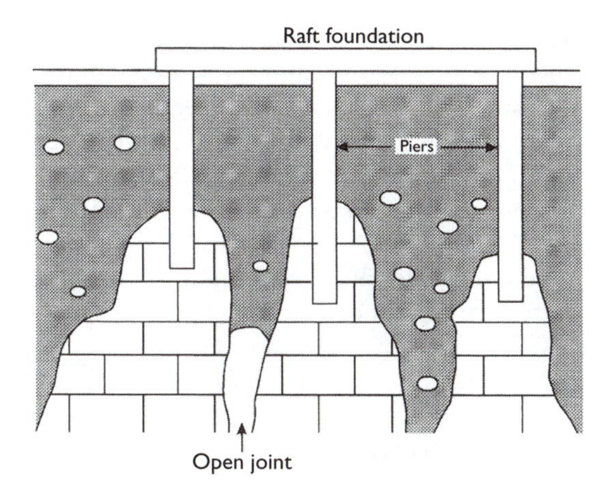

Figure 5.14 Raft foundation on piers founded in dolomite.

when the cover of unconsolidated deposits is thicker than 3.0 m. In situations where the cover ranges from 3.0 to 7.0 m in thickness the raft can be supported on piers, making sure that the piers are founded in bedrock. The structural loadings are transferred via an appropriate beam and pier arrangement (Fig. 5.14). If this type of foundation structure is used for founding larger heavier structures, it will be necessary to proof drill the pinnacles.

Piles should only be used in dolomitic areas when other foundation structures are not feasible. This is because large floaters of dolomite and steeply sloping dolomite pinnacles can deflect or damage piles, and the presence of cavities can make the installation of piles very difficult. It is necessary to prove the competence of the rock below founding level once the desired depth has been reached. Drilling and blasting techniques have been used to fracture rock prior to piling. Piling subsequently takes place through the fractured rock.

Van den Berg (1996) mentioned that large diameter augered, precast or displacement type piles are not recommended for dolomite due mainly to the presence of boulders and chert gravels in the soil mantle and the uneven topography of the bedrock. Conventional shaft sinking techniques can be used where heavy loads have to be carried by the foundations. Caissons also can be sunk in dolomitic areas. An advantage of caissons over shafts is that they can be sunk in poor ground conditions without fear of sidewall collapse as the cutting edge can always be kept close to the bottom of the excavation.

In some instances it may be possible to treat the ground. For instance, Partridge *et al.* (1981) referred to the use of the soil-cement arch method of foundation support (Fig. 5.15). The compacted layer of soil-cement bridges

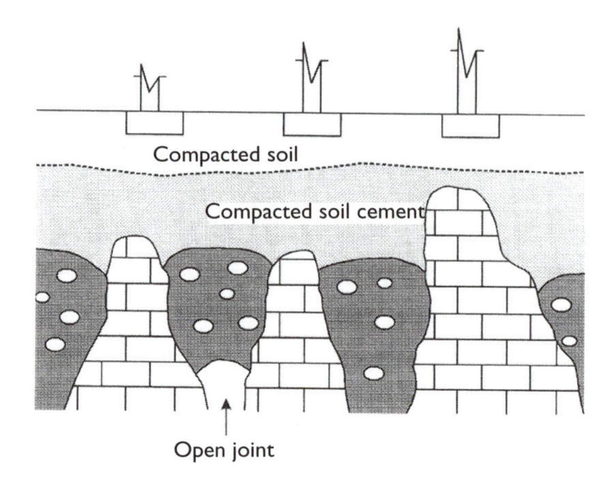

Figure 5.15 A compacted layer of soil cement arching between dolomite pinnacles.

between large open joints in dolomite and also inhibits downward movement of water. They alleged that such stabilized soil arches were most suitable where the depth of unconsolidated deposits was in excess of 3 m. The degree to which the properties of a clayey soil can be enhanced by cement, or indeed lime, depends not only on the quantity added but also on the amount and type of clay minerals present. Generally, however, an increase in unconfined compressive strength of several hundred per cent can be obtained by the addition of a few per cent, by weight, of either cement or lime to a clay soil (a rough rule of thumb is 1.0% for every 10% of clay minerals present; rarely do clay soils contain more than 80% clay minerals).

Dynamic consolidation has been used to compact overburden in some dolomitic areas. However, where chert gravel occurs at the surface it usually is more competent than the material beneath. Consequently, much of the energy expended by the compactive effort is absorbed by the chert gravel and so the material beneath is not compacted to the extent that it otherwise would be. In addition, where depth to bedrock varies widely, the compaction achieved will be laterally inconsistent.

The inverted filter method has been used to rehabilitate sinkholes that occur at shallow depth, that is, usually less than 3 m. This essentially consists of backfilling a sinkhole initially with boulders and soil–cement slurry to choke the throat of the sinkhole (Fig. 5.16). Once the slurry has set, any unstable material on the sidewalls of the sinkhole is removed by backhoe. Further backfilling of the sinkhole then takes place using a coarse gravel–sand mixture at the base, on top of the boulders. Progressively finer material, that is, coarse sand and then fine sand are placed to fill the sinkhole. The backfilling is carried out in 150 mm layers that are compacted

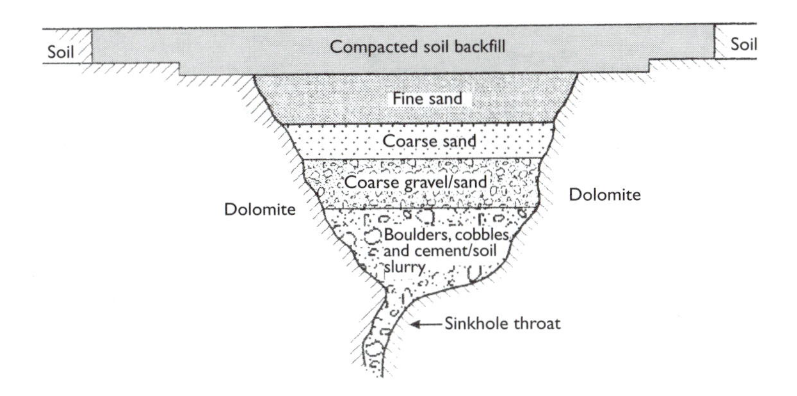

Figure 5.16 Rehabilitation of a sinkhole by the inverted filter method.

by small mechanical compactors. The soil around the sinkhole is excavated to a distance of 3 to 5 m beyond the perimeter of the sinkhole and to the depth of the soil-dolomite contact where the soil horizon is relatively thin. Soil then is placed and compacted over the excavated area and filled sinkhole. Anchored geogrid may be used to enhance the strength of the compacted soil.

Bulk grouts have been used to treat areas of potential subsidence risk. Thick grouts, injected at low grouting pressures, have been employed in order to avoid erosion of wad and to plug grykes. Gregory *et al.* (1988) referred to the use of grouting to protect a major road from the effects of sinkhole development. The grouting took place prior to dewatering of the Gemsbokfontein Compartment, in the Far West Rand, which commenced in 1986. Areas of potential high risk were identified. Grouting in these areas commenced in the hole where the dolomite or cavities occurred at greatest depth. Generally, primary holes were grouted at 20 m centres, this aiding the identification of zones that required further treatment in terms of secondary, tertiary or quaternary grouting, the latter resulting in a final hole spacing of 5 m. Monitoring was undertaken during the grouting programme. If this revealed that any subsidence had occurred because of erosion of wad by grout, then grouting was carried out in the centre of the area that had subsided. No sinkholes or notable subsidence depressions have developed in areas of high risk since dewatering began. Gregory *et al.* were able to show that the bedrock topography had a critical gradient below which grout movement would not occur. Furthermore, they maintained that the critical gryke width–depth ratio was about 0.4, with lesser gryke width–depth ratios being stable since material was able to bridge the gryke. Grouting also has been used to arrest subsidence under existing structures, thereby underpinning them.

5.6. Subsidence due to the abstraction of oil or gas

Subsidence due to the abstraction of oil occurs for the same reason as does subsidence associated with the abstraction of groundwater. In other words, pore pressures are lowered in the oil producing zones due to the removal of not only oil but also of gas and groundwater, and the increased effective load causes consolidation in compressible beds. Consolidation is likely to give rise to changes in the porosity and permeability of the reservoir rocks that can affect recovery efficiency and well productivity. It can deform well tubulars thereby creating operational problems and shortening well life. If the abstraction of oil is offshore, then subsidence of the sea floor can cause oil platforms to be lowered deeper into the sea and so create safety problems. Therefore, the problem of consolidation and associated subsidence must be addressed during the design and development stages of offshore projects if severe financial setbacks are to be avoided. In this regard, it is important to have a monitoring programme in place to assess subsidence development. For example, a formation consolidation monitoring tool is a wireline device that uses multiple gamma ray detectors to determine the locations of and precise distance between radioactive markers located in the formation or casing. An array of detectors is used to measure the distance between markers and so allow calculation of the consolidation that the formation has undergone.

One of the most notable and costly occurrences of land subsidence took place at Wilmington Oil Field, California. This area is underlain by approximately 1800 m of Miocene to Recent sediments consisting largely of sands, silts and shales. Subsidence was first noticed in 1940 after oil production had been under way for three years. Gilluly and Grant (1949) were among the first to demonstrate that the area of maximum subsidence showed a remarkable coincidence with the productive area of the oil field (Fig. 5.17). They also indicated that there was a very close agreement between the relative subsidence of the various parts of the oil field and the pressure decline, thickness of oil sand affected and the mechanical properties of the oil sands. By 1966 an elliptical area of over 75 km^2 had subsided more than 8.8 m, the largest amount of subsidence due to the withdrawal of fluids known to have occurred at that time in the world (Poland and Davis, 1969). By 1947 subsidence was occurring at a rate of 0.3 m $year^{-1}$ and had reached 0.7 m annually by 1951 when the maximum rate of withdrawal was attained (i.e. 140 000 barrels a day). Because of the seriousness of the subsidence, and the realization that it was due to declines in fluid pressures, remedial action was taken in 1957 by injecting water into and thereby repressurizing the abstraction zones. By 1962 this had brought subsidence to a halt in most of the field. In fact, Yerkes and Castle (1970) referred to more than 40 known examples of differential subsidence, horizontal displacement and surface

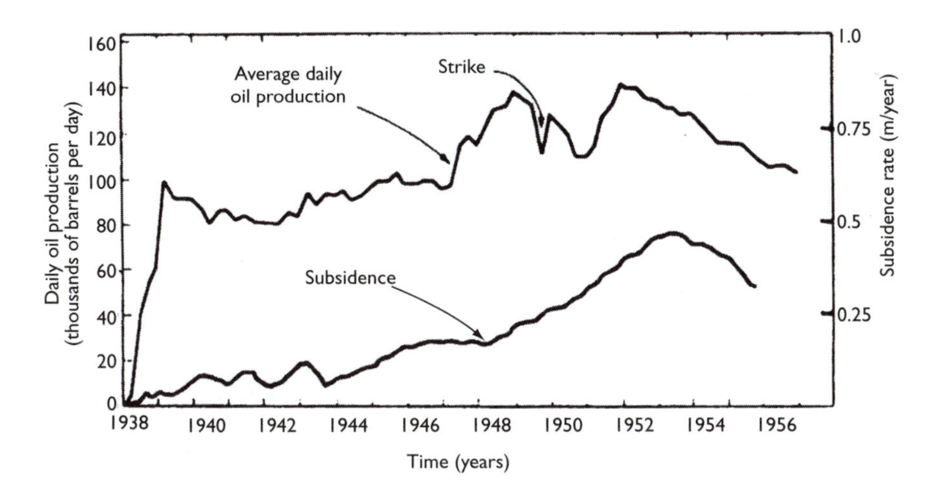

Figure 5.17 Oil production and rate of subsidence at BM 700, Wilmington oil field, California. (Modified after the United States Geological Survey and Steinbrugge, K.V. and Bush, V.R. 1958. Subsidence in Long Beach – Terminal Island – Wilmington, California, Pacific Fire Rating Bureau, San Francisco.)

fissuring/faulting, associated with 27 oil and gas fields in California and Texas. The maximum subsidences recorded in these fields were not comparable with that of Wilmington and generally were less than 1 m.

Yerkes and Castle (1970) also reported that centripetally directed horizontal displacements commonly accompany differential subsidence. Again the Wilmington Oil Field offers an example. There the maximum horizontal displacement exceeded 35% of the maximum vertical subsidence. It was located about halfway along the flanks of the subsidence basin from which it decreased progressively to zero in both directions, that is, towards the centre and periphery of the basin. A maximum horizontal displacement amounting to 3.66 m was recorded between 1937 and 1966, which gave rise to horizontal strains exceeding 1.2%. Between 1957, when repressurization commenced, and 1967, several survey stations along the eastern area of the subsidence basin (an area where vertical rebound approximating to 17% of total differential subsidence has taken place) recovered as much as 80% of their measured horizontal displacement.

Subsidence due to the abstraction of oil in Venezuela has been reported by Murria (1997). Production from the oilfields along the eastern coast of Lake Maricaibo began in the 1920s and up to 5.4 m of subsidence has occurred in unconsolidated sediments in the areas of most intensive production, and can be as much as 200 mm year^{-1}. The subsidence has necessitated the construction of dykes along the shore of the lake to protect the land from flooding.

More recent examples of subsidence due to the exploitation of oil have occurred in the North Sea. For example, there are six platforms in the Ekofisk Field that have been associated with subsidence of the seafloor. The Ekofisk Field comprises Eocene shales and Palaeocene deposits overlying Upper Cretaceous strata. The oil and gas occur in a crescent shaped reservoir in the Chalk and Palaeocene deposits located at a depth of up to 3000 m below the seabed. Oil was first abstracted in 1973 and the first evidences of subsidence are believed to date from 1978 when production peaked (this probably was a coincidence). Since the peak production of 620 000 barrels per day, the abstraction rate has fallen to 270 000 barrels a day. Subsidence was first reported in late 1984 when it was revealed that it varied between 1.8 and 2.5 m. By mid-1986 this had reached a maximum of 3.5 m and the subsidence basin had a diameter of 6 km. In addition to having to raise oil platforms, an attempt has been made to arrest further movement of the seafloor by injecting natural gas back into the reservoir. By the summer of 1985 up to 8 000 000 m^3 of natural gas had been pumped back into the reservoir deposits. The possibility of injecting water or nitrogen (both being cheaper than natural gas) has been investigated.

Spaceborne methods of detemining land subsidence have been referred to earlier and, for example, Fielding *et al.* (1998) used interferometric analysis of InSAR data in their study of the subsidence occurring in the San Joaquin Valley, California, due to the abstraction of oil from diatomite located at shallow depth. They showed that maximum subsidence rates were as high as 400 mm $year^{-1}$ and produced subsidence maps of the area from the satellite data obtained. Such data complements that gathered by ground based techniques, permits measurements to be made where access is difficult and aids identification of underlying causes.

The Groningen gas field in the Netherlands has been in production since 1965 and by 1983 the total quantity of gas abstracted exceeded 800×10^9 m^3. The gas reservoir occurs in sandstone belonging to the Permian Rotliegende formation. It is 70 m in thickness in the south increasing to 240 m in the north, occurs at a depth of around 3000 m, and extends over 900 km^2. Regular precision levelling surveys have been carried out since abstraction of gas commenced in order to monitor the amount of subsidence occurring at the ground surface. Contour maps showing the amount of subsidence have been produced from this data. Another method of subsidence monitoring has involved measuring the consolidation of near surface sediments by using borehole extensometers (Abidin *et al.*, 2001). Finally, movements of reservoir consolidation were taken by monitoring the relative displacement of radioactive bullets shot into the formation at regular intervals down a well (Schoonbeek, 1976). The initial levelling surveys indicated that subsidence was occurring at an amount considerably less than had been predicted. Furthermore, the consolidation coefficients of sediments determined in the laboratory turned out to be about three times those derived from field measurements. As a consequence, the mathematical model that had been developed to predict subsidence had to be modified (Geertsma and Van Opstal, 1973). The modified model has been used to predict the amount of subsidence up to the year 2050 when it is estimated that between 0.25 and 0.3 m will have occurred. Although not appreciable, this amount of subsidence still means that it is necessary to make provision for maintaining water management in the polder region of Groningen province. The model also was used to predict that the greatest amount of subsidence would occur before 1990. This prediction did not include any subsidence attributable to consolidation in the overlying Quaternary sediments. These average about 400 m in thickness and are consolidating at between 0.2 and 2 mm $year^{-1}$.

As referred to above, natural gas has been exploited in the Ravenna region, near Venice, in northern Italy. Natural gas also occurs in the Adriatic Sea some 4–40 km offshore from Venice. However, because of the high level of concern regarding the risk of subsidence and the possible effect that this could have on Venice, the project to develop this gas field was stopped by the Italian government in 1995. Subsequently, an investigation was undertaken to assess the possible amount of subsidence that would occur if the natural gas was exploited. This has involved both reservoir and subsidence modelling. According to Cassiani and Zoccatelli (2000), the results of the investigation indicate that no actual risk of subsidence exists at the coast. Nevertheless, if production goes ahead, then the natural gas reservoirs will be repressurized by the injection of water and an extensive monitoring programme will be put into effect. The latter will include precision levelling onshore, GPS monitoring of production platforms, the use of downhole radioactive markers to monitor consolidation, and the use of extensometers in shallow wells (i.e. less than 400 m depth) to record vertical movements.

Methane gas is dissolved in the groundwater of deep sandy aquifers beneath the Niigata area of Japan. Pumping groundwater from these deep sand layers, in order to extract the methane gas it contains, caused subsidence (Okumura, 1970). The piezometric level was lowered to a maximum of 44 m below sea level and resulted in considerable consolidation of the interbedded clay strata. The maximum rate of subsidence of 530 mm $year^{-1}$ was observed around the port area during 1958–59. Pumping restrictions were enforced from 1959 onwards and resulted in a marked recovery of the groundwater levels in the aquifers. In addition, Aoki (1976) reported that since 1973 water has been injected back into the gas reservoirs after the natural gas has been separated. Not only has this meant that the groundwater level has risen rapidly but rebound of the ground surface has occurred, the maximum amount measuring about 28 mm $year^{-1}$. However, this appears to be a temporary phenomenon.

5.7. Subsidence due to the abstraction of brine

Those deposits that readily go into solution, notably salt, can be extracted by solution mining. Salt occurs in nature either in the solid form as rock salt or halite, or in solution as brine. Rock salt occurs in sedimentary successions in beds varying from a few centimetres to hundreds of metres in thickness. Other evaporite minerals may be associated with rock salt. Most underground brines have been formed by groundwater dissolving either beds of rock salt or salt in saliferous beds. Brine abstracted by pumping may contain about 25% dissolved salt.

Salt has been worked in several areas of Britain and in some, like Cheshire, it has been obtained either directly by mining or indirectly by brine pumping (Cooper, 2002). Both these methods of extraction have involved subsidence, which in Cheshire was at times catastrophic. Consequently, subsidence due to salt extraction was an inhibiting factor, especially in certain parts of Cheshire, as far as major developments were concerned. This primarily was because of the unpredictable nature of the subsidence.

Rock salt in Cheshire is dissolved by circulating groundwater. As a result, salt generally is absent from within 60 m or so of the ground surface. Brines form locally immediately above the highest remaining bed of rock salt and the interface at which solution occurs or has occurred is referred to as a 'wet rockhead'. The complete sequence of saliferous beds is never present within wet rockhead areas, and the uppermost horizon is overlaid by sagged and brecciated mudstones. Where the saliferous strata are too deep to be affected by groundwater, the stratigraphical contact between these and the overlying rocks is described as 'dry rockhead'. Natural brine springs have been exploited since ancient times in Cheshire and were the basis of the medieval salt industry. At the end of the seventeenth century, however, the gradual lowering of the fresh water/brine interface led to shafts being sunk to pump the brine. Subsequently, natural or 'wild' brine was pumped from drillholes sunk to wet rockhead, which is at around 60–100 m depth. At present over 99% of the brine abstracted in Cheshire is produced by controlled solution mining, which involves salt being removed by solution from chambers located in the salt beds below dry rock-head so that there is no possibility of circulating groundwater dissolving the salt between the chambers. The individual chambers are large and may yield brine for 20–30 years. The chambers are set out on a grid pattern, averaging some 180 m apart, with an extraction ratio of about 20%. Basically, the controlled method involves pumping water down a steel tube into the chamber where dissolution of the salt occurs. The water is under sufficient pressure to raise the brine to the surface. The steel tubing is grouted into place to prevent leakage. Roof control of a chamber is effected by a cushion of air or oil and sonar techniques are used to control its size and shape. Once a chamber reaches the optimum size consistent with the stability of the roof rocks, abstraction ceases. The chambers then are used for the disposal of liquid waste from the salt industry. A typical chamber has a capacity of 500 000 m^3 for non-hazardous industrial wastes or 50 000 m^3 for difficult wastes. Controlled solution mining has resulted in no subsidence of any consequence.

The long-term behaviour of solution cavities depends on many factors. For a cavity that is located at sufficient distances from others, the main factors affecting its long-term behaviour are its size and shape, its depth and the *in situ* stress, as well as the mechanical properties of the rock salt and the internal pressure (Ghaboussi *et al.*, 1981).

Subsidence due to pumping of natural brine began in Cheshire about 1790 and after 1840, with the increasing production of brine, this type of subsidence became increasingly more important than subsidence due to salt mining. The most extensive wild brine pumping in Cheshire was carried out on the major natural brine runs (Bell, 1992). As wild brine was pumped to the surface in increasing amounts fresh unsaturated water was drawn to the beds of rock salt, hence giving rise to further solution, accelerating the formation of solution channels and associated subsidence. However, if pumping took place at a moderate rate, then the brine levels were not lowered to any significant extent. Active subsidence normally concentrated at the head and sides of a brine run where the fresh water first entered the system (Fig. 5.18(a)). Hence, serious subsidence could occur at considerable distances, anything up to 8 km, from pumping centres. As brine runs extended and widened the subsidence encroached into new areas and deepened existing depressions. It has been suggested that the subsidence troughs developed were somewhat similar to those developed by longwall mining of coal (Fig. 5.18(b)).

Subsidence due to wild brine pumping also gave rise to the formation of flashes. Flashes are water filled linear hollows that were developed in the superficial deposits (mainly tills) of wet rockhead areas. They may be about 10 m in depth by 50–70 m wide. They formed as a result of collapse above brine runs. Their sides are cambered, which is a surface manifestation of the associated subsidence curves. The flanks also may be interrupted by tension scars (c.f. subsidence faulting, Section 5.8) along which movements may occur (Fig. 5.19). Brine streams frequently merged and where they did, larger flashes were developed.

The worst subsidences were associated with bastard brine pumping that began in the 1860s and fortunately ceased in 1930. At the time bastard brine pumping commenced, many brine runs had failed due to the drainage of brine in wet rockhead areas into abandoned mines. As a result, the brine operators began to pump from the old mines. This hastened the solution of mine walls and pillars, bringing about collapse. Reckless driving of connections between partially flooded mines worsened the situation and culminated in a series of catastrophic collapses from 1870 onwards (Fig. 5.20).

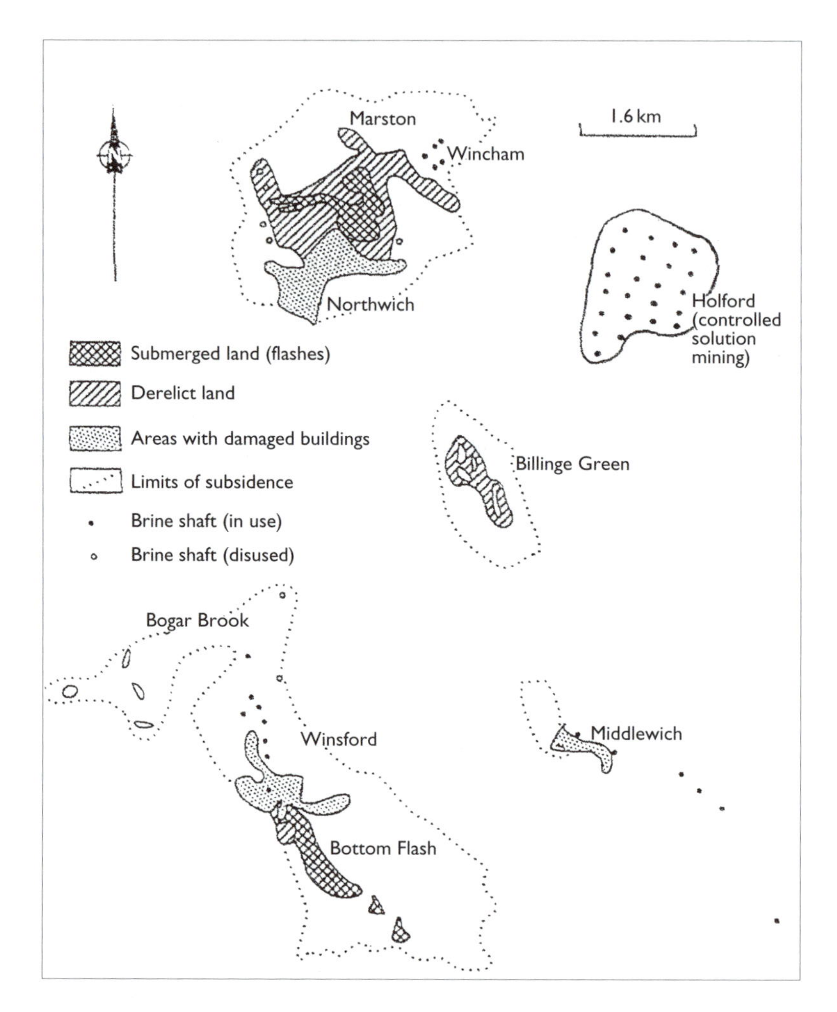

Figure 5.18a Subsidence and derelict land in Cheshire in the mid-1980s.

The geology of the deposits of rock salt, together with the character of the overlying strata and the location of brine pumps, influenced the extent and form that the subsidence took. Many of the subsidences were trough-shaped linear hollows with terraced sides. Indeed, many of the flashes were a consequence of this type of subsidence. Such features tended to occur where brine pumps were located more or less in line so that the brine runs merged into one broad channel. In areas where the overlying strata were stronger, collapse was not contemporaneous with removal of salt. Consequently, when collapse eventually occurred, a rapid subsidence of

Figure 5.18b Severe damage caused to farm buildings due to subsidence caused by the abstraction of salt brine, Cheshire, England.

Figure 5.19 Subsidence caused by wild brine pumping, Cheshire, England. Note the tension scar to the right of the fence and the flash in the top right-hand corner.

Figure 5.20 Catastrophic damage caused by bastard pumping of salt brine, Cheshire, England. (From Calvert, A.F. 1915. Salt in Cheshire, Spon, London.)

considerable dimension took place, rather than gradual subsidence. Where a brine run occurred beneath glacial deposits, especially sands, cavities were formed. These then collapsed to produce rather circular shaped small hollows at the surface, around 6 m or so in diameter. They made their appearance almost instantaneously but fortunately they occurred only infrequently. It would seem that when the sand collapsed it choked the brine run, which was diverted elsewhere. In this way intermittent localized subsidence was experienced over quite wide areas. This type of subsidence tended to be associated with areas in which the brine pumps had a more diffuse pattern so that brine runs presumably were not as concentrated as where pumps had a linear arrangement.

Although the distribution of linear subsidence channels can be fairly accurately defined, the stability of the ground immediately adjacent to these channels is still imperfectly understood. Although the potential subsidence can be estimated, the time taken for actual subsidence to match potential subsidence remains unknown. The ground movements developed by such subsidence could demolish buildings and it proved impossible to allot responsibility to any particular operator. Hence, two Compensation for Subsidence Acts (1891 and 1952) were passed that imposed levies on the operators in order to provide compensation funds for subsidence damage. The latter act primarily extended the compensation district (Fig. 5.21). As the time factor involved in subsidence and the exact dimensions of the area subjected to solution are unknown, it is difficult to ascertain whether subsidence is still occurring. It never really has been possible to predict when or where subsidence due to wild brine pumping was likely to occur or its

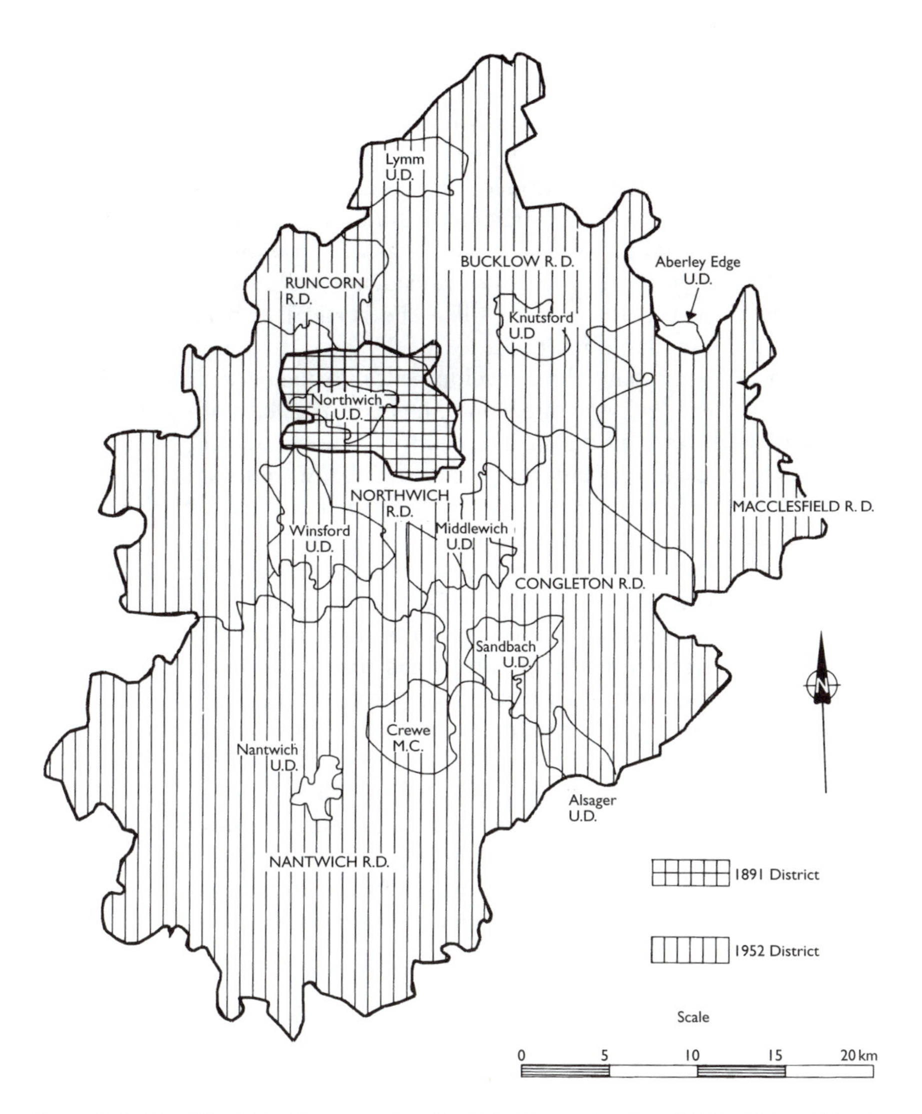

Figure 5.21 The Cheshire Compensation for Subsidence districts (U.D.=Urban District, R.D.=Regional District, M.C.=Metropolitan Council).

magnitude. Whenever any attempt was made to predict the amount and location of subsidence, then this could only be done by the extrapolation of past trends, a method that cannot guarantee success.

Because the exact area from which salt was extracted was not known, the magnitude of subsidence developed could not be related to the volume of salt worked. Consequently, there was no accurate means of predicting the amount of ground movement or strain. However, the rate at which subsidence occurred appears to have been related to the rate at which brine was extracted. Although it is not easy to tell, it seems that subsidence did not necessarily occur immediately after the removal of salt. For example, surface subsidence in some areas was found to continue for six months after the particular well thought to be responsible ceased to operate. Indeed, residual subsidence often continued for 1 or 2 years. The maximum strains developed by wild brine pumping could ruin buildings. What was worse as far as structural damage was concerned, was the formation of tension scars (small faults) on the convex flanks of subsidence hollows (Fig. 5.19). These features developed in the surface tills. The vertical displacement along a scar usually is less than a metre.

Subsidence meant that urban development in some parts of mid-Cheshire was severely curtailed. Although buildings constructed with a wooden or steel framework resting on a steel or reinforced concrete sill and capable of being jacked back into position after subsidence, were used to resist subsidence, their high cost acted as a deterrant to their construction. Reinforced concrete rafts also have been used for buildings in areas affected by subsidence. Services and communications also were damaged. For instance, in the late nineteenth and early twentieth centuries subsidence in Northwich, Cheshire, retarded the provision of paved streets, piped water supply, and main drainage and sewage disposal. Roads frequently were broken up by the formation of gullies and streets had to be raised periodically to prevent permanent flooding. Furthermore, the River Weaver, canalized in 1721, has been affected by subsidence for over 250 years, causing damage to banks, locks and bridges. The Trent-Mersey Canal was likewise affected. Rail communications in mid-Cheshire also have been affected by subsidence. The most notable example was on the main line between Crewe and Manchester at Elton where the ground surface was lowered by some 4.8 m between 1895 and 1960. Trains had to reduce speed over this section to around 40 km an hour and, more seriously, subsidence caused the destruction of the viaduct over the River Wheelock.

The legacy of damage caused by subsidence also includes damage to farmland, which ranged from deterioration of pasture through capillary movement of brine, to disturbance of the water table and natural drainage with associated flooding. Flooding caused by subsidence may be considered to be permanent dereliction of agricultural land, although flashes often have an amenity value. For example, they may provide recreational areas for sailing and fishing or reserves for wildlife.

Branston and Styles (2003) reported that subsidence in Northwich, Cheshire, was still going on and that in one area 230 mm of subsidence had occurred between July 1998 and July 2001. They undertook a series of microgravity surveys that indicated, in the area they were concerned with, that the subsidence was associated with a negative gravity anomaly. The latter was interpreted as a zone of lower density that was migrating.

Branston and Styles employed time lapse microgravity, which by using anomaly size and gradient can track the growth of cavities as they move towards the surface. In this way, they were able to demonstrate that the anomaly had grown in size over the period mentioned, maintaining that it extended from wet rock head at 23 m, upwards to within about 3 or 4 m of the surface. The low density of the anomaly was explained as a result of bulking due to fracturing as the cavity migrated towards the ground surface. Branston and Styles established that the anomaly was not related to old mine workings and therefore concluded that it was developed by dissolution of salt at wet rock head.

Subsidence similar to that described in Cheshire has occurred at Tuzla, Yugoslavia, according to Popovic (1970). Again the amount of ground movement and strain cannot be predicted with any degree of certainty. Wassman (1980) described subsidence that resulted from the solution mining of salt in the area around Hengelo in the Netherlands (Fig. 5.22). Cavities have been produced in the salt and they subsequently have become interconnected. Wassman quoted some 1.6 m of subsidence as having taken place in the centre of one subsidence trough and at Hengelo vertical subsidence amounting to about 50 mm year^{-1} was occurring. Again certain similarities with subsidence due to coal mining were observed. However, the duration over which the subsidence occurred differs and the area of influence at the surface in relation to depth of cavity was small. Furthermore, Wassman found that some cavities tended to collapse several years after pumping had ceased. In addition, brecciation has taken place in the overlying Red Beds of Bunter age (this mudstone loses its strength when wetted). Unfortunately, brine from the cavities in the underlying salt permeates into the mudstone, rising as a result of capillary action. Eventually, this weakened roof material collapses into the cavity. Void migration then takes place and, depending on the bulking factor of the rocks involved, the void may move into the overlying Tertiary clays. Pronounced subsidence occurs when this happens. The subsidence basin in these latter deposits is demarcated by an angle of influence of 45°, which extends upwards from the contact between the Red Beds and Tertiary clays. An additional cause of ground movement is attributable to consolidation that occurs in the brecciated mudstones. This is responsible for slowly decreasing, but long-lasting, subsidence.

Solution mining of rock salt has taken place in several parts of the United States where water is introduced into the salt formations by wells, thereby forming brine that then is recovered. Underground cavities are formed that are filled with brine. Johnson (1998) pointed out that these cavities may be 10–100 m in diameter and 10–600 m in depth. He further mentioned that these cavities can be joined hydraulically and so can have a long narrow shape. Two examples of cavity collapse were quoted by Johnson, namely, the Cargill Sink in Kansas and the Grand Saline Sink in Texas, where the

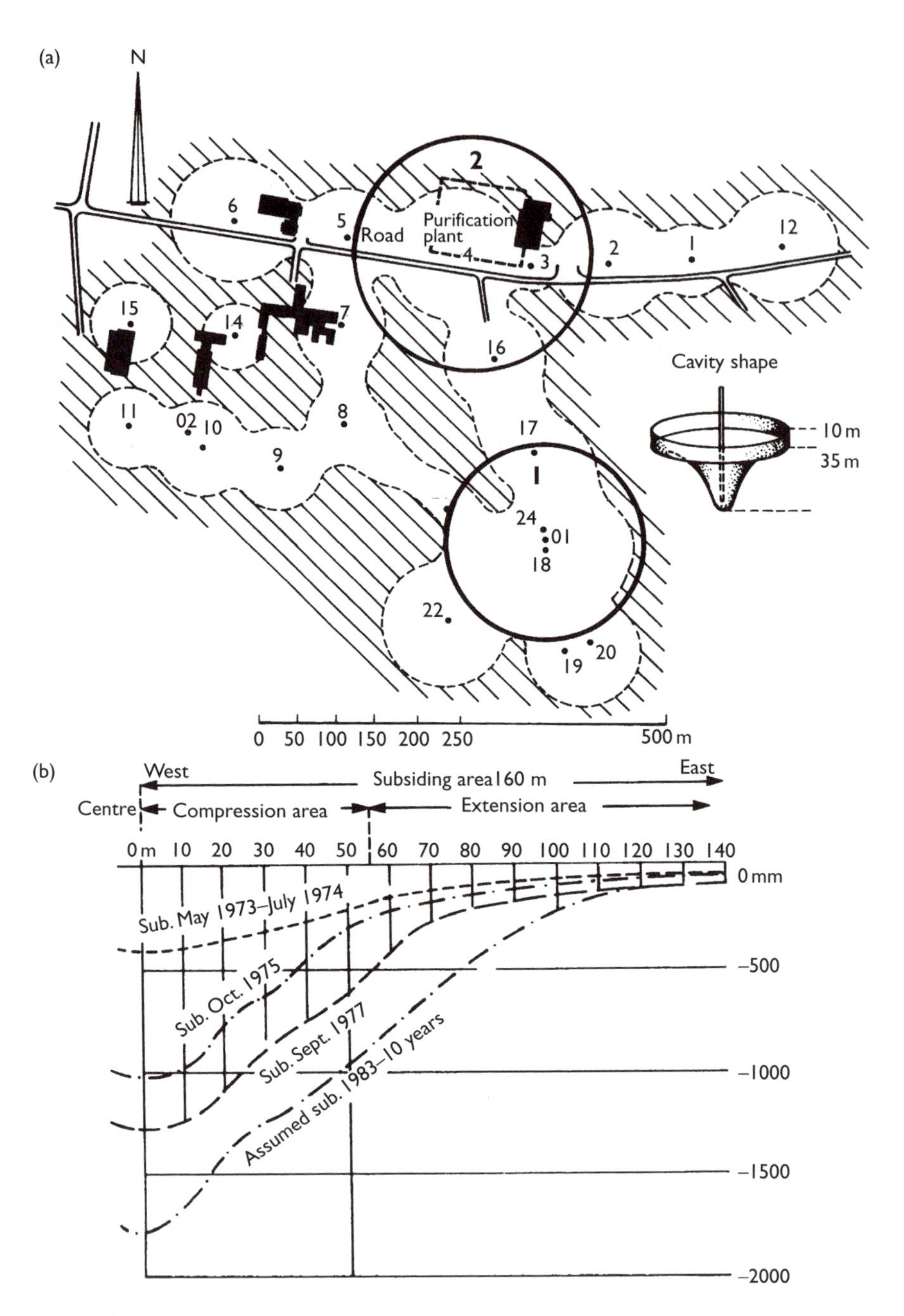

Figure 5.22 (a) Location of wells for the solution mining of salt and associated subsidence areas, Hengelo, The Netherlands. (b) Actual measured and predicted subsidence of Area 2 during 1973–83. (After Wassman, 1985; reproduced by kind permission of Kluwer Academic Publishers.)

roof spans exceeded their capacity to support the overlying rocks. However, Johnson indicated that most solution mining collapses in the United States were associated with cavities formed 50–100 years ago, modern day design of solution cavities more or less having eliminated this problem (c.f. controlled solution mining in Cheshire).

Deere (1961) described differential subsidence (resulting from the development of a concavo-convex subsidence profile), simultaneous horizontal displacements and surface faulting associated with the abstraction of sulphur from numerous wells in the Gulf region of Texas. The sulphur occurs at a depth of 397–488 m within the cap rock of a salt dome. The cap rock is overlain by sands, gravels, clays and clay-shales. Vertical and horizontal movements were noticed shortly after mining operations commenced. After nine months the maximum vertical subsidence amounted to 0.45 m and the diameter of the subsidence basin was approximately 600 m. During the first 31 months of operation, differential subsidence amounting to 1.75 m occurred over an elliptical area exceeding 5 km^2. The subsidence basin was centred directly above the narrow linear producing zone. A normal fault, around 650 m in length, with a downthrow of 0.1 m on the mining side and peripheral to the subsidence basin, running more or less parallel to the subsidence contours, developed suddenly during the fifth month of production (Lee and Strauss, 1970). By the thirty-first month the length of the fault had increased to 800 m and the displacement to 0.3 m. The fault dipped at about 40° directly inwards towards the top of the producing zone and it formed at a point of maximum surface tension (c.f. the tension scars associated with salt extraction referred to above). Mining of sulphur and potash in Louisiana has given rise to vertical subsidence ranging up to 9 and 6 m, respectively.

5.8. Ground failure and the abstraction of fluids

Ground failure often is associated with subsidence due to the abstraction of groundwater, oil, natural gas or brines. Because the withdrawal of groundwater exceeds that of other fluids, some of the most notable failures are found in such subsidence areas. For instance, ground failures associated with subsidence due to groundwater abstraction occur in at least 14 areas in six states in the United States. Some of the best examples of ground failure attributable to the withdrawal of groundwater are found in southern Arizona, the Houston-Galveston region of Texas and the Fremont Valley in California. By contrast, only four ground failures have been reported in the San Joaquin Valley in California, one of the most noteworthy areas of subsidence due to groundwater abstraction.

Two types of ground failure, namely, fissures and faults are recognized. Fissures may appear suddenly at the surface and the appearance of some may be preceded by the occurrence of minor depressions at the surface. Within a matter of a year or so most fissures become inactive. In some instances new fissures have formed in close proximity to older fissures. Rojas *et al.* (2002) proposed an approximate analysis of the fissuring process that could be used to predict the location of fissures. Their analysis indicated that tension cracks develop on the flanks of a subsiding basin where differential subsidence is occurring. They also maintained that with continuing differential subsidence fissures may evolve into faults. Faults develop as small scarps at the surface and their development depends on the thickness and compressibility of the soils involved, together with the rate at which the groundwater is abstracted.

Fissures vary in length from a few tens of metres to a few kilometres. The longest fissure zone in the United States is 3.5 km and lengths of hundreds of metres are typical (Holzer, 1980). Individual fissures commonly are not continuous but consist of a series of segments with the same trend. Exceptionally segments may form an *en* echelon pattern. Segments are not connected at the surface. Occasionally, a zone defined by several closely spaced parallel fissures occurs, the width of which may be of the order of 30 m. Secondary cracks frequently are associated with fissures. These cracks develop sub-parallel to the main fissure and usually are only a few metres long. They occur at distances up to 15 m from the parent fissure. Fissure traces range from linear to curvi-linear in outline. Fissures commonly intersect but do not cut through one another.

Fissure separations tend not to exceed a few tens of millimetres, the maximum reported being 64 mm. They, however, frequently are enlarged by erosion to form gullies 1–3 m in width and 2–10 m in depth (Fig. 5.23). The greatest measured depth is some 25 m.

Fissure development has caused an increasing problem in the alluvial basins of southern Arizona over several decades. On average groundwater levels have declined by some 45–120 m, accompanied by more than 0.3 m of subsidence. According to Larson (1986), subsidence generally begins when the decline in water level exceeds 30 m and with continued subsidence, fissures form at points of maximum horizontal tensile stress associated with maximum convex upward curvature of the subsidence profile (c.f. with the formation of tension scars on the subsidence curves developed by the abstraction of brine noted in Section 5.7). Certain conditions are believed to localize differential subsidence such as buried bedrock hills and scarps; changes in sedimentary facies; the hinge or zero line of subsidence; the edge of an advancing subsidence front; man-made changes in vertical loading; and neighbouring recharge mounds. Geophysical surveys (notably gravity surveys) indicate that most fissures in southern Arizona are associated with bedrock hills. The most critical depths for bedrock features would appear to be from 30 to 500 m because the greatest amount of consolidation of sediments occurs at this interval as the groundwater levels decline. In fact, most

Figure 5.23 A ground fissure in south central Arizona. The fissure is enlarged by erosion. (Courtesy of Dr T.L. Holzer; reproduced by kind permission of the United States Geological Survey.)

irregularities in the bedrock surface are inferred to occur at depths of less than 250 m. The association of fissure occurrence with variable aquifer thickness also suggests that differential consolidation is taking place near these fissures as groundwater levels fall. Theoretical estimates of horizontal strains thereby generated indicate that this mechanism is the dominant source of horizontal tension causing fissures (Jachens and Holzer, 1980). Tensile strains at fissures at times of their formation have ranged from 0.1% to 0.4%.

However, gravity and magnetic surveys in areas of the Lucerne Valley, California, as well as in some parts of south-eastern Arizona, where the fissures have polygonal patterns, according to Holzer (1986), did not indicate special subsurface conditions beneath the fissures. He suggested that these fissures probably are caused by tension induced by capillary stresses in the zone above the declining water table (c.f. the formation of desiccation cracks, which admittedly are on a smaller scale). Holzer further noted that in the Las Vegas Valley, Nevada, and the Fremont and San Jacinto valleys, California, some fissures are coincident with existing faults. Subsequently, Bell and Helm (1998) showed that around 80% of the fissures in the Las Vegas Valley occur within some 350 m of a fault trace. They argued that there was a relationship between Quaternary faults and subsidence related fissures, the faults providing pre-existing planes of weakness along which vertical subsidence may be localized. Sheng and Helm (1998) suggested four models to help explain the association of fissuring with groundwater abstraction. These models included an aquifer containing horizontal planes of weakness (i.e. bedding planes); an aquifer interrupted by a pre-existing fault; an aquifer above a ridge of bedrock; and an aquifer containing heterogeneities (i.e. abrupt changes of thickness or other types of beds within the aquifer).

Those faults that are suspected of being related to groundwater withdrawal are much less common than fissures. They frequently have scarps more than 1 km in length and more than 0.2 m high (Fig. 5.24). The longest such fault scarp in the United States is 16.7 km long and the highest scarp is around 1 m, both being in the Houston-Galveston region, which is that most affected by such faulting in the United States. In fact, an estimated 160 faults have been associated with subsidence in the Houston-Galveston region from 1906 to 1978, resulting in approximately 500 km of observed fault scarp generation. These fault scarps have been found to increase in height by dip-slip creep along the normal fault planes. Measured rates of vertical offset range from 4 to 60 mm annually, however, movement tends to vary with time. Although some short-term episodic movement has been reported, seasonal variations of offset that correlate both in magnitude and timing with seasonal fluctuations of water level are remarkably

Figure 5.24 Faulting induced by a decline in the water table due to groundwater abstraction in the Fremont Valley of southern California (Courtesy of Dr T.L. Holzer; reproduced by kind permission of the United States Geological Survey.)

widespread (Holzer, 1986). The land surface near the scarp may tilt, tilting being greatest near the scarp, but has been observed to extend as far as 500 m from the scarp.

The relationship between faulting and fluid abstraction (both of oil and gas, as well as groundwater) in the Houston-Galveston region is ambiguous. Many faults in this region are thought to act as hydrological barriers. Consequently, fluid production on one side of a fault may cause the piezometric surface to decline with attendant consolidation on that side of the fault and not the other. This differential consolidation may be translated to the surface as shear displacements along the faults. Buckley *et al.* (2003) have demonstrated by use of radar interferometry that differential subsidence is occurring along some faults and quoted the Long Point Fault in north west Houston as an example.

Van Siclen (1967) suggested that because faults appeared to form in all positions and orientations within subsidence bowls in the Houston-Galveston region and so crossed subsidence contours, then subsidence was not responsible for their formation. He therefore maintained that increases in vertical effective stress caused by falling piezometric surfaces are responsible for reactivating existing faults. Castle and Youd (1972) challenged Van Siclen's conclusions. Their alternative model suggested that surface faulting was the result of centripetally directed horizontal movements and radially orientated strain caused by consolidation of the sediments concerned. Several observations contradict the Castle and Youd model and favour fault control of the piezometric surface and differential consolidation. These include reversals in the direction of fault movement and increased subsidence on both the upthrown and downthrown sides with the fault remaining in a zone of minimal subsidence.

Kreitler (1977) maintained that in the Houston region all identified active faults with significant displacement occur in areas of intensive fluid production whilst there is little evidence of recent movement along faults in areas where fluids are not abstracted. He further argued that if these structures are followed along their strike from areas of non-production into areas of fluid production, then there is a change from a passive to an active fault condition.

In southern Arizona faults have been found to occur around the peripheries of subsidence basins. The downthrows of the high angled normal faults are on the side towards the centre of the basin and displacements are usually less than 1 m. The faults are shallow in depth, being confined to the zone where stresses due to lowering of the groundwater level occur (Holzer and Thatcher, 1979).

One of the most notable and most studied faults in this context is the Picacho Fault in Arizona. Faulting near Picacho began about 1961 and post-dated the commencement of the lowering of groundwater level and associated subsidence. Furthermore, it occurred along a zone in which there is no evidence of any faulting prior to groundwater level declines. It also appears that vertical displacements based on closely spaced bench marks near the fault are compatible with subsurface faulting restricted to the zone affected by the falling groundwater level. In other words, the Picacho Fault only extends to a depth similar to the thickness of alluvium affected by the decline in the piezometric surface. Finally, the fault undergoes seasonal displacements that suggests a seasonal variation in the stressing of the system due to seasonal changes in the amount of groundwater withdrawn. Faulting therefore is assumed to have been preceded by differential subsidence over a narrow zone, that is, a flexure, with significant subsidence occurring on both sides of the fault scarp (Holzer *et al.*, 1979). Indeed, fissuring initially was observed along the outcrop position of the fault in 1949. In the period from 1963 to 1977 subsidence took place over a basin of approximately 12 km in diameter, the maximum observed subsidence being 1.25 m. Ground movements were concentrated along the fault. Creep rates across the fault scarp were measured at approximately 60 mm year^{-1} during the initial phase of movement, decreasing to about 10 mm year^{-1} by the late 1970s.

A notable fissure system has developed in Shanxi Province, China, which Li *et al.* (2000) referred to as the Shanxi Graben System. They noted that hundreds of fissures have formed since the 1950s and that the fissures, together with subsidence, especially differential subsidence, have given rise to substantial damage to building, roads, canals and farmland. Displacement has taken place along some of the fissures and normal faults have been reactivated in the Quaternary deposits (Lee *et al.*, 1996). The ground rupturing that this has given rise to is manifested, for example, in Xian City where extensive damage to property has occurred along the paths of propagation (Fig. 5.25). Li *et al.* suggested that the fissures have more than one origin. Some fissures are related to the reduction in pore water pressures as accelerated groundwater withdrawal leads to a decline in the water table. This is particularly the case in Xian City where Li *et al.* pointed out that the groundwater levels were lowered by 10–15 m between 1975 and 1979. However, since then groundwater levels have stabilized and ground fissure expression has decreased accordingly. Other fissures are associated with faults that trend in a north easterly direction. Trenching has shown that these fissures extend into faults that are near the surface and therefore the fissures are regarded as a surface expression of movements along the faults. Yet other fissures, such as those in the Linfen and Yuncheng Basins, are associated with the collapse that has taken place in the loess as the groundwater levels have been lowered.

Fissuring has also been noted in loess like soils in western Saudi Arabia by Bankher and Al-Harthi (1999). There the groundwater level has declined as a result of pumping above the safe yield from the wadi aquifer. The subsidence that has taken place in unconsolidated metastable silty deposits has given rise to fine fissures above buried surface ridges. These fissures subsequently have been widened by floods. Monitoring of the ground

(a)

(b)

Figure 5.25 (a) Damage to a building caused by a ground fissure in Xian City, China. (b) A fissure in Xian City, China. The ground to the right of the fissure has been lowered by some 500 mm. The upper storey of the building immediately behind the wall had to be taken down due to the effects of the fissure.

surface using GPS has shown a good relationship between the subsidence of the ground and the decline in the water table.

5.9. Prediction of subsidence due to the abstraction of fluids

Obviously, it is necessary to determine the amount of subsidence that is likely to occur as a result of the withdrawal of fluids from the ground, as well as to estimate the rate at which it may occur. Unfortunately, a large number of prediction methods have been developed, some of which are relatively simple whilst others are complex. One of the probable reasons for this is that stratal sequences are different in different areas where subsidence has occurred and consequently different models have been devised. Nonetheless, a number of steps can be taken in order to evaluate the amount of subsidence that is likely to occur due to the abstraction of fluids from the ground. These include defining the *in situ* hydraulic conditions; computing the reduction in pore pressure due to removal of a given quantity of fluid; conversion of the reduction in pore pressure to an equivalent increase in effective stress; and estimating the amount of consolidation likely to take place in the formation affected from consolidation data and the increased effective load (Saxena and Mohan, 1979). In addition to depth of burial, the ratio between maximum subsidence and reservoir consolidation also should take account of the lateral extent of the reservoir in that small reservoirs that are deeply buried do not give rise to noticeable subsidence, even if undergoing considerable consolidation, whereas extremely large reservoirs may develop significant subsidence. The problem therefore is three-dimensional rather than one of simple vertical consolidation. Subsidence prediction in relation to the rate of groundwater pumping therefore should involve relating the consolidation model to a two- or three-dimensional hydrogeological model based on the groundwater flow equation. Variations in the hydraulic head in both time and space in response to groundwater abstraction are obtained from the hydrogeological model. These values then can be used to derive the time-dependent consolidation curve at any point in the system. This provides an indication of the amount of subsidence likely to occur.

Pottgens (1986) suggested that a simple rapid estimate of subsidence, S, for a disc-shaped reservoir could be obtained from the following expression:

$$S = \pm 2(1 - \nu) \times (1 - C)/(1 + C^2) \times m_v \times \sigma \times H \qquad (5.2)$$

where ν is Poisson's ratio, $C = z/R$, where z is the depth and R is the radius of the disc-shaped reservoir, m_v is the coefficient of volume compressibility, σ is the increase in load due to reservoir pressure reduction and H is the thickness of the reservoir.

The ratio of subsidence to head decline in coarse grained permeable beds of consolidating aquifer systems represents the ratio between these two factors over a common interval of time. This ratio reflects the change in thickness per unit change in effective stress and can be used to predict a lower limit for the amount of subsidence in response to a given increase in virgin stress (stress exceeding the past maximum). If pore pressures in the consolidating aquitards reach equilibrium with those in adjacent aquifers, then consolidation ceases and the subsidence-head decline ratio represents a true measure of the virgin compressibility of the system. Until equilibrium of pore pressure is attained, the ratio of subsidence to head decline is a transient value. Contours of this ratio for a given period of time can be plotted on a map and indicate the amount of head decline required to produce a particular magnitude of subsidence throughout the area concerned.

Like mining subsidence prediction methods, prediction methods associated with subsidence due to fluid abstraction also can be grouped into empirical, semi-empirical and theoretical categories. Empirical methods involve the extrapolation of available information such as the amount of subsidence, the amount of consolidation or the decline in fluid level being plotted against time in order to determine future trends (Figueroa Vega and Yamamoto, 1984). The semi-empirical approach also depends on the relationship between subsidence and related phenomena. For instance, Castle *et al.* (1969) found that subsidence in six oil fields in the United States varied more or less linearly with net production of fluid but the correlation between the decline in reservoir pressure and subsidence was poor. The most likely explanation of this poor correlation is that pressure decline as measured at individual producing wells probably is not representative of the average decline over a field.

As remarked earlier, the abstraction of fluid from the ground reduces the pore pressure that leads to a transfer of load to the granular skeleton and to its subsequent reduction in volume, primarily due to a decrease in void space. However, in trying to develop an explanation of the phenomenon one encounters the problems associated with multi-phase systems representing solids, liquids and gases, the properties of which must be inferred from statistical averages or from representative tests. The materials concerned include in their mechanical properties, the combined behaviour of their individual components (i.e. elasticity and plasticity of solids; viscosity of liquids; compressibility of gases; decay of organic matter; attraction and repulsion of ionic charges etc.). Hence, the mechanical properties are anisotropic, as well as dependent on stress history and time. Accordingly, such material is difficult, if not impossible, to deal with in a theoretical model of subsidence. Resort has to be made to simplifying assumptions in order to develop any type of model of subsidence prediction, in particular, the properties of the ground must be idealized. These include accepting the principle of effective stress, that displacement and deformation are small, that the soil skeleton behaves like a perfectly elastic solid within the range of linear elasticity, and that solid grains are incompressible (Gambolati *et al.*, 1998). Other simplifications may include the assumption that strata are horizontal; that flow in aquifers is horizontal whilst it is vertical in aquitards; and that subsidence primarily is due to consolidation of aquitards. Obviously, the greater the number of simplifications that are incorporated into a model, the more restricted its use becomes.

Subsidence above groundwater, oil or gas reservoirs frequently is treated as a mechanical problem involving the elastic behaviour of the reservoir and is analysed with the aid of a mathematical model. Unfortunately, the amount of data available frequently is limited, which means that, as remarked earlier, certain assumptions have to be made before the model can be applied. For example, Geertsma (1973) simulated the subsidence above the Groningen gas field by assuming a homogeneous and isotropic semi-infinite porous medium. The reservoir was assigned an idealized shape, that is, a horizontal circular cylinder of limited thickness. The ratio between maximum subsidence and reservoir consolidation was governed by the ratio between depth of burial and the lateral extent of the reservoir.

Finite element analysis frequently is used in modelling the amount of land subsidence an area is likely to undergo due to the abstraction of fluids. The finite element method has proved a useful mathematical tool to deal with problems in which material properties are inhomogeneous since material homogeneity only is required within elements of a small area (Shimizu, 1998). However, sufficient data must be available if finite element modelling is to provide reasonable results. If there is a lack of agreement between modelled and actual measured subsidence, then this normally is attributable to insufficient data being available for input into the model concerned. Lewis *et al.* (2003) used the finite element method to model the subsidence of the oil and gas reservoir in the Ekofisk Field in the North Sea, and produced a map of seabed subsidence that was in general agreement with that based on previous bathymetric surveys. Suzhou City in Jiangsu Province is one of the few cities in China that suffers severe ground subsidence. This subsidence is associated with the continuously increasing groundwater abstraction from a deep multilayered aquifer system in which three distinct mud formations occur. Chen *et al.* (2003) used a three-dimensional finite difference model to investigate the resultant ground subsidence. They claimed that the model outputs fitted reasonably well with the observed subsidence, which suggests that the model can reproduce the dynamic processes of groundwater flow and consolidation of the mud formations.

Gambolati and Freeze (1973) also developed a mathematical model to simulate subsidence due to groundwater withdrawal beneath Venice. First, with the aid of the finite element technique, the regional drawdowns in hydraulic head were determined in a two-dimensional vertical cross section in radial coordinates, using an idealized ten-layer representation of the geological conditions. Then, the values of the hydraulic head calculated for the aquifers were used as time-dependent boundary conditions in a set of one-dimensional vertical consolidation models solved by the finite difference technique and applied to a more refined representation of each aquitard. Subsequently, Gambolati *et al.* (1986, 1991) developed a method of analysis of subsidence due to the withdrawal of oil and gas from a reservoir overlain by layered anisotropic soils by using the finite element technique. They assumed a disc-shaped reservoir of uniform thickness, which underwent elastic deformation. The model consisted of layered anisotropic soil units that were characterized by five elastic constants. Alternating sands and clays were assumed to occur above the reservoir, their compressibility progressively decreasing with depth. They found that for a given geometry, depth of burial and fluid pressure decline, the subsidence was basically related to the compressibility of the oil-gas bearing strata and of the adjacent overlying/underlying clays. The depth at which a rigid basement occurred beneath a reservoir appeared to have only a limited influence on ground movement.

A geographical information system (GIS) based approach was used by Zhou *et al.* (2003) to predict regional groundwater flow and land subsidence in the Saga Plain, Japan. The system consists of three parts, namely, surface water hydrological cycle simulation, groundwater flow simulation and land subsidence simulation. Various data were used for a 21-year period to simulate subsidence and the results compared fairly with the observed results thereby allowing future land subsidence to be predicted assuming different groundwater abstraction scenarios. Such simulations provide information that will aid future subsidence mitigation decision making.

5.10. Mine water rebound

As noted earlier, rising water tables can give rise to problems. Similarly, rising mine water or mine water rebound, which occurs when mine dewatering ceases, can be responsible for problems, notably pollution of groundwater and surface water courses. Mine water is water that can be traced as having had contact with mine workings. Deep mining, in particular, has had to deal with the problem of dewatering of the workings. Early Roman mines in favourable locations utilized archimedean screws, rag-and-chain pumps or gravity drainage via tunnels (soughs) to discharge mine water into rivers or the sea but it was the development of the steam engine that enabled large volumes of water to be pumped from the mines (Davies, 1979). Frequently, deeper workings, especially in coal mines, became deliberately and complexly interlinked, often from one to another, to facilitate access, ventilation and mineral clearance. Hence, it became necessary to maintain a network of pumping stations utilizing old shafts or boreholes in order to protect mine workings from flooding.

Flows of water into mine workings come from the overlying rock masses, the workings acting as a drain. They also may be related to shafts, adits and other mine entries. Such flows may be consistent and similarly originate from the overlying rock masses. Water in old shallow mine workings is derived mainly from precipitation that has percolated into the ground. Fault zones also may contain significant volumes of water. However, the flow rates may be relatively small and are controlled by the permeability of the faulted rocks that they intersect. On the other hand, some faults may act as barriers to flow, for example, they may displace impermeable mudrocks with the fault zone being occupied primarily by clay gouge. Increasing pressures with depth on the one hand and the distance from mine entries on the other tend to mean that flows are controlled by discontinuities. In addition, mining and any associated subsidence frequently enlarge the gape of existing discontinuites in the overlying rock masses thereby increasing their storage capacity and permeability. Groundwater recovery occurs when a mine closes and pumping ceases. Flow rates reduce as the head of pressure controlling the flow reduces and with time tend to stabilize. Mine water rebound is likely to drive mine gas ahead of it in cases where the mines concerned contain gas.

Taking British coalfields as an example, there is at present a regional recovery of the groundwater levels (Younger and Adams, 1999; Younger, 2000). The complete closure of a coalfield and cessation of pumping ultimately results in flooding of the old workings and subsequent rebound of groundwater levels until equilibrium is reached. Groundwater rebound may flow into neighbouring mines or may rise and discharge at the surface from fault outcrops, mine entries, coal outcrops or fractured or permeable strata. In the Durham Coalfield, for instance, many years of pumping had dewatered the strata up to depths of 150 m and rebound is now taking place. The time scale of mine water rebound is difficult to determine. However, modelling studies of the Durham Coalfield undertaken by Sherwood and Younger (1994) indicated that if pumping were to cease completely, then mine water would rebound to ground level in 15 years in the southern part of the coalfield, increasing to 40 years in the northern part. In the Leicestershire Coalfield new seepages were predicted to appear by Smith and Colls (1996) at the surface six years after the last pumps were switched off. If the new water level intersects surface water courses, then uncontrolled mine water discharges will result.

The former British Coal Corporation supplied annual pumping data for each of its deep mine discharges and in 1988 it was estimated that 840 mega-litres a day of water were pumped from mines in England and Wales. Of this, 97% was discharged as waste into rivers and streams (Anon., 1998). Mine water pumping systems were designed to protect the deeper workings and are still active in some parts of Britain. For instance, the Woolley pumping station, South Yorkshire, which now is operated by the Coal Authority, has an output of around $152\ l\ s^{-1}$.

Mining subsidence is likely to have induced fractures in the overlying rocks, enhancing their hydraulic conductivity, and creating new pathways for mine water to migrate upwards through the strata on mine closure. For instance, eventual closure and cessation of pumping in the East Pennine Coalfield could result in mine water being driven up (via fractures, faults or inadequately sealed old shafts and drillholes) into the overlying Permo-Triassic aquifers due to the geometry of the interlinked workings coupled with high potential driving heads. This may cause pollution of potable water supply abstractions. Conversely, many abandoned mines now function as aquifers (Wood *et al.*, 1999).

Groundwater derived from near surface Coal Measures contains approximately 60–80% magnesium and calcium bicarbonates with some sodium sulphate and chloride. The dissolved salt concentration is usually less than 1000 mg l^{-1}. Groundwater from deeper seams contains higher proportions of chlorides. Groundwater that has been in contact with mine workings frequently has had its chemistry changed by reactions with iron pyrite and, if so, will have lower concentrations of alkalinity and higher concentrations of sulphate (see Chapter 8). Due to solution of pyrite oxidation products, mine water is commonly highly ferruginous, with an ochreous colour, and often has low pH values. Where mine water flows into surface water courses the latter may become grossly polluted, with extensive precipitation of ochreous ferric hydroxide on stream beds. Other metals may occur in solution, derived from natural sources and artificially introduced by mining equipment. For instance, Younger and Bradley (1994) mentioned aluminium hydroxide being precipitated from mine water discharges in the abandoned Durham Coalfield in Britain. Mine water discharging from abandoned metalliferous workings can contain high concentrations of metals that are toxic (Dumpleton *et al.*, 1996; see Case history 1, Chapter 4). Mine water also may become polluted with organic compounds derived from spills of lubricating and hydraulic oils underground (Glover, 1983). Connate water from Coal Measures sources are highly saline water and have been recorded from many Coal Measures rocks at depth.

Not all mine water discharges pose a threat to the environment. Some discharges provide dilution to otherwise poor quality streams or maintain flow rates during times of drought. This may be due to the fact that the groundwater in such instances is maintained at levels lower than the oxidized mineral horizon, or overlying carbonate rocks may have a buffering effect on originally acidic mine water.

Mine water rebound may be monitored in drillholes. Alternatively, conceptual models may be generated for individual mines and mining regions to predict the rate of mine water rebound and the possible date when surface discharge may occur (Fig. 5.26). Several software packages are now available to enable this to be undertaken (Burke and Younger, 2000; Dumpleton *et al.*, 2001). Models have tended to consider the general principles and simplistic flow over mining areas rather than the detailed and often complex flow regimes at individual mines. Mine water models require information on mine void space (no easy task, especially when no abandonment plans exist), and this sometimes can be estimated by digitizing the area of the workings, roadways and shafts, from the lowest to the highest point of a mined area. Generally, mine water rebound prediction rates have

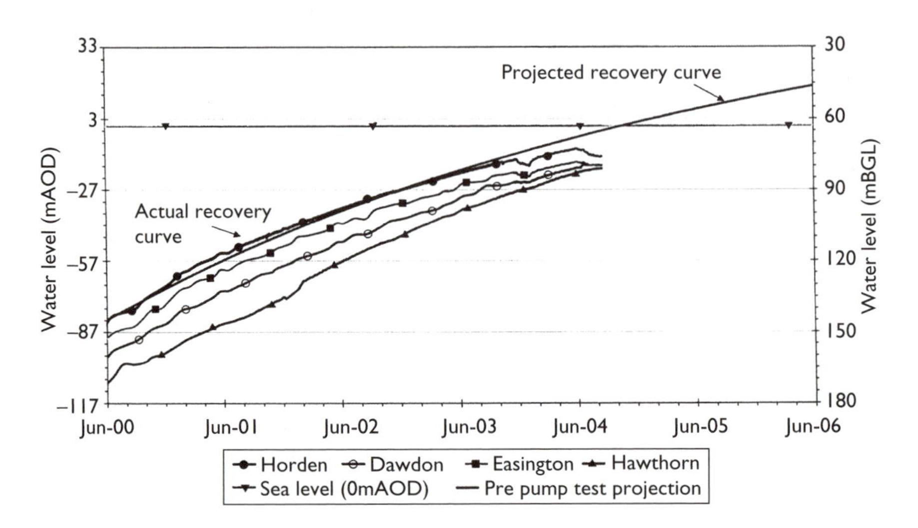

Figure 5.26 Projected recovery of mine water following mine abandonment and cessation of pumping at some mines in County Durham, England. (Reproduced by kind permission of the Coal Authority.)

appeared to be consistent with data obtained from mine water monitoring stations. Several factors may need to be considered when developing mine water rebound prediction models. These include:

- The type, thickness and extent of the overlying strata and the aquifers present.
- The amount and seasonal distribution of precipitation falling over the area concerned.
- The location of the water table.
- The zone of undisturbed or unmined strata involved.
- The zone of mining and area of influence.
- The presence and nature of faults and other rock mass discontinuities in the area under consideration.
- The mine entries, boreholes and wells in the area, together with groundwater abstraction and discharge.
- The nature and state of the mine workings.
- The sources of water and levels at which they enter mines, along with the hydraulic heads need to be identified and, if possible, estimated.

Flow paths from monitoring of mine water recovery in coalfields throughout Britain have shown that the mine water recovery process is progressive and that the recovery rates reduce exponentially. This, however, is dependent upon the bulk transmisstivity of the Coal Measures and the pressure heads driving the inflows. Considerable progress has been made in establishing monitoring points, notably in the Yorkshire and Durham Coalfields. Data on mine water levels in coalfields in Britain are now held by the Coal Authority on a Hydrolog database.

Obviously, mining activity has disrupted the hydrogeology in mining areas, sometimes so badly that it is difficult to predict where any mine water would emerge. However, McGinness (1999) pointed out that although individual coal mine closures in Britain were occurring that acid mine drainage associated with mine water rebound had not yet become a massive problem. This he noted was because pumping was continued so as to allow working to carry on at nearby mines. But now that whole coalfields have closed, acid mine drainage has become a serious problem. Although, as mentioned in Chapter 8, not all mine drainage is acidic, it may be ferruginous and contain heavy metals. It therefore is necessary to treat acid mine drainage to avoid pollution of water courses and groundwater in productive aquifers. Even so, the treatment of acid mine drainage by most current active and passive technologies produces a heavy metal containing residue that requires to be disposed of. Because of the heavy metal content, this may require special conditions for the disposal of the waste.

5.11. A note on groundwater lowering and pollution

Pollution can be regarded as impairment of water quality by the addition of chemical, physical or biological substances to an extent that does not necessarily create an actual public health hazard, but that it does adversely affect such water for domestic, agricultural or industrial use. The greatest danger of groundwater pollution generally is from surface sources. Areas with thin soil cover or where an aquifer is exposed as in the recharge area, are the most critical from the point of view of pollution potential. Concentrated sources of pollution are most undesirable because the self-cleansing ability of the soil/rock in area concerned is likely to be exceeded. As a result a raw pollutant may be able to enter an aquifer and travel some considerable distance from the source before being reduced to a negligible concentration.

When water is abstracted from a well the water table in the immediate vicinity is lowered and assumes the shape of an inverted cone that is referred to as a cone of depression. The steepness of the cone is governed by the soil/rock type(s) penetrated, it being flatter in highly permeable materials. The size of the cone depends on the rate of pumping, equilibrium being achieved when the rate of abstraction balances the rate of recharge. However, if abstraction exceeds recharge, then the cone of depression increases in size. As the cone of depression spreads, water is withdrawn from storage over a progressively increasing area of influence and the groundwater levels about the well continue to be lowered. Assuming relatively uniform conditions, the flow towards the well is greatest on the side nearest the source where the gradients are steepest. As pumping continues the proportion of water entering the cone of depression that is likely to be derived from a surface source increases progressively. The risk of pollution occurs when a surface source that itself is polluted is tapped. The list of water pollutants is almost endless and some of the more common surface sources of pollution include domestic, industrial, mining or agricultural wastes; polluted streams; excessive use of fertilizers, pesticides and/or herbicides; and volatile organic chemicals.

Induced infiltration occurs where a stream is connected hydraulically to an aquifer and lies within the area of influence of a well. If influent seepage of surface water is less than the amount required to balance the discharge from a well, the cone of depression spreads up and downstream until the drawdown and the area of the stream bed intercepted are sufficient to achieve the required rate of infiltration. If the stream bed has a high permeability, the cone of depression may extend over only part of the width of the stream. Conversely, if it has a low permeability the cone may expand across and beyond the stream. If pumping is continued over a prolonged period, a new condition of equilibrium is established with essentially steady flows. Most of the abstracted water then is derived from the surface source. Induced infiltration is significant from the point of view of groundwater pollution in two respects. First, the new condition of equilibrium that is established may involve the reversal of the non-pumping hydraulic gradients, particularly when groundwater levels have been significantly lowered by groundwater abstraction. This may result in any pollutants travelling in the opposite direction from that which they normally do. Second, surface water resources are often less pure than the underlying groundwater so induced infiltration introduces the danger of pollution. However, whether or not induced infiltration gives rise to pollution depends upon the quality of the surface water source, the nature of the aquifer, the quantity of infiltration involved and the intended use of the abstracted groundwater.

Excessive lowering of the water table along a coast leads to saline intrusion, the saltwater entering an aquifer via submarine outcrops thereby displacing fresh water. However, the fresh water still overlies the saline water and continues to flow from the aquifer to the sea. The problem of saline intrusion usually starts with the abstraction of groundwater from a coastal aquifer, which leads to the disruption of the equilibrium condition. Generally, saline water is drawn up towards a well and this is sometimes termed upconing. This is a dangerous condition that can occur even if the aquifer is not overpumped and a significant proportion of the fresh water flow still reaches the sea. A well may be ruined by an increase in salt content even before the actual 'cone' reaches the bottom of the well. This is due to leaching of the interface between salt and fresh water by the fresh water. The encroachment of saltwater may extend for several kilometres inland leading to the abandonment of wells. The first sign of saline intrusion is likely to be a progressively upward trend in the chloride concentration of water obtained from the affected wells. Chloride levels may increase from a normal value of around 25 mg l^{-1} to something approaching 19 000 mg l^{-1}, which is the concentration in sea water. Overpumping is not the only cause of saltwater encroachment, continuous pumping or inappropriate location and design of wells also may be contributory factors. In other words, the saline-fresh water interface is the result of a hydrodynamic balance, hence if the natural flow of fresh water to the sea is interrupted or significantly reduced by abstraction, then saline intrusion is almost certain to occur. Although it is difficult to control saline intrusion and to effect its reversal, the encroachment of saltwater can be checked by maintaining a fresh water hydraulic gradient towards the sea, which means that there is a flow of fresh water towards the sea. This gradient can be maintained naturally, or by some artificial means such as artificial recharge to form a groundwater mound between the coast and the area where abstraction is taking place. However, the technique requires an additional supply of clean water. Alternatively, an extraction barrier, that is, a line of wells parallel to the coast, which abstracts encroaching salt water before it reaches the protected inland wellfield can be employed. The abstracted water will be brackish and therefore generally is pumped back into the sea.

5.12. Case history 1

The mine at West Driefontein, Far West Rand, South Africa, initiated a dewatering programme in 1956, pumping some 35 Ml day^{-1}. During 1959 serious delays were caused in the tubular ball mill sections of the reduction works where, almost daily, the main bearings came under stress due to misalignment of the axles. At the same time the thickener and agitation tanks showed signs of tilting. Subsidence movements were suspected as being responsible and it was decided to sink a shaft to investigate. Two tunnels were excavated from the shaft towards a fault that was thought to be the cause of the instability. These encountered a mud filled conduit, material from which moved into the tunnels. Investigations were still in progress when on 12 December 1962, a huge sinkhole (55 m in diameter) engulfed the three-story crusher plant taking 29 men with it (Fig. 5.27). There was practically no warning, collapse occurring suddenly. The sinkhole subsequently was filled with material from an adjoining waste rock dump, this being completed by 10 January 1963. A gravity survey had suggested that there was a subsurface karst valley of substantial size beneath the plant and that it probably was coincident with a fault in the dolomite. Prior to the construction of the crusher plant the site had been consolidation grouted by the injection of cement and the area around the plant had been hard surfaced to prevent the ingress of water.

Figure 5.27 A sinkhole which engulfed a three-storey crusher plant at West Driefontein Mine. Far West Rand, and claimed 29 lives.

Subsequently, Jennings *et al.* (1965) maintained that the appearance of the sinkhole at the surface at West Driefontein was influenced by pumping of groundwater from the mine causing the water table to be lowered by some 120 m. They went on to suggest that there was considerable evidence to indicate that sinkhole occurrence was associated with the collapse of an arch or dome that bridged a cavern. Lowering of the water table had led to fissures in the roofs of caverns in the dolomite being enlarged by dissolution due to percolating water. Avens (i.e. chimneys) subsequently developed in the roofs and grew upwards into the overlying mantle of soil (Fig. 5.28). Hence, arches above caverns had been seriously weakened and thinned by material spalling from them. In this way upward migration of the void eventually led to a sinkhole appearing at the surface. In such instances, the cavern must be large enough to accommodate the material that falls, otherwise upward migration of the void will cease due to it becoming choked with fallen residual material that bulks. Alternatively, fallen material may be removed by groundwater flowing through the cavern and into discontinuities in the dolomite opened by dissolution.

5.13. Case history 2

The Wheal Jane and South Crofty were the two last, of more than 50 mines, to be worked in the Gwennap Mining District of Cornwall, England. There

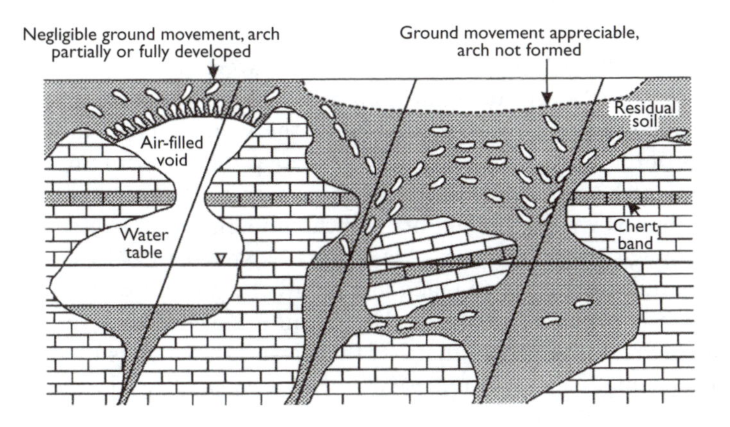

Figure 5.28 Section through the ground showing the conditions leading to a void liable to give rise to a sinkhole (on left) and conditions leading to caving and subsidence (on right). (After Jennings *et al.*, 1965; reproduced by kind permission of the University of Toronto Press.)

mining for tin and other metals dates back to at least the seventeenth century, continuing until March 1991 when closure became inevitable due to the low price for tin.

The mines exploited complex ores and veins of fine grained tin in a sulphide rich matrix. The ore minerals were associated with quartz porphyry dykes intruded into the Killas Mudstones during a phase of regional metamorphism associated with the emplacement of the Cornish granite batholith during the Hercynian Orogeny. The main lodes at Wheal Jane trend north-east to south-west and dip at approximately 45° to the north-west. They typically include minerals such as cassiterite (tin, SnO_2), chalcopyrite (copper, $CuFeS_2$), pyrite (iron, FeS_2), wolframite (tungsten, $(Fe, Mn)WO_4$) and arsenopyrite (arsenic, FeAsS), with smaller amounts of native silver (Ag), galena (PbS) and other minerals. Although the porosity of the Killas rocks is low, their permeability is enhanced due to the density of faults and joints in the rock mass and the high degree of weathering and hydrothermal alteration to which it has been subjected. The mine workings provide a preferential flow for the groundwater. Wheal Jane was one of the wettest mines per tonne of ore worked in the world. Dewatering of the mine complex occurred to a depth of about 450 m below ground level and caused a cone of depression in the water table, although the actual extent of the cone is not known. The majority of the mine workings drain mine water into the Carnon River and associated surface water courses. The entire catchment area contains numerous abandoned mine shafts and adits, and has been worked in the past for alluvial deposits. The water pressures encountered during drilling were as high as 1.38 MPa and 1215 Ml water were pumped to the surface at a cost of over £1 M year^{-1}. The hydrogeology of the mine was further complicated by being linked to the neighbouring Mount Wellington Mine. Both of these mines were worked simultaneously from 1979 to 1991 producing around 360 000 tonnes per year.

Before the closure of Wheal Jane Mine the water quality in the streams and rivers of the Carnon catchment zone were poor and contained high concentrations of heavy metals and acid mine water. Upon the closure of the mine the pumps were switched off and the underground assets were recovered. Pumping stopped in June 1991 and by November 1991 the water table had recovered sufficiently to reach just below the ground surface and minor discharge was observed in the nearby Carnon River. This river discharges 3 km downstream into the estuary of the River Fal, which is used for sailing and shell fishing. This acid mine water was very acidic, with pH values in the range of 2–3. It also contained high concentrations of arsenic, cadmium, copper, iron, lead and zinc (particularly copper, zinc and lead).

The mine complex subsequently began to fill with water, as rebound occurred, and large areas of oxidized minerals were exposed to mine water leaching. Concerns were expressed that once the mine was flooded then acid mine water could discharge into the local rivers and water courses (Hamilton *et al.*, 1994a). A monitoring programme was therefore established. This consisted of a survey of mine workings, interconnections, adits and shafts; a water level survey of the Wheal Jane shaft; measurements of mine water rise; collection of mine water samples from the mine complex; surveys of water level in shafts, wells and drillholes not associated with the mine; surveys of private and licensed groundwater sources; and baseline water quality surveys (Hamilton *et al.*, 1994b). Early recovery was rapid initially but this slowed down and became irregular due to the filling of various void spaces and different levels. Discharge occurred on 17 November 1991, from Wheal Jane adit into the Carnon River. The monitoring programme accurately determined the location and timing of the discharge. The rate of discharge was measured at between 5000 and 40 000 m^3 day^{-1} and also was related to rainfall events. Samples collected from the six mine shafts showed variations in the heavy metal content and acidity due to variations in the metal content of the mine, and length of exposure and the pattern of water flow. The pH varied from 2.5 to 6.0 but in general was around 3.0. Attempts were made to control the water level by pumping water from the mine shafts into tanks where lime was added to raise the pH and precipitate any toxic metals. Storms and poor weather in early 1992 caused the pumping and treatment to be stopped. Within nine days of the cessation of pumping the water pressures had increased sufficiently to burst the concrete plug built to seal the adit to the mine. Approximately 320 Ml of polluted water flowed into the estuary discolouring it, turning it a deep orange, and depositing toxic metals. Cadmium levels reached 600 mg l^{-1} compared to a British drinking water quality standard of less than 0.01 mg l^{-1}. Public

Table 5.3 Variations in the chemical quality of mine water from Wheal Jane mine, Cornwall (units in parts mg l^{-1} except pH) (From Hamilton *et al.*, 1994b and Dodds-Smith *et al.*, 1995)

Year	*1992*	*1995*
pH	2.6–3.1	3.5
Iron	1720–1900	300
Zinc	1260–1700	120

concerns led to the resumption of pumping until a solution to the problem was found.

As noted, when Wheal Jane Mine closed the mine flooded and the mine waters emerged at the ground surface in January 1992, causing huge environment impact. It was the presence of the iron in the mine water that drew a large amount of attention to this incident since it turned the Fal Estuary to a highly visible bright orange. The quality of the initial mine water influx has improved with time. A marked decline in heavy metal content can be attributed to the initial flush of water carrying sulphate salts that had accumulated within the mine when it was working and dry. This is illustrated in Table 5.3.

As mentioned, the mine water at Wheal Jane is pumped from the shaft and treatment involves the addition of lime and flocculants to encourage the precipitation of the finer particles and heavy metals. The result is a low density sludge that contains approximately 2% solids. This is deposited in the tailings lagoon. The treatment of the mine water by the addition of lime required quarrying limestone in Derbyshire and its transportation to Cornwall, a distance of over 400 km. Waste may then have to be transported to landfill sites. In the long-term this may not be sustainable environmentally and passive treatment methods, that is, the use of reed beds were investigated.

References

Abidin, H.Z., Djaja, R. and Darmawan, D. 2001. Land subsidence of Jakarta (Indonesia) and its geodetic monitoring system. *Natural Hazards*, **23**, 365–387.

Allis, R.G. 2000. Review of subsidence at the Wairakei field, New Zealand. *Geothermics*, **29**, 455–478.

Anon. 1980. Subsidence – a geological problem with a political solution. *Civil Engineering, American Society of Civil Engineers*, May, 60–63.

Anon. 1998. *Assessment of Groundwater Quality in England and Wales*. Department of the Environment, Her Majesty's Stationery Office, London.

Aoki, S. 1976. Land subsidence in Niigata. *Proceedings of the Second International Symposium on Land Subsidence*, Anaheim, International Association Hydrological Sciences, UNESCO Publication No. 121, 10–12.

Bankher, K.A. and Al-Harthi, A.A. 1999. Earth fissuring and land subsidence in western Saudi Arabia. *Natural Hazards*, **20**, 21–42.

Bell, F.G. 1988. Ground movements associated with the withdrawal of fluids. In: *Engineering Geology of Underground Movements*, Engineering Geology Special Publication No. 5, Bell, F.G., Culshaw, M.G., Cripps, J.C. and Lovell, M.A. (eds), Geological Society, London, 363–376.

Bell, F.G. 1992. Salt mining and associated subsidence in mid-Cheshire, England, and its influence on planning. *Bulletin Association of Engineering Geologists*, **29**, 371–386.

Bell, J.W. and Helm, D.C. 1998. Ground cracks and Quaternary faults in Nevada: hydraulic and tectonic. In: *Land Subsidence Case Studies and Current Research*, Special Publication No. 8, Association Engineering Geologists, Borchers, J.W. (ed.), Star Publishing Company, Belmont, CA, 165–173.

Bell, J.W., Amelung, F., Ramelli, A.R. and Blewitt, G. 2002. Land subsidence in Las Vegas, Nevada, 1935–2000: new geodetic data show evolution, revised spatial patterns, and reduced rates. *Environmental and Engineering Geoscience*, **8**, 155–174.

Bezuidenhout, C.A. and Enslin, J.F. 1969. Surface subsidence and sinkholes in the dolomitic area of the Far West Rand, Transvaal, South Africa. *Proceedings of the First International Symposium on Land Subsidence*, Tokyo, International Association Hydrological Sciences, UNESCO Publication No. 88, Paris, 482–495.

Branston, M.W. and Styles, P. 2003. The application of time lapse microgravity for the investigation and monitoring of subsidence at Northwich, Cheshire. *Quarterly Journal of Engineering Geology and Hydrogeology*, **36**, 231–244.

Brook, C.A. and Allison, T.L. 1986. Fracture mapping and ground subsidence susceptibility modeling in covered karst terrain: the example of Dougherty County, Georgia. *Proceedings of the Third International Symposium on Land Subsidence*, Venice, International Association of Hydrological Sciences, Publication No. 151, 595–606.

Buckley, S.M., Rosen, P.A., Hensley, S. and Tapley, B.D. 2003. Land subsidence in Houston, Texas, measured by radar interferometry and constrained by extensometers. *Journal of Geophysical Research B, Solid Earth*, **108**, 1–13.

Bull, W.B. and Poland, J.F. 1975. *Land Subsidence due to Groundwater Withdrawal in California. Part 3: Interrelations of Water Level Change, Change in Aquifer System Thickness, and Subsidence.* Professional Paper 437–9, United States Geological Survey, Government Printer, Washington, DC.

Burke, S.P. and Younger, P.L. 2000. Groundwater rebound in the South Yorkshire Coalfield: a first approximation using the GRAM model. *Quarterly Journal of Engineering Geology and Hydrogeology*, **33**, 149–160.

Buttrick, D.B. and Calitz, F. 1995. *Guidelines for the Application of the Method of Scenario Supposition Risk Evaluation Technique for the Assessment of the Stability of Sites underlain by Dolomite*, Council for Geoscience, Pretoria.

Buttrick, D.B. and Van Schalkwyk, A. 1995. The method of scenario supposition for stability evaluation of sites on dolomitic land in South Africa. *Journal of South African Institution of Civil Engineers*, **37**, 9–14.

Buttrick, D.B., Van Schalkwyk, A., Kleywegt, R.J. and Watermeyer, R.B. 2001. Proposed method for dolomite land hazard and risk assessment in South Africa. *Journal of South African Institution of Civil Engineers*, **43**, 27–36.

Carbognin, L., Gatto, P., Mozzi, G., Gambolati, G. and Ricceri, G. 1976. New trend in the subsidence of Venice. *Proceedings of the Second International Symposium on Land Subsidence*, Anaheim, International Association of Hydrological Sciences, UNESCO Publication No. 121, 65–81.

Carillo, N. 1948. Influence of artesian wells on the sinking of Mexico City. *Proceedings of the Second International Conference on Soil Mechanics and Foundation Engineering*, Rotterdam, **2**, 156–159.

Carminati, E. and Martinelli, G. 2002. Subsidence rates in the Po Plain, northern Italy: the relative impact of natural and anthropogenic causation. *Engineering Geology*, **66**, 241–255.

Cassiani, G. and Zoccatelli, C. 2000. Subsidence risk in Venice and nearby areas, Italy, owing to offshore gas fields: a stochastic analysis. *Environmental and Engineering Geoscience*, **6**, 115–128.

Castle, R.O. and Youd, T.L. 1972. Discussion: the Houston fault problems. *Bulletin of the Association of Engineering Geologists*, **9**, 57–68.

Castle, R.O., Yerkes, R.F. and Riley, F.S. 1969. A linear relationship between liquid withdrawal and oil field subsidence. *Proceedings of the First International Symposium on Land Subsidence*, Tokyo, International Association Hydrological Sciences, Publication No. 81, **1**, 167–173.

Chen, C., Pei, S. and Jiao, J.J. 2003. Land subsidence caused by groundwater exploitation in Suzhou City, China. *Hydrogeology Journal*, **11**, 275–287.

Chi, S.C. and Reilinger, R.E. 1984. Geodetic evidence for subsidence due to groundwater withdrawal in many parts of the United States of America. *Journal of Hydrology*, **67**, 152–82.

Chilingarian, G.V., Donaldson, E.C. and Yen, T.F. 1995. *Subsidence due to Fluid Withdrawal*. Elsevier, Amsterdam.

Cliff, M.I. and Smart, P.C. 1998. The use of recharge trenches to maintain groundwater levels. *Quarterly Journal of Engineering Geology*, **31**, 137–145.

Cooper, A.H. 2002. Halite karst geohazards (natural and man-made) in the United Kingdom. *Environmental Geology*, **42**, 505–512.

Davies, O. 1979. *Roman Mines in Europe*, Arno Press, New York.

De Bruyn, I.A. and Bell, F.G. 2001. The occurrence of sinkholes and subsidence depressions in the Far West Rand and Gauteng Province, South Africa. *Environmental and Engineering Geoscience*, 7, 281–295.

Deere, D.U. 1961. Subsidence due to mining – a case history from the Gulf region of Texas. *Proceedings of the Fourth Symposium on Rock Mechanics, Bulletin Mining Industries Experimental Station, Engineering Series*, Hartman, H.L. (ed.), Pennsylvania State University, Pittsburg, PA, 59–64.

Delflache, A.P. 1979. Land subsidence versus head decline in Texas. In: *Evaluation and Prediction of Subsidence*. Proceedings of the Speciality Conference American Society Civil Engineers, Gainsville, Saxena, S.K. (ed.), New York, 320–331.

Dodds-Smith, M.E., Payne, C.A. and Gusek, J.J. 1995. Reedbeds at Wheal Jane. Mining Environment Management, Sept., 22–24.

Dumpleton, S., Donnelly, L.J. and Young, B. 1996. A Survey of Force Crag Mine, near Keswick, Cumbria. Technical Report WE/96/34, British Geological Survey Nottingham.

Dumpleton, S., Robins, N.S., Walker, J.A. and Merrin, P.D. 2001. Mine water rebound in South Nottinghamshire; risk evaluation using 3-D visualisation and predictive modelling. *Quarterly Journal of Engineering Geology and Hydrogeology*, **34**, 307–319.

Fielding, E.J., Blom, R.G. and Goldstein, R.M. 1998. Rapid subsidence over oil-fields measured by SAR interferometry. *Geophysical Research Letters*, **25**, 3215–3218.

Figueroa Vega, G.E. 1976. Subsidence of the city of Mexico: a historical review. *Proceedings of the Second International Symposium on Land Subsidence*, Anaheim, International Association of Hydrological Sciences, UNESCO Publication No. 121, 35–38.

Figueroa Vega, G.E. and Yamamoto, S. 1984. Techniques of prediction of subsidence. In: *Guidelines to Studies of Land Subsidence due to Groundwater Withdrawal*, Poland, J.F. (ed.), UNESCO, Paris, 88–117.

Gabrysch, R.K. 1969. Land surface subsidence in the Houston-Galveston region, Texas. *Proceedings of the First International Symposium on Land Subsidence*, Tokyo, International Association of Hydrological Sciences, UNESCO Publication No. 88, **1**, 43–54.

Gabrysch, R.K. 1976. Land subsidence in the Houston-Galveston region, Texas. *Proceedings of the Second International Symposium on Land Subsidence*, Anaheim. International Association of Hydrological Sciences, UNESCO Publication No. 121, 16–24.

Gambolati, G. and Freeze, R.A. 1973. Mathematical simulation of subsidence of Venice. *Water Resources Research*, **5**, 721–733.

Gambolati, G., Gatto, P. and Ricceri, G. 1986. Land subsidence due gas-oil removal in layered anisotropic soils by a finite element model. *Proceedings of the Third International Symposium on Land Subsidence*, Venice, International Association Hydrological Sciences, Publication No. 151, 29–41.

Gambolati, G., Ricceri, G., Bertoni, Brighenti, G. and Vuillermin, E. 1991. Mathematical simulation of subsidence of Ravenna. *Water Resources Research*, **27**, 2899–2918.

Gambolati, G., Teatini, P. and Bertoni, W. 1998. Numerical prediction of land subsidence over Dosso degli Angeli gas field, Ravenna, Italy. In: *Land Subsidence Case Studies and Current Research*, Special Publication No. 8, Association of Engineering Geologists, Borchers, J.W. (ed.), Star Publishing Company, Belmont, CA, 229–238.

Geertsma, J. 1973. A basic theory of subsidence due to reservoir compaction: the homogeneous case. *Geologie en Mijnbouw*, **28**, 43–62.

Geertsma, J. and Van Opstal, G. 1973. A numerical technique for predicting subsidence above compacting reservoirs, based on the nucleus strain concept. *Geologie en Mijnbouw*, **28**, 63–74.

Ghaboussi, J., Ranken, R.E. and Hendron, A.J. 1981. Time dependent behavior of solution caverns in salt. *Proceedings of the American Society of Civil Engineers, Journal of Geotechnical Engineering Division*, **107**, 1379–1401.

Gilluly, J. and Grant, U.S. 1949. Subsidence in the Long Beach area, California. *Bulletin of the Geological Society of America*, **60**, 461–560.

Glover, H.G. 1983. Mine water pollution – an overview of problems and control strategies in the United Kingdom. *Water Science Technology*, **15**, 39–70 (Pretoria).

Gregory, B.J., Venter, I.S. and Kruger, L.J. 1988. Grouting induced ground movements. In: *Engineering Geology of Underground Movements*, Engineering

Geology Special Publication No. 5, Bell, F.G., Culshaw, M.G., Cripps, J.C. and Lovell, M.A. (eds), the Geological Society, London, 153–157.

Gubellini, A., Lombardini, G. and Russo, P. 1986. Application of high precision levelling and photogrammetry to the detection of the movements of an architectonic complex produced by subsidence in the town of Bologna. *Proceedings of the Third International Symposium on Land Subsidence*, Venice, International Association Hydrological Sciences, Publication No. 151, 257–267.

Hagerty, T.J. and Lippert, K. 1982. Rising groundwater – problem or resource. *Ground Water*, **20**, No. 2, 217–223.

Hamilton, R.M., Bowen, G.G., Postlethwaite, N.A. and Dussek, C.J. 1994a. The abandonment of Wheal Jane, a tin mine in south west England. *Proceedings of the Fifth International Mine Water Congress, Mine Water and the Environment*, Nottingham, 543–551.

Hamilton, R.M., Taberham, J., Waite, R.R.J., Cambridge, M., Coulton, R.H. and Hallewell, M.P. 1994b. The development of a temporary treatment solution for the acid mine water discharge at Wheal Jane. *Proceedings of the Fifth International Mine Water Congress, Mine Water and the Environment*, Nottingham, 643–656.

Holzer, T.L. 1980. Faulting caused by groundwater level declines, San Joaquin Valley, California. *Water Resources Research*, **16**, 1065–1070.

Holzer, T.L. 1986. Ground failure caused by groundwater withdrawal from unconsolidated sediments. *Proceedings of the Third Symposium on Land Subsidence*, Venice, International Association Hydrological Sciences, Publication No. 151, 747–756.

Holzer, T.L. 1998. The history of the aquifer drainage model. In: *Land Subsidence Case Studies and Current Research*, Special Publication No. 8, Association of Engineering Geologists, Borchers, J.W. (ed.), Star Publishing Company, Belmont, CA, 7–12.

Holzer, T.L. and Thatcher, W. 1979. Modeling deformation due to subsidence faulting. In: *Evaluation and Prediction of Subsidence*. Proceedings of Speciality Conference American Society Civil Engineers, Gainsville, Saxena, S.K. (ed.), New York, 349–357.

Holzer, T.L., Davis, S.N. and Lofgren, B.N. 1979. Faulting caused by groundwater abstraction in south central Arizona. *Journal of Geophysical Research*, **84**, 603–612.

Jachens, R.C. and Holzer, T.L. 1980. Geophysical investigations of ground failure related to groundwater withdrawal, Picacho basin, Arizona. *Ground Water*, **17**, No. 6, 574–585.

Jammel, S.E. 1986. The Winter Park sinkhole and Central Florida sinkhole type subsidence. *Proceedings of the Third International Symposium on Land Subsidence*, Venice, International Association of Hydrological Sciences, Publication No. 151, 585–594.

Jennings, J.E., Brink, A.B.A., Louw, A. and Gowan, G.D. 1965. Sinkholes and subsidences in the Transvaal Dolomite of South Africa. *Proceedings of the Sixth International Conference on Soil Mechanics and Foundation Engineering*, Montreal, **1**, 51–54.

Johnson, K.S. 1998. Land subsidence above man-made salt-dissolution cavities. In: *Land Subsidence Case Studies and Current Research*, Special Publication No. 8,

Association of Engineering Geologists, Borchers, J.W. (ed.), Star Publishing Company, Belmont, CA, 385–392.

Kleywegt, R.L. and Enslin, J.F. 1973. The application of the gravity method to the problem of ground settlement and sinkhole formation in dolomite on the Far West Rand. *Proceedings of the International Symposium on Sinkholes, Sinkholes and Engineering Geological Problems related to Soluble Rocks*, Hannover, T2/0–T3/15.

Kreitler, C.W. 1977. Fault control of subsidence, Houston, TX. *Ground Water*, **15**, No. 3, 203–214.

Larson, M.K. 1986. Potential for fissuring in the Phoenix, Arizona, USA, area. *Proceedings of the Third International Symposium on Land Subsidence*, Venice, International Association of Hydrological Sciences, Publication No. 151, 291–300.

Lee, C.F., Zhang, J.M. and Zhang, Y.X. 1996. Evolution and origin of the ground fissures in Xian, China, *Engineering Geology*, **43**, 45–55.

Lee, K.L. and Strauss, M.E. 1970. Prediction of horizontal movements due to subsidence over mined areas. *Proceedings of the First International Symposium on Land Subsidence*, Tokyo, International Association of Hydrological Sciences, UNESCO Publication No. 88, **2**, 515–522.

Lewis, R.W., Makurat, A. and Pao, W.KS. 2003. Fully coupled modeling of seabed subsidence and reservoir compaction of North Sea oil fields. *Hydrogeology Journal*, **11**, 142–161.

Li, J. and Helm, D.C. 2001. Using an analytical solution to estimate the subsidence risk caused by ASR applications. *Environmental and Engineering Geoscience*, **7**, 67–79.

Li, Y., Yang, J. and Hu, X. 2000. Origin of ground fissures in the Shanxi Graben System, northern China. *Engineering Geology*, **55**, 267–275.

Lofgren, B.N. 1968. *Analysis of Stress Causing Land Subsidence*. United States Geological Survey, Professional Paper 600-B, 219–225.

Lofgren, B.N. 1979. Changes in aquifer system properties with groundwater depletion. In: *Evaluation and Prediction of Subsidence*. Proceedings of Speciality Conference American Society Civil Engineers, Gainsville, Saxena, S.K. (ed.), New York, 26–46.

McGinness, S. 1999. Treatment of acid mine drainage. *House of Commons Library*, Research Paper 99/10, London.

Marsh, T.J. and Davies, P.A. 1983. The decline and partial recovery of groundwater levels below London. *Proceedings of the Institution of Civil Engineers*, Part 1, **74**, 263–276.

Meinzer, O.E. 1928. Compressibility and elasticity of artesian aquifers. *Journal of Geology*, **23**, 263–291.

Mes, M.J., Landau, H., Luttenberger, C. and Gustavsen, K. 1998. Automatic GPS subsidence measurements at offshore oil and gas production platforms, Ekofisk, North Sea. In: *Land Subsidence Case Studies and Current Research*, Special Publication No. 8, Association of Engineering Geologists, Borchers, J.W. (ed.), Star Publishing Company, Belmont, CA, 417–425.

Murria, J. 1997. Subsidence due to oil extraction in Venezuela: problems and solutions. *Proceedings of the International Symposium on Engineering Geology and the Environment*, Athens, Marinos, P.G., Koukis, G.C., Tsiambaos, G.C. and Stournaras, G.C. (eds), A.A. Balkema, Rotterdam, **1**, 899–904.

Narasimham, T.N. and Goyal, K.P. 1984. Subsidence due to geothermal fluid withdrawal. In: *Man-induced Land Subsidence*, Holzer, T.L. (ed.), Reviews in Engineering Geology VI, Geological Society of America, New York.

Okumura, T. 1970. Analysis of land subsidence in Niigata. *Proceedings of the First International Conference on Land Subsidence*, Tokyo, International Association of Hydrological Science. UNESCO Publication No. 88, **1**, 130–143.

Ortega-Guerrero, A., Rudolf, D.L. and Cherry, J.A. 1999. Analysis of long-term land subsidence near Mexico City: field investigations and predictive modeling. *Water Resources Research*, **35**, 3327–3341.

Pan, K.L. 1997. Controlling land subsidence in Taiwan. *Proceedings of the International Symposium on Engineering Geology and the Environment*, Athens, Marinos, P.G., Koukis, G.C., Tsiambaos, G.C. and Stournaras, G.C. (eds), A.A. Balkema, Rotterdam, **1**, 943–948.

Partridge, T.C., Harris, G.M. and Diesil, V.A. 1981. Construction upon dolomites in the south western Transvaal. *Bulletin of International Association of Engineering Geology*, No. 24, 135–135.

Phien-wej, N., Giao, P.H. and Nutalaya, P. 1998. Field experiment of artificial recharge through a small well with reference to land subsidence control. *Engineering Geology*, **50**, 187–201.

Poland, J.F. 1972. Subsidence and its control. *American Association of Petroleum of Geologists*, Memoir 18, 50–71.

Poland, J.F. and Davis, G.H. 1969. Land subsidence due to withdrawal of fluids. In: *Reviews in Engineering Geology*, Varnes, D.J. and Kiersch, G.A. (eds), Geological Society of America, **2**, 187–269.

Poland, J.F., Lofgren, B.E., Ireland, R.L. and Pugh, R.G. 1975. *Land Subsidence in the San Joaquin Valley, California, as of 1972*. Professional Paper 437-H, United States Geological Survey, Government Printer, Washington, DC.

Pollard, W.S. and Johnson, K.H. 1984. Groundwater subsidence model: a management tool. *Proceedings of the American Society of Civil Engineers, Journal of Water Resources Planning and Management Division*, **110**, 323–332.

Popovic, M. 1970. Deformations of soil surface caused by salt exploitation in Tuzla and their influence on structures. *Proceedings of the Second International Congress on Rock Mechanics*, Belgrade, 34–45.

Pottgens, J.J.E. 1986. Ground movements caused by mining in the Netherlands. *Proceedings of the Third International Symposium on Land Subsidence*, Venice, International Association Hydrological Sciences, Publication No. 151, 651–664.

Pratt, W.E. and Johnson, D.W. 1926. Local subsidence of the Goose Creek oil field. *Journal of Geology*, **34**, 577–590.

Rau, J.L. and Nutalaya, P. 1982. Geomorphology and land subsidence in Bangkok. In: *Applied Geomorphology*, Craig, R.G. and Craft, I.L. (eds), Allen and Unwin, London, 181–201.

Riley, F.S. 1986. Developments of borehole extensometry. *Proceedings of the Third International Symposium on Land Subsidence*, Venice, International Association of Hydrological Sciences, Publication No. 151, 169–186.

Rojas, E., Arzate, J. and Arroyo, M. 2002. A method to predict the group fissuring and faulting caused by regional groundwater decline. *Engineering Geology*, **65**, 245–260.

Sato, H.P., Abe, K. and Ootaki, O. 2003. GPS-measured land subsidence in Ojiya City, Niigata Prefecture, Japan. *Engineering Geology*, **67**, 379–390.
Saxena, S.K. and Mohan, A. 1979. Study of subsidence due to withdrawal of water below an aquitard. In: *Evaluation and Prediction of Subsidence*. Proceedings of Speciality Conference, American Society of Civil Engineers, Gainsville, Saxena, S.K. (ed.), New York, 332–338.
Schoonbeek, J.B. 1976. Land subsidence as a result of natural gas extraction in the province of Groningen. *Journal of Society Petroleum Engineers, American Institute of Mining Engineers*, Paper No SPE 5751, 1–20.
Scott, R.F. 1979. Subsidence – a review. In: *Evaluation and Prediction of Subsidence*. Proceedings of Speciality Conference, American Society of Civil Engineers, Gainsville, Saxena, S.K. (ed.), New York, 1–25.
Sheng, Z. and Helm, D.C. 1998. Multiple steps of earth fissuring caused by groundwater withdrawal. In: *Land Subsidence Case Studies and Current Research*, Special Publication No. 8, Association of Engineering Geologists, Borchers, J.W. (ed.), Star Publishing Company, Belmont, CA, 149–154.
Sherwood, J.M. and Younger, P.L. 1994. Modelling groundwater rebound after coalfield closure: an example from County Durham, UK. *Proceedings of the Fifth International Minewater Congress*, Nottingham, **2**, 769–777.
Shimizu, M. 1998. Application of a large-strain finite element model in predicting land subsidence due to the variation of groundwater levels. In: *Land Subsidence Case Studies and Current Research*, Special Publication No. 8, Association of Engineering Geologists, Borchers, J.W. (ed.), Star Publishing Company, Belmont, CA, 239247.
Smith, J.A. and Colls, J.J. 1996. Groundwater rebound in the Leicestershire Coalfield. *Journal of Chartered Institution of Water and Environmental Management*, **10**, 280–289.
Stiros, S.C. 2001. Subsidence of the Thessaloniki (northern Greece) coastal plain, 1960–1999. *Engineering Geology*, **61**, 243–256.
Swanson, A.A. 1998. Land subsidence in the San Joaquin Valley, updated to 1995. In: *Land Subsidence Case Studies and Current Research*, Special Publication No. 8, Association of Engineering Geologists, Borchers, J.W. (ed.), Star Publishing Company, Belmont, CA, 75–79.
Swart, C.J.U., Stoch, E.J., Van Jaarsveld, C.F. and Brink, A.B.A. 2003. The Lower Wonderfontein Spruit: an expose. *Environmental Geology*, **43**, 635–653.
Terzaghi, K. 1925. *Erdbaumechanik auf Bodenphysikalischer Grunlage*. Deuticke, Vienna.
Terzaghi, K. 1943. *Theoretical Soil Mechanics*. Wiley, New York.
Tolman, C.F. and Poland, J.F. 1940. Ground-water, salt-water infiltration and ground-surface recession in Santa Clara Valley, Santa Clara County, California. *Transactions of American Geophysical Union*, **21**, 23–34.
Trudgill, S.T. and Viles, H.A. 1998. Field and laboratory approaches to limestone weathering. *Quarterly Journal of Engineering Geology*, **31**, 333–341.
Van den Berg, J.P. 1996. Construction methods on dolomitic land. *Seminar on Development of Dolomitic Land*, Pretoria, 11/1–11/18.
Van Rooy, J.L. 1989. A new proposed classification system for dolomitic areas south of Pretoria. In: *Contributions to Engineering Geology*, South African Institution of Engineering Geologists and Geological Survey of South Africa, **1**, Government Printer, Pretoria, 57–65.

Van Schalkwyk, A. 1998. Legal aspects of development on dolomitic land in South Africa. *Environmental Geology*, **36**, 167–169.

Van Siclen, D.C. 1967. The Houston fault problem. *Proceedings of the Third Meeting, American Institute of Professional Geologists*, Texas Section, Dallas, 9–31.

Venter, I.S. and Gregory, B.J. 1987. A risk assessment in dolomitic terrain – a case history. In: *Planning and Engineering Geology*, Special Publication in Engineering Geology No. 4. Culshaw, M.G., Bell, F.G., Cripps, J.C. and O'Hara, M. (eds), the Geological Society, London, 329–334.

Wagener, F. von M. 1985. Dolomites. *The Civil Engineer in South Africa*, **27**, 395–407.

Wassmann, T.H. 1980. Mining subsidence in Twente, east Netherlands. *Geologie en Mijnbouw*, **59**, 225–231.

Wolmarans, J.F. 1996. Sinkholes and subsidences on the Far West Rand. *Seminar on the Development of Dolomitic Land*, Pretoria, 6/1–6.12.

Wood, S.C., Younger, P.L. and Robins, N.S. 1999. Long term changes in the quality of polluted mine water discharges from abandoned underground mines in Scotland. *Quarterly Journal of Engineering Geology*, **32**, 69–79.

Yerkes, R.F. and Castle, R.O. 1970. Surface deformation associated with oil and gas field operation in the United States. *Proceedings of the First International Symposium on Land Subsidence*, Tokyo, International Association Hydrological Sciences, UNESCO Publication No. 88, Paris, **1**, 55–66.

Younger, P.L. 2000. Holistic remedial strategies for short and long term water pollution from abandoned mines. *Transactions of the Institution of Mining and Metallurgy*, Section A. Mining Technology, **109**, A131–A144.

Younger, P.L. and Adams, R. 1999. *Predicting Minewater Rebound*. Technical Report W179, Environmental Agency, Her Majesty's Stationery Office, London.

Younger, P.L. and Bradley, K.F. 1994. Application of geochemical mineral exploration techniques to the cataloguing of problematic discharges from abandoned mines in north east England. *Proceedings of the Fifth International Congress*, Nottingham, 857–871.

Zhou, G., Esaki, T. and Mori, J. 2003. GIS-based spatial and temporal prediction system development for regional land subsidence hazard mitigation. *Environmental Geology*, **44**, 665–678.

Chapter 6

Quarrying and surface mining

6.1. Introduction

Before any mineral deposit is worked, there has to be an investigation to assess its economic feasibility, that is, to determine whether there is sufficient quantity of acceptable material to make extraction worthwhile (Smith and Collis, 2001). The investigation should provide data relating to the thickness and extent of the deposit, the depth over which it occurs, any variation within it, and the geological conditions, notably the geological structure and hydrogeology, involved.

Most surface mining methods are large scale, involving removal of massive volumes of material, including overburden, to obtain the mineral deposit. Huge amounts of waste can be produced in the process. Surface mining also can cause noise and disturbance, leave scars on the landscape and may pollute the air with dust. Nonetheless, surface mining is important for the local, and usually national, economy. Therefore, any decision to open or extend a quarry or pit must take into consideration how any local residents, the landscape and the environment will be affected. In this context, landscaping schemes and reclamation work should now form part of any planning proposal for the development of a quarry or open-pit.

6.2. Quarrying and open-pit mining

A number of other factors, in addition to those mentioned above, have to be taken into account when a feasibility study is undertaken for a quarry or open-pit. These include the stripping ratio, that is, the ratio of overburden to deposit. This can determine the limit of working in terms of the economic operation of a quarry or open-pit in that a mineral occurring at a depth beyond the maximum stripping ratio will either have to remain unworked or be mined by underground methods. Open-pit mining is used to work ore bodies that occur at or near the surface where the orebody is steeply dipping or occurs in the form of a pipe. Initially, the overburden is stripped to expose the ore. Mining and stripping then continue in a carefully phased

Figure 6.1 Open-pit workings of coal at Lethbridge, British Columbia, note the overturned fold running through the benches.

manner. Coal also is mined in open-pits under certain geological and topographical conditions. For instance, coal mining on a large scale is carried out by open-pit methods at Lethbridge in the Canadian Rocky Mountains (Fig. 6.1).

The hydrogeological conditions determine whether dewatering is necessary during the operation of a quarry or open-pit and if it is, then the quantity of water that has to be removed affects the economics of working. Accordingly, groundwater levels should be recorded in drillholes so that a contoured map of the groundwater surface can be produced. This provides an indication of whether or not the base of the proposed quarry or open-pit will extend below the water table and, if so, the possible depth to which it could extend. If the planned base extends beneath the water table, then this will affect groundwater flow, which could extend outside the boundary of the excavation. In such a situation, it is necessary to determine whether any users of groundwater in the immediate neighbourhood will be affected by a reduction in supply or by the supply being polluted.

Another factor that has to be taken into account during the working life of such an operation is the stability of the slopes that will be produced (Fig. 6.2). This is influenced by the nature of the rock mass(es) concerned, the geological structure and the hydrogeological conditions. Slope stability will affect the limits of the excavation and bench dimensions where benches are employed (Anon., 1987). Open joints in rock masses facilitate

Figure 6.2 Slides in Sandsloot open-pit, Northern Province, South Africa. (a) Planar slide. (b) Wedge slide.

weathering and generally aid slope failure. Fissure zones usually represent zones of weakness along which rock masses may have been altered to appreciable depth by rising hot fluids or gases, or may have undergone weathering. Faults that traverse the area in which excavation is to be made may contribute to the instability and failure of faces and benches. This is principally because of the greater freedom afforded to the rock masses to move along fault planes. In particular, if a fault intersects a prominent joint or bedding plane in such a way that it produces a wedge that daylights into the quarry or pit, then this is likely to slide. Obviously, slope stability must be taken into account when a quarry or pit is being designed (Bye and Bell, 2001).

Once a decision has been made to work a deposit, then a further number of factors have to be considered. These include how the deposit is going to be worked, which type of equipment is going to be employed during the operation of the quarry or open-pit, how any overburden is going to be removed and disposed off, how any excess water will be dealt with, and how environmental factors such as noise, dust and visual impacts will be monitored and managed.

6.3. Methods of working quarries and open-pits

In some cases when a sedimentary rock mass such as limestone or sandstone is worked for building or dimension stone, then it may be obtained by splitting along bedding and/or joint surfaces by using a wedge and feathers. Another method of quarrying such rock types for building stone consists of drilling a series of closely spaced holes (often with as little as 150 mm spacing between them) in line in order to split a large block from the face. Stone may also be cut from a quarry face by using a chain saw or a wire saw (the stone is cut by sand fed between the wire and the rock) or by a diamond impregnated wire saw (Fig. 6.3(a)). Flame torch cutting has been used primarily for winning granitic rocks (Fig. 6.3(b)). It is claimed that this technique is the only way of cutting stone in areas of high stress relief.

If explosive is used to work building stone, then the blast should only weaken the rock along joint and/or bedding planes, and not fracture the material. The object is to obtain blocks of large dimension that can be sawn to size. Hence, the blasting pattern and amount of charge (black powder) are very important. Every effort should be made to keep wastage and hair cracking of rock to a minimum.

Drilling and blasting are used to work aggregate in quarries and ore in open-pits. Two of the principal methods of drilling and blasting used in quarries and open-pits are well-hole blasting and benching. Single face with well-hole blasting is the standard method for working quarries. Here a line of holes is drilled parallel to the face and the holes usually are inclined, often between 10° and 15°, for safety reasons. Sometimes multiple rows of holes may be used. The height of the face brought down is greater than that in benching and tends to vary between 15 and 35 m. As the name suggests, in benching the face is worked in a series of levels or benches, the maximum height of each bench generally not exceeding 10 m. Individual benches can be worked simultaneously, the rows of blastholes being drilled parallel to the free face. Single or multiple rows of holes may be used. In open-pit mining the orebody is blasted in a series of benches, the walls of which are steeply dipping (Fig. 6.4). Rock surrounding the orebody may have to be removed in order to maintain stable slopes in the pit. This represents waste and, together with the waste produced by processing the ore, has to be disposed off.

If drilling is not carried out properly, then blasts are unable to provide material having the characteristics required for subsequent operations. Optimum drilling therefore is a prerequisite of optimum blasting (Hagan, 1986). Rotary percussion drills are designed for rapid drilling in rock. The technique is most effective in brittle materials since it relies upon chipping the rock. Holes with a diameter of 100 mm cater for most normal requirements. As the diameter of blastholes is increased, so the charge is increased.

Figure 6.3 (a) Extraction of marble by diamond impregnated saw at Estremos, Portugal. (b) Working granite with a flame torch in Cornwall, England.

Figure 6.4 Rudny open-pit iron ore mine in Kazakhstan, one of the largest open-pit mines in the world.

Ground vibrations may present a problem when charges are fired. However, vibrations generally can be reduced to an acceptable level by, for example, using short-delay blasting. As the diameter of a blasthole increases, the burdens and spacings are increased, so that the discontinuities in the rock mass become more significant in terms of fragmentation.

Depending on the pattern of holes and firing technique, subgrade drilling, that is, drilling blastholes some 300 to 900 mm below the excavation grade of a quarry, usually is necessary to ensure that the area being blasted is broken down to grade. It is only when there is a notable parting, such as a prominent bedding plane that is coincident with the base of the quarry, that subgrade drilling is unnecessary.

If a rock mass is to be blasted efficiently it must be capable of transmitting the explosive energy some distance from the blastholes. When this does not happen, then the rock immediately surrounding the hole is pulverized and some of the area between holes is not fractured. The early movement of the rock face is most important for an efficient blast, that is, one that achieves good fragmentation per drillhole with minimum quantity of explosive (Fourney, 1993). Rock breakage in blasting, apart from the character of the rock itself, especially the fracture index, depends largely on the relation between the burden and the hole spacing, as well as the time of ignition between the holes (Muller, 1997). Obviously, efficient blasting should produce rock fragments sufficiently reduced in size so that they can be easily loaded without resort to secondary breakage (Latham *et al.*, 1999). Accordingly, blastholes should be drilled accurately to the requisite pattern and proper depth to ensure satisfactory fragmentation.

The quantity of a particular explosive required to blast a certain volume of rock is difficult to estimate since it depends upon the strength, toughness, grain size, density and incidence of discontinuities within the rock mass. Consequently, no rigid sequence for progressive resistance to blasting offered by different rock types can be formulated. Nevertheless, it can be said that an explosive with the greatest energy and concentration is required for removing very hard rock; in medium hard rocks a high-velocity detonation produces a shattering effect; a medium to high explosive can be used in medium to hard laminated rocks; and the greatest efficiency is obtained with fairly bulky explosive in soft to medium rocks. Table 6.1 affords some idea of the amount of charge in relation to burden for primary blasting. Over-charging can give rise to air-blast and fly-rock (see later), this usually occurring when a weaker zone in a rock mass is undetected prior to firing.

Calculation of the burden and charge also must take account of the toe section since this is confined at the back and base. In other words, there needs to be a higher concentration of charge at the base than in the column, whilst at the top of the hole no charging is required. Hence, the concentration of explosive in the bottom of a hole generally is approximately 2.5 times greater than in the column sections. The basal charge may be regarded as extending from the bottom of a hole to a point above the floor level equal to the burden. The column charge alternates with stemming

Table 6.1 Typical charges and burdens for primary blasting by shot-hole methods (After Sinclair, 1969)

Minimum finishing diameter of hole (mm)	*Cartridge diameter (mm)*	*Depth of hole (m)*	*Burden (m)*	*Spacing (m)*	*Explosive charge (kg)*	*Rock yield (tones)*	*Blasting ratio*	*Tonnes rock per metre of drillhole*
25	22	1.5	0.9	0.9	0.3	3	10.0	2
35	32	3.0	1.5	1.5	1.8	19	10.5	6.3
57	50	6.1	2.4	2.4	9.5	97	10.2	15.9
75	64	9.1	2.7	2.7	18.0	180	10.0	19.8
75	64	12.2	2.7	2.7	25.0	245	9.8	20.1
75	64	18.3	2.7	2.7	36.3	365	10.1	20.0
100	83	12.2	3.7	3.7	43.0	430	10.0	35.3
100	83	30.5	3.7	3.7	43.0	1120	10.7	36.7
170	150	18.3	6.7	6.7	104.3	2235	10.3	122.1
170	150	30.5	6.7	6.7	216.5	3660	10.1	120.0
230	200	21.3	7.6	7.6	363.0	3350	10.2	157.3
230	200	30.5	7.6	7.6	476.3	4670	9.8	153.1

(often fine chippings from drillholes) and extends from there to a point below the crest equal to the burden.

6.4. Strip mining and opencasting

Stratified mineral deposits that occur at or near the surface such as coal or sedimentary iron ore in relatively flat terrain generally are mined by strip or opencast mining. The deposit usually is either horizontal or gently dipping and normally is within 60 m of the surface. All the strata overlying the mineral deposit are removed, frequently by dragline, and placed in stockpiles (Fig. 6.5(a)). Bucket-wheel excavators sometimes are used for stripping (Fig. 6.5(b)). In fact, Scoble and Muftuoglu (1984) proposed a diggability index to provide a guide for the selection of equipment for use in surface mines. The parameters that were used in this system included the degree of weathering, the unconfined compressive strength, the joint spacing and the bed separation. Use of the joint spacing and bedding separation provides a measure of block size.

When necessary, blasting is used to break overburden above the mineral deposit. The mineral deposit itself may have to be drilled and blasted, once exposed, prior to its removal by conventional loading and haulage equipment (Fig. 6.6). The deposition of spoil in mined out areas means that very little advanced stripping is necessary and that mining activity can be located in a relatively small area. Because the face is exposed for a relatively short time, this allows it to have a steeper slope than otherwise.

When the overburden is removed from the working face in long parallel strips, and placed in long spoil heaps in the worked out area, this is referred to as strip mining (Fig. 6.7). Parallel rows of spoil are produced as the face moves on. Soil is removed before the overburden and is placed in separate piles so that it can be used in the rehabilitation process. Rehabilitation of the spoil heaps can proceed before the mining operation ceases, the spoil heaps being regraded to fit into the surrounding landscape. Hence, the short life span of the spoil heaps means that they can be maintained at their natural angle of repose. Once the spoil has been regraded, it is covered with soil, fertilized if necessary and seeded.

In Britain opencast coal mining involves the exploitation of shallow seams at sites normally from 10 to 800 ha in area. This usually involves creating a box-cut to reveal the coal. Once the coal has been extracted, the box-cut is filled with overburden from the next cut. Maximum depths of working are about 100 m, with stripping ratios up to 25 : 1. Working is advanced on a broad front with face lengths of 3–5 km not being uncommon (Fig. 6.8). Draglines may be used for overburden removal along with face shovels and dump trucks, and scrapers. The topsoil is removed separately by a scraper and placed in dumps around the site for use in subsequent restoration. Similarly, the subsoil is stripped and stored temporarily

Figure 6.5 (a) A dragline working Athabasca Tar Sands, Fort McMurray, Alberta. (b) A bucket-wheel excavator working at Taiba Phosphate Mine, Senegal. (Reproduced by kind permission of International Mining Consultants Limited.)

in separate dumps about the site, again for restoration purposes. Face shovels are used to excavate the initial box-cut and for subsequent forward reduction of overburden. The material from the box-cut is placed above ground in a suitable position in relation to void filling so that later

Figure 6.6 A dump truck being loaded by an electrically operated shovel. (Reproduced by kind permission of International Mining Consultants Limited.)

Figure 6.7 Strip-mining at Jayant mine, Singrauli Coalfield, India.

Figure 6.8 Opencast working of coal at St Aiden's opencast site, Yorkshire, England, with partly restored area at the top of the photograph.

rehandling is minimized. Draglines excavate to lower seams in a progressive strip cut behind the face shovels. Drilling and blasting is carried out where necessary.

Many of the large opencast mines in India are worked by draglines, with individual faces often reaching 2–4 km in length. The reach of the draglines, which is around 45 m, and the extraction of the coal along the face known as the 'high wall' creates a working face from 60 m to 100 m high. The dragline operations leave behind pillars of coal 6–10 m wide along their base, 18–20 m high and around 2 m wide at the top. These pillars retain the tip material that has been dumped by the draglines (Fig. 6.9).

Fires at times are encountered in opencast coal mines (see Chapter 10). Spontaneous combustion of the coal may fill an opencast mine with a haze of gas and smoke. This creates low visibility and difficult working conditions in the upper parts of the mine. Potentially hazardous conditions to

Figure 6.9 Pillars of coal left to retain overburden, Jayant Opencast Mine, Singrauli Coalfield, India.

health may exist at the base of an opencast mine. This is caused by the presence of carbon dioxide, which being heavier than air sinks to the bottom of the pit.

6.5. Blasting and vibrations

One of the problems that blasting can give rise to is that of air-blast and another is fly-rock. Air-blast refers to vibrations in the air that are generated by blasting, an explosion causing a diverging shock wave front that rapidly reduces to the speed of sound. The air-blast then is propagated through the atmosphere at the speed of sound. Hence, air-blast consists of an initial concussion wave of a few milliseconds duration, rising rapidly to a peak and then falling off more slowly. The principal factors influencing the effects of air-blast include the type and quantity of explosive used; the method of initiation; the degree of confinement; the atmospheric conditions; the local geological conditions and topography; and the distance from and condition of nearby buildings. Damage is closely related to the concussion wave and so air-blast usually is characterized by the peak overpressure created (Table 6.2). Panes of glass are damaged before any other type of damage but normal blasting results in general overpressures well below those that can cause a pane of correctly installed glass to fail. Any cracks in plaster usually are brought about by flexing in a building

Table 6.2 The effects of overpressure on buildings

Overpressure $g\,cm^{-2}$	*Effect on buildings*
Under 2	Usually no effect but causes nuisance
2–4	Loose windows rattle
4–7	Failure of badly installed windows
7–52	Failure of correctly installed windows begins
52–140	All window panes fail
Over 140	Plaster cracks begin

due the waves created by the air-blast. Nonetheless, nuisance can be caused by overpressures well below those that give rise to damage. The effects of air-blast can be reduced by the correct design of blasting rounds, avoiding the detonation of large unconfined charges and considering atmospheric conditions prior to blasting (e.g. avoid adverse wind directions, low cloud, temperature inversions). A pre-blast survey of buildings close to the site should be carried out and a monitoring programme should be established at certain points around the site. Obviously, wherever possible, blasting should not take place at inconvenient times such as at night, weekends or public holidays. If blasting occurs at a regular time, then this should reduce the element of surprise.

Fly-rock is essentially the product of uncontrolled venting of the gases that are developed when a charge is initiated that may cause rock fragments to be thrown some distance from the face. This is especially hazardous if a zone in the rock face being blasted is weaker so that the charge in effect becomes more effective. The weaker zone may be due to the presence of weathered material or to the greater frequency or significance of discontinuities, which remained unnoticed during drilling operations. In practice, fly-rock can occur regardless of the precautions taken because the factors that cause fly-rock are difficult to identify, measure and control. According to Davies (1995), fly-rock can be regarded as short range when the maximum horizontal travel distance and elevation are normally less than 300 m and 150 m respectively. Models have been developed to provide estimates of this type of fly-rock (Roth, 1979). However, if as mentioned, there is some abnormality in the rock formation or in the blast, then fly-rock may travel much further than 300 m. This type of fly-rock is referred to as wild fly-rock, which in terms of blasting is related to incorrect blast design and poor blasting practice. The factors responsible can include damaged or unstable holes, insufficient burden, blastholes that are too closely spaced, inadequate stemming, incorrect firing sequence or delay period, misfire, accidental initiation or poor site clearance. Davies suggested that danger zones should be established around blast sites and that these should be based on risk assessment where a stand-off distance cannot be accommodated easily. Danger zones are established either to exclude injury to people and/or damage to property, or to limit blasting specifications. Danger zones have tended to be established on a basis of consequences (e.g. the maximum distance of fly-rock projection or expected ground vibration plus a defined safety margin). Such an approach is satisfactory where the requirement for distance can be accommodated readily. Increasingly, according to Davies, sites are being developed in potentially sensitive areas and danger zones are imposing constrains on blasting specifications. The establishment of danger zones on a basis of risk management takes into consideration both consequences and frequency of occurrence. In this way distances can be optimized with respect to acceptable levels of risk.

The three most commonly derived quantities relating to vibration are amplitude, particle velocity and acceleration. Of the three, particle velocity appears to be the one most closely related to damage in the frequency range of typical blasting vibrations. Edwards and Northfield (1960) defined three categories of damage attributable to vibrations. Threshold damage refers to widening of old cracks and the formation of new ones in plaster, and the dislodgement of loose objects. Minor damage is that which does not affect the strength of the structure, it includes broken windows, loosened or fallen plaster and hairline cracks in masonry. Major damage seriously weakens the structure, it includes large cracks, shifting of foundations and bearing walls, and distortion of the superstructure caused by settlement and walls out of plumb. They proposed that a vibration level with a peak particle velocity of 50 mm s^{-1} could be regarded as safe as far as structural damage was concerned, 50–100 mm s^{-1} would require caution, and above 100 mm s^{-1} would present a high probability of damage occurring (Fig. 6.10(a)). The United States Bureau of Mines subsequently lent support to the idea that 50 mm s^{-1} provides a reasonable safety from the possibility of damage (Nicholls *et al.*, 1971). However, Oriard (1972) maintained that no single value of velocity, amplitude or acceleration could be used indiscriminately as a criterion for limiting blasting vibrations. In particular, low vibration levels may disturb sensitive machinery and in this case it is impossible to specify a limit of ground velocity; hence each instance should be separately assessed. Moreover, blasting vibrations at 50 mm s^{-1} peak particle velocity, in terms of human response, are regarded as highly unpleasant or intolerable (Nicholls *et al.*, 1971). Indeed, Crandell (1949) previously had concluded that the average person would consider a vibration to be 'severe' at about one-fifth the level that might damage structures and the threshold of subjective perception has been variously placed from as low as 0.5 mm s^{-1} to 10 mm s^{-1} (Roberts, 1971). People react more unfavourably to large amplitude vibrations of long duration than to low amplitude short duration vibrations of the same intensity. It is likely that this sensitivity is increased in the low frequency range of 3–10 Hz. The duration of the operation and the frequency of occurrence of the blasts are almost as important as the level of the physical effects. Experience has shown that the great majority of complaints and even law suits against blasting operations are due to irritation and that subjective response to vibrations normally leads a person to react strongly long before there is any likelihood of damage occurring to his/her property (Fig. 6.10(b)).

When dealing with high levels of shock and vibration, the time history of the motion and the characteristic response of the structure concerned to the

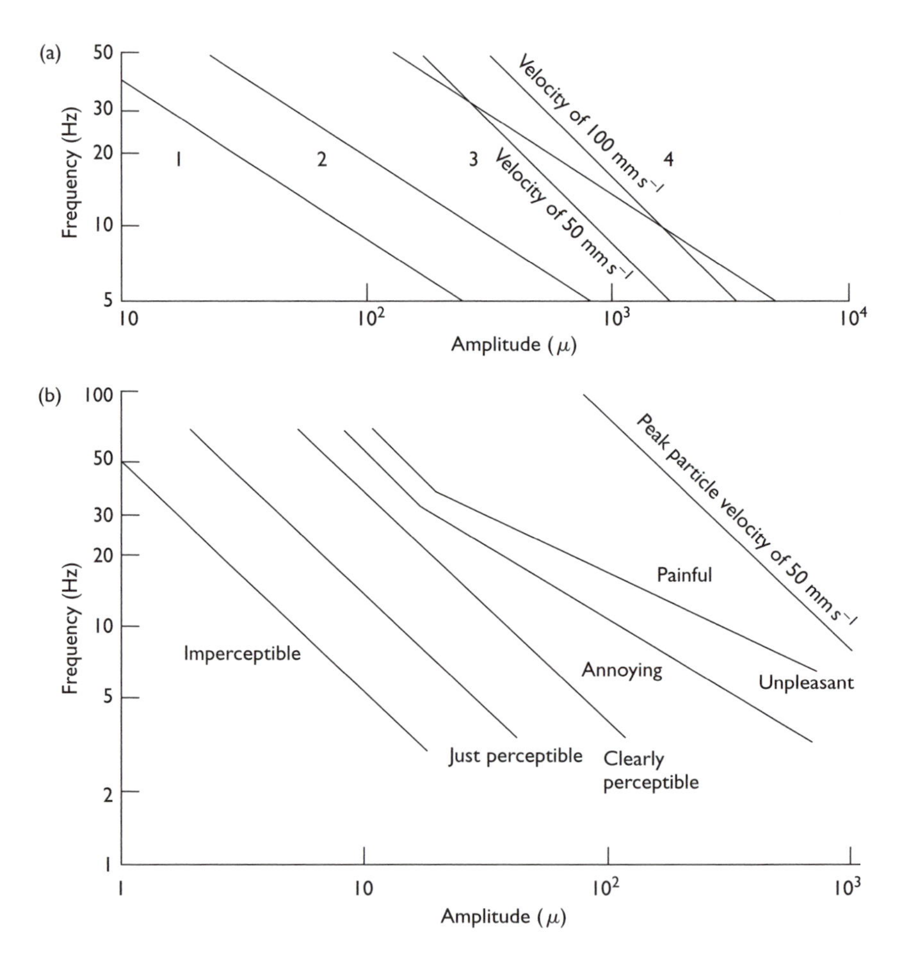

Figure 6.10 (a) Possible damage to buildings for frequencies between 5 and 50 Hz, the range most frequently encountered in buildings: 1 = no damage; 2 = possibility of cracks; 3 = possible damage to load bearing structural units; 4 = damage to load bearing units. (b) Human sensitivity to vibration. If the peak particle velocity of 50 mm s^{-1} is taken as the threshold of damage, then the Figure indicates that humans react strongly long before there is any reason to anticipate damage to property.

type of motion imposed become increasingly important. For instance, a structure with a slow response, such as a tall chimney, when subjected to vibrations with large amplitude, low frequency and long duration, would come closer to the resonant response of the structure. This would be more dangerous than vibrations with small amplitude, high frequency and short duration, even though both may have the same acceleration or velocity. Because of the dependence of response on frequency, conservative limits should be accepted when applying single values of velocity or acceleration as criteria for different types of structures subjected to different kinds of motion.

Vibrations associated with blasting generally fall within the frequency range of 5–60 Hz. The types of vibration depend on the size of the explosive charge; the delay time sequence; the spatial pattern of the blastholes; the volume of the ground set into vibration; the attenuation characteristics of the ground; and the distance from the blast (Blair, 1999). A small explosive charge generates a low vibration with relatively high frequency and relatively low amplitude. By contrast, a large explosive charge produces a vibration with relatively low frequency and relatively high amplitude. The shock waves are attenuated with distance from the blast, the higher frequencies being maintained more readily in dense rock masses. In other words, these pulses are rapidly attenuated in unconsolidated deposits that are characterized by lower frequencies.

Vibrographs can be placed in locations considered susceptible to blast damage in order to monitor ground velocity. A record of the blasting effects compared with the size of the charge and the distance from the point of detonation normally is sufficient to reduce the possibility of damage to a minimum (Fig. 6.11). A pre-blast survey informs owners as to the condition of their buildings before the commencement of blasting and offers some protection to the contractor against unwarranted claims for damage.

The effects of vibration due to blasting operations can be reduced by

- Time dispersion, which includes the use of delay intervals in ignition. Delay intervals mean that shock waves generated by individual blasts are mutually interfering, thereby reducing vibration. For instance, Duvall (1964) noted that a delay interval of around 9 ms reduced ground vibrations and Oriard (1972) suggested that even shorter delay intervals were important as far as the control of vibration and potential damage was concerned. Blair and Armstrong (1999) maintained that the use of electronic delay detonators can control the blasting sequence and so the vibrations in resonant structures such as houses. However, where the geological conditions are variable or complex, then it may not be possible to control the vibration frequency adequately.
- Spatial dispersion, which involves the pattern and orientation of the blastholes.

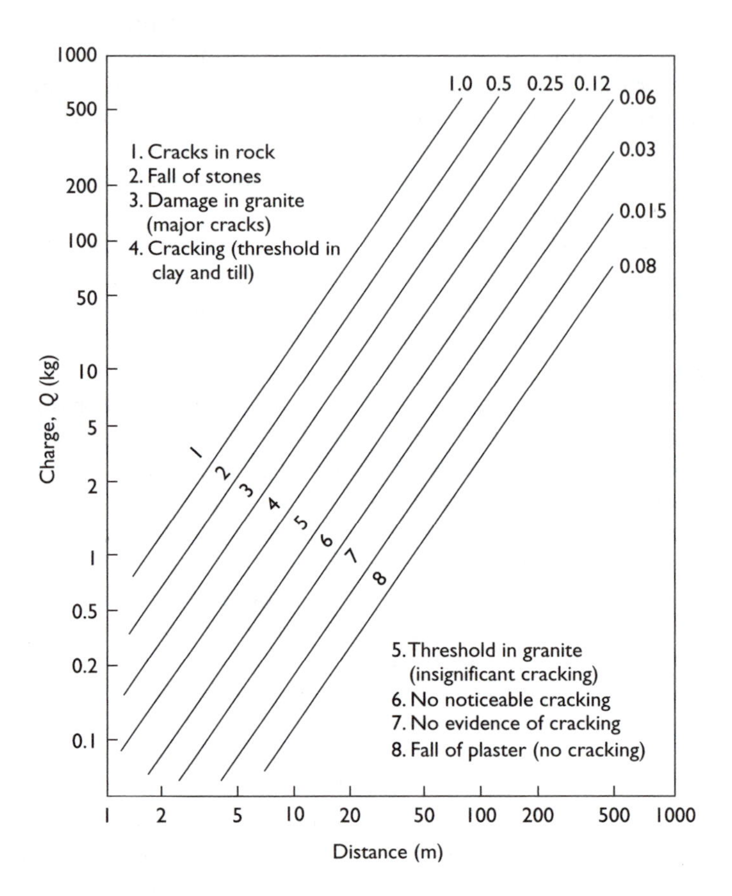

Fig. 6.11 Damage as a function of distance for various charge levels. Numbers 3 to 8 inclusive describe damage expected in normal houses. (Modified after Langefors and Kihlstrom, 1962.)

- The way in which the charge is distributed in the blasthole, the diameter of the hole and the depth of lift.
- The confinement of the charge, which involves the type of stemming/decking and the powder factor, as well as the width of burden and the spacing of the blastholes.

6.6. Slope stability in rock masses

Sliding on steep slopes in hard unweathered rock (defined as rock with an unconfined compressive strength of 35 MPa and over) depends primarily on the incidence, orientation and nature of the discontinuities present. It is only on very high slopes or in weak rock masses that failure of intact material becomes significant. Data relating to the spatial relationships between discontinuities affords some indication of the modes of failure that may occur and information relating to the shear strength along discontinuities is required for use in stability analysis. Indeed, one of the most important aspects of rock slope analysis is the systematic collection and presentation of geological data so that it can be incorporated into and readily evaluated in stability analyses (Bye and Bell, 2001).

Other factors apart, the maximum height that can be developed safely in a rock slope is roughly proportional to its shearing strength, that is, the stronger the rock, the steeper the slopes that may be cut into it. For instance, excavations in fresh massive plutonic igneous rocks such as granite can be left more or less vertical after the removal of loose fragments. Quarrying usually is straightforward in stratified rocks that are horizontally bedded and slopes can be determined with some degree of certainty. However, slopes may have to be modified in accordance with how the dip and strike directions are related to an excavation when it occurs in inclined strata. The most stable excavation in dipping strata is one in which the strata dip into the face. Conversely, if the strata daylight into a quarry or open-pit face (i.e. the dip of the bedding and therefore potential failure planes are into and less than that of the slope), then there is the potential for a slide to occur (Lisle, 2004). This is most critical where the strata dip at angles between 30° and 70°. If the dip exceeds 70° and there is no alternative to working against the dip, then the face should be developed parallel to the bedding planes for safety reasons. Sedimentary sequences in which thin layers of shale, mudstone or clay are present may have to be treated with caution, especially if the bedding planes dip at a critical angle. Indeed, wherever weaker strata underlying stronger rocks are exposed on excavation, then undermining of the latter is likely to occur as the former are removed by agents of denudation. Ultimately, this action will produce a rock fall or slide.

In a bedded and jointed rock mass, if the bedding planes are inclined, the critical slope angle depends upon their orientation in relation to the slope and the orientation of the joints. The relation between the angle of shearing resistance, ϕ, along a discontinuity, at which sliding will occur under gravity, and the inclination of the discontinuity, α, is important. If $\alpha < \phi$ the slope is stable at any angle, whilst if $\phi < \alpha$, then gravity will induce movement along the discontinuity surface and the slope will not exceed the critical angle, which will have a maximum value equal to the inclination of the discontinuities. It must be borne in mind, however, that rock masses generally are interrupted by more than one set of discontinuities.

Hard rock masses are liable to sudden and violent failure if their peak strength is exceeded in an excessively steep or high slope. On the other hand, soft materials, which exhibit small differences between peak and residual strengths, tend to fail by gradual sliding. The relative sensitivity of the factor of safety to the variation in importance of each parameter that influences the stability of slopes depends initially on the height of the slope. For example, Richards *et al.* (1978) graded each parameter concerned, in order of importance with respect to their effect on the factor of safety, in relation to slopes with heights of 10,100 and 1000 m. The results are shown in Table 6.3 and Figure 6.12. The heights and angles of slopes in hard rocks can be estimated roughly from Figure 6.12. The joint inclination is always the most important parameter as far as slope stability is concerned. Friction is the next most important parameter for slopes of medium and large height, whereas unit weight is more important for small slopes than friction. Cohesion becomes less significant with increasing slope height whilst the converse is true as far as the effects of water pressure are concerned. Knill (1994) showed that the maximum measured slope angle of faces in quarries in the igneous and metamorphic rocks of the Malvern Hills, England, was reduced with increasing height. He further distinguished between short-term (working) and long-term (abandoned) conditions in that the cohesive strength was reduced between the former and the latter. This Knill attributed to weathering of infill along weaker shear zones and means that stable slopes have a lower angle in abandoned quarries than in working quarries. Once abandoned, the quarry faces continue to fail with rock falls contributing to scree slopes that cover part of the faces, thereby reducing the overall slope angle.

The shear strength along a joint is attributable mainly to the basic frictional resistance that can be mobilized on opposing joint surfaces.

Table 6.3 Sensitivity of the factor of safety to various parameters (After Richards *et al.*, 1978)

(a) Parameters affecting the factor of safety for a given slope height and angle

Parameter	*Probable range of magnitude*
Unit weight	15–30 kN m^{-3}
Cohesion	0–300 kPa
Water pressure, h_w/H	0 – H m
Friction angle	0–60°
Joint inclination	10–50°

(b) Order of importance of parameters

Rank	*Slope height*		
	10 m	*100 m*	*1000 m*
1	Joint inclination	Joint inclination	Joint inclination
2	Cohesion	Friction angle	Friction angle
3	Unit weight	Cohesion	Water pressure
4	Friction angle	Water pressure	Cohesion
5	Water pressure	Unit weight	Unit weight

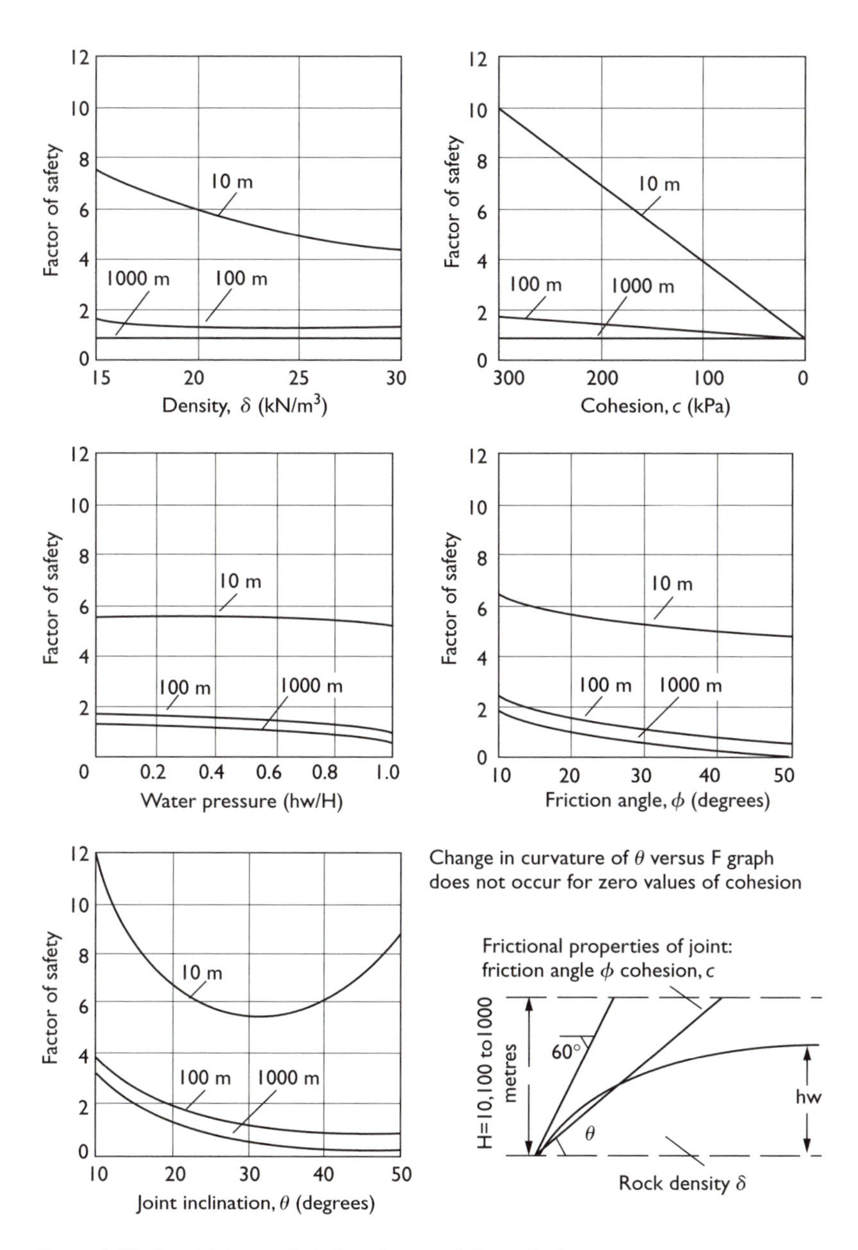

Figure 6.12 Sensitivity analysis for slope stability calculations.

Normally, the basic friction angle, ϕ_b, approximates to the residual strength along the discontinuity. An additional resistance is consequent upon the roughness of the joint surface. Shearing at low normal stresses occurs when the asperities are overriden, at higher confining conditions and stresses they are sheared through. Barton (1974) proposed that the shear strength, τ, of a joint surface could be represented by the following expression:

$$\tau = \sigma_n \tan [\text{JRC} \log_{10}(\text{JCS}/\sigma_n) + \phi_b] \tag{6.1}$$

where σ_n is the effective normal stress, JRC is the joint roughness coefficient and JCS is the joint compressive strength. The shear strength along a discontinuity also is influenced by the presence and type of any fill material, and by the degree of weathering undergone along the discontinuity.

According to Hoek and Bray (1981), in most hard rock masses neither the angle of friction nor the cohesion are dependent upon moisture content to a significant degree. Consequently, any reduction in shear strength is attributable almost solely to a reduction in normal stress across the failure plane and it is water pressure, rather than moisture content, which influences the strength characteristics of a rock mass.

The principal types of failure that are generated in rock slopes are rotational, translational and toppling modes (Fig. 6.13). Rotational failures normally only occur in structureless overburden, highly weathered material or very high slopes in closely jointed rock. They may develop either circular or non-circular failure surfaces. Circular failures take place where rock masses are intensely fractured, or where the stresses involved over-ride the influence of the discontinuities in the rock mass. Relict jointing may persist in highly weathered materials, along which sliding may take place. These failure surfaces are often intermediate in geometry between planar and circular slides.

There are three kinds of translational failures, namely plane failure, active and passive block failure and wedge failure. Plane failures are a common type of translational failure and occur by sliding along a single plane that daylights into a slope face (Figs 6.2(a), 6.13). When considered in isolation, a single block may be stable. Forces imposed by unstable adjacent blocks may give rise to active and passive block failures (Fig. 6.13). In wedge failure two planar discontinuities intersect, the wedge so formed daylighting into the face (Figs 6.2(b), 6.13). In other words, failure may occur if the line of intersection of both planes dips into the slope at an angle less than that of the slope.

Toppling failure generally is associated with steep slopes in which the jointing is near vertical. It involves the overturning of individual blocks and therefore is governed by discontinuity spacing as well as orientation. The likelihood of toppling increases with increasing inclination of the discontinuities (Fig. 6.13). Water pressure within discontinuities helps promote the development of toppling.

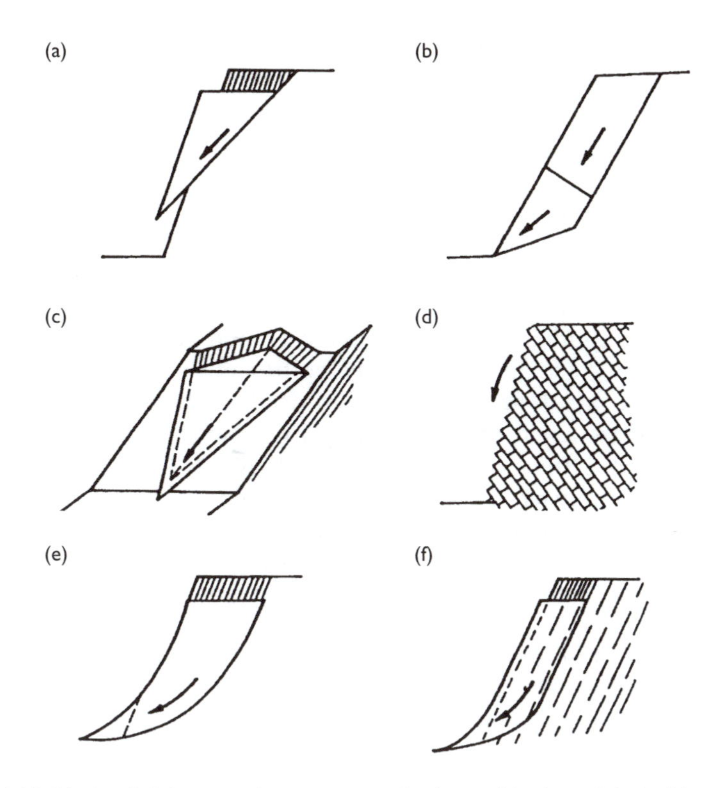

Figure 6.13 Idealized failure mechanisms in rock slopes (a) plane failure (b) active and passive block failure (c) wedge failure (d) toppling failure (e) circular failure (f) noncircular failure.

6.7. Placer deposits and mining

A placer is a surficial mineral deposit formed by mechanical concentration of mineral particles from weathered and eroded material. Usually, this natural gravity separation is brought about by running water such as streams or rivers but it also can be by marine, lacustrine, aeolian or even glacial action. Placers also can be formed as residual deposits or as eluvial deposits (Evans, 1997). The minerals that are concentrated must be freed from their source rock, must have a high density, be chemically resistant to weathering and possess mechanical durability. Such minerals include gold, platinum, tin, diamonds and other gemstones, and heavy minerals such as ilmenite, rutile, zircon and monazite. Most placer deposits are small and frequently ephemeral, forming at the surface of the Earth and so being subject to erosion and removal before being preserved by burial. Placer deposits have formed throughout geological time but most are of Tertiary or Quaternary age. Nonetheless, older placer deposits do occur that have been buried and lithified, one of the most notable examples being the gold deposits of the Witwatersrand, South Africa.

6.7.1. Small-scale alluvial mining

Alluvial mining is used to develop alluvial deposits in which valuable minerals occur in silt, sand and gravel of stream, river, lake or beach deposits. Panning is the simplest form of small-scale alluvial mining in which the miner scoops sediment and water into a shallow pan and shakes it with a rotary motion to wash out the silt, sand and gravel. The heavy minerals such as gold concentrate in the bottom of the pan. The sluice box is a larger device working on the same principle and consists of a shallow trough with riffles across its base. Sediment is washed through the box and the heavy minerals are trapped behind the riffles.

The individual impact of each small alluvial mine is usually not great but the combined effects of numerous operations may be disastrous to the environment. As a result, the environmental effects associated with alluvial mining often include damage to natural drainage, contamination of land and water, the formation of polluted wastelands, reduction of grazing land, deterioration of water quality and the destruction of natural habitats. One of the most notable effects of these activities is the siltation of rivers and dams. Siltation of rivers means that their importance as a means of conveyance is reduced, and that they become more prone to flooding and that fish stocks are depleted.

Flooding of land and disruption of the flow of rivers also is caused by the placement of obstacles in a river, which form part of extraction and processing, causing flooding at times of heavy rain. Waste water from the recovery processes often drain into rivers causing them to become contaminated. Rivers also can become polluted by tailings being washed into them. In areas of moderate to high relief, the removal of vegetation and stockpiling of waste or tailings on slopes can result in the generation of landslides. The creation of wastelands by alluvial mining means that the land is unable to grow and sustain crops, and that it cannot be used for building. In addition, these wastelands may contain contaminants and tailings left behind by the alluvial mining. Small-scale alluvial mining requires firewood for fuel and this, in turn, causes deforestation. In Zimbabwe, for example, approximately 4 million tonnes of timber are used each year for fuel, which equates to about 100 000 trees. Areas that have been stripped of vegetation begin to fill with water following prolonged heavy rainfall. They also become more prone to soil erosion.

Another environmental problem associated with small-scale alluvial mining of gold is the use of mercury for amalgamation of the gold (see Chapter 10). The amalgam is heated, which allows some mercury vapour to escape. The process means that mercury is dispensed into the atmosphere and water courses where it is transformed into methyl mercury (CH_3Hg). Mercury is a potent toxin and can cause severe health problems in humans and animals. For example, in Victoria Fields, Tanzania, approximately 6 tonnes of mercury are used annually for gold mining and much gaseous mercury has been released into the atmosphere (Ikingura *et al.*, 1997). A total mercury level of more than 48.3 ppm was found in the hair of gold miners and local inhabitants. The highest value was 953 ppm, which compares with a level of 50 ppm and above that can lead to Minamata disease. Approximately 78% of water samples analysed in the Lake Victoria goldfields contained mercury in concentrations in excess of 1 g l^{-1} of the acceptable limits for drinking water.

Significant alluvial gold deposits occur in Papua New Guinea and are mined at several places. Alluvial mining is concentrated along the Porgera River (Apte, 2001). The low-grade orebody is crushed and the heavy residue containing gold is recovered by panning, the waste being washed into the rivers. The gold subsequently is extracted from the residue by mercury amalgamation, the mercury–gold amalgam then is heated and processed to release the gold and the mercury vapour escapes into the air. It has been estimated that 2 g of mercury are released into the environment for each gram of gold recovered. Several cases of chronic mercury poisoning of miners and local inhabitants have been reported.

Environmental disasters caused by alluvial mining have prompted government intervention and legislation. Any mining companies involved are responsible for the post-alluvial mining restoration, and the safe treatment and disposal of mining waters and tailings after mining has ceased. In Malaysia many former alluvial tin mining areas have been successfully converted into housing estates whilst others have been developed as tourist attractions. For instance, a mining pool was converted into a recreational lake at Clearwater Sanctuary Club, Perak, and many old mines have been restored for their historical value. Replanting programmes have been introduced to protect the soil against further erosion and fertilizers added to the soil to encourage farming. Non-poisonous mine tailings and waste have been used to infill voids in the ground surface and contaminated ground has been cleaned up.

6.7.2. Dredge mining

Dredge mining is used to recover minerals from alluvial, marine or aeolian (dune) deposits (Fig. 6.14(a)). In alluvial and marine deposits, it involves the underwater excavation of a deposit, usually carried out from a floating vessel that may incorporate processing and waste disposal facilities. The body of water may be natural or man-made.

The deposits that contain the minerals must be diggable and able to hold water. For example, a pit usually may be excavated in the deposit and filled

Figure 6.14 (a) Dredge mining of heavy minerals, Richards Bay, Natal, South Africa. (b) Some 20 years after restoration of a dredge mined area at Richards Bay, Natal, South Africa. (Courtesy of Richards Bay Minerals Limited.)

with water so forming a pond. The dredge excavates the deposit and pumps the material to the separation plant, which may be separate from the dredge. Heavy minerals (e.g. ilmenite, rutile, cassiterite, chromite or scheelite) are concentrated and pumped ashore for further processing. Others such as gold or diamonds are collected. The tailings from the separation plant are placed in the mined out area of the pond for subsequent rehabilitation. Hence, the dredge, separation plant (if separate) and pond move on, excavating at one end of the pond and depositing waste at the other.

The flexibility of waste disposal allows different types of topography to be created after mining has passed on, so that restoration can be very successful (Fig. 6.14(b)). The topsoil is stripped prior to mining and placed initially in stockpiles for use during restoration. However, once mining is in progress soil can be transported directly from the area ahead of the dredging operation to the waste that is being landscaped. The scope for establishing vegetation is governed by the moisture holding capacity of the waste, which is usually sand or sand–gravel mixtures.

6.8. Gravel, sand and clay pits

Gravels and sands represent sources of coarse and fine aggregate respectively. The voids of a gravel deposit are rarely empty, being occupied by sand, silt or clay material. The composition of a gravel deposit reflects not only the type of rocks in the source area but it also is influenced by the agent(s) responsible for its formation and the climatic regime in which it was or is being deposited. Furthermore, relief influences the character of a gravel deposit, for example, under low relief gravel production is small and the pebbles tend to be chemically inert residues such as vein quartz, quartzite, chert and flint. By contrast, high relief and rapid erosion yield coarse immature gravels. Be that as it may, a gravel achieves maturity much more rapidly than does a sand under the same conditions.

Gravels and sands may occur as river deposits, river terrace deposits, beach deposits, raised beach deposits or fluvio-glacial deposits. Sand may also accumulate as wind blown deposits. The composition of a river gravel deposit reflects the rocks of its drainage basin. Sorting takes place with increasing length of river transportation, the coarsest deposits being deposited first, although during flood periods large fragments can be carried great distances. Thus, river deposits may possess some degree of uniformity as far as sorting is concerned. However, like fluvio-glacial gravels, river gravels tend to be bimodal, the principal mode being in the gravel grade, the secondary in the sand grade. Naturally, differences in gradation occur in different deposits within the same river channel but the gradation requirements for aggregate are generally met with or they can be made satisfactory by a small amount of processing. Moreover, as the length of river transportation increases softer material is progressively eliminated, although in a complicated river system with many tributaries new sediment is being added constantly. Deposits found in river beds are usually characterized by rounded particles. This is particularly true of gravels. River transportation also roughens the surfaces of pebbles.

If any gravel and/or sand operation is located within the floodplain of a river, then consideration should be given to the possibility that this could affect a flood event. This is particularly the case if gravel and sand working induce changes in the river bed, which could affect bank stability, accelerate erosion and change river morphology. If pits are flooded, then the steeper slopes may collapse as floodwaters move through and sediment is likely to be deposited in the pit.

River terrace deposits are similar in character to those found in river channels. The pebbles of terrace deposits may possess secondary coatings, frequently of calcium carbonate, due to leaching and precipitation. The longer the period of post-depositional weathering to which a terrace deposit is subjected, the greater is the likelihood of its quality being impaired.

Gravels and sands of marine origin are used as natural aggregate. The winnowing action of the sea leads to marine deposits being relatively clean and uniformly sorted. For the latter reason these deposits may require some blending. The particles are generally well rounded with roughened surfaces. Gravels and sands that occur on beaches generally contain deleterious salts and therefore require vigorous washing. By contrast, much of the salt may have been leached out of the deposits found on raised beaches. One factor that must be taken into account in any proposal to work marine gravels and sands is the role they play in coastal protection, for example, will removal of such deposits lead to increasing coastal erosion and increased likelihood of marine inundation?

Glacial deposits are poorly graded, commonly containing an admixture of boulders and rock flour. As a consequence, they are usually of limited value as far as aggregate is concerned. Conversely, fluvio-glacial deposits are frequently worked for this purpose. These deposits were laid down by melt waters that issued from bodies of ice. They take the form of eskers, kames and outwash fans (Fig. 6.15). Fluvio-glacial gravels may contain a variety of rock types, depending on the nature of the glacial material in the ice sheet from which they were derived. The influence of water on these sediments means that they have undergone varying degrees of sorting. They may be composed of gravels or, more frequently, of sands. The latter are often well sorted and may be sharp, thus forming ideal building material.

Wind blown sands are uniformly sorted. They are composed predominantly of well rounded quartz grains that have frosted surfaces.

In a typical gravel pit the material is dug from the face by a mechanical excavator. This loads the material into trucks or onto a conveyor that transports it to the primary screening and crushing plant (Fig. 6.16). After crushing the material is further screened and washed. This sorts the gravel

Figure 6.15 A kame being worked for sand, north of Lillehammer, Norway.

Figure 6.16 Screening gravel and sand at a pit in north Derbyshire, England.

into various grades and separates it from the sand fraction. The latter usually is sorted into coarser and finer grades, the coarser is used for concrete and the finer is preferred for mortar. Because gravel deposits are highly permeable, if the water table is high, then a gravel pit will flood. The gravels then have to be worked by dragline or by dredging from a floating pontoon. One possible advantage of wet working is that some silt and clay is washed out during the extraction process and the dredged material may be sufficiently liquid to be transferred to the processing plant by pipeline.

Sands are worked in a similar manner to gravels. They are used for building purposes to give bulk to concrete, mortars, plasters and renderings. For example, sand is used in concrete to lessen the void space created by the coarse aggregate. A sand consisting of a range of grade sizes gives a lower proportion of voids than one in which the grains are of uniform size. Indeed, grading is probably the most important property as far as the suitability of sand for concrete is concerned (Anon., 1992). In any concrete mix, consideration should be given to the total specific surface of the coarse and fine aggregates, since this represents the surface that has to be lubricated by the cement paste to produce a workable mix. Poorly graded sands can be improved by adding the missing grade sizes to them, so that a high quality material can be produced with correct blending. The commonest deleterious materials present in sand are fine silt and clay particles, coaly and organic material, and carbonate material such as fine shell fragments.

It is alleged that generally sand with rounded particles produces slightly more workable concrete than sand consisting of irregularly shaped particles. Sands used for building purposes are usually siliceous in composition and should be as free from impurities as possible. As far as concrete is concerned, sands should contain less than 3%, by weight, of silt or clay since these need a high water content to produce a workable concrete mix. This, in turn, leads to shrinkage and cracking on drying. Furthermore, clay and shaley material tend to retard setting and hardening, or they may spoil the finished appearance of concrete.

Sufficient quantities of suitable raw material must be available at a site before a clay pit can be developed. The volume of suitable mudrock must be determined, as must the amount of waste, that is, the overburden and unsuitable material within the sequence that is likely to be extracted (Bell, 1992). The first stages of the investigation are topographical and geological surveys, followed by a borehole/drillhole programme. This leads to a lithostratigraphic and structural evaluation of the site. It also should provide data on the position of the water table and the stability of the slopes that will be produced during excavation of the pit.

The suitability of a raw material for brick making is determined by its physical, chemical and mineralogical character, and the changes that occur when it is fired. The ideal raw material should possess moderate plasticity, good workability, high dry strength, total shrinkage on firing of less than 10% and a long vitrification range. Although bricks can be made from most mudrocks, the varying proportions of the different clay minerals have a profound effect on the processing and on the character of the fired brick. Those brick clays that contain a single predominant clay mineral have a shorter temperature interval between the onset of vitrification and complete fusion, than those consisting of a mixture of clay minerals. Also, the type and proportions of clay minerals present influence the amount of shrinkage that occurs when fired, those mudrocks containing predominantly expansive clay minerals being the most likely to shrink and distort. As far as the unfired properties of the raw materials are concerned the non-clay minerals present act mainly as a diluent, but they may be of considerable importance in relation to the fired properties. The non-clay material also may enhance the working properties, for instance, colloidal silica improves the workability by increasing plasticity and the presence of quartz, in significant amounts, gives strength and durability to a brick. However, as the proportion of quartz increases, the plasticity of the raw material decreases.

Once the clay material has been extracted it is usually placed in a stockpile, which thereby affords a degree of mixing of the raw material. It then is left to weather for up to a year. Weathering helps to break down the mudrock and therefore makes subsequent crushing and screening easier. In addition, the material in the stockpile acquires a relatively uniform moisture content, which means that the subsequent addition of water to the raw material in a pugmill is more easy to control. Weathering also helps the breakdown of pyrite, a common constituent in clays, into sulphate and iron compounds. In fact, too high a percentage of pyrite or other iron bearing minerals gives rise to rapid melting that can lead to difficulties on firing. Some of the sulphates may be leached out of the stockpile. Any soluble sulphates that remain will dissolve in the water used to mix the clay material. During drying and firing they may form a white scum on the surface of a brick. Barium carbonate may be added to render such salts insoluble and so prevent scumming.

Gravels, sands and clays, as well as some of the softer rocks such as shales, can be excavated by digging machines. The diggability of ground is of major importance in the selection of excavating equipment and depends primarily upon its intact strength, bulk density, bulking factor (intact density per unit volume/disturbed density per unit volume) and natural moisture content. At present there is no generally accepted quantitative measure of diggability, assessment usually being made according to the experience of the operators. However, a fairly reliable indication can be obtained from similar excavations in the area concerned and the behaviour of the ground excavated in trial pits. Nevertheless, attempts have been made to evaluate the performance of excavating equipment in terms of seismic velocity (Fig. 6.17). It would appear that most earthmoving equipment operates most effectively when the seismic velocity of the ground is less than

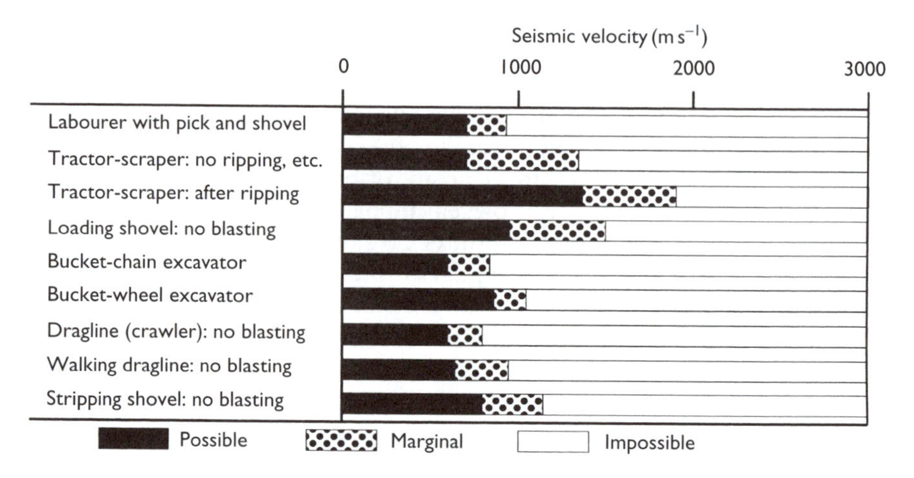

Figure 6.17 Assessment of diggability in terms of seismic velocity.

Figure 6.18 Working the Oxford Clay with a bucket-chain excavator for brick making at Whittlesey, near Peterborough, England.

1 km s^{-1} and will not function above 1.8 km s^{-1}, but in areas of complex geology seismic evaluation may prove difficult if not impossible.

With their long reach and dumping radius draglines are especially suited to bulk excavation below track level. Draglines can be used in loose soils or for soft ground and inundated areas, for they are designed to stand above the working area on firm ground and move backwards as the excavation proceeds. They also can be used to excavate badly fractured, weathered and soft rocks. Draglines can work to appreciable depths, although they cannot cut vertical faces. One of the advantages of large draglines is that they can pile earth adjacent to the excavation when earth needs to be returned subsequently as backfill. The two most important types of continuous excavators are the bucket-chain excavator and the bucket-wheel excavator (Atkinson, 1971). For example, clay deposits in large pits may be worked by a bucket-chain excavator. This consists of a series of toothed grabs arranged along a continuous chain. The chain and grabs are placed alongside the working face, the grabs making a continuous cut and discharging the clay at the top of the face (Fig. 6.18). The method produces an even mix of the clay material and leaves a clean, and therefore relatively stable, face. Loading shovels are only suitable for cuts up to a few metres in depth. They are capable of handling hard dense soils or badly fragmented ground but they have a poor sub-grade digging capability. The face shovel is the most powerful of the front-end excavators and is designed primarily for bulk excavation above track level, working from the bottom of the excavation upwards. It is most efficient when operating against a high face in firm to hard clay soil, densely packed sand or gravel with clay binder.

6.9. Slope stability in soils

Displacement in soil, usually along a well-defined plane of failure, occurs when shear stress rises to the value of shear strength of the soil. The shear strength of the material along the slip surface is reduced to its residual value so that subsequent movement can take place at a lower level of stress.

A slope of 1 : 1.5 generally is used when excavating dry gravel or sand, this more or less corresponding to the angle of repose, that is, 35°–45°. This means that an excavation in coarse grained soil will be stable, irrespective of its height, as long as the slope is equal to the lower limit of the angle of internal friction, provided that the slope is suitably drained. As far as sands are concerned, their packing density is important. For example, densely packed sands that are very slightly cemented may have excavated faces with high angles that are stable. Failure on a slope composed of coarse grained soil involves the translational movement of a shallow surface layer. The slip is often appreciably longer than it is in depth.

The most frequently used gradients in many clay soils vary between 30°and 45°. The stability of slopes in clay soil depends not only on its strength and the angle of the slope but also on the depth to which the excavation is taken and on the depth of a firm stratum, if one exists, not far below the base level of the excavation. In stiff fissured clays, the fissures appreciably reduce the strength below that of intact material. Thus, reliable estimation of slope stability in stiff fissured clays is difficult. Generally, steep slopes can be excavated in such clay soils but their excavation means that fissures open due to the relief of residual stress and there is a change from negative to positive pore water pressures along the fissures, the former tending to hold the fissures together. This change can occur within a matter of days or hours. Not only does this weaken the clay but it also permits more significant ingress of water, which means that the clay is softened. Irregular shaped blocks may begin to fall from the face and slippage may occur along well-defined fissure surfaces that are by no means circular. Slopes in stiff fissured clays can be excavated at about 35°. Although this will not prevent slips, those that occur are likely to be small.

Clay soils, especially in the short-term conditions, may exhibit relatively uniform strength with increasing depth. As a result, slope failures, particularly short-term failures, may be comparatively deep-seated, with roughly circular slip surfaces. This type of failure is typical of relatively small slopes. Slides on larger slopes are often non-circular failure surfaces following bedding planes or other weak horizons.

In slopes excavated in fissured overconsolidated clay, although stable initially, there is a steady decrease in the strength of the clay towards a long-term residual condition. During the intermediate stages swelling and softening, due to the dissipation of residual stress and the ingress of water along fissures that open on exposure, take place. Large strains can occur locally due to the presence of the fissures. Considerable non-uniformity of shear stress along a potential failure surface and local overstressing leads to progressive slope failure, with initial sliding occurring at a value below peak strength.

6.10. Groundwater and surface mining

Groundwater can represent a problem during surface mining and its removal can prove costly. Not only does groundwater make working conditions more difficult, but piping, uplift pressures and flow of water into the pit can lead to erosion and failure of the sides. Collapsed material has to be removed and the damage has to be made good. Sub-surface water is normally under pressure, which increases with increasing depth below the water table. Under high pressure gradients soils and weakly cemented rock can disintegrate. Excessive flow of groundwater can lead to pits flooding. For instance, flow of groundwater from springs into the Clayhanger Brick Pit near Walsall, Staffordshire, England, which was some 12.5 m in depth, was eventually so strong that the pump used to remove it could not cope and the pit flooded. Hence, data relating to the groundwater conditions should be obtained prior to the commencement of operations.

Groundwater flow in rock masses is controlled primarily by the discontinuities present rather than by permeation through the intact rock. Since discontinuities generally fall into 'sets' having a certain orientation, the permeability of a rock mass varies with direction. Since measurements of permeability represent spot values in the field, an adequate description of the geology of the rock mass and its discontinuity pattern is essential before an assessment of groundwater conditions can be made or means of controlling groundwater designed. Geological features such as faults or dykes may have a significant influence on groundwater flow patterns. For example, faults containing clay gouge impede flow and may cause high groundwater pressures behind them. Alternatively, shear zones may form zones of high permeability, concentrating groundwater flow and effectively draining the rock on either side. The geological structure and topography will indicate the potential for recharge of the rocks that are to be exposed in a quarry (Fig. 6.19).

The permeability of rock masses should be determined from detailed studies of drillhole core, information on water or air loss during drilling and field permeability tests. Groundwater pressures may adversely affect the stability of rock masses by reducing their shear strength along discontinuities and by producing hydraulic driving forces in steeply dipping discontinuities that are not subject to shear failure (e.g. tension cracks). Hence, the groundwater pressure distribution in a slope must be known before the effect of groundwater on stability can be evaluated. This means that measurements have to be taken with piezometers at representative locations. Piezometric heads within rock masses often vary considerably from one location to another and a sufficient number of piezometer tips must be installed to determine the overall conditions. Tips should be located with reference to the local geology, and especially in relation to intersecting sets of discontinuities, and must be properly sealed in the drillhole both above and below the tip so that measurements of piezometric head represent only conditions encountered at the tip. For this reason, open holes rarely give reliable results. There must be a sufficient number of piezometer installations to allow for malfunction of individual instruments, as well as to give a reasonable statistical assessment of piezometric conditions. Analytical methods to determine the groundwater pressure distribution within a slope from permeability and piezometric measurements have included graphical flow net sketching and electrical resistance analogues but today these involve computer analyses.

In particular, artesian conditions can cause serious trouble in quarries and pits, and therefore if such conditions are suspected it is essential that both the position of the water table and the piezometric pressures should be determined before work commences. Otherwise excavations that extend

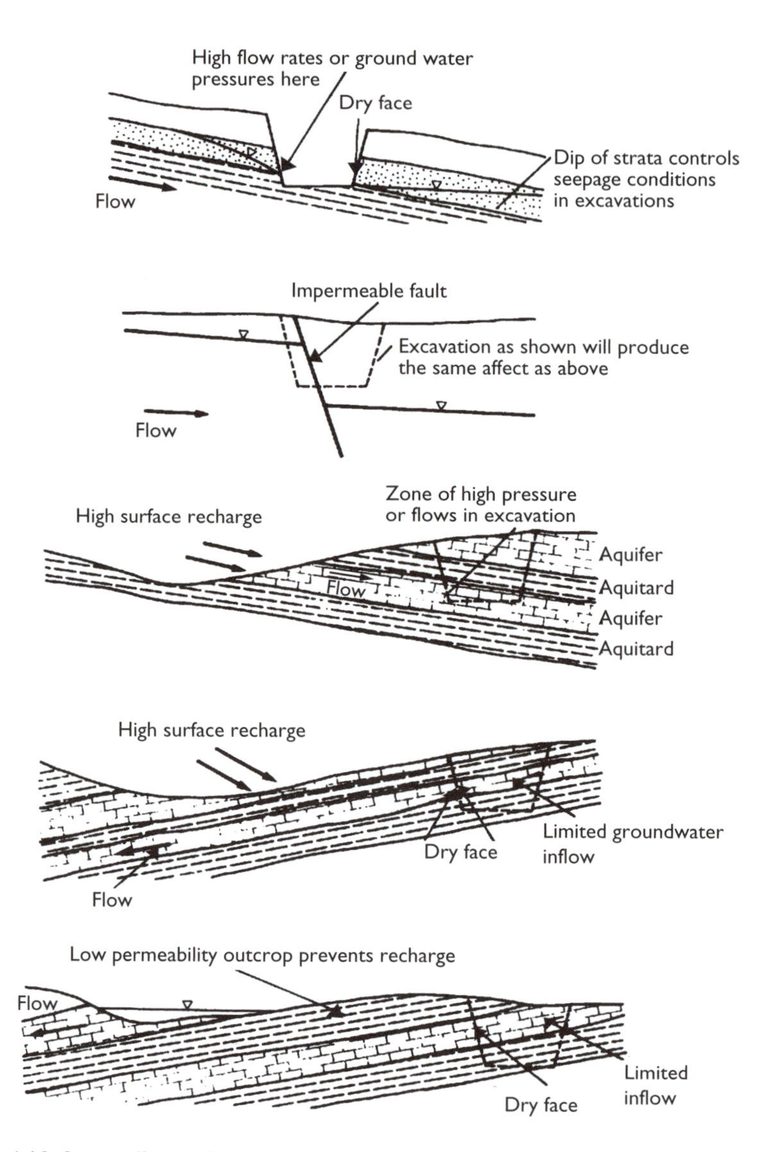

Figure 6.19 Some effects of geological structure on groundwater conditions.

close to strata under artesian pressure may be severely damaged due to blowouts taking place in their floors. Such action may cause slopes to fail and could lead to the abandonment of the pit. Moore and Longworth (1979), for instance, recorded a failure in a clay pit, 29 m in depth, excavated in Oxford Clay at Whittlesey, Cambridgeshire, England. The failure was brought about by a build-up of hydrostatic pressure in an underlying aquifer (either the Cornbrash or Blisworth Limestone located at depths of 6 and 11 m respectively below the floor of the pit). This initially gave rise to a heave of some 150 mm, and then ruptured the surface clay, thereby allowing the rapid escape of approximately 7000 m^3 of groundwater. The floor of the pit then settled up to 100 mm. Sites where such problems are likely to be encountered should be dealt with by employing dewatering techniques.

The most commonly used method of groundwater control is drainage, which by reducing the detrimental effects of groundwater, can increase the stability of quarry or pit slopes. The maximum benefit is obtained by reducing groundwater pressures within the slope prior to movements taking place during excavation and while the rock mass still possesses its undisturbed strength. The most probable failure zone must be determined in order to delimit the zone of the rock mass that needs to be drained. Drainage of a slope includes such measures as drainage trenches constructed down and/or along the slope face; horizontal or near horizontal drain holes drilled into the slope face; and vertical pumped wells drilled behind the slope crest or on the slope face (Fig. 6.20). The method

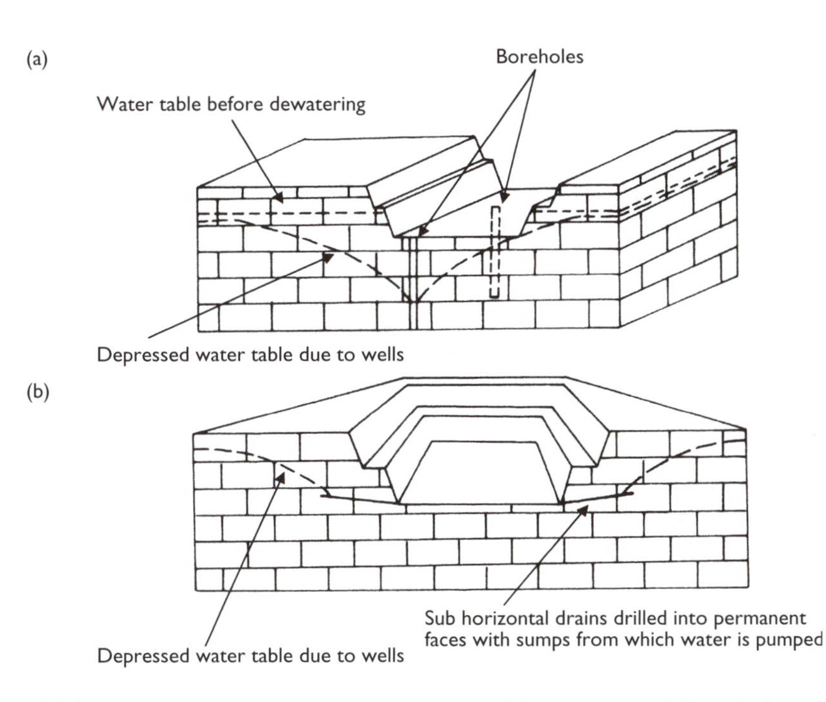

Figure 6.20 Dewatering in quarries and opencast pits (a) using vertical beoreholes or wells (b) using subhorizontal drains.

chosen to drain an excavated slope depends on slope geometry, as well as permeability of the slope material. Effectiveness of drainage installations can be enhanced by controlling surface water before it infiltrates the slope. This usually is accomplished by constructing trenches at the top and toe of the slope and, where necessary, surface drainage over the slope (Anon., 1987). Control of water is particularly important when individual benches are drained by horizontal holes in order to prevent water from percolating into the bench below. Each row of drain outlets should issue into a properly graded interceptor or trench, lined if necessary with impermeable material to prevent erosion.

In some cases, it may be more expedient to prevent the groundwater from reaching a critical zone by creating a cut-off that retains the groundwater behind it and allows natural drainage to occur within the slope. Cut-offs are advantageous where internal erosion of soft material may be caused by increased groundwater flows induced by drainage measures. In rock masses cut-offs can be formed by the construction of a grout curtain, or in soft rocks by excavating a trench and backfilling with concrete. On occasions, grout curtains have been used to prevent or control large flows of groundwater from conduits in karstic carbonate rocks into quarries. For example, Bruce *et al.* (2001) referred to an inflow of about 140 000 $l\,min^{-1}$ that suddenly occurred in the floor of an operational quarry in West Virginia, the flow originating from a river some 450 m distant and moving via karstic features in the limestone concerned. The head differential was over 50 m. As pumping would have been prohibitively expensive, it was decided to construct a two line grout curtain, some 340 m long and 70 m in depth, within 20 m of the river bank. The curtain had to be re-grouted in places due to water under high pressure removing clay fill that occupied conduits in the karstified limestone. Alternatively, a row of pumped vertical wells can be sunk to form a well curtain that intercepts flow towards a quarry and provides relatively dry conditions on its downstream side.

Some of the worst conditions are met in quarries that have to be taken below the water table. In such cases the water level may be lowered by dewatering. The method adopted for dewatering depends upon the permeability of the ground and its variation within the stratal sequence; the depth of base level of the excavation below the water table; and the piezometric conditions in underlying horizons. Pumping from a sump within the excavation generally can be achieved where the rate of inflow does not lead to instability of the sides or base of an excavation. Ditches are dug in the floor of the quarry, then lined or filled with gravel, in order to lead water to a sump or sumps, which must be deep enough to ensure that the quarry is drained. Each sump requires its own pump and the method is only capable of lowering the water table by up to approximately 8 m. Sump pumping cannot dewater confined aquifers and may not be able to cope with flow from aquifers with large storage capacities or that transmit copious quantities of groundwater. In such situations internal drainage measures, as mentioned earlier, may provide an answer. Alternatively, groundwater may be removed by abstraction via vertical wells to form interconnected cones of depression, thereby lowering the water table. A bored filter well consists of a perforated tube surrounded by an annulus of filter media and the operational depth may, in theory, be unlimited. Generally, wells are placed in a 600 mm diameter hole. Having drilled the hole to a sufficient depth within the aquifer (in the case of a thin aquifer the well bore is often taken some 1–2 m into the impervious stratum beneath the aquifer), the well-screen is lowered into place and the appropriate filter medium then is placed. An electric submersible pump is lowered into each well on its own riser pipe and connected to a common discharge main. Deep wells are particularly suited to multi-layer aquifers, as well as to the control of groundwater under artesian and subartesian conditions.

6.11. Reclamation of old quarries and pits

Abandoned quarries and pits have been restored or reclaimed for a variety of uses, where ground conditions are suitable. These include for landfill disposal, use as nature reserves and recreational areas, sites of special scientific interest, occasionally for railway or tram museums, and larger quarries have been used for housing development, restored to farmland and developed as golf courses. Some quarries have been afforested, especially spoil heaps associated with quarries. In future quarry reclamation may include sustainable biomass energy production, that is, the growth of crops such as willow that can be harvested for power generation. Abandoned gravel and sand pits often have high water tables and therefore have become flooded. These areas frequently are converted to marinas. Old bricks pits located in thick formations of clay frequently are used for the disposal of domestic waste and then subsequently reclaimed. In addition, old quarries and pits may contain a range of interests from geology to industrial archaeology to wildlife, which can form part of a restoration programme.

Abandoned quarries, pits and associated spoil heaps are particularly conspicuous and can be difficult to rehabilitate into the landscape. For instance, many of the large modern quarries in the chalk escarpment of south-east England, that are associated with cement works have a dramatic impact on the landscape. Once abandoned such large quarries can have a blighting affect on both the local environment and economy. Any old buildings associated with former workings represent another form of dereliction. In fact, some former quarries such as the Pitstone works in Buckinghamshire, are large enough for the re-development proposal to include the provision of housing and community buildings. The floors of such other large quarries have been restored to farming.

According to Bailey and Gunn (1991) reclamation strategies for hard rock quarries have been surprisingly limited. In the past quarries have been screened from view by planting trees around them or, as mentioned, used for waste disposal where conditions allow. Some large old quarry voids have been used for storage or for the location of industrial units. Such uses usually require that the quarry faces are treated to reduce the potential for rock fall. This involves removal of loose rock from quarry faces and possibly stabilization measures. It also may involve material being placed at the base of faces, and then being covered with soil and planted.

Quarry restoration often involves an attempt to simulate natural landforms and in this way to reduce the visual obstructiveness of a quarry and help blend it into the landscape. Accordingly, an assessment of the surrounding landscape should be undertaken prior to a quarry restoration scheme being designed. The form, scale, proportion, texture and colour of landform simulations should be realistic, as well as being adapted to the quarry setting. Obviously, the continuous sharp linear crest of a quarry should be done away with (e.g. rounded by mechanical excavators or by shallow blasting) and the straight faces of quarries or quarry benches broken into buttresses and headwalls so that the contrast between the quarry and surrounding landscape is reduced. In the case of quarries in limestone areas in particular, Bailey and Gunn (1991) suggested that landform replication represented the best method of quarry reclamation. Landform replication involves the construction of landforms and associated habitats similar to those of the surrounding environment (Fig. 6.21). Restoration blasting along quarry faces can be used to replicate landforms such as scree slopes and to produce multi-faceted slope sequences with rock buttresses. The design of any restoration blasting requires consideration of the degree of fragmentation needed to form a scree of suitable size and that the scree is correctly positioned by blasting. It also requires that the rock slopes remaining are stable. Several methods have been used to minimize or eliminate rock face breakage due to blasting. These include line drilling, pre-splitting or the use of lower velocity explosives. In some instances, rock faces may need to be scaled in order to remove potentially unstable or overhanging material. Quarry waste or different grades of crushed limestone can be used to dress some scree slopes and weathered limestone material from the upper horizons of a quarry can be used to aid the growth of characteristic limestone flora. Hydroseeding may be used to encourage the growth of such species. With time the walls of quarries can revegetate in such a way as to replicate natural escarpments and it is now believed that abandoned quarries have conservation potential for endangered species of flora and fauna (Wheater and Cullen, 1997).

Abandoned quarries, especially limestone quarries, may contain a range of interesting geological features and fossils, and as such be preserved as sites of special scientific interest. A pretty unique development has taken place at Tout Quarry on the Isle of Portland, England. There the Portland Sculpture Trust has developed a tradition since 1983 of producing carvings and sculptures on the quarry walls. Carving and sculpture courses are run

(a)

(b)

Figure 6.21 Abandoned quarries in limestone, Derbyshire, England, that have undergone landform replication. (Courtesy of Dr John Cripps.)

by the Trust, as well as educational visits to develop an appreciation of the skills of quarrymen, masons and carvers working in Portland stone.

Because of the size of many open-pits, for example, Bingham Canyon copper mine in Utah is some 800 m in depth and covers nearly 8 km^2, it being one of the largest man-made excavations in the world, there is normally not enough waste material available to backfill them. Accordingly, the principal objectives after cessation of mining operations generally are to ensure that the walls of the pit are stable, and that the waste dumps and tailings lagoons are rehabilitated. It is possible that an open-pit could be used for the disposal of other waste if the hydrogeological conditions permit. Perhaps open-pits could be left to fill with water and then could act as a source of supply and/or be used for recreational purposes. Bingham Canyon copper mine, although still a working mine, is also a tourist attraction. Open-pit coal mines are reclaimed by either rehabilitating and then planting trees on slopes or by backfilling with mine waste (Fig. 6.22).

Figure 6.22a Slopes in the process of being restored and planted with trees, near Lethbridge, British Columbia.

Figure 6.22b Restoration by end-tipping of waste into old open-pit coal workings, near Lethbridge, British Columbia.

Figure 6.22c Long-horn sheep dining on grass planted on a restored area, near Lethbridge, British Columbia.

The vast majority of opencast sites are restored to agriculture or sometimes to forestry (Fig. 6.23). Nevertheless, due to increasing pressures on land use and the proximity between some sites and urban areas, backfilled opencast sites have been used for industrial and housing developments or may be used for local amenities such as country parks and golf courses. Restoration of opencast sites normally begins before closure, with worked out areas behind the excavation front being filled with rock waste. This means that the final contours can be designed with less spoil movement than if the two operations were undertaken separately. Furthermore, more soil for spreading can be conserved when restoration and coal working are carried out simultaneously. Because of high stripping ratios (often 15 : 1 to 25 : 1), there usually is enough spoil more or less to fill the void. If a site is to be restored for agricultural use or for forestry, then this takes place without control of compaction. However, if a site is to be built over, then settlement is likely to be a problem without proper compaction. In fact, varying amounts of settlement have been noted at restored opencast sites. Differential settlement can occur across a site and especially at the fill-solid interface. Such settlement can induce cracking in buildings, and affect roads and services.

A further factor that should be considered at an opencast site where the water table was lowered by pumping in order to provide dry working conditions in the pit, is the affect that a rising water table will have on the fill. This can lead to a second phase of surface settlement after the initial phase that occurs under the self-weight of the backfill. The latter generally is completed largely within two or three years. Significant settlements of

Figure 6.23a Rehabilitation of the Las Mercedes opencast bauxite mine providing land for forestry, Pedernales, Dominican Republic, West Indies; showing abandoned mine workings and remnant limestone pillars (upper), reforestation and restoration (lower).

Figure 6.23b Restoration of the Westfield Opencast workings for coal, UK. The area beyond the current opencast workings has been rehabilitated and restored to agricultural land.

opencast backfill can occur when the partially saturated fill becomes saturated by rising groundwater after pumping has ceased. For example, Charles *et al.* (1993) referred to an opencast site in Northumberland, England, where the waste was backfilled without any systematic compaction. They recorded that when the water table rose, then some 0.33 m of settlement occurred where the fill was 63 m in depth. Blanchfield and Anderson (2000) maintained that settlement due to wetting collapse is more significant than that due to the self-weight of the backfill, which can give rise to difficulties in settlement prediction. Therefore, as remarked, if a site is to be built upon after restoration, then the waste should be properly compacted. However, the subsequent development of a site frequently has been made after an opencast area has been worked out and backfilled. As a consequence, such areas require investigation prior to any building development (Anon., 1983). Consequently, Charles (1993) recommended that if a former opencast site is to be built over but the backfill was not properly compacted, then collapse compression could be brought about prior to development by inundation. However, he further noted that in many instances flooding would be difficult to carry out in a controlled and effective way. Alternatively, the ground could be pre-loaded or subjected to deep compaction, or the structures could be designed to accommodate any subsequent ground movements.

Alternatively, where backfilled opencast coal workings exceed 30 m in depth, because greater settlements may occur, Kilkenny (1968) recommended that the minimum time that should elapse before development takes place should be 12 years after restoration is complete. He noted that settlement of opencast backfill appeared to be complete within 5–10 years after the operation. Kilkenny undertook comprehensive observations of the opencast restored area at Chibburn, Northumberland, England, which was 23–38 m in depth. These revealed that the ultimate settlement amounted to approximately 1.2% of fill thickness and that some 50% of the settlement was complete after two years and 75% within five years. In shallow opencast fills, that is, up to 20 m in depth, settlements of up to 75 mm have been observed. Greater settlements may occur if there are movements of groundwater through permeable fill, including a rising water table, as noted above, as a result of breakdown of point-to-point contacts in the shale or sandstone components of the fill. In such circumstances settlement may continue over a greater length of time than noted by Kilkenny. In fact, at some sites in the Midlands of England settlement has continued for over 30 years after the sites were restored.

Restoration of other opencast mineral workings can be dealt with in a similar way to those associated with coal. Another mineral in Britain that was extracted on a large scale by opencast mining was sedimentary iron ore, for example, such deposits were worked near Scunthorpe, Lincolnshire, and near Corby in Northamptonshire. In the latter area Penman and Godwin (1975) noted that maximum rates of settlement occurred immediately after the construction of two storey semi-detached houses on an old opencast site that had been backfilled. Settlement decreased to small rates after about four years. They suggested that two of the causes of settlement in this fill were creep, which is proportional to log time, and partial inundation. Similar conclusions were reached by Sowers *et al.* (1965). The houses at Corby were constructed 12 years after the fill was placed (cf. the comments in the previous paragraph). The amount of damage that they suffered was relatively small and was attributable to differential settlement; it was not related to the type of foundation structure used. These were either strip footings or reinforced concrete rafts with edge beams.

The extraction of sand and gravel deposits in low-lying areas along the flanks of rivers frequently means that the workings extend beneath the water table. On restoration it is not necessary to fill the flooded pits completely. Partial filling and landscaping can convert such sites into recreational areas offering such facilities as sailing, fishing and other water sports. They also can act as nature reserves and bird sanctuaries. It is necessary to carry out a thorough survey of flooded workings, with soundings being taken so that accurate plans and sections can be prepared. The resultant report then forms the basis for the design of the measures involved in rehabilitation.

As noted earlier, former clay pits are frequently used for the disposal of domestic and industrial waste, and once filled such areas may be landscaped and developed. For example, controlled tipping of domestic and industrial waste takes place in brick pits, in the Oxford Clay, as they are abandoned at Stewartby, Bedfordshire, England. Pits that have been excavated in thick deposits of clay like those at Stewartby, usually offer sites that are relatively impermeable and therefore will contain the leachate that is produced from the waste as it decomposes. In this way, groundwater pollution is avoided. Furthermore, clay material is available at such sites for the construction of cellular landfills and impermeable cappings, which reduce leachate production and further inhibit the possibility of groundwater pollution. However, the biochemical decomposition of domestic and other putrescible refuse in a landfill produces gas consisting primarily of methane. Methane production can constitute a dangerous hazard because methane is combustible, and in certain concentrations is explosive (5–15% by volume in air), as well as asphyxiating. In many instances landfill gas is able to disperse safely to the atmosphere from the surface of a landfill. However, when a landfill is completely covered with a clay capping of low permeability in order to minimize leachate generation, the potential for gas to migrate along unknown pathways increases and methane migration may present a hazard. Furthermore, there are unfortunately cases on record of explosions occurring in buildings due to the ignition of accumulated methane derived from landfills near to or on which the buildings were built (Williams and Aitkenhead, 1991). Accordingly, planners of residential developments ideally should avoid landfill sites. Obviously, the identification of possible migration pathways in the assessment of a landfill site is important, especially where residential property is nearby. Measures to prevent migration of the gas include impermeable barriers and gas venting. Flooded clay pits also are reclaimed as marinas, nature reserves and bird sanctuaries. Some have been designated sites of special scientific interest.

6.12. Case history I

The Eden Project near St Austell, Cornwall, England, represents one of the most ambitious and innovative attempts to reclaim a surface pit. The site is located in an old china clay pit, up to 60 m in depth and covering some 15 ha, the floor of which was a complex mixture of clay and rock around which were heaps of spoil waste. The object was to construct three giant conservatories or biomes to house plants from the humid tropical regions, the subtropical regions such as the Mediterranean and South Africa, and the temperate regions such as Britain (Fig. 6.24). Various facilities are provided for the many visitors, including an education centre.

Figure 6.24 Eden Project, showing biomes to the right of centre. (Courtesy of Mrs Susan Mills.)

Muck shifting to create a level floor for the pit involved moving some 1.8 million tonnes of material. Thirteen metres was removed from the tops of the spoil heaps and placed in the bottom of the pit, thereby raising it between 17 m and 20 m. Band drains were placed in the clayey material occupying the floor of the pit to facilitate its drainage and settlement prior to placement and compaction of the fill. The foundations for the biomes consist of reinforced concrete, 1.5 m in depth and 2 m wide. The biomes are huge dome-shaped structures that consist of triple glazed hexagonal panels made of ethyl tetra fluoro-ethylene, which is lightweight, strong, has better insulation properties than glass, and is highly transparent to ultra violet light and so is not degraded by sunlight. As the slopes of the pit were unstable, they were cut back to safe angles and terraces were excavated. In addition, 2000 rock anchors were used, some of which were up to 12 m long, to help retain the slopes. Finally, the slopes were hydro-seeded.

The base of the pit is some 15 m below the water table so that a state-of-the-art drainage and pumping scheme had to be installed to deal with the water. In other words, the water is filtered by a layer of matting beneath the fill in the pit, the water then flowing via drains into a sump of 22 m in depth. This water, together with the water from the site, on average amounts to 22 $l\,s^{-1}$, and is used to irrigate the plants, as well as for sanitary purposes.

Unfortunately, there was no soil available of the quality and in the quantity needed locally. Accordingly, a decision was taken to manufacture the 85 000 tonnes of soil that was required. The local mine waste, sand and reject clay provided the basis for the soil and a series of trial mixes were carried out in order to determine the best physical structure for the artificial soil. The organic matter was provided by composted bark fines from the local forestry industry and from composted domestic green waste. Obviously, different types of soil were produced for the different regions. For example, the soil mix for the humid tropical regions has a high organic content, together with sand and clay, that for the Mediterranean regions has more sand so that it holds less moisture whilst that for the Fynbos plants from South Africa consists of composted bark and sand. The type and amount of nutrients are added according to the type of plants concerned. The area outside the biomes was landscaped.

6.13. Case history 2

Old sand and gravel workings to the west of Brighouse in West Yorkshire, England, were in need of rehabilitation (Bell and Genske, 2000). The site was approximately 16.3 ha in extent. It was about 1.2 km long and 0.4 km wide, at its widest point. The site had been an active sand and gravel pit located in the valley of the River Calder, the valley containing a main road, a canal and a railway line.

Approximately 200 000 tonnes of sand and gravel had been extracted annually from the workings. Work had commenced at the eastern end of the site, where there was between 150 mm and 230 mm of topsoil, and 2–2.5 m of subsoil to remove before the sand and gravel could be excavated. As the site extended westwards, the subsoil increased up to 6 m in thickness, which meant that excavation became deeper. Normally, if there is more than 1 m^3 of overburden to remove in order to win 1 m^3 of sand and gravel it is not worked as it is unprofitable. However, as the site was in an area where much development was taking place, there was ready sale locally of the material for aggregate.

The overburden on the site had been removed by a scraper, towed behind a dozer, and a dragline was used to excavate the sand and gravel. As the site lies in the bottom of a valley, there is a high water table and much of the excavation was below water. To reduce the amount of pumping needed, two berms had been constructed across the workings from north to south, splitting the site into three approximately equal parts (Fig. 6.25). It was only necessary to keep the water level down in the active pit. The silt–clay fraction that was removed from the sand and gravel by washing had been piped into a settling pond.

When the sand and gravel was worked out, it was decided to convert the site into a marina. The western part of the site was substantially higher than

Figure 6.25 Gravel pit undergoing restoration as a marina near Brighouse, West Yorkshire, England.

the east and so the site had to be graded, excavation being done on a cut and fill basis by scrapers. The two berms that ran across the flooded workings were removed to form a lake and the material from the berms was used to construct several islands. The lake was connected to the canal in the valley bottom by a lock at its south eastern corner and the water level in the lake is compatible with that of the canal.

The whole area was landscaped. After the site was regraded, a layer of subsoil, 150 mm thick, was spread over it and on top of this was laid a layer of topsoil, again 150 mm thick. The soil was ploughed and harrowed at least twice, the second pass being made at right angles to the first. Grass seed then was sown over the whole area. Trees were planted alongside the main road adjoining the site and around the lake, and flower beds and small shrubs set out along the sides of the site roads and around the marina. Grass was sown and trees planted on the artificial islands in the lake. The lake and islands now provide a sanctuary for wildlife, and the lake was stocked with fish.

Access to the marina was provided at the north-east and north-west corners of the site from the major road in the area. About half way along the northern bank of the site, adjacent to the main road, were some disused farm buildings. As they were in a good state of repair, they were incorporated into the scheme by converting them into a boathouse and restaurant. These buildings command an excellent view over the marina. A cafe and snackbar were built alongside. An access road was laid from the boathouse to the edge of the lake, and a launching area and jetty were provided at the lakeside.

A large area on the northern bank of the lake was paved and grassed to provide a caravan park. All the usual amenities, such as a site shop, washing and toilet facilities, a launderette and refuse disposal areas were provided. Services, such as water supply, electricity and sewage disposal were installed. In the south-west corner of the site an outdoor swimming pool was constructed, and alongside it there is a paddling pool. Children are catered for by the provision of an adventure playground.

References

Anon. 1983. *Fill. Part 1: Classification and Load Carrying Characteristics*. Digest 274, Building Research Establishment, Watford, Hertfordshire.

Anon. 1987. *Hydrogeology and Stability of Excavated Slopes in Quarries*. Department of the Environment, Her Majesty's Stationery Office, London.

Anon. 1992. *Specification for Aggregates from Natural Sources for Concrete*. BS 882, British Standards Institution, London.

Apte, S.C. 2001. *Tracing Mine Derived Sediments and Assessing Their Impact Downstream of the Porgera Gold Mine*. Report No. ET/IR385, Centre for Advanced Analytical Chemistry and Energy Technology, Commonwealth Scientific and Industrial Research Organization (CSIRO), Adelaide.

Atkinson, T. 1971. Selection of open pit excavation and loading equipment. *Transactions of the Institution of Mining Metallurgy*, **80**, Section A, Mining Industry, A101–A129.

Bailey, D. and Gunn, J. 1991. Landform replication as an approach to the reclamation of limestone quarries. In: *Land Reclamation, an End to Dereliction*, Davies, M.C.R. (ed.), Elsevier Applied Science, London, 96–105.

Barton, N. 1974. Estimating the shear strength of rock joints. *Proceedings of the Third International Congress Rock Mechanics* (ISRM), Denver, **2**, 219–225.

Bell, F.G. 1992. An investigation of a site in Coal Measures for brickmaking materials: an illustration of procedures. *Engineering Geology*, **32**, 39–52.

Bell, F.G. and Genske, D.D. 2000. Restoration of derelict mining sites and mineral workings. *Bulletin of Engineering Geology and the Environment*, **59**, 173–185.

Blair, D.P. 1999. Statistical models for ground vibration and airblast. *Fragblast*, **3**, 335–364.

Blair, D.P. and Armstrong, I.W. 1999. Spectral control of ground vibration using electronic delay detonators. *Fragblast*, **3**, 303–334.

Blanchfield, R. and Anderson, W.F. 2000. Wetting collapse in opencast coal mine backfill. *Proceedings of the Institution of Civil Engineers, Geotechnical Engineering*, **143**, 139–149.

Bruce, D.A., Taylor, R.P. and Lolcama, J. 2001. The sealing of a massive water flow through karstic limestone. *Proceedings of the Speciality Conference, Foundations and Ground Improvement*, Blacksburg, Virginia, Geotechnical Special

Publication No. 113, American Society of Civil Engineers, Reston, Virginia, 160–174.

Bye, A.R. and Bell, F.G. 2001. Stability assessment and slope design at Sandsloot open pit, South Africa. *International Journal of Rock Mechanics and Mining Science*, **38**, 449–466.

Charles, J.A. 1993. *Building on Fill: Geotechnical Aspects*. Building Research Establishment Report, Watford, Hertfordshire.

Charles, J.A., Burford, D. and Hughes, D.B. 1993. Settlement of opencast backfill at Horsley 1973–1992. *Proceedings of the Conference on Engineered Fills*, Newcastle upon Tyne, Clarke, B.G., Jones, C.J.F.P. and Moffat, A.I.B. (eds), Thomas Telford Press, London, 429–440.

Crandell, F.J. 1949. Ground vibrations due to blasting and its effect on stress meters. *Journal of Boston Society Civil Engineers*, **36**, 222–225.

Davies, P.A. 1995. Risk-based approach to setting of flyrock 'danger zones' for blast sites. *Transactions of the Institution of Mining and Metallurgy*, Section A, Mining Industry, **79**, A96–A100.

Duvall, W.I. 1964. *Design Requirements for Instrumentation to Record Vibrations Produced by Blasting*. Report Investigation No. 6487, United States Bureau of Mines, Washington, DC.

Edwards, A.T. and Northfield, R.D. 1960. Experimental studies of the effects of blasting on structures. *The Engineer*, **210**, 539–546.

Evans, A.M. 1997. *An Introduction to Economic Geology and Its Environmental Impact*. Blackwell Science Ltd., Oxford.

Fourney, W.L. 1993. Mechanisms of rock fragmentation by blasting. In: *Comprehensive Rock Engineering*, Vol. 4, Brown, E.T., Fairhurst, C. and Hoek, E. (eds), Pergamon Press, Oxford, 39–69.

Hagan, T.N. 1986. The influence of some controllable blast parameters upon muck-pile characteristics and open pit mining costs. *Proceedings of the Conference on Large Open Pit Mining*, Australian Institute Mining Metallurgy/Institute Engineers, 123–132.

Hoek, E. and Bray, J.W. 1981. *Rock Slope Engineering*. Institution of Mining and Metallurgy, London.

Ikingura, J.R., Mutakyahwa, M.K.D. and Kahatano, J.M.J. 1997. Mercury and mining in Africa with special reference to Tanzania. *Water, Air and Soil Pollution*, **97**, 223–232.

Kilkenny, W.M. 1968. *A Study of the Settlement of Restored Opencast Coal Sites and their Suitability for Development*. Bulletin No. 38, Department of Civil Engineering, University of Newcastle upon Tyne, Newcastle upon Tyne.

Knill, J.L. 1994. Quarry slope stability and landscape preservation in the Malvern Hills, UK. In: *Geological and Landscape Preservation*, O'Halloran, D., Green, C., Harley, M., Stanley, M. and Knill, J.L. (eds), Geological Society, London, 287–290.

Langefors, U. and Kihlstrom, B. 1962. *The Modern Technique of Rock Blasting*. Wiley, New York.

Latham, J.-P., Munjiza, A. and Lu, P. 1999. Rock fragmentation by blasting – a literature study of research in the 1980's and 1990's. *Fragblast*, **3**, 193–212.

Lisle, R.J. 2004. Calculation of the daylight envelop for plane failures of rock slopes. *Geotechnique*, **54**, 279–280.

Moore, J.F.A. and Longworth, I.T. 1979. Hydraulic uplift at the base of a deep excavation in Oxford Clay. *Geotechnique*, **29**, 35–46.
Muller, B. 1997. Adapting blasting technologies to the characteristics of rock masses in order to improve blasting results and reduce blasting vibrations. *Fragblast*, **1**, 361–378.
Nicholls, H.R., Johnson, C.F. and Duvall, W.I. 1971. *Blasting Vibrations and Their Effects on Structures*. United States Bureau Mines, Bulletin 656, Washington, DC.
Oriard, L.L. 1972. Blasting operations in the urban environment. *Bulletin of Association Engineering Geologists*, **9**, 27–46.
Penman, A.D.M. and Godwin, E.W. 1975. Settlement of experimental houses on land left by opencast mining at Corby. In: *Settlement of Structures*, British Geotechnical Society, Pentech Press, London, 53–61.
Richards, R.L., Leg, O.M.M. and Whittle, R.A. 1978. Appraisal of stability in rock slopes. In: *Foundation Engineering in Difficult Ground*, Bell, F.G. (ed.), Butterworths, London, 449–512.
Roberts, A. 1971. Ground vibrations due to quarry blasting and other sources – an environmental factor in rock mechanics. *Proceedings of the Twelfth Symposium on Rock Mechanics*, Rolla, Missouri, American Institute of Mining Engineers, New York, 427–456.
Roth, J.A. 1979. A model for the determination of flyrock range as a function of shot conditions. *United States Bureau of Mines*, Open File Report, 77–81.
Scoble, M.J. and Muftuoglu, Y.V. 1984. Derivation of a diggability index for surface mine equipment selection. *Mining Science and Technology*, **1**, 305–322.
Sinclair, J. 1969. *Quarrying, Opencast and Alluvial Mining*. Elsevier, London.
Smith, M.R. and Collis, L. (eds). 2001. *Aggregates: Sand, Gravel and Crushed Rock Aggregates for Construction Purposes*, Third Edition, Engineering Geology Special Publication No. 17, the Geological Society, London.
Sowers, G.F., Williams, R.C. and Wallace, T.S. 1965. Compressibility of broken rock and settlement of rock fills. *Proceedings of the Sixth International Conference on Soil Mechanics and Foundation Engineering*, Montreal, **2**, 561–565.
Wheater, C.P. and Cullen, W.R. 1997. The flora and invertebrate fauna of abandoned limestone quarries in Derbyshire, United Kingdom. *Restoration Ecology*, **5**, 77–84.
Williams, G.M. and Aitkenhead, N. 1991. Lessons from Loscoe: the uncontrolled migration of landfill gas. *Quarterly Journal of Engineering Geology*, **24**, 191–208.

Chapter 7

Waste materials from mining and their disposal

7.1. Introduction

Mine wastes result from the extraction of metalliferous and non-metalliferous deposits. In the case of metalliferous mining, high volumes of waste are produced because of the low or very low concentrations of metal in the ore. In fact, mine wastes represent the highest proportion of waste produced by industrial activity, billions of tonnes being produced annually. Waste from mines has been deposited on the surface in spoil heaps or tailings impoundments, which disfigure the landscape. Such waste can be inert or contain hazardous constituents but generally is of low toxicity. The chemical characteristics of mine waste and waters arising therefrom depend upon the type of mineral being mined, as well as the chemicals that are used in the extraction or beneficiation processes. Because of its high volume, mine waste historically has been disposed of at the lowest cost, often without regard for safety and often with considerable environmental impacts. Catastrophic failures of spoil heaps and tailings dams, although uncommon, have led to the loss of lives.

Mining waste at times is reworked. This is the result of improved extraction techniques, particularly in the case of metals, and more favourable economic conditions. For instance, waste tips from the old gold mines around Johannesburg, South Africa, have been reworked for their gold content. The mining of galena in the Pennines of Britain produced notable volumes of waste containing calcite, barite and fluorspar. Some of the spoil heaps subsequently were reworked to extract the barite, which has been used as an additive for drilling fluids in the offshore oil industry; and to extract fluorspar for use in the steel industry. Similarly, many colliery spoil heaps have been reprocessed to extract the coal content that was not removed by earlier, less advanced mineral processing techniques.

As mentioned, mining waste can be inert or contain hazardous constituents but generally is of low toxicity. Nonetheless, in some areas where metals were mined in the past, because little regard was given to the disposal of waste, relatively high concentrations of heavy metals can represent an environmental problem. For example, Smith and Williams (1996) referred to the United States Superfund Site near Kellogg, Idaho, where over twenty lead, zinc and silver mines had been operational in the district since the late 1880s, and mine and mill waste were discarded primarily into the South Fork of the Coeur d'Alene River. One area within this Superfund Site, namely, Smelterville Flats required urgent attention since the wastes, which contained various heavy metals, were a source of airborne particulates. Risk assessments undertaken by the United States Environmental Protection Agency (USEPA) had suggested that a major threat to human health was posed by the ingestion of airborne tailings and smelter wastes, and elevated levels of lead had been found in the blood of children who lived in the neighbourhood. Furthermore, the wastes also had degraded the quality of the groundwater. According to Smith and Williams, reclamation of Smelterville Flats was imperative if the objectives of the CERCLA were to be met. They suggested that typical reclamation procedures for such wastes included removing them to an engineered disposal facility; stabilizing them *in situ* either chemically or biochemically; or selective *in situ* leaching or remining and reprocessing them. After sampling the wastes and carrying out a statistical analysis to determine metal content, Smith and Williams suggested that selective treatment was a viable option. Selective treatment of wastes can reduce clean-up costs by avoiding handling or processing wastes that possess lower levels of contamination.

The character of waste rock from metalliferous mines reflects that of the rock hosting the metal, as well as the rock surrounding the orebody. The type of waste rock disposal facility depends on the topography and drainage of the site, and the volume of waste. Van Zyl (1993) referred to the disposal of coarse mine waste in valley fills, side-hill dumps and open piles. Valley fills normally commence at the upstream end of a valley and progress downstream, increasing in thickness. Side-hill dumps are constructed by the placement of waste along hillsides or valley slopes, avoiding natural drainage courses. Open piles tend to be constructed in relatively flat-lying areas. Obviously, an important factor in the construction of a spoil heap is its slope stability, which includes its long-term stability (Hughes and Clarke, 2003). Acid mine drainage from spoil heaps is another environmental concern.

7.2. Basic properties of coarse discard associated with colliery spoil heaps

Spoil heaps associated with coal mines represent ugly blemishes in the landscape and have a blighting effect on the environment (Fig. 7.1). They consist of coarse discard, that is run-of-mine material that reflects the various rock types that are extracted during mining operations. As such, coarse discard contains varying amounts of coal that has not been separated

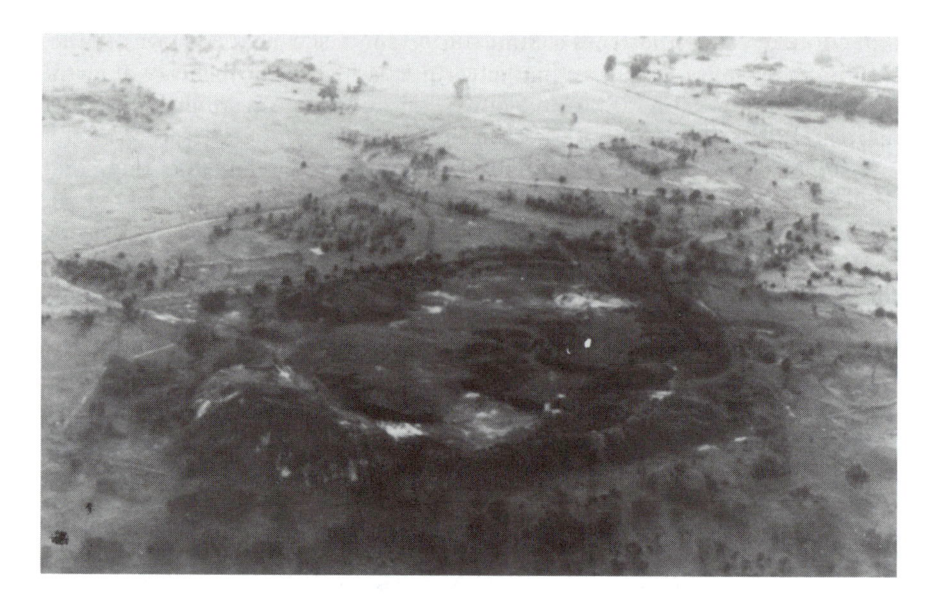

Figure 7.1 A colliery spoil heap in the Witbank Coalfield, South Africa, from which coal has been partially recovered and which shows signs of having been burnt.

by the preparation process. Obviously, the characteristics of coarse colliery discard differ according to the nature of the spoil. The method of tipping also appears to influence the character of coarse discard. In addition, some spoil heaps, particularly those with relatively high coal contents, may be burnt, or still be burning, and this affects their mineralogical composition (Taylor, 1973). However, investigations by Spears *et al.* (1970) in the unburnt section of the 50-year-old Yorkshire Main tip, England, showed that below the intensely weathered skin, which was around 1 m in thickness, that chemical activity did not extend beyond a depth of three metres.

The chemical composition of spoil material reflects that of the mineralogical composition. Free silica may be present in concentrations up to 80% and above, and combined silica in the form of clay minerals may range up to 60%. Concentrations of aluminium oxide may be between a few per cent and 40% or so. Calcium, magnesium, iron, sodium, potassium and titanium oxides may be present in concentrations of a few per cent. Lower amounts of manganese and phosphorus also may be present, with copper, nickel, lead and zinc in trace amounts. The sulphur content of fresh spoils is often less than 1% and occurs as organic sulphur in coal, and in pyrite.

In Britain illites and mixed-layer clays are the principal components of unburnt spoil in English and Welsh tips (Taylor, 1975). Although kaolinite is a common constituent in Northumberland and Durham, it averages only

10.5% in the discard of other areas. Quartz exceeds the organic carbon or coal content, but the latter is significant in that it acts as the major diluent, that is, it behaves in an antipathetic manner towards the clay mineral content. Sulphates, feldspars, calcite, siderite, ankerite, chlorite, pyrite, rutile and phosphates average less than 2%.

Pyrite is a relatively common iron sulphide in some of the coals and argillaceous rocks of the Coal Measures. It also is an unstable mineral, breaking down quickly under the influence of weathering. The primary oxidation products of pyrite are ferrous and ferric sulphates, and sulphuric acid (see Chapter 8). Oxidation of pyrite within spoil heap waste is governed by the access of air that, in turn, depends upon the particle size distribution, amount of water saturation and the degree of compaction. However, any highly acidic oxidation products, which may form may be neutralized by alkaline materials in the waste material.

The moisture content of spoil increases with increasing content of fines. It also is influenced by the permeability of the material, the topography and climatic conditions. Generally, it falls within the range 5–15% (Table 7.1).

The range of specific gravity depends on the relative proportions of coal, shale, mudstone and sandstone in the waste, and tends to vary between 1.7 and 2.7. The proportion of coal is of particular importance, the higher the coal content, the lower the specific gravity. The bulk density of material in spoil heaps shows a wide variation, most material falling within the range 1.5–2.5 Mg m^{-3}. Of course, the bulk density may vary within an individual deposit of mine waste. Low densities are mainly a function of low specific gravity. Bulk density tends to increase with increasing clay content.

The argillaceous content influences the grading of spoil, although most spoil material would appear to be essentially granular in the mechanical

Table 7.1 Examples of soil properties of coarse discard (After Bell, 1996)

Property	*Yorkshire Main*	*Brancepath*	*Wharncliffe*
Moisture content (%)	8.0–13.6	5.3–11.9	6–13 (7.14)[a]
Bulk density (Mg m^{-3})	1.67–2.19	1.27–1.88	1.58–2.21
Dry density (Mg m^{-3})	1.51–1.94	1.06–1.68	1.39–1.91
Specific gravity	2.04–2.63	1.81–2.54	2.16–2.61
Plastic limit (%)	16–25	Non-plastic –35	14–21
Liquid limit (%)	23–44	23–42	25–46
Permeability (m s^{-1})	$1.42–9.78 \times 10^{-6}$		
Size $<$ 0.002 mm (%)	0.0–17.0	} Most material of	2.0–20
Size $>$ 2.0 mm (%)	30.0–57.0	} sand size range	38–67
Shear strength (ϕ')	31.5–35.0°	27.5–39.5°	29–37°
Shear strength (c')	19.44–21.41 kPa	3.65–39.03 kPa	16–40 kPa

Note
a Average value.

sense. In fact, as far as the particle size distribution of coarse discard is concerned there is a wide variation, often most material may fall within the sand range but significant proportions of gravel and cobble range also may be present (Fig. 7.2(a)). Indeed, at placement coarse discard very often consists mainly of gravel-cobble size but subsequent breakdown on weathering reduces the particle size. Once buried within a spoil heap, coarse discard undergoes little further reduction in size. Hence, older and surface samples of spoil contain a higher proportion of fines than those obtained from depth.

The liquid and plastic limits can provide a rough guide to the engineering characteristics of a soil. In the case of coarse discard, however, they are only representative of that fraction passing the British Standard 425 μm BS sieve, which frequently is less than 40% of the sample concerned. Nevertheless, the results of these consistency tests suggest a low to medium plasticity, whilst in certain instances spoil has proved to be virtually non-plastic (Fig. 7.2(b)). Plasticity increases with increasing clay content.

The shear strength parameters of coarse discard do not exhibit any systematic variation with depth in a spoil heap so are not related to age, that is time dependent. As far as effective shear strength of coarse discard is concerned, the angle of shearing resistance usually varies from 25° to 45°. The angle of shearing resistance and therefore the strength increases in spoil that has been burnt. With increasing content of fine coal, the angle of shearing resistance is reduced. Also, as the clay mineral content in spoil increases, its shear strength decreases.

The shear strength of discard within a spoil heap, and therefore its stability, is dependent upon the pore water pressures developed within it. Pore water pressures in spoil heaps may be developed as a result of the increasing weight of material added during construction, by seepage though the heap of natural drainage or by mine water rebound. High pore water pressures usually are associated with fine grained materials that have a low permeability and high moisture content. Thus, the relationship between permeability and the build-up of pore water pressures is crucial. In fact, in soils with a coefficient of permeability of less than 5×10^{-9} m s^{-1} there is no dissipation of pore water pressures, whilst above 5×10^{-7} m s^{-1} they are completely dissipated. The permeability of colliery discard depends primarily upon its grading and its degree of compaction. It tends to vary between 1×10^{-4} and 5×10^{-8} m s^{-1}, depending upon the amount of degradation in size that has occurred.

The most significant change in the character of coarse colliery discard brought about by weathering is the reduction of particle size. The extent to which breakdown occurs depends upon the type of parent material involved and the effects of air, water and handling between mining and placing on the spoil heap. After a few months of weathering the debris resulting from sandstones and siltstones usually is greater than cobble size.

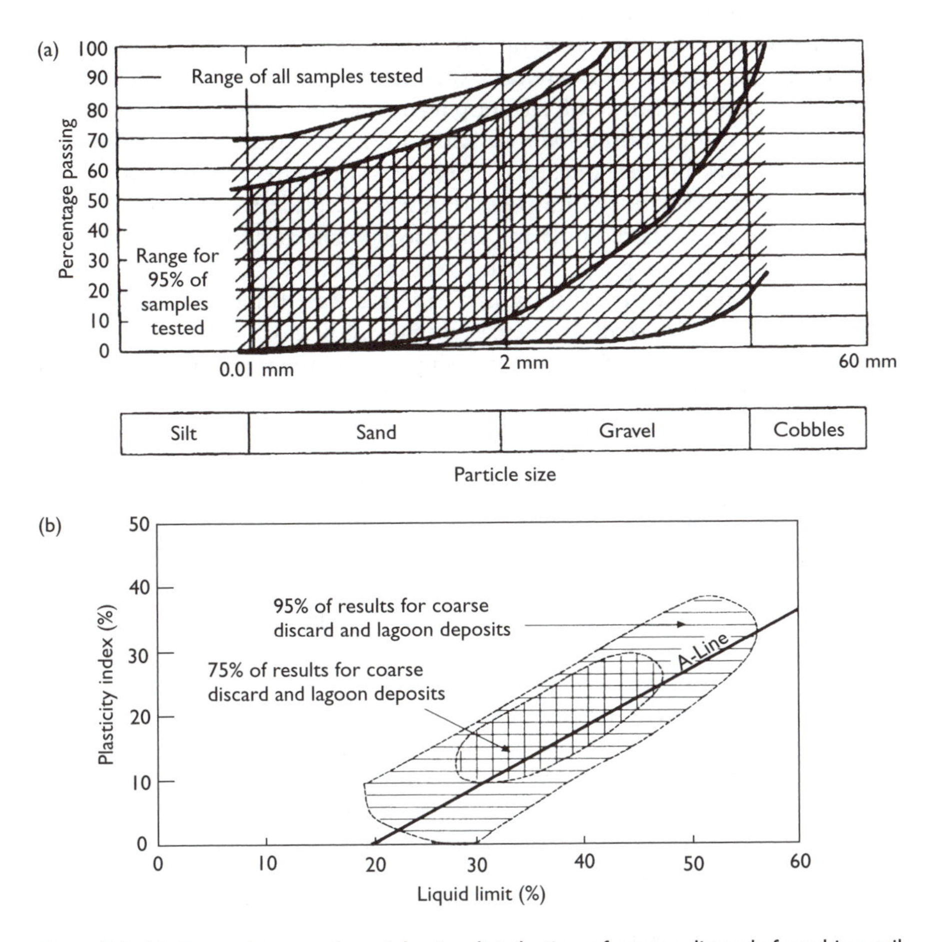

Figure 7.2 (a) General range of particle size distribution of coarse discards found in spoil heaps in Britain. (b) General range of plasticity characteristics found in spoil heaps in Britain. (After Anon., 1973; reproduced by kind permission of the Coal Authority.)

After that, the degradation to component grains takes place at a very slow rate. Mudstones, shales and seatearth exhibit rapid disintegration to gravel size. Although coarse discard may reach its level of degradation within a matter of months, with the degradation of many mudstones and shales taking place within days, once it is buried within a spoil heap it suffers little change. When spoil material is burnt, it becomes much more stable as far as weathering is concerned.

Taylor (1988) maintained that disintegration of British Coal Measures mudrocks takes place as a consequence of air breakage after a sufficient number of cycles of wetting and drying. In other words, if mudrocks undergo desiccation, then air is drawn into the outer pores and capillaries as high suction pressures develop. Then, on saturation, entrapped air is pressurized as water is drawn into the rock by capillarity. Such slaking therefore causes the fabric of the rock to be stressed. The size of the pores is more important than the volume of the pores, as far as the development of capillary pressure is concerned. Capillary pressure is proportional to surface tension. Previously, however, Badger *et al.* (1956), had contended that air breakage occurred only in Coal Measures shales that were mechanically weak, whilst the presence of dispersed colloidal material appeared to be a general cause of disintegration. They also found that the variation in disintegration of different shales in water usually was not connected with the total amount of clay colloid or the variation in the types of clay mineral present. It rather was controlled by the type of exchangeable cations attached to the clay particles and the accessibility of the latter to attack by water that, in turn, depended on the porosity of the shale. Air breakage could assist this process by presenting new surfaces of shale to water. Subsequently, Dick and Shakoor (1992) maintained that the influence of clay minerals on slaking diminishes in those mudrocks that contain less than 50% clay size particles, and that microfractures become the dominant lithological characteristic controlling durability. In such cases, slaking is initiated along microfractures.

Intraparticle swelling was considered by Taylor and Spears (1970) to exert a major control on the durability of mudrocks from the Coal Measures in Britain that contained significant amounts of expandable mixed-layer clay minerals. What is more, they suggested that those mudrocks that exhibited the greater breakdown were found to contain a higher exchangeable Na ion content in the mixed-layer clay. In fact, the presence of expandable mixed-layer clay is thought to be one of the factors promoting the breakdown of argillaceous rocks, particularly if Na^+ is a prominent interlayer cation. Present evidence nevertheless suggests that in England, colliery spoils with high exchangeable Na^+ levels are restricted. Sodium may be replaced by Ca^{++} and Mg^{++} cations originating from sulphates and carbonates in the waste, but dilution of spoil by more inert rock types is possibly the most usual reason for the generally low levels of sodium. Moreover, after undergoing the preparation process and exposure in the new surface layer of a spoil heap prior to burial, disintegration will have been largely achieved. However, it appears that the low shear strength of certain fine grained English spoils is associated with comparatively high exchangeable Na^+ levels.

7.3. Colliery spoil heap material and combustion

Spontaneous combustion of carbonaceous material, frequently aggravated by the oxidation of pyrite, is the most common cause of burning spoil.

It can be regarded as an atmospheric oxidation (exothermic) process in which self-heating occurs. Coal and carbonaceous materials may be oxidized in the presence of air at ordinary temperatures, below their ignition point. Generally, the lower rank coals are more reactive and accordingly more susceptible to self-heating than coals of higher rank.

Oxidation of pyrite at ambient temperature in moist air leads, as mentioned earlier, to the formation of ferric and ferrous sulphate, and sulphuric acid. This reaction also is exothermic. When present in sufficient amounts, and especially when in finely divided form, pyrite associated with coaly material, increases the likelihood of spontaneous combustion. When heated, the oxidation of pyrite and organic sulphur in coal gives rise to the generation of sulphur dioxide. If there is not enough air for complete oxidation, then hydrogen sulphide is formed.

The moisture content and grading of spoil are also important factors in spontaneous combustion. At relatively low temperatures an increase in free moisture increases the rate of spontaneous heating. Oxidation generally takes place very slowly at ambient temperatures but as the temperature rises, oxidation increases rapidly. In material of large size the movement of air can cause heat to be dissipated whilst in fine material the air remains stagnant, but this means that burning ceases when the supply of oxygen is consumed. Consequently, the best conditions for spontaneous combustion occur when the grading is intermediate between fine and coarse material, and hot spots may develop under such conditions. These hot spots may have temperatures around 600°C or occasionally up to 900°C (Bell, 1996). Furthermore, the rate of oxidation generally increases as the specific surface of particles increases.

Spontaneous combustion of colliery spoil may give rise to long-term smouldering, as well as to subsurface cavities in spoil heaps, the roofs of which may be incapable of supporting a person. Burnt ashes also may cover zones that are red-hot to appreciable depths. Anon. (1973) recommended that during restoration of a spoil heap that a probe or crane with a drop-weight could be used to prove areas of doubtful safety that are burning. Badly fissured areas should be avoided and workmen should wear lifelines if they walk over areas not proved safe. Any area that is suspected of having cavities should be excavated by dragline or drag scraper rather than allow plant to move over suspect ground.

When steam comes in contact with red-hot carbonaceous material, water-gas is formed, and when the latter is mixed with air, over a wide range of concentrations, it becomes potentially explosive. If a cloud of coal dust is formed near burning spoil when reworking a heap, then this also can ignite and explode. Damping with a spray may prove useful in the latter case.

Noxious gases are emitted from burning spoil. These include carbon monoxide, carbon dioxide, sulphur dioxide and, less frequently, hydrogen sulphide. Each may be dangerous if breathed in certain concentrations that may be present at fires on spoil heaps (Table 10.2). The rate of evolution of these gases may be accelerated by disturbing burning spoil by excavating into or reshaping it. Carbon monoxide is the most dangerous since it cannot be detected by taste, smell or irritation and may be present in potentially lethal concentrations. By contrast, sulphur gases are readily detectable in the aforementioned ways and usually are not present in high concentrations. Even so, when diluted they still may cause distress to persons with respiratory ailments. Nonetheless, the sulphur gases are mainly a nuisance rather than a threat to life. In certain situations a gas monitoring programme may have to be carried out. Where danger areas are identified, personnel should wear breathing apparatus.

The problem of combustion in spoil material has sometimes to be faced when reclaiming old tips (Fig. 7.3). Spontaneous combustion of coal in colliery spoil can be averted if the coal occurs in an oxygen deficient atmosphere that is humid enough with excess moisture to dissipate any heating that develops. Cook (1990), for example, described shrouding a burning spoil heap with a cover of compacted discard to smother the existing burning and prevent further spontaneous combustion. Anon. (1973) also recommended blanketing and compaction, as well as digging out, trenching, injection with non-combustible material and water, and water spraying as methods by which spontaneous combustion in spoil material may be controlled. Bell (1996) described the injection of a curtain wall of pulverized

Figure 7.3 Levelling spoil over a hot spot caused by spontaneous combustion, Barnsley, England.

fuel ash, extending to original ground level, about hot spots, as a means of dealing with hot spots.

7.4. Restoration of spoil heaps

The configuration of a spoil heap depends upon the type of equipment used in its construction and the sequence of tipping the waste. The shape, aspect and height of a spoil heap affects the intensity of exposure, the amount of surface erosion that occurs, the moisture content in its surface layers and its stability. The mineralogical composition of coarse discard from different coal mines obviously varies but pyrite frequently occurs in the shales and coaly material present in spoil heaps. When pyrite weathers it gives rise to the formation of sulphuric acid, along with ferrous and ferric sulphates and ferric hydroxide that lead to acidic conditions in the weathered material. Such conditions do not promote the growth of vegetation. Indeed, some spoils may contain elements that are toxic to plant life. To support vegetation a spoil heap should have a stable surface in which roots can become established, must be non-toxic and contain an adequate and available supply of nutrients. The uppermost slopes of a spoil heap frequently are devoid of near-surface moisture. Hence, spoil heaps are often barren of vegetation.

Restoration of a spoil heap represents an exercise in large scale earth-moving. Since it invariably will involve spreading the waste over a larger area, this may mean that additional land beyond the site boundaries has to be purchased. Where a spoil heap is very close to the disused colliery, spoil may be spread over the latter area. This will involve the burial or removal of derelict colliery buildings and may be the treatment of old mine shafts or shallow old workings (Johnson and James, 1990). Water courses may have to be diverted, as may services, notably roads.

Landscaping of spoil heaps frequently is to allow them to be used for agriculture or forestry. This type of restoration is generally less critical than when structures are to be erected on the site since bearing capacities in the former case are not so important and steeper surface gradients are acceptable. Most spoil heaps offer no special handling problems other than the cost of regrading and possibly the provision of adjacent land, so that the gradients on the existing site can be reduced by the transfer of spoil to the adjacent land. Some spoil heaps, however, present problems due to spontaneous combustion of coaly material that they contain. A spoil heap that is burning or that is so acidic that polluted waters are being discharged requires special treatment. Sealing layers have been used to control spontaneous combustion and polluted drainage, the sealing layers being well compacted.

Surface treatments of spoil heaps vary according to the chemical and physical nature of the spoil and the climatic conditions. Preparation of the surface of a spoil heap also depends upon the use to which it is to be put subsequently. Where it is intended to sow grass or plant trees, the surface layer should not be compacted. Drainage plays an important part in the restoration of a spoil heap and should take account of erosion control during landscaping. If the spoil is acid, then this can be neutralized by liming. The chemical composition will influence the choice of fertilizer used and in some instances spoil can be seeded without the addition of top soil, if suitably fertilized. More commonly topsoil is added prior to seeding or planting.

7.5. Spoil heaps and stability

The stability of a spoil heap is influenced by the material of which it is composed, its height, the gradients and lengths of its slopes, and the nature and rate of erosion to which it is exposed that, in turn, is influenced by the climatic regime. Occasionally landslips may be associated with spoil heaps. In particular, old colliery spoil heaps were formed by tipping the waste without any compaction, which means that their stability is adversely affected if the material becomes saturated. Flow slides can develop under such conditions and these are more likely to claim lives than rotational slides, although both can damage property. According to Garrard and Walton (1990) poor drainage and foundation preparation account for many dangerous incidents related to spoil heaps. Failure to remove weak weathered foundation material or to place a drainage layer prior to tipping, along with too rapid a rate of construction or removal of the toe have accounted for many failures. Garrard and Walton noted that most incidents have been rotational or curvilinear failures whereas flow slides involving liquefaction have not been so common but where they occurred the failures were large and material travelled appreciable distances. Failures are more likely to occur after periods of prolonged heavy rainfall (i.e. nearly 1.75 times as likely after over 135 mm of rain has fallen the previous month). Garrard and Walton provided an outline of the data that should be collected by a site investigation prior to the construction of a spoil heap and made recommendations relating to its design to help ensure its stability.

Siddle *et al.* (1996) reviewed 23 rapid failures of spoil heaps in the South Wales Coalfield that occurred during the early and mid-twentieth century, and caused damage to both property and infrastructure, as well as flooding due to blocking of streams. These spoil heaps were constructed with no control over waste disposal and were commonly placed on valley sides. The failures referred to primarily took the form of flow slides but debris slides and two failures brought about by outbursts of groundwater also have taken place. The flow slides would appear to have been triggered by rotational or non-circular sliding of the spoil and, in some instances, of the underlying soils and rock. Debris slides probably resulted from rapid disposal of wet spoil over the face of a heap, which caused undrained loading of the spoil at the toe of the heap. In addition, debris flow was a secondary mechanism of failure at five sites. Siddle *et al.* noted that these failures tended to happen in those areas of the Coalfield with the highest relief and the highest rainfall. They also noted, however, that antecedent rainfall conditions had varied as far as the failures were concerned. Most failures were associated with active tipping faces but one flow slide occurred at a spoil heap some four years after it had been abandoned. Siddle *et al.* suggested that self-compaction in older spoil heaps gave rise to an average increase in dry density that was sufficient to reduce the potential for flow sliding to occur. They went on to add, however, that one of the outbursts occurred about fifteen years after the spoil heap concerned was no longer operative and that this indicates the potential for such hazards to take place on older heaps, particularly where subsurface drainage is impeded.

The residents of the South Wales mining village of Cilfynydd, located downslope of a colliery spoil heap, known as the Albion Lower Tip, were put under notice of evacuation in February 1990, following movements of the heap material (Maddison, 2000). The heap is located on the west-facing hillside of the Taff Valley, approximately 20 km north-west of Cardiff. It was constructed in two phases between 1886 and 1920 and between 1920 and the closure of the colliery in 1966. The heap was constructed on the site of a degraded landslide and had experienced past movements. It also had a history of combustion and burning. In 1975 the spoil heap was regraded as part of a reclamation scheme. Thick beds of sandstone crop out below the heap and form the steep hillside beyond it, and mudrocks interbedded with coal seams and sandstones crop out at the bottom of the heap. The strata are faulted and dip gently to the north-north-west. Overlying superficial deposits consist of about 15 m of colluvium and glacial till, which merge with the landslip material. Unconfined groundwater conditions and different temporary water tables contributed to the instability of the spoil. Fissures, the dilation of joints, heave and displacements were observed in the 1980s at the top of the regraded heap. High artesian pressures were found to exist within the heap and these were remediated by drilling 17 emergency wells and 150 m long sub-horizontal bored drains as temporary measures to relieve the groundwater pressures and instability of the spoil. In 1989 ground movements were identified as resulting from a non-circular deep-seated slide of the toe of the slope and heap, with secondary failures extending upslope. Ground movements increased in 1990 following a period of prolonged and intense winter rainfall and elevated groundwater levels. The heap was placed under 24 hours surveillance, however, due to the control exerted on the heap by the drainage measures no evacuation of nearby houses was necessary.

The worst disaster in Britain due to the failure of a spoil heap occurred at Aberfan, in South Wales (Fig. 7.4). The spoil heaps from the local colliery had been built on the steep valley slopes above the village. On 21 October 1966, about 107 000 m^3 of spoil, representing one-third of the heap, flowed into the village (Penman, 2000). The Pennant Sandstone in which the valley is carved is well jointed and the joints had been affected by mining subsidence. This allowed rain falling on the upper slopes to be discharged as springs on the lower slopes. Accordingly, water issued beneath the toe of the spoil heap eroding material. This had given rise to intermittent slips previously. Heavy rains two days prior to the fatal slide meant that spoil became saturated and, with its shear strength very much reduced, the spoil became liable to flow, which it subsequently did with disastrous consequences (Bishop, 1973). The resultant mudflow invaded a school and several houses, killing 116 children and 28 adults. Over two years later work began on removing the spoil to a new site. The spoil

Figure 7.4 Mudflow from a spoil heap at Aberfan, South Wales, October 1966.

was placed in a number of terraces and extensive drainage works were undertaken.

7.6. Waste disposal in tailings dams

Tailings are fine grained residues that result from crushing rock that contains ore or are produced by the washeries at collieries. As far as metalliferous deposits are concerned, mineral extraction usually takes place by wet processes in which the ore is separated from the parent rock by flotation, dissolution by cyanide, washing etc. Consequently, most tailings originate as slurries, in which form they are transported for hydraulic deposition. The water in tailings may contain certain chemicals associated with the metal recovery process such as cyanide in tailings from gold mines, and heavy metals in tailings from copper-lead-zinc mines. Tailings also may contain sulphide minerals like pyrite that can give rise to acid mine drainage (or acid rock drainage; AMD, ARD). The particle size distribution, permeability and resistance to weathering of tailings affect the process of acid generation (Bell and Bullock, 1996). Acid drainage also may contain elevated levels of dissolved heavy metals. Accordingly, contaminants carried in the tailings represent a source of pollution for both ground and surface water, as well as soil. Tailings storage facilities therefore represent one of the most notable environmental liabilities associated with mining operations. As such, care should be taken in all aspects of design, construction, operation, inspection, surveillance, review and management. Furthermore, Martin *et al.* (2002) emphasized that all facets of tailings facility stewardship should be considered from pre-feasibility studies onward to what happens when closure takes place.

7.6.1. *Tailings dam construction*

The first dams to retain tailings frequently were built across a stream channel with only limited provision for large floods and so few of these dams survived. Very little engineering input was involved in the construction of these early dams and a construction procedure was developed like the upstream method (see later). First, a low dyked impoundment was filled with hydraulically deposited tailings and then incrementally raised by constructing low berms above and behind the dyke of the previous level. The first departure from the traditional upstream method of construction, according to Martin *et al.* (2002), followed the failure of the Barahona tailings dam in Chile when, in 1928, a large earthquake caused the dam to fail, the resulting catastrophic flow slide killing more than fifty people. In fact, earthquake induced liquefaction associated with tailings dams remains a key consideration in their design (Anon., 1998). The Barahona Dam was replaced by a more stable downstream dam (see later), in which cyclones were used during its construction to obtain coarser sized material. With the availability of high capacity earthmoving equipment from the 1940s onwards, it became possible to construct tailings dams of compacted earthfill.

As can be inferred from the earlier discussion, tailings are deposited as slurry usually in specially constructed tailings dams, the construction of which generally takes place in stages. Tailings dams are of two main types, that is, embankment dams that are constructed across a valley impounding the tailings upstream of the dam and embankment dams that encircle the tailings (Fig. 7.5(a)). Blight (1997) referred to these two types of enclosures as valley dams and ring dykes. Ring dykes or dams may be several kilometres in length and are constructed on relatively flat ground (Fig. 7.5(b)). Both types of embankment dams may be constructed to their full height before the tailings are discharged. As such, they are constructed in a similar manner to earth fill dams used to impound reservoirs. They may be zoned with a clay core and have filter drains. Such dams are best suited to tailings impoundments with high water storage requirements. However, tailings dams usually consist of raised embankments, that is, the construction of the dam is staged over the life of the impoundment. Raised embankments consist initially of a starter dyke that normally is constructed of earth fill from a borrow pit. This dyke may be large enough to accommodate the first 2 or 3 years of tailings production. A variety of materials can be used subsequently to complete such embankments including earth fill, mine waste or tailings themselves. Tailings dams consisting of tailings can be constructed using the upstream, centreline or downstream method of construction (Vick, 1983; Fig. 7.6). In the upstream method of construction, tailings are discharged from spigots or small pipes to form the impoundment (Fig. 7.7). This allows separation of particles according to size, with the coarsest particles accumulating in the centre of the embankment beneath the spigots and the finer particles being transported down the beach. Alternatively, cycloning may be undertaken to remove coarser particles from tailings so that they can be used in embankment construction. Centreline and downstream construction of embankments use coarse particles separated by cyclones for the dam. For equivalent embankment heights, water retention embankment dams and downstream impoundments require approximately three times more fill than an upstream embankment. A centreline embankment would require about twice as much fill as an upstream embankment of similar height. The design of tailings dams must pay due attention to their stability both in terms of static and dynamic loading (Anon., 1998). Vick mentioned that the upstream construction method has some limitations that include control of the phreatic surface within the dam, limited storage water capacity and susceptibility of tailings to seismic liquefaction. In order to obtain a low phreatic surface, which is necessary to ensure stability, a low pond condition must be maintained, sedimentation

(a)

(b)

Figure 7.5 (a) Part of a large tailings facility at Athabasca Tar Sands, Fort McMurray, Alberta, showing dam, beach and lagoon. (b) A ring tailings dam at a tin mine in North West Province, South Africa.

upstream of the dam should be layered, that is, alternating sandy and silty beach deposits, and underdrainage should be provided beneath the dam.

Centreline and downstream constructed tailings dams usually are considered more robust than upstream tailings dams. Failures have been the result

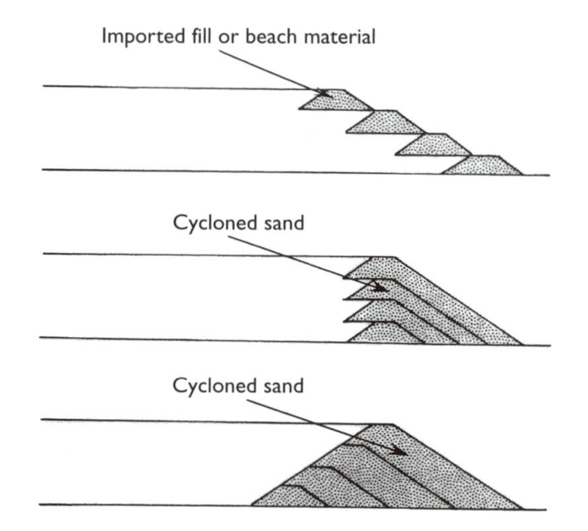

Figure 7.6 Upstream, centreline and downstream methods of embankment construction for tailings dams. (Modified after Vick, 1983.)

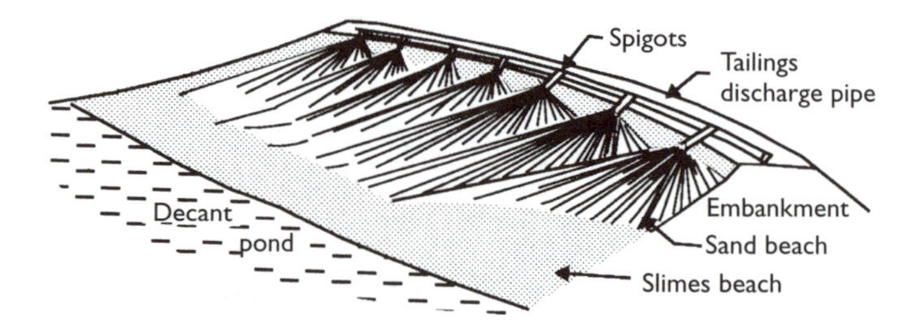

Figure 7.7 Tailings discharge by spigotting.

of earthquakes, high saturation levels, steep slopes, poor water control in the lagoon, poor construction techniques incorporating fines in the dam shell, static liquefaction and failures of embedded decant structures (Martin *et al.*, 2002). Considerable attention has been given to improving traditional upstream dam construction to improve stability under both static and dynamic conditions. Hence, a number of important design features may be incorporated into modern dams. These include underdrainage, either as finger drains or blanket drains, to lower the pheatic level in the dam shell; beaches compacted to some minimum width to provide a stable dam shell; and slopes designed at a lower angle than was used for many failed dams. Beaches are compacted by tracking with bulldozers, which also are used for pushing up berms for support of spigot lines. Slopes generally are set at 3 horizontal to 1 vertical or flatter, depending on other measures incorporated into the design. Steeper slopes without an adequate drained and/or compacted beach create the potential for spontaneous static liquefaction.

The slopes of hydraulically deposited material, according to Blight (1994a), assume slopes that are geometrically similar almost irrespective of the height and the length of the slopes. Hence, hydraulic fill beaches can be regarded as developing a master profile that can be useful in the design of tailings dams. In other words, Blight maintained that for a given concentration of solids, particle size distribution and rate of flow a beach slope will conform to a certain master profile that can be simulated in a flume in a laboratory. Once the master profile has been determined, then the shape of the top surface of the proposed dam can be modelled for hydrological design and calculation of the volume that can be stored. The beach profiles can be fitted between the predicted area of the lagoon and the areas where deposition occurs when the positions of the penstock(s) and the necessary freeboard have been decided upon. Blight further noted that if deposition is in the particle sorting regime, then this allows advantage to be taken of the gradient of particle size down the beaches, thereby providing for the lowering of the phreatic surface due to the increasing permeability between the edge of the lagoon and the outer slopes of the tailings dam.

Erosion of the outer slopes of tailings dams should be minimized by erosion control measures such as berms, surface drainage, riprap or geofabrics and where appropriate the rapid establishment of vegetation (Anon., 1989). Blight and Amponsah Da Costa (1999) pointed out that climate has an influence on erosion control measures in that water and wind erosion differ in effectiveness between humid, semi-arid and arid regions. Furthermore, vegetation is less easy to establish on slopes in arid regions so that a cover of riprap over a slope may be more appropriate. They undertook an evaluation of the cost effectiveness of various slope protection measures and found that a cover of 300 mm of fine rock was most effective.

7.6.2. Deposition of tailings

The tailings slurry can be discharged under water in the impoundment or it may be disposed of by subaerial deposition. In subaerial disposal the slurry is discharged from one or more points around the perimeter of the impoundment with the slurry spreading over the floor to form beaches or deltas. The free water drains from the slurry to form a pond over the lowest area of the impoundment. The discharge points can be relocated around the dam to allow the exposed solids to dry and thereby increase in density. Alternatively, discharge can take place from one fixed point (Fig. 7.8).

The deposition of tailings in a dam may lead to the formation of a beach or mudflat above the water level. When discharged, the coarser particles in

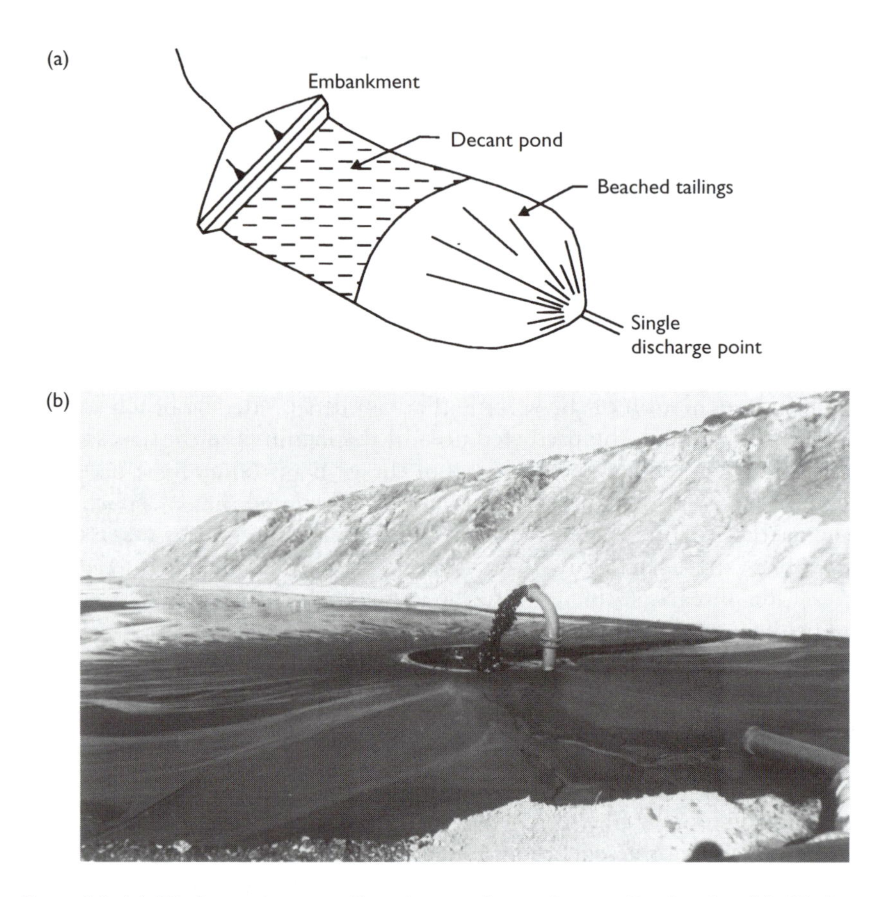

Figure 7.8 (a) Discharge into a tailings impoundment from a fixed point. (b) Discharge of slurry from a colliery washery at a coal mine in the Witbank Coalfield, South Africa.

tailings settle closer to the discharge point(s) with the finer particles being deposited further away (Blight, 1994b). The amount of sorting that takes place is influenced by the way in which the tailings are discharged, for instance, high volume discharge from one point produces little sorting of tailings. On the other hand, discharge from multiple points at moderate rates gives rise to good sorting. After being deposited, the geotechnical properties of tailings such as moisture content, density, strength and permeability are governed initially by the amount of sorting that occurs, and subsequently by the amount and rate of consolidation that takes place. For example, if sorting has led to material becoming progressively finer as the pool is approached, then the permeability of the tailings is likely to decrease in the same direction. The moisture content of deposited tailings can vary from 20% to over 60% and dry densities from 1 to 1.3 Mg m^{-3}. At the end of deposition the density of the tailings generally increases with depth due to the increasing self-weight on the lower material. The coarser particles that settle out first also drain more quickly than the finer material, which accumulates further down the beach, and so develop shear strength more quickly than the latter. These variations in sorting also affect the permeability of the materials deposited.

The quantity of tailings that can be stored by a dam of a given volume is dependent upon the density that can be achieved. The latter is influenced by the type of tailings, the method by which they are deposited, whether they are deposited in water or subaerially, the drainage conditions within the dam and whether or not they are subjected to desiccation. Ideally, the pool should be kept as small as possible and the penstock inlet should be located in the centre of the pool. This depresses the phreatic surface, thus helping to consolidate the tailings above the latter. Blight and Steffen (1979) referred to the semi-dry or subaerial method of tailings deposition whereby in semi-arid or arid climates a layer of tailings is deposited in a dam and allowed to dry before the next layer is placed. Since this action reduces the volume of the tailings, it allows more storage to take place within the dam. However, drying out can give rise to the formation of desiccation cracks in the fine discard that, in turn, can represent locations where piping can be initiated. In fact, Blight (1988) pointed out that tailings dams have failed as a result of desiccation cracks and horizontal layering of fine grained particles leading to piping failure. The most dangerous situation occurs when the ponded water on the discard increases in size and thereby erodes the cracks to form pipes that may emerge on the outer slopes of the dam.

The permeability and rheological properties of tailings are, according to East and Morgan (1999), the principal factors influencing the design of effective drainage systems at a tailings impoundment. Tailings are commonly transported as slurries via a pipeline to the impoundment where the solid particles settle out of the liquid fraction, so that, as noted, a pond or lagoon often occurs within a tailings dam. For example, slurries produced by gold processing usually contain between 25% and 50% solid particles by weight. However, slurry tends to be added to the saturated tailings waste within the dam faster than the excess pore water pressures can be dissipated. This means that undrained tailings dams are difficult to reclaim once they become inactive because working surfaces for light earth moving equipment are costly to establish until appreciable consolidation has taken place, which may take several years. In addition, ponding and saturated tailings can adversely affect the stability of the embankments and therefore are important factors to be considered in the design of tailings dams. Consequently, removal of pond water and reduction of pore water pressures, together with internal drainage of the embankment should be incorporated into the design of a tailings dam. Removal of pond water can be achieved by decanting via towers, by pumping or by sloping filter drains. Underdrainage can help reduce pore water pressures in the waste and can consist of a simple system of finger drains on the one hand or a drainage blanket of sand or gravel with internal collection pipework on the other. Alternatively, geotextile and geonet sandwiches can be used for underdrainage. Underdrainage conveys water to collection centres. However, whether such measures are put in place depends upon the size of the impoundment, which influences cost, and the location.

7.6.3. Seepage from tailings dams

The rate at which seepage occurs from a tailings dam is governed by the permeability of the tailings and the ground beneath the impoundment. Climate and the way in which a tailings dam is managed also have some influence on seepage losses. In many instances, because of the relatively low permeability of the tailings compared with the ground beneath, a partially saturated flow condition will occur in the foundation. Nonetheless, the permeability of tailings can vary significantly within an impoundment depending on the nearness to discharge points, the degree of sorting, the amount of consolidation that has taken place, the stratification of coarser and finer layers that has developed, and the amount of desiccation that the discard has undergone.

Seepage losses from tailings dams that contain toxic materials can have an adverse effect on the environment. For example, with the advent of larger gold mining operations, and the almost universal use of sodium cyanide as an essential part of gold extraction, it became necessary to construct impervious impoundments to contain the cyanide solutions. Although cyanide is unstable in most forms, breaking down on exposure to air, it can be persistent and migrate over long distances in groundwater. Hence, in order to protect aquifers where tailings impoundments are not located on impermeable soils or bedrock, then it may be necesssary to design and construct a liner beneath tailings over the base of the impoundment. An illustration of pollution has been provided by Sharma and Al-Busaidi (2001) who described a serious pollution problem associated with a tailings dam in Oman. Between 1982 and 1994, 11 million tonnes of sulphide rich tailings from a copper mine and 5 million m^3 of sea water (used in the processing of the copper ore) were disposed of within an unlined tailings dam. The resultant pollution plume extended some 14 km downstream of the tailings dam and the total dissolved solids, TDS, of the groundwater in some monitoring wells increased from around 1000 mg l^{-1} in 1984 to some 55 000 mg l^{-1} in 1996. In fact, it is usually necessary to operate a dam as a closed or controlled system from which water can be released at specific times. It therefore is necessary to carry out a water balance determination for a dam, and any associated catchments such as a return water reservoir to assess whether and/or when it may be necessary to release water from or make up water within the system. Because of variations in precipitation and evaporation, water balance calculations should be undertaken frequently. If provision has been made for draining the pool of a dam, then after tailings deposition has ceased it should be possible to maintain a permanent water deficit that will reduce seepage of polluted water from the dam.

According to Fell *et al.* (1993), one of the most cost effective methods of controlling seepage loss from a tailings dam is to cover the whole floor of the impoundment with tailings from the start of the operation (Fig. 7.9). This cover of tailings, provided it is of low permeability, will form a liner. Tailings normally have permeabilities between 10^{-7} and 10^{-9} m s^{-1} or less. Nevertheless, Fell *et al.* mentioned that a problem could arise when using tailings to line an impoundment, that is, if a sandy zone develops near the point(s) of discharge, then localized higher seepage rates will occur if water covers this zone. This can be avoided by moving the points of discharge or by placing fines. Alternatively, a seepage collection system can be placed beneath the sandy zone prior to its development. Clay liners also represent an effective method of reducing seepage from a tailings facility. The permeability of properly compacted clay soils usually varies between 10^{-8} and 10^{-9} m s^{-1}. However, clay liners may have to be protected from drying out, with attendant development of cracks, by placement of a sand layer on top. The sand on slopes may need to be kept in place by using geotextiles. Geomembranes have tended not to be used for lining tailings facilities, primarily because of their cost. However, if used, then it is normal practice to provide a drainage layer above the geomembrane to reduce the head of

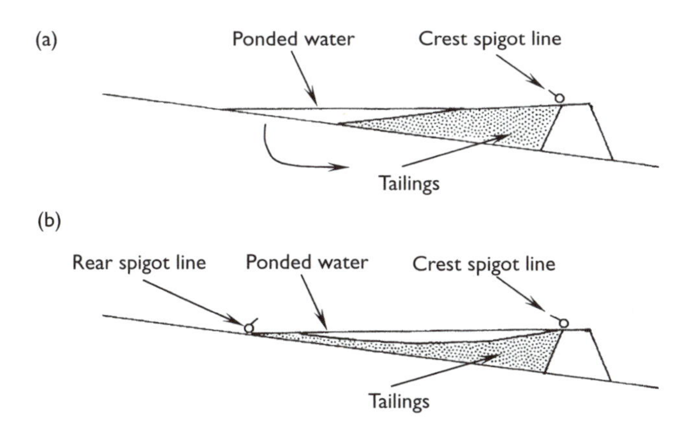

Figure 7.9 Control of seepage by spiggotting procedures. (a) Major seepage at water-foundation contact. (b) Foundation sealing by rear spiggotting of tailings. (Modified after Vick, 1983.)

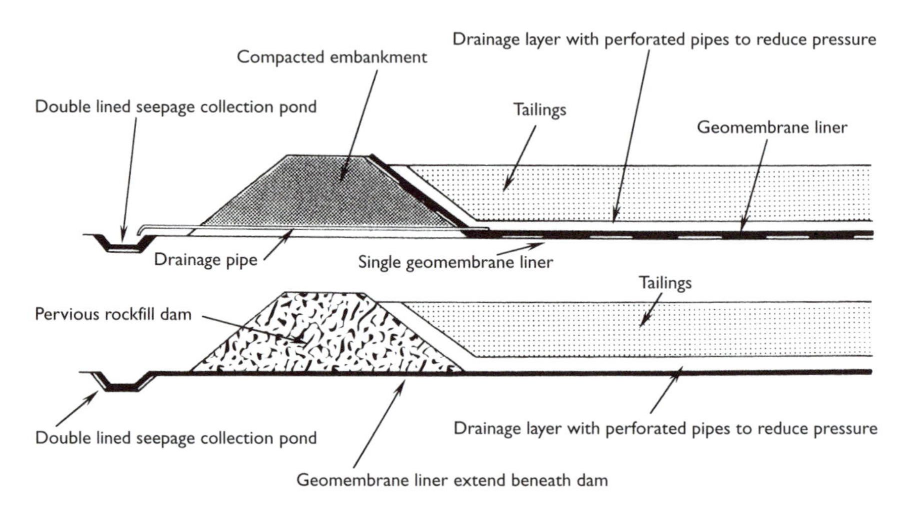

Figure 7.10 Sections of lined impoundments with underdrains. (After Martin *et al.*, 2002; reproduced by kind permission of the International Institution for Environment and Development.)

pressure on the liner and to minimize seepage through any imperfections in the liner. Another benefit of a drainage layer is that it lowers pore water pressures in the tailings, thereby affording them higher strength. The drainage layer typically consists of at least 300 mm of granular material, with perforated pipes set out at intervals within the layer (Martin *et al.*, 2002). The pipes drain water from the base of the tailings and discharge it to a seepage recovery pond (Fig. 7.10). Filter drains may be placed at the base of tailings, with or without a clay liner. They convey water to collection dams. A toe drain may be incorporated into the embankment. This will intercept seepage that emerges at that location. Where a tailings dam has to be constructed on sand or sand and gravel, a slurry trench may be used to intercept seepage water but slurry trenches are expensive. Bowell *et al.* (1999) reviewed various chemical methods of containing mine waste, especially in relation to water quality.

7.6.4. Acid mine drainage and the character of tailings impoundments

Where acid mine drainage (or acid rock drainage) is associated with tailings impoundments it may be necessary to operate AMD collection and treatment systems (see Chapter 8) both during the working life and after closure of the mine. It is, however, far easier to prevent AMD in the first place than to control it. Impervious covers can be placed over tailings to prevent ingress of air and water. Multiple-layer covers, incorporating impervious zones, pervious capillary barriers and topsoil to encourage the growth of vegetation have been developed. Pastore (2003) indicated that organic waste such as de-inking residues from paper recycling, where about 30% of waste is composed of secondary cellulose, calcite and metakaolinite and leaves, bark, sawdust and wood shavings from wood mills have the potential to reduce oxygen diffusion when used as a liner cover. Nevertheless, covers have been found to present a risk of long-term cracking or erosion, and to be ineffective in excluding air, so are often less favoured solutions than submergence from the geochemical standpoint at sites where it is feasible to maintain a submerged condition.

Some of the principal ways to reduce the potential for acid mine drainage from sulphide bearing tailings include submergence by flooding the tailings at closure, treatment of tailings to create non-acid generating covers, and evapotranspiration covers in arid regions. Although flooding is considered one of the best available technologies for long-term storage of acid generating tailings, flooded impoundments may create a risky legacy. There is, however, a concern that oxidation of sulphides may still occur (Peacey *et al.*, 2002). Lime can be added to neutralize acidity. The more traditional closure configuration for tailings impoundments has been to draw down water lagoons as completely as possible, in order to reduce the potential for dam failure by overtopping or erosion. To raise water levels in impoundments formed by high dams could present considerable long-term risk. One of the reasons that closed tailings impoundments generally have proved to be more stable than operating impoundments is the relatively more drained condition of the former that do not include a large water lagoon. The flooded closure scenario represents an undrained condition that does not allow this improvement in physical stability to develop, so the risk does not decrease with time. For instance, Vick (1999) described the implications for dam safety associated with the use of submergence in Ontario compared with Peru. In Ontario, the topography is relatively subdued, tailings dams are constructed to relatively low heights, the terrain is geologically stable not being subject to frequent earthquakes, and the hydrology is well understood. In other words, conditions are favourable towards maintaining flooded tailings impoundments in a safe condition. Conversely, in Peru the topography is frequently mountainous, tailings dams are constructed to great heights, the terrain is subject to frequent great earthquakes, and the hydrology is not well understood. As a result, the likelihood of failure of a flooded tailings impoundment in Peru is likely to be higher than in Ontario.

In order to avoid the necessity of flooding impoundments, non-reactive covers for tailings can be placed on the top of the impoundment during the last few years of operation. Pyrite can be removed to a level where the tailings can be made non-acid generating by the relatively inexpensive installation of some additional flotation capacity. The upper non-acid generating tailings placed on top can be left as a wide beach for dam safety, while the underlying mass of potentially acid generating tailings remains saturated below the long-term water table in the impoundment. Normally, the small amount of pyrite removed by flotation can be disposed of as a separate tailings stream, placed in the deepest part of the impoundment where it can be left flooded.

As noted, submergence cannot be considered as an option in arid regions due to the dry climate. In such regions a single cover constructed of naturally occurring materials is used increasingly. During precipitation events the cover stores moisture, subsequently releasing it as evaporation and transpiration. By so doing, the cover prevents infiltration from penetrating into the underlying tailings. Such covers are termed evapotranspiration or store and release covers. To be effective, an evapotranspiration cover must have sufficient fines content (especially silt) to have a relatively low hydraulic conductivity so as to provide a high moisture retention capacity and to maximize evaporation and transpiration. Second, they should not have an excessively high degree of plasticity such that they would be subject to desiccation cracking. Third, they should possess sufficient thickness to store infiltration from wet periods and prevent percolation into the underlying tailings. A problem that may arise when tailings are impounded in arid regions is the precipitation of salts on the surface of the tailings leading to the formation of thin salt crusts that reduce the rate of evaporation. Newman and Fahey (2003) referred to this in relation to gold mining in western Australia where ore processing frequently uses groundwater that can have salinities that approach solution saturation. In such cases, quantification of the rate of evaporation from the drying tailings surface is important when assessing disposal strategies.

Wilson (2001) referred to the use of co-disposal of tailings and waste rock for mines with acid mine drainage problems. This methodology is focused on disposing of waste rock and tailings as a single waste stream, instead of as two separate waste streams, one wet (tailings) and the other dry (waste rock). Sulphidic waste rock may have high shear strength but may be highly susceptible to AMD. Tailings are much less pervious to water and atmospheric oxygen, but when deposited in conventional slurry impoundments have a very low shear strength, and are susceptible to liquefaction and related catastrophic flow slide behaviour. The fine grained nature of tailings promotes water retention and saturation, which inhibits oxidation. By encapsulating waste rock in tailings, the beneficial qualities of each type of waste (high shear strength of waste rock and low permeability of tailings) can be realized. The resulting single tailings stream is more stable physically and is more chemically stable. In addition, Wilson maintained that co-disposal of tailings and waste rock may have particularly promising potential for the construction of a low permeability cover for tailings lagoons and waste rock dumps. Morris and Williams (1997) described the co-disposal of colliery waste at Jeebropilly Mine, south east Queensland. Significant segregation of the waste occurs with the formation of distinct upper and lower segments, which are dominated by coarse and fine material respectively. The coarse material drains rapidly so that its surface is able to support traffic more or less immediately after deposition.

7.6.5. Tailings dam failure

Hundreds of tailings dams have failed around the world, often resulting in damage to land and infrastructure, and to loss of life (Anon., 1994; Anon., 2001). In particular, the generation of a wave of slurry is caused by the catastrophic failure of a tailings dam. These waves typically travel at velocities of between 8 and 40 km hr^{-1}. Typical types of failure modes of tailings dams include, first, weak foundations and the development of a failure plane within the soil or rock beneath the foundation of a dam. Second, during earthquakes, induced cyclic loading of tailings slurries and the materials used for the construction of the dam may liquefy. As a consequence, large parts of the impounded tailings may be released as a series of waves causing catastrophic downstream destruction. In parts of the Chilean Andes, for example, a region subject to frequent earthquake activity, the location of tailings dams across steep upstream mountain slopes is no longer acceptable for the construction of tailings dams. Third, the excessive rise in the level of pond water may cause the failure of tailings dams even if overtopping does not occur. The rise in water level can be brought about by increased and prolonged precipitation, snow melt or inappropriate water management. If the phreatic surface in the embankment rises this may cause it to fail. If water level rise causes overtopping and breaching of the crest of the dam, then this may lead to its failure in a relatively short time. Fourth, piping is caused by seepage within or beneath a dam. This causes erosion along the flow path that may result in the general failure of the dam. Fifth, if a tailings dam is raised too quickly, then this may generate excessive pore pressures within the dam that can result in its failure.

As noted, failure of a dam can lead to catastrophic consequences. For example, failure of the tailings dam at Buffalo Creek in West Virginia after heavy rain in February 1972, destroyed over 1500 houses and cost 118 lives. Static liquefaction caused a tailings impoundment to fail at the Sullivan Mine in Canada in 1991, resulting in a flow slide. Fortunately, according to Davies (2003), a second tailings dam contained the flow. The dam concerned had been built on older tailings that were placed as beach material below water. The failure occurred with the initiation of shear stresses in the foundation tailings that exceeded the shear strength. Pore water pressures rose as the silty sand was strained, which impeded drainage thereby leading to liquefaction. The failure of the Merriespruit gold tailings dam in South Africa in 1994 resulted in 17 deaths. This was essentially an upstream constructed tailings dam with little freeboard and relatively saturated beach below water. A massive failure of the north wall occurred after a heavy rainstorm. Overtopping as a result of insufficient freeboard provision, together with poor pool control, led to static liquefaction and the development of a flow slide once enough toe material had been eroded. More than 600 000 m^3 of tailings and 90 000 m^3 of water were released. The slurry travelled about 2 km and covered nearly 500 000 m^2. Fourie *et al.* (2001) suggested that high void ratios in some parts of the dam could have meant that these zones were in a metastable condition, overtopping and erosion of the dam wall exposing such zones. This, they maintained, resulted in liquefaction of the tailings and consequent flow failure. The tailings dam at Aznalcollar/Los Frailes mine 45 km west of Seville, Spain, was breached in 1998, flooding approximately 4600 ha of land along the Ros Agrio and Guadiamar rivers with around 5.5 million m^3 of acidic water and 1.3×10^6 m^3 of heavy metal bearing tailings. Most of the deposited tailings and around 4.7×10^6 m^3 of contaminated soils were removed to the Aznalcollar open-pit during the clean-up operation, removing enough spill deposited sediment to achieve pre-spill metal (Ag, As, Cd, Cu, Pb, Sb, Tl and Zn) concentrations in the surface sediment (Hudson-Edwards *et al.*, 2003).

Davies *et al.* (2000) suggested that every tailings dam failure was predictable in hindsight, and that in all cases over the preceding 30 years the necessary knowledge was available to prevent failure at both the design and operational stages but that the knowledge was not used. In other words, failures commonly occur when one or more aspects of design and construction, and/or operation are deficient.

Blight (1997) reviewed the failures of five tailings dams in South Africa and concluded that the nature of the mudflow that issued from them on failure was closely associated with the condition of the ground surface over which the tailings moved, that is, whether it was wet or dry. All five dams examined were of the ring dyke type and Blight pointed out that this type of dam, because of its appreciable length and frequent variability in foundation conditions, tends to be more susceptible to failure than valleys dams, although the latter because of their topographic situation are likely to be more catastrophic when they fail (see Case history 2). Moreover, tailings are often distributed around the circumference of a ring dam by open channel flow. As a consequence, deposition does not occur at the same time all around the perimeter of the dam, and so the crest elevation and freeboard generally vary around the perimeter. Hence, a ring dam may be at greater risk of overtopping than a valley dam because of the existence of zones of lower elevation around its perimeter. This is especially the case during times of unusually wet weather. Be that as it may, overtopping represents a danger to both types of dams if excessive volumes of water are stored in the lagoon, so reducing freeboard to a dangerous level, and there is poor control over the operation of the dam. For instance, dams may be susceptible to shear failure, if due to poor operational control, rates of rise in height of the dam become excessive so that the phreatic level rises or the outer slopes of the dam are oversteepened, thereby reducing the factor of safety against shear failure. When a ring dam is breached a significant body of tailings moves through the breach and often spreads out in lobe-like fashion (Fig. 7.11). For example, when tailings escape over wet ground, then they tend to form a lobate shaped or bottle-necked mudflow. The ground may be wetted because of a period of rainfall or by water spilling through the breach from the lagoon. If, however, the tailings move over a dry surface, then Blight contended that they were not likely to move far and a mudflow is unlikely to develop. Obviously, ring dams are more likely to fail in wet than in dry weather.

Even if the impounding dam remains stable, the facility can undergo failure from the environmental point of view. For instance, a tailings facility can give rise to significant dust problems that impact upon the quality of surrounding air. In addition, a tailings facility can have an impact upon groundwater or surface water quality because of seepage from the facility. This is especially the case if the preceding hydrological and hydrogeological

Figure 7.11 Failure of the Merriespruit tailings dam, South Africa, showing lobate flowslide. (Courtesy of Professor Dick Stacey & SRK Consulting South Africa.)

investigations have been inadequate. Hence, if this type of failure is to be avoided, then the geochemistry of the tailings and the lagoon water has to be understood, as well as the hydrological/hydrogeological conditions, and must be taken account of during the design, construction, operation and closure of a tailings facility.

7.6.6. Other forms of tailings

Until recently the only form of tailings for most tailings facilities, according to Martin *et al.* (2002), was the conventional segregating pumpable slurry with water contents of well over 100% (Fig. 7.12). There are, however, a number of situations in which dewatered tailings systems can be of advantage to a mining operation. Be that as it may, dewatered tailings systems have less application for larger operations for which tailings lagoons also must serve as water storage reservoirs. This is particularly the case where water balances must be managed to store annual snow melt run-off in order to provide water for all year operation.

The development of large capacity vacuum and pressure belt filter technology has afforded the opportunity for disposing of tailings in a dewatered state, rather than as a conventional slurry. The basic elements of

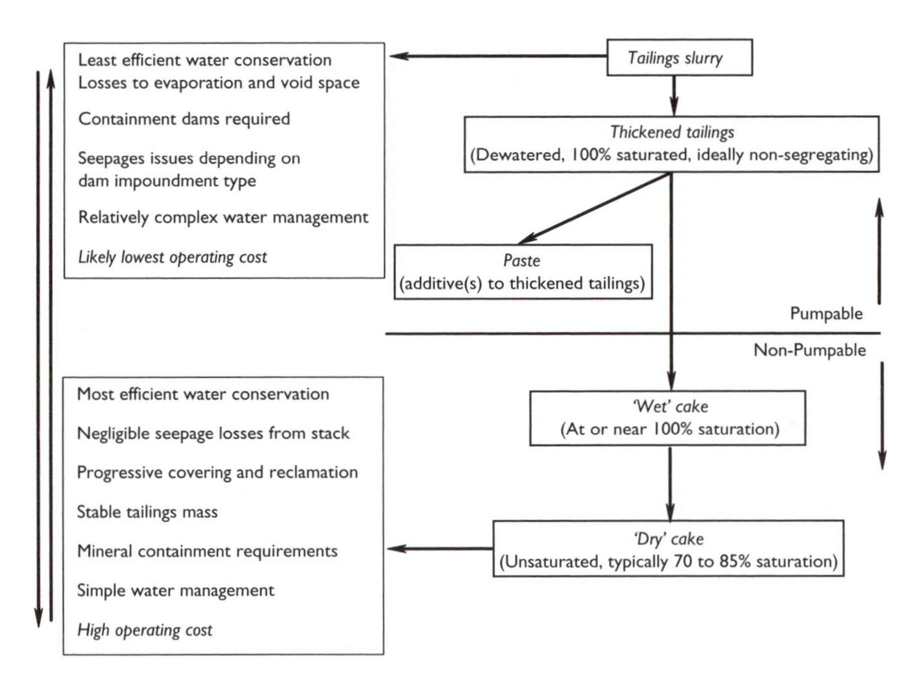

Figure 7.12 Classification of tailings by degree of dewatering. (After Martin *et al.*, 2002; reproduced by kind permission of the International Institution for Environment and Development.)

filtered tailings management have been described by Davies and Rice (2001). As indicated, filtering can be carried out using a pressure or vacuum force system. Drums, horizontally or vertically stacked plates and horizontal belts are the most common filtration plant configurations. Pressure filtration can be carried out on a much wider spectrum of materials although vacuum belt filtration is probably the most logical for larger scale operations. Tailings can be dewatered in a filter plant to less than 20% moisture content for tailings with specific gravities around 2.7. At these moisture contents, the material can be transported by conveyor or truck, and placed, spread and compacted to form an unsaturated, dense and stable tailings stack (often termed a 'dry stack') requiring no dam for retention. Whilst the technology is at present considerably more expensive per tonne of tailings stored than is the case for conventional slurry facilities, it has particular advantages in some situations, for instance, in arid regions where water conservation is an important issue. Davies and Rice quoted an example of such system at the La Coipa silver and gold operation in the Atacama desert region of Chile. A daily tailings production of 18 000 tonnes is dewatered, conveyed to the tailings site and stacked by a radial mobile conveyor system. The vacuum filter system was selected for this site because of the need to recover dissolved gold from solution, but it also is advantageous for water conservation and for stability of the tailings deposit in this region of high seismicity. Alternatively, dry stacking proves of value in cold regions where water handling is difficult in winter. Again Davies and Rice quoted an example at Raglan nickel operation in the arctic region of northern Quebec, where a dewatered tailings system, using truck transport, is in operation. The system also is intended to provide a solution for potential acid generation, as the tailings stack will become permanently frozen. Moreover, filtered tailings stacks require a smaller footprint for tailings storage, as they have a much lower bulking factor, are easier to reclaim and have a much lower long-term liability in terms of structural integrity and potential environmental impact.

Several other tailings disposal technologies have been introduced into the mining industry. For instance, the development of improved thickener technology has led to advances in paste tailings for use in underground backfill operations. Paste tailings are essentially the whole tailings thickened to a dense slurry, that is, dewatered tailings with little or no water bleed that are non-segregating. Cement can be added to the thickened tailings and the material pumped underground to use as ground support in mined out stopes. In addition to cement, PFA and ground granulated blast furnace slag (GGBS) can be added to improve strength (e.g. 3–6% by weight). In addition, paste can be stacked at the surface, being placed in stable configurations. Cement or other additives are not added routinely for surface disposal because of the cost. Nonetheless, the addition of 1% by weight of cement increases the neutralizing potential by an order of magnitude and raises the pH of a paste, which can lead to the immobilization of some metals through mineral precipitation. As noted, the use of virtually water-free tailings storage can eliminate the need for large dams and remove the associated liability that such structures pose to the environment. Moreover, there is very little free water to generate leachate and the relatively low permeability (10^{-7} m s^{-1} and less) of the tailings limits infiltration, resulting in reduced seepage volume present in deposited paste. The preparation of the paste allows the material to be engineered with the use of additives that can enhance the beneficial properties of the paste. The Bulyanhulu gold mine in Tanzania was the first mine in the world to adopt a total paste solution for all its tailings, being commissioned in 2001 (Newman *et al.*, 2001). Some of the paste is to be used underground as backfill, with the remainder being stacked at the surface.

Thickened tailings are paste without the additives. Thickened disposal is a technique that has been implemented in a few operations, most notably and successfully in the arid regions of Australia. The tailings may be thickened to the extent that they may be discharged from one or several discharge points to form a non-segregating mass of tailings with little or no lagoon. They form a conical mass but still require retaining dams.

Spearing *et al.* (2001) have suggested that the introduction of micro-air bubbles into dewatered tailings would permit them to be transported as a foam medium rather than a slurry. Hence, a dry stack tailings deposit could be developed using foam technology instead of trucks or conveyors to transport the tailings. The air could be removed at the point of discharge by the addition of a de-foaming agent or by compaction of the tailings. Transport costs for filtered tailings operations make up a very significant portion of the operating costs. If lower cost transport methods such as foam technology are viable, then the economics of dry stack tailings disposal becomes more favourable than is presently the case.

7.6.7. Reclamation of tailings impoundments

Those tailings deposits that remain after mining operations have ceased often present a post-mining liability. Unfortunately, many tailings impoundments currently undergoing remediation were designed and constructed using methods that are inappropriate by present day standards and contain inherent deficiencies that must be remediated. The minimization of liability after closure can be brought about by stabilization of the impoundment/deposit, provision of hydrological control, control of seepage, integration of created landforms (i.e. during reclamation) into the landscape, and minimization of the need for water treatment, surveillance and monitoring (Jakubick *et al.*, 2003). Upgrading of tailings dams may include the construction of internal cut-off walls, durable downstream drainage blankets and berms, and the installation of upgraded diversion and spillway systems to achieve long-term control of water levels in tailings impoundments.

The objectives of rehabilitation of tailings impoundments include their long-term mass stability, long-term stability against erosion, prevention of environmental contamination and return of the area to productive use. Normally, when the discharge of tailings comes to an end the level of the phreatic surface in the embankment falls as water replenishment ceases. This results in an enhancement of the stability of embankment slopes. However, where tailings impoundments are located on slopes, excess run-off into the impoundment may reduce embankment stability or overtopping may lead to failure by erosion of the downstream slope. The minimization of inflow due to run-off by judicious siting is called for when locating an impoundment that may be so affected. Diversion ditches can cater for some run-off but have to be maintained, as do abandonment spillways and culverts. Accumulation of water may be prevented by capping the impoundment, the capping sloping towards the boundaries. Erosion by water or wind can be impeded by placing riprap on slopes and by the establishment of vegetation on the waste. The latter also will help to return the impoundment to some form of productive use (Troncoso and Troncoso, 1999). However, tailings may contain high concentrations of heavy metals and be very low in plant nutients, with low pH values. Accordingly, the application of lime and/or fertilizer may be necessary to establish successful plant growth. Even then, the number of species that can be grown initially without their growth being stunted may be limited. Where long-term potential for environmental contamination exists, particular precautions need to be taken. For example, as the water level in the impoundment declines, the rate of oxidation of any pyrite present in the tailings increases, reducing the pH and increasing the potential for heavy metal contamination. In the case of tailings from uranium mining, radioactive decay of radium gives rise to radon gas. Diffusion of radon gas does not occur in saturated tailings but after abandonment radon reduction measures may be necessary. In both these cases a vegetated clay cover can be placed over the tailings impoundment to prevent leaching of contaminants or to reduce emission of radon gas (Gatzweiler *et al.*, 2001).

Once the discharge of tailings ceases, the surface of the impoundment is allowed to dry. Drying of the decant pond may take place by evaporation and/or by drainage to an effluent plant. Desiccation and consolidation of the slimes may take tens of years. In the latter case, the dissipation of pore water pressures may be hastened by the installation of band drains. They may be installed through the full depth of the tailings or only placed in the upper few metres to allow the surface to consolidate to a sufficient extent to allow access for placement of a cover. The desiccated crust may be reinforced by the placement of geogrids, over which a cover is laid. The final cover can be laid once the surface is firm enough to support equipment.

The objectives of the final cover are to provide a medium for the establishment of vegetation, to reduce erosion, to prevent surface flow from coming in contact with the tailings and to reduce infiltration. The ultimate surface topography of the remediated tailings facility should not allow ponding to occur and long-term settlement should not alter the slope of the drainage patterns on the surface of the cover or within the cover (Jakubick *et al.*, 2003). Covers are sensitive to erosion on moderately to steeply inclined slopes (1 in 20 or steeper), which are common for tailings dams. Slopes may be interrupted by berms, placed at intervals of 25–50 m, in order to reduce the potential for rill and gully development. Slope drainage is collected on the berms by erosion resistant (i.e. lined) ditches that convey the drainage from the berm and dam. Mine tailings can be stabilized by the addition of lime, PFA and aluminium (around 5%, 10% and 110 mg kg^{-1} respectively) to improve their mechanical properties, especially their strength and reduce their permeability (Mohamed *et al.*, 2002). However, such treatment depends on availability of the materials and generally may be of restricted use because of cost. Lime also can be used to fix heavy metals in tailings thereby inhibiting or preventing their release. Eusden *et al.* (2002) referred to the use of soluble phosphate and lime for stabilization of heavy metals in tailings at Leadville, Colorado. The tailings were associated with the Wolftone and Maid of Erin mines where the dominant minerals are galena (PbS), cerrusite ($PbCO_3$), pyromorphite ($(PbCl)Pb_4(PO_4)_3$) plumbojarosite [$Pb\{Fe_3(OH)_6(SO_4)_2\}_2$] and chalcophanite [$(Zn,Mn,Fe)Mn_2O_5.nH_2O$]. The soluble heavy metals were converted to insoluble metal phosphates, the treatment causing bulk mineralogical changes and the formation of reaction rinds, dominated by Ca and P, around offending particles. At Fort McMurray, Alberta, an aqueous fines suspension, referred to as mature fine tails (MFT), is produced during the extraction of bitumen from the Athabasca Tar Sands (Matthews *et al.*, 2002). These tailings, because of their geotechnical properties, require long-term storage in secure impoundments. Accordingly, they are mixed with a coarse tailings stream and a coagulant is added to form a slurry. This releases water rapidly once deposited and binds the MFT in the coarse tailings.

7.7. Case history I

Two old colliery spoils heaps at Wharncliffe-Woodmoor near Barnsley, South Yorkshire, England, offer examples of reclamation to improve the local amenity and to provide a site for light industrial development. The site consisted of two distinct areas, namely, a western area comprising 19 ha of agricultural land, and an area containing the spoil heaps occurred on the eastern part. The latter covered 29.5 ha and contained two spoil heaps, one 45 m in height, the other 25 m high, and a large area occupied by old tailings ponds, as well as pithead buildings and stock grounds. The smaller heap and pithead buildings were separated from the main heap and tailings ponds by a canal that carried compensation water from a local reservoir to the River Dearne. As a consequence the canal had to be diverted prior to reclamation work commencing.

The larger spoil heap had an estimated volume of 3 000 000 m^3 and consisted mainly of shale, some of which had been burnt. The small spoil heap to the north of the canal was composed of similar material and its volume was less than 1 000 000 m^3. Spoil had been spread over other areas and had been tipped over some of the tailings ponds. However, there was no evidence of burning in the latter spoil.

The geology of the area consists of Middle Coal Measures in which shales, mudstones and sandy shales alternate with sandstones, notably, the Woolley Edge Rock, the Oaks Rock and the Mexborough Rock. The general dip is towards the east or east-north-east and the area contains a number of faults. A large patch of till covers the Coal Measures in this area.

The dominant mineral in the unburnt shaley material was quartz. Illite was the most important clay mineral, with kaolinite usually averaging less than 5% or 10%. Mica and chlorite tended to occur in similar amounts to that of kaolinite. Feldspars, sulphates, pyrite and carbonate occurred in trace amounts.

The moisture content of the spoil material tended to increase with increasing content of fines. It also was influenced by the permeability of the material, the topography and the climatic conditions. Generally, it fell within the range 6–13%. The bulk density of the unburnt spoil showed a wide variation, with most of the material having values between 1.5 and 2.5 Mg m^{-3}. Low densities were mainly a function of low specific gravities. These obviously were influenced by the relative proportions of mudrock, sandstone and coal in the waste. In particular, the higher the coal content, the lower was the specific gravity. As far as the particle size distribution of the coarse discard was concerned, it mostly fell within the sand range although significant proportions of gravel and cobble size were present.

Although the pyrite content of the spoil heaps was in trace amount, nonetheless its presence could have a significant effect. As mentioned earlier, pyrite breaks down rapidly under the influence of weathering to produce, amongst other things, sulphuric acid. If the oxidation products are not neutralized by alkaline materials in the waste material, then the waste material is acidic. It therefore probably will be deficient in plant nutrients and so will require careful treatment if vegetation is to be established, notably dozing with lime. In fact, there was little vegetation cover over the tipped areas, that which did occur being found at the base of the spoil heaps. However, the results of pH tests on the waste material gave values around 6.5–6.8. Hence, the material was only very slightly acidic.

The site investigation included the production of topographic plans of the site from aerial photographs and sinking boreholes in the old tailings ponds and spoil heaps. The tailings ponds appeared to be quite stable, so that their level could be raised by covering with waste material during the reclamation process.

Reclamation of a colliery spoil heap is essentially a large scale exercise in earthmoving and the coarse discard frequently is spread over a larger area than that occupied by the heap. The first part of the work involved stripping over 1 000 000 m^3 in order to reduce the height of the spoil heaps. The purpose of the levelling was to achieve a pleasant contoured landform within the site, related to the topography of the adjoining land (Fig. 7.13). This was brought about not only by reducing the height of the spoil heaps, but also by regrading and rounding off the steeper slopes. The other areas, such as those occupied by the pithead buildings, as well as the tailings ponds, were buried by spoil. These areas also were sloped. In this way a gently sloping ridge, valley and hill topography was created on the site.

This work unfortunately encountered problems with hot spots. These were normally around 600°C but at some locations temperatures were as

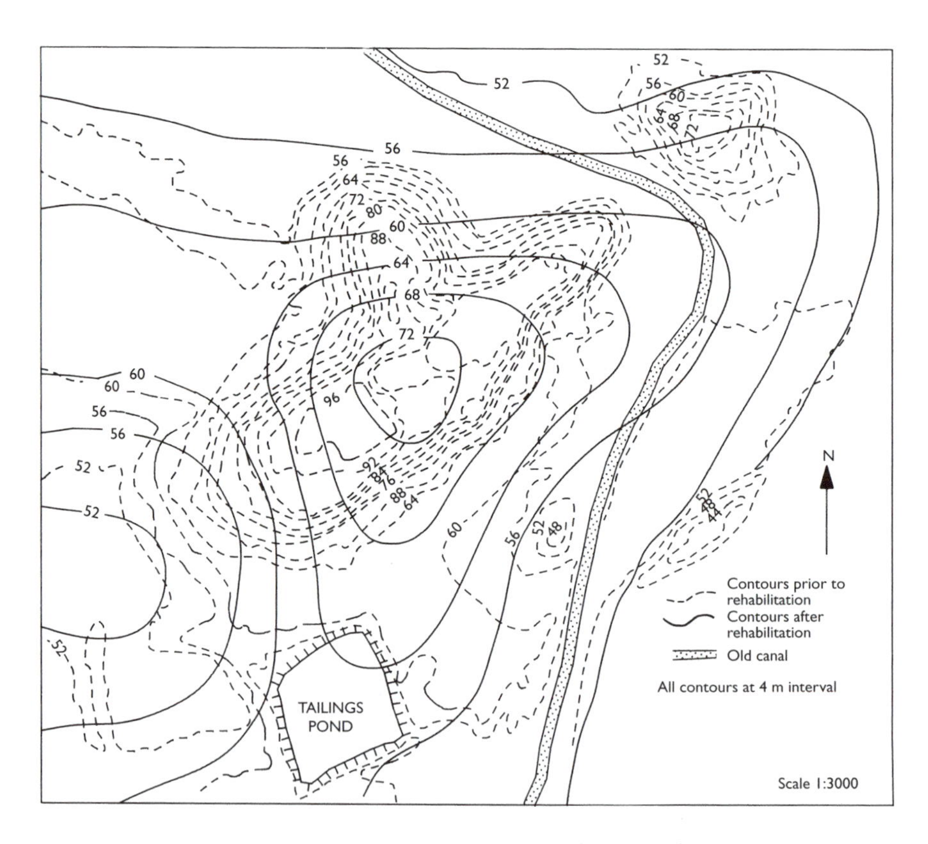

Figure 7.13 Lay-out of spoil heaps showing contours before and after reclamation.

Figure 7.14 Compacting spoil with a vibratory roller.

high as 900°C. The heat often caused engines on the scrapers to cease and rescue of both operator and machine often proved hazardous. In addition, the tyres on scrapers frequently melted. Where levelling was still in progress and hot spots were present, it was decided to only remove a layer of 300 mm in thickness and then compact the spoil with a vibratory roller (Fig. 7.14). This reduced the temperature by about 50% after two hours. Levelling then could commence again. When hot spots were found in areas where the tip material had been lowered to its finished level, 450 mm of clay was spread and compacted on top of the hot spot. Then a layer of shale was spread and compacted over the clay. Boreholes were sunk into these hot spots to determine whether or not they were cooling. Those spots where the temperatures had not dropped within a year were enclosed by an injected curtain wall of pulverized fuel ash that extended down to original ground level. Fortunately, no subsurface cavities in the spoil, due to spontaneous combustion, were met with on this site. Noxious gases were not a problem.

Drains, in the form of open ditches containing pipes and filled with gravel, were constructed after levelling was completed and led into existing streams or culverts (Fig. 7.15). An open ditch also was placed around the site to catch any surface run-off.

The next operation involved harrowing the spoil, after which lime was applied at 7.5 tonnes ha^{-1}. Once liming was complete, subsoil and then topsoil were applied. This was obtained from the agricultural land in the

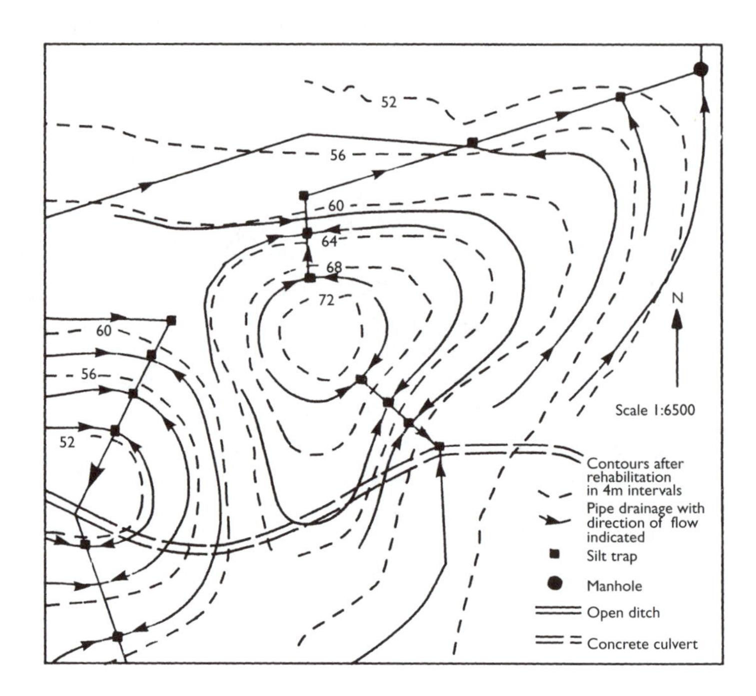

Figure 7.15 Drainage for rehabilitated spoil heap.

Figure 7.16 Bakery constructed in the north west part of the site.

western part of the site that had been stripped and stockpiled prior to spoil being spread over it. Fertilizers were added to the ground prior to seeding.

Part of the site was developed for a large bakery and most of the rest was used as a golf course (Fig. 7.16). The latter was mainly grassed, although a large number of trees were planted on the course. Trees also were planted to enhance the character of the landscape, to reduce erosion on steeper slopes and to screen parts of the site. In all, some 20 000 trees were planted over a three year period. The area now represents a source of employment and a popular local amenity rather than an unpleasant eyesore.

7.8. Case history 2

At around 12:30 on 19 July 1985, a tailings dam collapsed near Stava, northern Italy, which resulted in the loss of 269 lives and the destruction of much property (Genevois and Tecca, 1993). In fact, according to Chandler and Tosatti (1995), this was the worst failure of its type in Europe. The collapse of the dam liquefied the tailings and gave rise to the generation of a mudflow that moved downhill initially at about 30 km hr^{-1} into the small village of Stava, some 500 m distant, where it destroyed two hotels and numerous houses. As the mudflow progressed it became more fluid, moving faster (up to 90 km hr^{-1}) and it flowed a further 3 km down the Tesero Valley causing death and destruction in its wake, then continued into the Stava Valley as far as the confluence with the River Avisio, where it came to a halt (Fig. 7.17).

The tailings disposal facility consisted of two adjacent dams, one upstream of the other (Fig. 7.18). The tailings were derived from the Prestavel Mine that worked fluorite. Collapse occurred in the upper basin, which released its material into the lower impoundment causing it also to collapse. Chandler and Tosatti (1995) suggested that the primary cause of failure was due to the fact that the dams were constructed with unacceptably low factors of safety and that the failure was triggered by a blocked decant pipe located within the tailings of the upper facility. Other factors that may have been involved in the failure include a high rate of lagoon filling after 1982 and high pore water pressure in the dam foundation soils with increasing vertical and horizontal displacements taking place from 1975 onwards. These factors must be taken into consideration along with the steep downstream slope of the dam and the poor geotechnical properties of the hydrocycloned tailings material (Table 7.2; Genevois and Tecca, 1993).

The lower tailings dam was constructed in 1962, the ground consisting of water bearing, highly heterogeneous morainic soils consisting of boulders, cobbles and gravels set in a finer matrix, overlying dolostone bedrock at depth. The groundwater level tends to fluctuate in these soils. Fluvio-glacial deposits also were present in the area and were used in part of the construction of the dams. Construction involved the site being cleared of trees. The

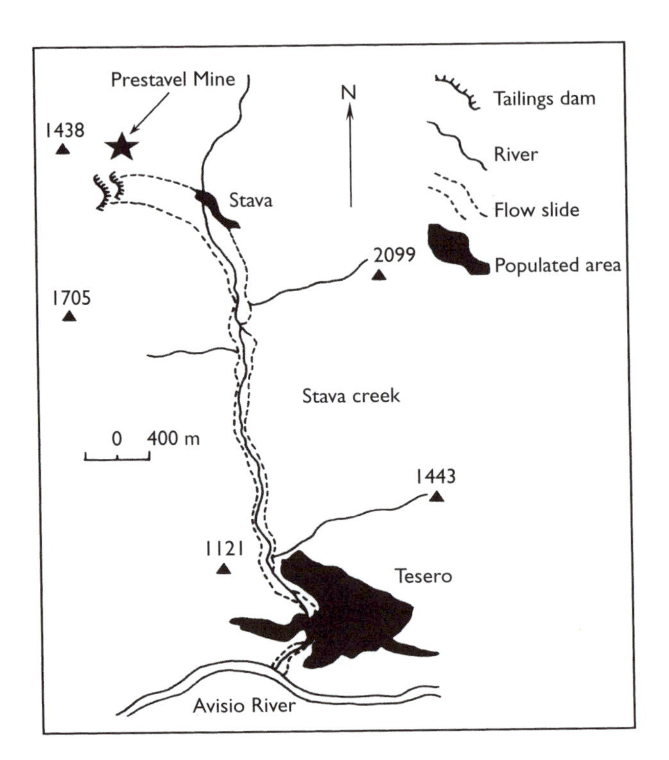

Figure 7.17 Location of Stava tailings dams and the area of Stava Creek Valley swept by the flowslide. (After Genevois and Tecca, 1993.)

site was not well drained and no attempt was made to enhance the ground conditions except for a crude form of soil nailing with wooden stakes in the dam foundation zone. Neither was there a drainage blanket of granular material placed beneath the dam to act as an underdrain. The dam itself was constructed by the upstream method (see earlier) and a hydrocyclone was used to separate the sandy fraction from the tailings. The lowermost part of the dam was constructed of local material and then the sandy fraction was used to finish off the rest of the dam. The tailings were pumped into the lagoon, where they settled out and the excess water was removed via a decant pipe. The draw-off point from the latter was situated towards the upper part of the lagoon and was progressively extended upslope as the dam was raised. Consequently, a low lagoon position was not maintained that, as referred to earlier, is necessary to ensure stability in tailings facilities constructed by the upstream method. In addition to there being no drainage blanket, natural run-off was not diverted away from the lagoon by a diversion ditch.

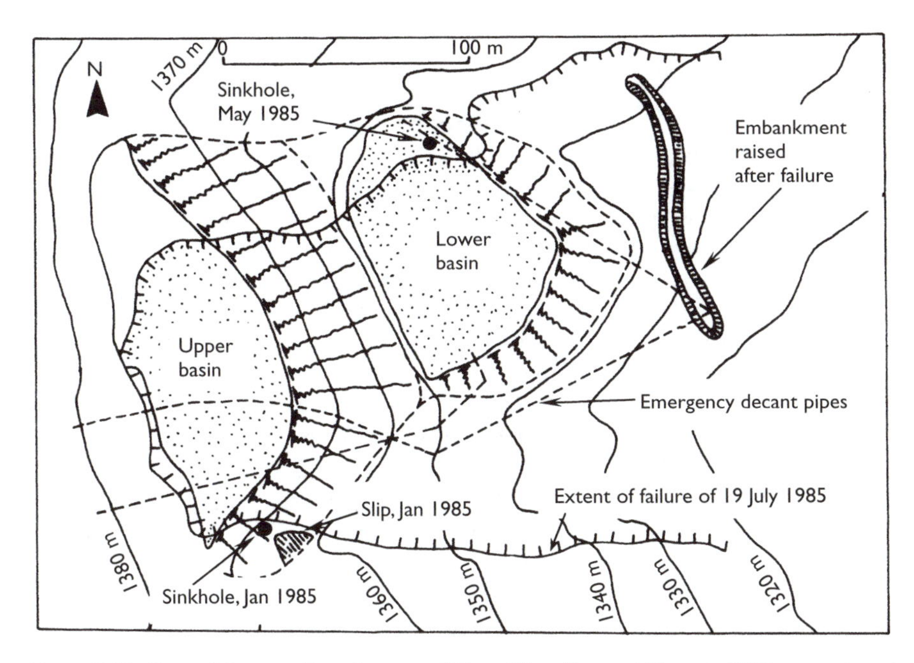

Figure 7.18 Plan of Stava tailings lagoons. (After Chandler and Tosatti, 1995; reproduced by kind permission of Thomas Telford Publishing.)

Table 7.2 Some physical properties of colliery and mine tailings

Property	*Colliery tailings*[a]	*Metal mine tailings*[b]
Specific gravity	1.94	2.86
Natural moisture content (%)	34.8	35
Void ratio	0.82	—
Bulk density (Mg m^{-3})	1.46	18.4
Dry density (Mg m^{-3})	1.04	15.5
Clay content (%)	16	—
Plastic limit (%)	23	17
Liquid limit (%)	39	29
Plasticity index (%)	16	12
Angle of friction (°)	24.5	36.2[c]
Cohesion (kPa)	2.3	0

Notes
a After Taylor (1984), mean values.
b After Genevois and Tecca (1993) and Chandler and Tossati (1995).
c Remoulded samples.

The upper dam was constructed in 1969 and again no foundation treatment was given to the soft wet soils of the foundation area of the dam, in which seepages occurred, nor was underdrainage provided. The upper dam was constructed by the centreline method, initially of local materials and then of hydrocycloned sandy tailings. Unfortunately, the downstream toe of the dam extended over the tailings of the lower lagoon. Two decant pipes were constructed and progressively raised as the level of the dam was raised, with their discharge points being sealed as the level of the tailings were raised. Initially, the dam height was planned for 19 m with an average downstream slope of some 39°, but then this was to be extended to 34 m at a similar downstream slope angle. A berm, 4 m in width, was constructed at 19 m, thereby giving an average downstream slope angle of 35°. The upper part of the dam was constructed by the upstream method. However, the dam had only been raised to a height of 28.5 m by the time failure occurred, the uppermost part of the dam being finished with fluvio-glacial material rather than hydrocycloned sandy tailings. Neither of the dams were ever monitored to record their performance.

A small slip occurred on the southern part of the upper dam in January 1985 from which seepage occurred, which was followed by the development of a sinkhole in the outer bank of the dam. The slip and sinkhole may have been caused by leakage from the nearby decant pipe that had been frozen at its outfall. Seepage continued at the slip until March. In May a sinkhole developed in the lower dam. Repairs involved diverting the decant water from the upper lagoon around the southern and northern part of the lower lagoon and re-routing the lower decant pipe downslope of the lower dam. The lagoons were brought back into operation in July when tailings again were pumped into the upper lagoon, from which decant water was pumped into the lower, which then refilled. The failure took place four days later.

References

Anon. 1973. *Spoil Heaps and Lagoons: Technical Handbook*. National Coal Board, London.

Anon. 1989. *Tailings Dams Safety Guidelines*. Bulletin 74, International Committee on Large Dams (ICOLD), Paris.

Anon. 1994. *Tailings Dam Incidents*. United States Committee on Large Dams, Denver, Colorado.

Anon. 1998. *Tailings Dams and Seismicity – Review and Recommendations*. Bulletin 98, International Committee on Large Dams (ICOLD), Paris.

Anon. 2001. *Tailings Dams – Risk of Dangerous Occurrences, Lessons Learnt from Practical Experiences*. United Nations Environmental Programme (UNEP) and International Committee on Large Dams (ICOLD), Bulletin 121, Paris.

Badger, C.W., Cummings, A.D. and Whitmore, R.L. 1965. The disintegration of shale. *Journal Institute Fuel*, **29**, 417–423.

Bell, F.G. 1996. Dereliction: colliery spoil heaps and their rehabilitation. *Environmental and Engineering Geoscience*, **2**, 85–96.

Bell, F.G. and Bullock, S.E.T. 1996. The problem of acid mine drainage, with an illustrative case history. *Environmental and Engineering Geoscience*, **2**, 369–392.

Bishop, A.W. 1973. The stability of tips and spoil heaps. *Quarterly Journal of Engineering Geology*, **6**, 335–376.

Blight, G.E. 1988. Some less familiar aspects of hydraulic fill structures. In: *Hydraulic Fill Structures*, Van Zyl, D.J.A. and Vick, S.G. (eds), American Society of Civil Engineers, Geotechnical Special Publication No. 21, New York, 1000–1064.

Blight, G.E. 1994a. The master profile for hydraulic fill tailings beaches. *Proceedings of the Institution of Civil Engineers, Geotechnical Engineering*, **107**, 27–40.

Blight, G.E. 1994b. Environmentally acceptable tailings dams. *Proceedings of the First International Congress on Environmental Geotechnics*, Edmonton, Carrier, W.D. (ed.), BiTech Publishers Ltd., Richmond, B.C., 417–426.

Blight, G.E. 1997. Destructive mudflows as a consequence of tailings dyke failures. *Proceedings of the Institution of Civil Engineers, Geotechnical Engineering*, **125**, 9–18.

Blight, G.E. and Amponsah Da Costa, F. 1999. Improving the erosional stability of tailings dam slopes. *Proceedings of the Sixth International Conference on Tailings and Mine Waste '99*, Fort Collins, Colorado, A.A. Balkema, Rotterdam, 197–206.

Blight, G.E. and Fourie, A.B. 1999. Leachate generation in landfills in semi-arid climates. *Proceedings of the Institution of Civil Engineers, Geotechnical Engineering*, **137**, 181–187.

Blight, G.E. and Steffen, O.K.H. 1979. Geotechnics of gold mining waste disposal. In: *Current Geotechnical Practice in Mine Waste Disposal*, American Society of Civil Engineers, New York, 1–52.

Bowell, R.J., Williams, K.P., Connelly, R.J., Sadler, P.J.K. and Dodds, J.E. 1999. Chemical containment of mine waste. In: *Chemical Containment of Waste in the Geosphere*, Metcalfe, R. and Rochelle, C.A. (eds), Special Publication No. 157, the Geological Society, London, 213–240.

Chandler, R.J. and Tosatti, G. 1995. The Stava tailings dam failure, Italy, July 1985. *Proceedings of the Institution of Civil Engineers, Geotechnical Engineering*, **113**, 67–79.

Cook, B.J. 1990. Coal discard-rehabilitation of a burning heap. In: *Reclamation, Treatment and Utilization of Coal Mining Wastes*, Rainbow, A.K.M. (ed.), A.A. Balkema, Rotterdam, 223–230.

Davies, M.P. 2003. Tailings impoundment failures: are geotechnical engineers listening? *Geotechnical News*, **21**, No. 3, 31–36.

Davies, M.P. and Rice, S. 2001. An alternative to conventional tailings management – 'dry stack' filtered tailings. *Proceedings of the Eighth International Conference on Tailings and Mine Waste '01*, Las Vegas, A.A. Balkema, Rotterdam, 411–420.

Davies, M.P., Martin, T.E. and Lighthall, P.C. 2000. Tailings dam stability: essential ingredients for success. In: *Slope Stability in Surface Mining*, Hustrulid, W., McCarter, M. and Van Zyl, D.J.A. (eds), Society for Mining, Metallurgy and Exploration, Denver, Colorado, 365–377.

Dick, J.C. and Shakoor, A. 1992. Lithological controls of mudrock durability. *Quaterly Journal of Engineering Geology*, **25**, 31–46.

East, D.E. and Morgan, D.J.T. 1999. The use of geomembranes in the design of mineral waste storage facilities. *Proceedings of the Sixth International Conference on Tailings and Mine Waste* '99, Fort Collins, Colorado, A.A. Balkema, Rotterdam, 371–376.

Eusden, J.D., Gallagher, L. and Eighmy, T.T. 2002. Petrographic and spectrographic characterization of phosphate-stabilized mine tailings from Leadville, Colorado. *Waste Management*, **22**, 117–135.

Fell, R., Miller, S. and de Ambrosio, L. 1993. Seepage and contamination from mine waste. *Proceedings of the Conference on Geotechnical Management of Waste and Contamination*, Sydney, Fell, R., Phillips, A. and Gerrard, C. (eds), A.A. Balkema, Rotterdam, 253–311.

Fourie, A.B., Blight, G.E. and Papageorgiou, G. 2001. Static liquefaction as a possible explanation for the Merriespruit tailings dam failure. *Canadian Geotechnical Journal*, **38**, 707–719.

Garrard, G.F.G. and Walton, G. 1990. Guidance on design and inspection of tips and related structures. *Transactions of the Institution of Mining and Metallurgy*, Section A, Mining Industry, **99**, A115–A124.

Gatzweiler, R., Jahn, S., Neubert, G. and Paul, M. 2001. Cover design for radioactivity and AMD-producing mine waste in the Ronneburg area, Eastern Thuringia. *Waste Management*, **21**, 175–184.

Genevois, R. and Tecca, P.R. 1993. The tailings dams of Stava (nothern Italy); an analysis of the disaster. *Proceedings of the International Conference on Environmental Management, Geowater and Engineering Aspects*, Wollongong, Chowdhury, R.N. and Sivakumar, M. (eds), A.A. Balkema, Rotterdam, 23–36.

Hudson-Edwards, K.A., Macklin, M.G. and Jamieson, H.E. 2003. The impact of tailings dam spills and clean-up operations on sediment and water quality in river systems: the Ros Agrio-Guadiamar, Aznalcollar, Spain. *Applied Geochemistry*, **18**, 221–239.

Hughes, D.B. and Clarke, B.G. 2003. Surface coal mining and the reclamation of tips, landfills and quarries – some geotechnical case studies from northern England. *International Journal of Surface Mining, Reclamation and Environment*, **17**, 67–97.

Jakubick, A.T., McKenna, G. and Robertson, A.M. 2003. Stabilisation of tailings deposits: international experience. http://www.geoconsultants.com/papers/Sudbury2003_Jakubick_McKenna_AMR.pdf

Johnson, A.C. and James, E.J. 1990. Granville colliery land reclamation/coal recovery scheme. In: *Reclamation and Treatment of Coal Mining Wastes*, Rainbow, A.K.M. (ed.), A.A. Balkema, Rotterdam, 193–202.

Maddison, J.D. 2000. Albion Colliery tip and landslide: deep drainage by bored drains. In: *Landslides and Landslide Management in South Wales*, Siddle, H.J., Bromhead, E.N. and Bassett, M.G. (eds), Geological Series No. 18, National Museum of Wales, Cardiff, 100–103.

Martin, T.E., Davies, M.P., Rice, S., Higgs, T. and Lighthall, P.C. 2002. *Stewartship of Tailings Facilities*. Mining, Minerals and Sustainable Development, International Institution for Environment and Development, Report No. 20, Ottawa.

Matthews, J.G., Shaw, W.H., MacKinnon, M.D. and Cuddy, R.G. 2002. Development of composite tailings technology at Syncrude. *International Journal of Surface Mining, Reclamation and Environment*, **16**, 24–39.

Mohamed, A.M.O., Hossein, M. and Hassani, F.P. 2002. Hydro-mechanical evaluation of stabilized mine tailings. *Environmental Geology*, **41**, 749–759.

Morris, P.H. and Williams, D.J. 1997. Co-disposal of washery wastes at Jeebropilly colliery, Queensland, Australia. *Transactions of the Institution of Mining and Metallurgy*, Section A, Mining Industry, **106**, A25–A29.

Newman, P., White, R. and Cadden, A. 2001. Paste – the future of tailings disposal? http://www.golder.com/archive/skelleftea2001pastedisposal.pdf

Newman, T.A. and Fahey, M. 2003. Measurement of evaporation from saline tailings storages. *Engineering Geology*, **70**, 217–233.

Pastore, E.L. 2003. Impacts de recouvrements de residus organique sur des residus miniers reactifs. *Bulletin Engineering Geology and the Environment*, **62**, 269–277.

Peacey, V., Yanful, E.K., Li, M. and Patterson, M. 2002. Water cover over mine tailings and sludge: field studies of water quality and resuspension. *International Journal of Surface Mining, Reclamation and Environment*, **16**, 289–303.

Penman, A.D.M. 2000. The Aberfan flow slide, Taff Valley. In: *Landslides and Landslide Management in South Wales*, Siddle, H.J., Bromhead, E.N. and Bassett, M.G. (eds), Geological Series No. 18, National Museum of Wales, Cardiff, 62–69.

Sharma, R.S. and Al-Busaidi, T.S. 2001. Groundwater pollution due to a tailings dam. *Engineering Geology*, **60**, 235–244.

Siddle, H.J., Wright, M.D. and Hutchinson, J.N. 1996. Rapid failures of colliery spoil heaps in the South Wales Coalfield. *Quarterly Journal of Engineering Geology*, **29**, 103–132.

Smith, M.L. and Williams, R.E. 1996. Examination of methods for evaluating remining a mine waste site. Part 1: Geostatistical characterization methodology. Part 2: Indicator kridging for selective remediation. *Engineering Geology*, **43**, 11–21; 23–30.

Spearing, A., Millette, D. and Gay, F. 2001. Foam technology. *Mining Magazine*, No. 8 (August), 69–70.

Spears, D.A., Taylor, R.K. and Till, R. 1970. A mineralogical investigation of a spoil heap at Yorkshire Main Colliery. *Quarterly Journal of Engineering Geology*, **3**, 239–252.

Taylor, R.K. 1973. Compositional and geotechnical characteristics of a 100-year-old colliery spoil heap. *Transaction of the Institution of Mining and Metallurgy*, Section A, Mining Industry, **82**, A1–A14.

Taylor, R.K. 1975. English and Welsh colliery spoil heaps – mineralogical and mechanical relationships. *Engineering Geology*, 7, 39–52.

Taylor, R.K. 1984. *Composition and Engineering Properties of British Colliery Discards*. National Coal Board, London.

Taylor, R.K. 1988. Coal Measures mudrocks: composition, classification and weathering processes. *Quarterly Journal of Engineering Geology*, **21**, 85–99.

Taylor, R.K. and Spears, D.A. 1970. The breakdown of British Coal Measures rocks. *International Journal of Rock Mechanics Mining Science*, **7**, 481–501.

Troncoso, J.H. and Troncoso, D.A. 1999. Rehabilitation of abandoned deposits of mineral residues. *Proceedings of the Sixth International Conference on Tailings and Mine Waste* '99, Fort Collins, Colorado, A.A. Balkema, Rotterdam, 693–700.

Van Zyl, D.J.A. 1993. Mine waste disposal. In: *Geotechnical Practice for Waste Disposal*, Daniel, D.E. (ed.), Chapman and Hall, London, 269–287.

Vick, S.G. 1983. *Planning, Design and Analysis of Tailings Dams*. Wiley, New York.

Vick, S.G. 1999. Tailings dam safety – implications for the dam safety community. *Proceedings of the Annual Conference of Canadian Dam Safety Association*, Sudbury, 1–12.

Wilson, G.W. 2001. Co-disposal of tailings and waste rock. *Geotechnical News*, **19**, No. 2, 44–49.

Chapter 8

Mine effluents and acid mine drainage

Different types of process effluents and waste waters are produced as a result of mining. These may arise due to the extraction process, by the subsequent preparation of the mineral that is mined, from the disposal of spoil or from stockpiles. The rock mass from which the groundwater involved is derived, the mineralogical character of the material mined and the spoil, and the preparation processes employed all affect the type of effluent produced. Drainage problems also result from past mining operations.

8.1. Effluents associated with coal mines

Generally, the major pollutants associated with coal mining are suspended solids, dissolved salts (especially chlorides), acidity and iron compounds (Bell and Kerr, 1993). When suspended solids enter a small stream they may cause the water to become turbid and may deposit a fine layer of sediment over the bed of the stream, both of which adversely affect aquatic life. However, the character of drainage from coal mines varies from area to area and from coal seam to coal seam. Hence, mine drainage waters are liable to vary in both quality, and also in quantity, sometimes unpredictably, as the mine workings develop. Nonetheless, drainage water, for example, from coal mines in Britain can be classified as hard, alkaline, moderately saline, highly saline, alkaline and ferruginous, and acidic and ferruginous (Best and Aikman, 1983; Table 8.1). Colliery discharges have little oxygen demand, the biochemical oxygen demand (BOD) normally being very low. Elevated levels of suspended matter are associated with most coal mining effluents, with occasionally high values being recorded. Although not all mine waters are highly mineralized, a high level of mineralization is typical of many coal mining discharges and is reflected in the high values of electrical conductivity. Highly mineralized mine waters usually contain high concentrations of sodium and potassium salts, and mine waters that do not contain sulphate may contain high levels of strontium and barium. Similarly, not all mine waters are ferruginous, and in fact some are of the highest quality and can be used for potable supply. Nonetheless, mine

Table 8.1 Composition of coal mine drainage waters (From Bell and Kerr, 1993)

Column number	*1*	*2*	*3*	*4*	*5*	*6*
Approximate percentage of waters in each class	55%	25%	10%	7%	7%	2%
Quality	Hard alkaline		Moderately saline	Alkaline and ferruginous	Acidic and ferruginous	Highly saline
pH value	7.8	6.8	8.2	6.9	2.9	7.5
Alkalinity, $mg\,l^{-1}$ $CaCO_3$	260	850	240	340	Nil	190
Calcium, $mg\,l^{-1}$	75	28	90	190	125	2560
Magnesium, $mg\,l^{-1}$	90	17	40	130	90	720
Dissolved iron, $mg\,l^{-1}$	0.1	0.5	0.1	25	122	0.6
Suspended iron, $mg\,l^{-1}$	0.1	2	0.1	21	0.1	0.2
Manganese, $mg\,l^{-1}$	0.1	0.1	0.1	6	7	0.9
Chloride, $mg\,l^{-1}$	180	200	3400	42	50	30 800
Sulphate, $mg\,l^{-1}$	170	210	250	1720	1250	350

waters are commonly high in iron and sulphates. The low pH values of many mine waters are commonly associated with highly ferruginous discharges. In some associated streams the pH value is less than 4.0, the iron concentration is greater than several hundred milligrams per litre and the sulphates exceed one thousand milligrams per litre. Iron in mine water draining from coal mines may be in the more stable ferrous form underground but it may oxidize in the presence of oxygen to ferric iron whereby it may form an orange precipitate termed ochre. Ochre often is seen as deposits at points of discharge and coating stream beds.

The high level of dissolved salts that often is present in mine waters represents the most intractable water pollution problem connected with coal mining. This is because dissolved salts are not readily susceptible to treatment or removal. In some situations this option is not available. The range of dissolved salts encountered in mine water is variable, with electrical conductivity values up to 335 000 $\mu S\,cm^{-1}$ and chloride levels of 60 000 $mg\,l^{-1}$ being recorded (Woodward and Selby, 1981). Some average values for various coal mining effluents associated with the Nottinghamshire Coalfield, England, are given in Table 8.2.

The principal groups of salts in mine discharge waters are chlorides and sulphates. The former occur in the groundwater lying in the confined aquifers between coal seams in most coalfields in Britain, South Wales being a notable exception. These salts are released into the workings by mining operations. In general, the salinity increases with depth below the surface and with distance from the outcrop or incrop. The concentration of ions in mine waters conform to established ratios that are remarkably consistent

Table 8.2 Average quality characteristics of coal mining effluents in the Nottinghamshire Coalfield, England (From Bell and Kerr, 1993)

Type of effluent	*BOD (ATU) ($mg\,l^{-1}$)*	*Suspended solid ($mg\,l^{-1}$)*	*Chloride ($mg\,l^{-1}$)*	*Electrical conductivy ($\mu S\,cm^{-1}$)*	*Minimum pH*	*Other potential contaminants*
Mine waters	2.1	57	4900	14 000	3.5	Iron barium nickel, aluminium sodium sulphate
Drainage from coal stocking sites	2.6	128	600	2200	2.2	Iron, zinc
Spoil tip drainage	3.1	317	1600	4100	2.7	Iron, zinc
Coal preparation plant discharges	2.1	39	1500	4200	3.2	Oil from flotation chemicals
Slurry lagoon discharges	2.4	493	2000	6100	3.8	

throughout British coalfields. The more saline waters contain significant concentrations of barium, strontium, ammonium and manganese ions. Dissolved sulphates occur only in trace concentrations in waters of confined aquifers.

Mine drainage waters at the point of discharge almost invariably contain sulphates that are either present in the groundwater lying in the more shallow unconfined aquifers, or are generated in the workings by the action of atmospheric oxygen on pyrite. The physical changes such as delamination, bedding plane separation, fault reactivation and fissuring of rock masses caused by mining permit air to penetrate a much larger surface area than the immediate boundaries of the working faces and associated roadways. These changes also alter the hydrogeological conditions within a coalfield and allow wider movement of groundwater through the rock masses than existed prior to mining. These factors mean that groundwater comes in contact with a large surface area of rock, and is exposed to atmospheric oxidation. Hence, the groundwater becomes contaminated. Pyrite may occur in relatively high concentrations, particularly in the upper part of a coal seam or in associated black shales. As noted previously, the primary oxidation products of pyrite are ferrous and ferric sulphates, and sulphuric acid. Sulphates and sulphuric acid react with clay and carbonate minerals to form secondary products including manganese and aluminium sulphates. Further reactions with these minerals and incoming waters give rise to tertiary products such as calcium and magnesium sulphates. Generally, the stratal waters are sufficiently alkaline to ensure that only the tertiary products appear in the discharge at the surface (columns 1, 2, 3 and 6 of Table 8.1). Exceptionally both primary and secondary products may appear in waters from intermediate depths (column 4 of Table 8.1). The primary oxidation products tend to predominate in very shallow workings liable to leaching by meteoric water (column 5 of Table 8.1).

Precipitation of metal hydroxides occurs as acid mine water is neutralized by water into which it flows. Acidic ferruginous mine water also may contain high concentrations of aluminium that precipitates as hydroxide as the pH value rises on entering a receiving body of water, giving a milky appearance to the water. Concentrations of heavy metals may be high in some acid waters and may exert a toxic effect.

The prevention of pollution of groundwater by coal mining effluent is of particular importance. Movement of pollutants through strata is often very slow and is difficult to detect. Hence, effective remedial action is often either impractical or prohibitively expensive. Because of this there are few successful recoveries of polluted aquifers. In coalfields of the east Midlands and south Staffordshire, England, colliery spoil was often tipped on top of the Sherwood Sandstone, which is the second most important aquifer in Britain. Although modern tipping techniques may render spoil impervious, surface water run-off can leach out soluble salts, especially chloride. This may result in the loss of up to 1 tonne of chloride per hectare of exposed spoil heap per annum under average rainfall conditions (spoil heaps may extent to many hundreds of hectares in area). The run-off from these spoil heaps may discharge directly into drainage ditches or to land around the periphery of the heap and infiltrate into the aquifer. Figure 8.1 shows the isopleths for chloride ion concentration in the groundwater of the Sherwood Sandstone in the concealed part of the Nottinghamshire Coalfield in the 1980s. It also indicates the locations of the working collieries at that time and so demonstrates the relationship between elevated chloride ion level in the groundwater and mining activity. Such pollution of an aquifer can be alleviated by lining the beds of influent streams that flow across the aquifer or by providing pipelines to convey mine discharge to less sensitive water courses that do not flow across the aquifer.

Old adits often are unmapped and unknown, and even currently discharging ones are not always immediately evident. An examination of the catchment data is often the only way that such discharges come to light. Discharges from old adits and mine mouths are usually gravity flows. The number of gravitational discharges from abandoned mines in the Fife coalfield, Scotland, are shown in Figure 8.2 and the associated discharges are listed in Table 8.3.

8.2. Acid mine drainage

The term acid mine drainage (AMD) or acid rock drainage (ARD) is used to describe drainage resulting from natural oxidation of sulphide minerals

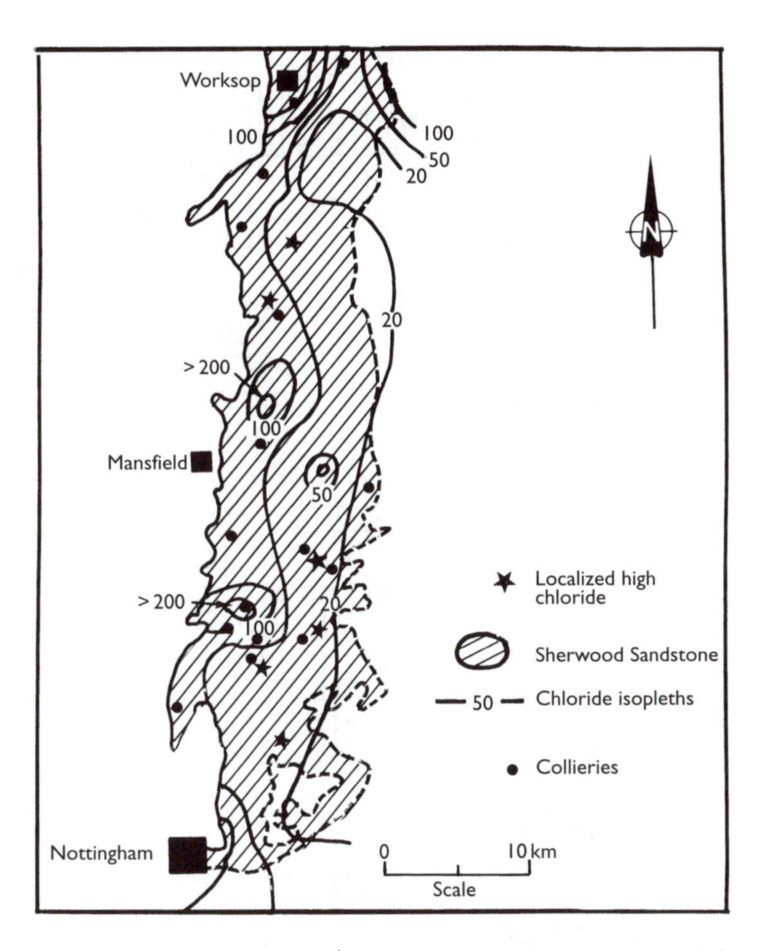

Figure 8.1 Chloride contours (in mg l^{-1}) for groundwater in the Sherwood Sandstone aquifer of the concealed Nottinghamshire Coalfield, England, in the late 1970s/early 1980s. (From Bell and Kerr, 1993.)

that occur in mine rock or waste that are exposed to air and water. This is a consequence of the oxidation of sulphur in the material to a higher state of oxidation and, if aqueous iron is present and unstable, the precipitation of ferric iron with hydroxide occurs. It can be associated with underground workings, with opencast workings, with spoil heaps, with tailings ponds or with mineral stockpiles (Brodie *et al.*, 1989). For instance, Sracek *et al.* (2004) have provided a recent account of the characteristics of acid mine drainage associated with a waste rock pile at Mine Doyon, Quebec.

Acid mine drainage is responsible for problems of water pollution in major coal and metal mining areas around the world. However, it will not occur if the sulphide minerals are non-reactive or if the rock contains

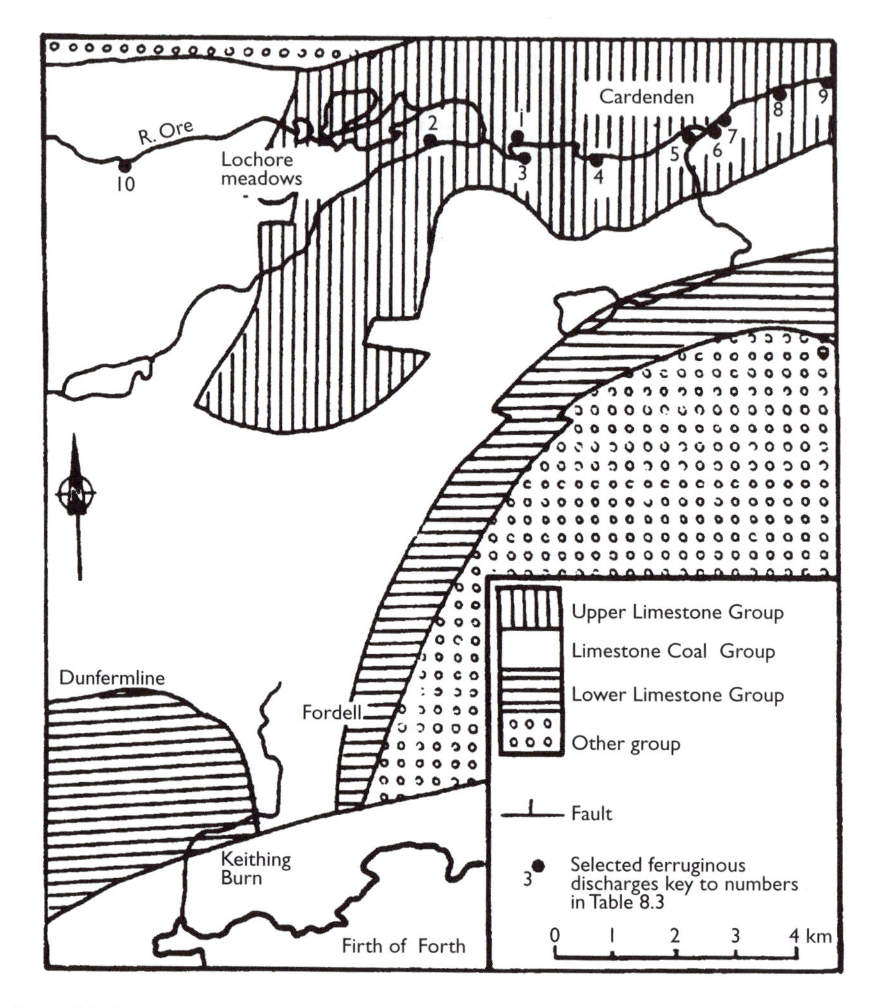

Figure 8.2 Location map of places referred to in Table 8.3 along the River Ore, Fifeshire Coalfield, Scotland, scale in kilometres. (From Bell and Kerr, 1993.)

sufficient alkaline material to neutralize the acidity. In the latter instance, the pH value of the water may be near neutral but it may carry elevated salt loads, especially of calcium sulphate. The character and rate of release of acid mine drainage is influenced by various chemical and biological reactions that occur at the source of acid generation (see later). Moreover, the development of acid mine drainage is time dependent and at some mines may evolve over a period of years. If acid mine drainage is not controlled it can pose a threat to the environment since acid generation can lead to elevated levels of heavy metals and sulphate in the water that obviously have a

Table 8.3 Examples of iron loading along River Ore (From Bell and Kerr, 1993)

Location	*Average flow* ($l\,s^{-1}$)	*Average iron* ($mg\,l^{-1}$)	*Average loading* ($kg\,day^{-1}$)	*Receiving stream*	*Origin*
1 Inchgall	0	—	—	Ore	Coal outcrop
2 Glencraig	1	—	—	Fitty Burn	Fissure?
3 W Colquhally	22	8	15	Fitty Burn	Drainage adit
4 Minto Pit	38	4	13	Ore	Drainage adit
5 Cardenden Gas Works	8	2	1	Den Burn	Coal outcrop
6 Cardenden Sewage Works upstream	4	11	4	Ore	Seepage
7 New Garden No 1	10	13	11	Ore	Pit shaft?
8 New Garden No 2	5	11	5	Ore	Borehole
9 Cluny	16	49	68	Ore	Pit shaft/ borehole
10 Blairenbathie	36	5	15	Ore	Pit shaft

Table 8.4 Examples of acid mine drainage groundwater

pH	*EC*	*NO_3*	*Ca*	*Mg*	*Na*	*K*	*SO_4*	*Cl*	*TDS*
6.30	444	11.8	401	289	418	18.5	2937	26.7	4114
5.9	421	4.9	539	373	202	23.8	3275	73.5	4404
3.0	3130	29.4	493	3376	228	12.7	47 720	86.9	51 946
2.9	2370	64.8	627	621	102	6.9	14 580	33.4	16 039
2.8	1680	19.6	509	561	70	0.7	27 070	53.5	28 284
3.2	857	7.4	614	309	103	12.5	2857	173.8	4078
1.9	471	0.1	174	84	247	7	3250	310	4844
2.4	430	0.1	114	48	326	9	1610	431	2968
2.9	377	0.1		61	278	6	1256	324	2490
2.3	404	0.1	113	49	267	4	2124	353	3364

Note
Except for pH and EC ($mS\,m^{-1}$) all units are expressed in $mg\,l^{-1}$.

detrimental effect on its quality (Table 8.4). This can have a serious impact on the aquatic environment, killing fish and amphibians, as well as vegetation (Fig. 8.3).

Generally, acid mine drainage from underground mines occurs as point discharges (Geldenhuis and Bell, 1997). A major source of acid mine drainage may result from the closure of a mine (Bell *et al.*, 2002). When a mine is abandoned and dewatering by pumping ceases, the water level rebounds and groundwater reoccupies the strata (Younger, 1995). However, the workings often act as drainage systems so that the water does not rise to its former level. Consequently, a residual dewatered zone remains

Figure 8.3 Dying vegetation as a result of acid mine drainage contaminating the ground.

that is subject to continuing oxidation. Groundwater may drain to the surface from old drainage adits, river bank mine mouths, faults, springs and shafts that intercept strata in which water is under artesian pressure. Nonetheless, it may take a number of years before this happens. Burke and Younger (2000), and Dumpleton *et al.* (2001) described modelling methods used to assess such problems in two coalfield areas in England. Mine water quality is determined by the hydrogeological system and the geochemistry of the upper mine levels. Hence, in terms of a working mine it is important that groundwater levels are monitored to estimate rebound potential. In addition, records should be kept of the hydrochemistry of the water throughout the workings, drives and adits so that the potential for acid generation on closure can be assessed.

The large areas of fractured rock exposed in opencast mines or open-pits can give rise to large volumes of acid mine drainage. Even when abandoned, slope deterioration and failure can lead to fresh rock being exposed, allowing the process of acid generation to continue. Where the workings extend beneath the surrounding topography, the pit drainage system leads to the water table being lowered. Increased oxidation can occur in the dewatered zone.

Spoil heaps represent waste generated by the mining operation or waste produced by any associated smelting or beneficiation. As such they vary from waste produced by subsurface mining on the one hand to waste produced by any associated processing on the other. Consequently, the sulphide content of the waste can vary significantly. Acid generation tends to occur in the surface layers of spoil heaps where air and water have access to sulphide minerals.

Tailings deposits that have a high content of sulphide represent another potential source of acid generation (see Chapter 7). However, the low permeability of many tailings deposits together with the fact that they commonly are flooded means that the rate of acid generation and release is limited. Consequently, the generation of acid mine drainage can continue to take place long after a tailings deposit has been abandoned.

Mineral stockpiles may represent a concentrated source of acid mine drainage. Major acid flushes commonly occur during periods of heavy rainfall after long periods of dry weather. Heap-leach operations at metalliferous mines include, for example, cyanide leach for gold recovery and acid leach for base metal recovery. Spent leach heaps can represent sources of acid mine drainage, especially those associated with low pH leachates.

8.3. Acid generation

Certain conditions including the right combination of mineralogy, water and oxygen are necessary for the development of acid mine drainage. Such conditions do not always exist. Consequently, acid mine drainage is not found at all mines with sulphide bearing minerals. The ability of a particular mine rock or waste to generate net acidity depends on the relative content of acid generating minerals and acid consuming or neutralizing minerals. Acid waters produced by sulphide oxidation of mine rock or waste may be neutralized by contact with acid consuming minerals. As a result the water draining from the parent material may have a neutral pH value or negligible acidity despite ongoing sulphide oxidation. If the acid consuming minerals are dissolved, washed out or surrounded by other minerals, then acid generation continues. Where neutralizing carbonate minerals are present, metal hydroxide sludges, such as iron hydroxides and oxyhydroxides are formed. Sulphate concentrations generally are not affected by neutralization unless mineral saturation with respect to gypsum is attained. Hence, sulphate sometimes may be used as an overall indicator of the extent of acid generation after neutralization by acid consuming minerals.

Oxidation of sulphide minerals may give rise to the formation of secondary minerals after some amount of pH neutralization if the pH is maintained near neutral during oxidization (Table 8.5). However, other minerals may form instead of or in addition to those listed in Table 8.5. This will depend upon the extent of oxidation, water chemistry and the presence of other minerals such as aluminosilicates. The secondary minerals that are developed may surround the sulphide minerals and in this way reduce the reaction rate.

The primary chemical factors that determine the rate of acid generation include pH value; temperature; oxygen content of the gas phase if saturation is less than 100%; concentration of oxygen in the water phase; degree of saturation with water; chemical activity of Fe^{3+}; surface area of exposed metal sulphide; and chemical activation energy required to initiate acid generation. In addition, the biotic micro-organism, *Thiobacillus ferrooxidans* may accelerate reaction by its enhancement of the rate of ferrous iron oxidation. It also may accelerate reaction through its enchancement of the rate of reduced sulphur oxidation. *Thiobacillus ferrooxidans* is most active in waters with a pH value around 3.2. If conditions are not favourable, the bacterial influence on acid generation will be minimal.

As remarked, acid mine drainage occurs as a result of the oxidation of sulphide minerals, notably pyrite, contained in either mine rock or waste when this is exposed to air and water. In the case of pyrite, the initial reaction for direct oxidation, either abiotically or by bacterial action, according to Lundgren and Silver (1980) is:

$$2FeS_2 + 2H_2O + 7O_2 \rightarrow 2FeSO_4 + 2H_2SO_4 \quad (8.1)$$

Subsequent biotic and abiotic reactions that lead to the final oxidation of pyrite by ferric ions (indirect oxidation mechanism), can be represented as follows:

$$4FeSO_4 + O_2 + 2H_2SO_4 \rightarrow 2Fe_2(SO_4)_3 + 2H_2O \quad (8.2)$$

$$Fe_2(SO_4)_3 + 6H_2O \rightarrow 2Fe(OH)_3 + 3H_2SO_4 \quad (8.3)$$

$$4Fe^{2+} + O_2 + 4H \rightarrow 4Fe^{3+} + 2H_2O \quad (8.4)$$

$$FeS_2 + 14Fe^{3+} + 8H_2O \rightarrow 15Fe^{2+} + 2SO_4 + 16H \quad (8.5)$$

Reaction (8.1) shows the initiation of pyrite oxidation, either abiotically (auto-oxidation) or biotically. As noted earlier, the biotic micro-organism primarily responsible for acid water pollution from mine waste is *Thiobacillus ferrooxidans*. *Thiobacillus ferrooxidans* converts the ferrous

Table 8.5 Summary of common sulphide minerals and their oxidation products (After Brodie *et al.*, 1989; reproduced by kind permission of Robertson Geoconsultants Inc)

Mineral	*Composition*	*Aqueous end products of complete oxidation*[a]	*Possible secondary minerals formed at neutral pH after complete oxidation and neutralization*[b]
Pyrite	FeS_2	Fe_3, SO_4^2, H	Ferric hydroxides and sulphates; gypsum
Marcasite	FeS_2	Fe_3, SO_4^2, H	Ferric hydroxides and sulphates; gypsum
Pyrrhotite	Fe_3S	Fe_3, SO_4^2, H	Ferric hydroxides and sulphates; gypsum
Smythite, Greigite	Fe_3S_4	Fe_3, SO_4^2, H	Ferric hydroxides and sulphates; gypsum
Mackinawite	FeS	Fe_3, SO_4^2, H	Ferric hydroxides and sulphates; gypsum
Chalcopyrite	$CuFeS_2$	Cu_2, Fe_3, SO_4^2, H	Ferric hydroxides and sulphates; copper hydroxides and carbonates; gypsum
Chalcocite	Cu_2S	Cu^2, SO_4^2, H	Copper hydroxides and carbonates; gypsum
Bornite	Cu_5FeS_4	Cu^2, Fe_3, SO_4^2, H	Ferric hydroxides and sulphates; copper hydroxides and carbonates; gypsum
Arsenopyrite	FeAsS	Fe_3, AsO_4^2, SO_4^2, H	Ferric hydroxides and sulphates; ferric and calcium arsenates; gypsum
Realgar	As_2S_2	AsO_4^2, SO_4^2, H	Ferric and calcium arsenates; gypsum
Orpiment	As_2S_3	AsO_4^2, SO_4^2, H	Ferric and calcium arsenates; gypsum
Tetrahedrite and Tennenite	$Cu_{12}(Sb,As)_4S_{13}$	Cu^2, SbO^3, AsO_4^2, SO_4^2, H	Copper hydroxides and carbonates; calcium and ferric arsenates; antimony materials, gypsum
Molybdenite	MoS_2	MoO_4^2, SO_4^2, H	Ferric hydroxides; sulphates; molybdates; molybdenum oxides; gypsum
Sphalerite	ZnS	Zn^2, SO_4^2, H	Zinc hydroxides and carbonates; gypsum
Galena	PbS	Pb^2, SO_4^2, H	Lead hydroxides, carbonates, and sulphates; gypsum
Cinnabar	HgS	Hg^2, SO_4^2, H	Mercuric hydroxide; gypsum
Cobaltite	CoAsS	Co^2, AsO_4^2, H	Cobalt hydroxides and carbonates; ferric and calcium arsenates; gypsum
Niccolite	NiAs	Ni_2, AsO_4^2, SO_4^2, H	Nickel hydroxides and carbonates; ferric, nickel and calcium arsenates; gypsum
Pentlandite	$(Fe,Ni)_9S_8$	$Fe_3Ni_2SO_4^2H$	Ferric and nickel hydroxides; gypsum

Notes

a Intermediate species such as ferrous iron (Fe^{2+}) and $S_2O_3^{2+}$ may be important.

b Depending on overall water chemistry other minerals may form with or instead of the minerals listed above.

iron of pyrite to its ferric form. The formation of sulphuric acid in the initial oxidation reaction and concomitant decrease in the pH make conditions more favourable for the biotic oxidation of the pyrite by *Thiobacillus ferrooxidans*. The biotic oxidation of pyrite is four times faster than the abiotic reaction at pH 3.0 (Pugh *et al.*, 1984). The presence of *Thiobacillus ferrooxidans* also may accelerate the oxidation of sulphides of antimony, arsenic, cadmium, cobalt, copper, gallium, lead, molybdenium, nickel and zinc.

Hence, the development of acid mine drainage is a complex combination of inorganic and sometimes organic processes and reactions. In order to generate severe acid mine drainage ($pH < 3$) sulphide minerals must create an optimum micro-environment for rapid oxidation and must continue to oxidize long enough to exhaust the neutralization potential of the rock.

8.4. Prediction and control of acid mine drainage

Accurate prediction of acid mine drainage is required in order to determine how to bring it under control. The objective of acid mine drainage control is to satisfy environmental requirements using the most cost effective techniques. The options available for the control of polluted drainage are greater at proposed rather than existing operations as control measures at working mines are limited by site specific and waste disposal conditions. For instance, the control of acid mine drainage that develops as a consequence of mine dewatering is helped by the approach taken towards the site water balance. In other words, the water resource management strategy developed during mine planning will enable mine water discharge to be controlled and treated, prior to release, or to be re-used. The length of time that the control measures require to be effective is a factor that needs to be determined prior to the design of a system to control acid mine drainage.

Prediction of the potential for acid generation involves the collection of available data and carrying out static tests and kinetic tests. A static test determines the balance between potentially acid generating and acid neutralizing minerals in representative samples. One of the frequently used static tests is acid-base accounting. Acid-base accounting allows determination of the proportions of acid generating and neutralizing minerals present. As mentioned earlier, the mineral components considered as acid generating are the sulphides, the most common of which are pyrite and pyrrhotite. Carbonates and hydroxides are regarded as neutralizing minerals. Feldspars may be considered potentially acid consuming minerals but are less effective than calcite. Obviously, a sample will produce net acidity in a period of time if the acid potential (AP) exceeds the neutralizing potential (NP), that is, if the net neutralization potential (NP – AP) is negative. Initially, the pH of a sample paste is measured to assess the natural pH value of the material, as well as to determine if acid generation has occurred prior to analysis.

Generally, if the pH of the paste is less than 5, then there probably has been acid generation in the sample. Next the total sulphur content of the sample is measured and the maximum potential acidity as sulphuric acid is calculated from the sulphur content. Finally, the neutralization potential is found by using a base titration procedure of a pre-acidified sample. However, static tests cannot be used to predict the quality of drainage waters and when acid generation will occur. If potential problems are indicated, the more complex kinetic tests should be used to obtain a better insight of the rate of acid generation. Kinetic tests involve weathering of samples under laboratory or on-site conditions in order to confirm the potential to generate net acidity, determine the rates of acid formation, sulphide oxidation, neutralization, metal dissolution and to test control techniques. The static and kinetic tests provide data that may be used in various models to predict the effect of acid generation and control processes beyond the time frame of kinetic tests. An evaluation of static tests as used to predict the potential for acid drainage generation has been provided by Adam *et al.* (1997).

There are three key strategies in acid mine drainage management, namely, control of the acid generation process, control of acid migration, and collection and treatment of acid mine drainage (Connelly *et al.*, 1995). Control of acid mine drainage may require different approaches, depending on the severity of potential acid generation, the longevity of the source of exposure and the sensitivity of the receiving waters. Mine water treatment systems installed during mine operation may be adequate to cope with both operational and long-term post-closure treatment with little maintenance. On the other hand, in many mineral operations, especially those associated with abandoned workings, the long-term method of treatment may be different from that used while a mine was operational. Hence, there may have to be two stages involved with the design of a system for treatment of acid mine drainage, one for during mine operation and another for after closure (Cambridge, 1995).

Obviously, the best solution is to control acid generation, if possible. Source control of acid mine drainage involves measures to prevent or inhibit oxidation, acid generation or containment leaching. If acid generation is prevented, then there is no risk of the resultant contaminants entering the environment. Such control methods involve the removal or isolation of sulphide material, or the exclusion of water or air. The latter is much more practical and can be achieved by the placement of a cover over acid generating material such as waste or air-sealing adits in mines. Bussiere *et al.* (2004) maintained that covers with capillary barrier effects (CCBE) were considered to be one of the most effective methods of controlling acid mine drainage from mine wastes. They considered the use of CCBE on low sulphide tailings to control drainage from tailings. A series of leaching column tests were undertaken from which Bussiere *et al.* concluded that

CCBEs with a moisture retaining layer can limit the production of acid mine drainage from acid generating tailings beneath, maintaining the leachate pH near neutrality. In addition, Bussiere *et al.* found that these covers also reduced the heavy metal content of acid mine drainage, for example, the amounts of zinc, copper and iron were reduced to less than 1%. Bussiere *et al.* (2003) indicated that the hydraulic behaviour of layered CCBEs is influenced by the inclination of the slope. As the upper part of a slope usually contains less groundwater than the lower part, it is less efficient than the lower part as far as limiting gas migration (i.e. oxygen) is concerned. Gandy and Younger (2003) referred to the use of a clay cover to reduce the availability of oxygen to sulphide minerals in a spoil heap that had been generating acid mine drainage. A monitoring programme was established after placement of the cover and this showed that both the acid generating potential of the spoil and the concentration of contaminants, notably iron and sulphate, in the discharge water decreased. Monitoring programmes of groundwater are important, especially when a mine has been closed (Younger and Robins, 2002).

Migration control is considered when acid generation is occurring and cannot be inhibited. Since water is the transport medium, control relies on the prevention of water entry to the source of acid mine drainage. Water entry may be controlled by diversion of surface water flowing towards the source by drainage ditches; prevention of groundwater flow into the source by interception and isolation of groundwater (this is very difficult to maintain over the long-term); and prevention of infiltration of precipitation into the source by the placement of cover materials but again their long-term integrity is difficult to ensure.

Release control is based on measures to collect and treat acid mine drainage. In some cases, especially at working mines, this is the only practical option available. Collection requires the collection of both ground and surface water polluted by acid mine drainage, and involves the installation of drainage ditches, and collection trenches and wells.

In general, methods of treatment of acid mine drainage fall into two main groups, namely, active and passive treatment systems. Cambridge (1995) pointed out that conventional active treatment of mine water requires the installation of a treatment plant, continuous operation in a closely controlled process and maintenance. Hence, the capital and operational costs of active treatment are high. According to Younger (2000), the classic approach to active treatment of acidic and/or ferruginous mine drainage involves three steps. First, oxidization, usually by a simple cascade, that helps convert soluble ferrous iron to the less soluble ferric state, as well as permitting the pH value to rise by venting carbon dioxide, where present, until equilibration with the atmosphere has been achieved. Second, dosing with alkalis, generally hydrated lime [$Ca(OH)_2$] or less often with caustic soda ($NaOH$), raises the pH value, thereby lowering the solubility of most problematic metals, and providing hydroxyl ions for rapid precipitation of metal hydroxides. Third, accelerated sedimentation normally is brought about by using a clarifier or lamellar plate thickener, frequently aided by the addition of flocculants and/or coagulants. Acid mine water treated with active systems tends to produce a solid residue that has to be disposed of in tailings lagoons. This sludge contains metal hydroxides. However, according to Cambridge, the long-term disposal of such sludge in tailings lagoons is not appropriate. Alternatively, sludges could be placed in hazardous waste landfills but such sites are limited.

Treatment processes have concentrated on neutralization to raise the pH and precipitate metals. Lime or limestone are commonly used to treat acid mine waters, this raises their pH value, although offering only a partial solution to the problem (Zinck and Aube, 2000). Adding lime or limestone also causes iron and other metals to precipitate out of solution. The decant water from the settling process is usually of a high enough quality for it to be discharged directly into surface water courses. However, there are drawbacks associated with liming, first, it may be expensive to maintain due to scaling. Second, the pH needed to remove metals such as manganese may cause the remobilization of other metal hydroxides such as that of aluminium. Third, there may be problems with the long-term disposal of sludge, which is produced in large volume. Other alkalis such as sodium hydroxide may be used, but are more expensive although they do not produce such dense sludges as does lime. The sludges recovered from alkali neutralization, followed by sedimentation and consolidation, are of relatively low density (2–5% dry solids), but drain fairly well. Bodurtha and Brassard (2000) suggested that powdered slag from the steel making industry potentionally could be useful in neutralizing acid mine drainage due to its high content of basic minerals. From their investigation, they noted that only 60% by volume of the slag dissolved but that the process still had the same neutralizing capacity of calcite. Furthermore, Bodurtha and Brassard suggested that the transfer from saturation to pH driven dissolution would be brought about by armouring of the grain surface by the precipitation of iron oxides.

The oxidation of pyrite can be eliminated by flooding a mine with water. However, it obviously is not possible to flood an active mine. It also is not possible to flood abandoned mine workings that are above the water table and that are drained by gravity. Tailings that are acid generating can be covered with water. More sophisticated active treatment methods involve osmosis (waste removal through membranes), electrodialysis (selective ion removal through membranes), ion exchange (ion removal using resin), electrolysis (metal recovery with electrodes) and solvent extraction (removal of specific ions with solvents). A state-of-the-art review of active systems has been provided by Coulton *et al.* (2003).

Passive systems try to minimize input of energy, materials and manpower, and so reduce operational costs. In other words, they should utilize naturally available energy sources such as topographic gradient, microbial metabolic energy, photosynthesis and chemical energy, and need infrequent maintenance to operate successfully (Hedin *et al.*, 1994). Such treatment involves engineering a combination of low maintenance biochemical systems (e.g. anoxic limestone drains, subsurface reactive barriers to treat acidic and metalliferous groundwater, aerobic reed beds and anaerobic wetlands, and rock filters for aerobic treatment of ferruginous waters). A wetland treatment system, in particular, can be integrated aesthetically into the surrounding environment (Fig. 8.4). If not properly managed, however, the generation of ochreous sludge may overwhelm a wetland. This is brought about by the reed beds not being able to transform the dissolved iron to iron sulphides but instead enhancing the oxidization process to precipitate iron as hydroxyferrous sulphates. Passive systems do not produce large volumes of sludge, the metals being precipitated as oxides or sulphides in the substrate materials.

Nuttall and Younger (2000) mentioned that certain difficulties had been met with in Britain when passive systems had been used to treat hard net alkaline mine waters with elevated dissolved zinc. These waters are not amenable to sulphidization in compost based systems like anoxic limestone drains as zinc sulphide does not precipitate readily at neutral pH. They suggested that closed system limestone dissolution, with concomitant precipitation of zinc carbonate minerals offers a means of passive treatment of such waters. This is brought about by using sealed beds of limestone gravel in which zinc is removed as smithsonite ($ZnCO_3$). Such a closed system requires venting after a few hours retention time to restore the partial pressure of carbon dioxide to relatively aggressive values.

Nuttall (2003) described a pump and treat system that was constructed at Whittle Colliery, Northumberland, England, where the system receives groundwater pumped from a purpose sunk drillhole that extends into the drift for the mine. In this way, mine water levels are kept below predicted decant points, which would otherwise give rise to discharges at the surface, and mine water can be treated in a designated setting. The passive system consists of an aeration chamber, two settling lagoons and three aerobic reed beds. Nuttall remarked that the system proved very successful in removing iron. A somewhat similar system using a drillhole to pump groundwater from the Six Bells Colliery in South Wales has been described by Jarvis *et al.* (2003). There two settlement lagoons and a wetland were used to treat the mine water. As most of the iron present in the mine water is in the ferrous state, a hydrogen peroxide dosing system also was installed. This was to ensure effective oxidation of the iron and limit hydrogen sulphide odours around the discharge. The effect of dosing with hydrogen peroxide was to reduce the mean concentration of iron in the effluent to 5 $mg\,l^{-1}$, which will reduce further as the wetland matures. Although acid mine drainage from the former Wheal Jane tin mine in Cornwall, England, is remediated using conventional lime treatment, a pilot scheme has been established that uses

Figure 8.4 Taff Merthyr mine water treatment scheme, South Wales. (a) When Taff Merthyr, Deep Navigation and Trelewis collieries closed in 1990, mine water rebound resulted in discharge from two shafts at Taff Merthyr Colliery causing pollution of a 3 km stretch of the Bargoed Taff, a tributary of the River Taff. The treatment scheme is an aeration and settlement system to facilitate oxidation and settlement of iron. A series of aerobic surface flow reed beds follow the aeration and settlement treatment. There are 16 wetland cells with a fall of 1 m between cells to re-aerate the water by means of cascades. The sludge in the settlement tanks needs to be emptied every five years, the parallel arrangement of the system allowing one tank to continue in operation while the other is emptied. (b) Reed beds following aeration and settlement. (Reproduced by kind permission of the Coal Authority.)

a passive system. According to Hallberg and Johnson (2003), the latter consists of 5 aerobic wetlands, 1 compost reactor (anaerobic cell) and 10 algal ponds or rock filters. The aerobic cells facilitate the removal of iron by oxidation of ferrous iron and hydrolysis of ferric iron, as well as bringing about the co-precipitation of arsenic with ferric iron. Chalcophilic metals (e.g. zinc and cadmium) are removed as sulphides in the compost bioreactor, which also generates alkalinity. The rock filters encourage the removal of manganese by oxygenic photosynthesis increasing the pH of the mine water and thereby promoting oxidation and precipitation of manganese, and by mineralization of dissolved organic carbon compounds in water draining from the compost bioreactor. Furthermore, there is the option to pre-dose the acid mine drainage flowing into the passive system with lime to a predetermined pH or to treat by using an anoxic limestone drain.

A number of novel treatments of acid mine water were reviewed by Davison (1990). These included the use of lime products, fertilizers and bulk organic matter. Calcium carbonate is the most readily available lime product but calcium hydroxide, sodium carbonate and sodium hydroxide are more soluble. Davison noted that when the neutralizing capacities of calcium and sodium based lime products were compared their efficiency was found to depend on the initial pH value and the calcium content of the water being neutralized. Davison went on to discuss the possibility of generating base by the addition of fertilizer in terms of considering base generating reactions associated with the decomposition of organic carbon. He pointed out that theoretically the addition of phosphate fertilizer means that each mole of phosphate added generates 94 moles of base so that practically 10–30 moles should be achievable. The production of base is governed by what proportion of organic material that is formed undergoes aerobic or anaerobic decomposition. Even if organic material is not a particularly efficient neutralizing agent it often is freely available, which could make its use a viable option. For example, Davison suggested that sewage sludge could be used to treat acid waters. Subsequently, Younger and Rose (2000) also mentioned the use of sewage sludge as a means of treating acidic mine waters. Sulphates are removed from the mine water by bacterial sulphate reduction.

Choi and West (1995) referred to the use of phosphate [francolite, $Ca_{10-x-y}Na_xMg_x(PO4)_{6-2}(CO_3)_zF_{0.4z}F_2$] pebbles as a possible method of treating acid mine drainage at Friar Tuck abandoned mine in south-west Indiana. Laboratory tests showed that, depending on flow rate, phosphate pebbles removed up to 1200 $mg\,l^{-1}$ of ferric iron, 8600 $mg\,l^{-1}$ of sulphate and 800 $mg\,l^{-1}$ of aluminium in three weeks. Their removal is inversely proportional to the flow rate and the flow rates concerned varied from 1.17×10^{-4} to 1.05×10^{-3} litres per minute per kilogram of phosphate pebble. In addition, the pH value of the acid mine drainage increased from around 2.1 and 2.2 to 3.2. This was due to the carbonate present in the francolite. The Ca ion reacts with the sulphate in the mine water, with gypsum being precipitated. Choi and West maintained that if a phosphate pebble system is to work successfully, then the ratio of Fe^{3+}/Fe_{total} should be high. Hence, aeration ponds are required to oxidize ferrous to ferric iron. They also indicated that the particle size is important, recommending that the pebble size should be 50 mm and that it should be clean. A smaller size will stop reacting with the mine water as it will become plugged with gypsum precipitate.

Natural zeolites have been used to remove heavy metals from acid mine drainage. The filtering abilities of zeolites offer a versatile and environmental friendly option to capture most contaminants found in water systems. Natural zeolites can perform these functions due to their high ion exchange capacity, adsorption–desorption energies and ability to be modified. For instance, when dry the zeolites adsorb relatively high volumes of liquid (e.g. 100 g of zeolite absorbs 70 g of liquid). Zeolites have an open regular crystalline framework that generates an electric field that interacts, attracts and binds various cations and, after modification, anions. They can remove ammonium (NH_4^+) and metal cations such as Pb, Cu, Cd, Zn, Co, Cr, Mn and Fe from acid mine drainage, with the recovery of ammonium and some metals (Pb, Cu) being as high as 97%. Lindsay *et al.* (2002) described the use of zeolites as a remediation option for acid mine drainage from mines on the west coast of South Island, New Zealand. The primary aim at these sites was to reduce the Fe content entering the streams from mine drainage and raise the pH to natural background levels of between 4 and 5.

Another relatively new and promising method of dealing with acid mine drainage, according to Fytas and Bousquet (2002), is to coat the sulphide minerals with silicates. Pyrite is treated with a solution containing H_2O_2, sodium silicate and a buffering agent. The H_2O_2 oxidizes a small amount of the pyrite thereby producing ions of ferric iron. These ions subsequently react with the silicate ions to form ferric hydroxide-silica, which is precipitated on the surface of the pyrite to give a protective coating that prevents oxidation.

Due to the impact on the environment of acid mine drainage regular monitoring is required. The major objectives of a monitoring programme developed for acid mine drainage are, first, to detect the onset of acid generation before acid mine drainage develops to the stage where environmental impact occurs. If required, control measures should be put in place as quickly as possible. Second, it is necessary to monitor the effectiveness of the prevention-control-treatment techniques and to detect whether the techniques are unsuccessful at the earliest possible time.

Banks *et al.* (2002) noted that it is not unusual to find water draining from pyrite rich mines or wastes that are circum-neutral or even alkaline. These waters may have modest loadings of heavy metals. This may be explained as a consequence of the mine water not coming into contact with sulphide minerals, possibly because of the low content of sulphide minerals or them being coated with non-acidic material; of large pyrite grain size limiting the oxidation rate; of oxygen not coming into direct contact with sulphide minerals or influent water being highly reducing; or of subsequent neutralization of acidic minerals by other minerals or highly alkaline waters. In fact, after an investigation of discharges from coal mines in Britain, Rees *et al.* (2002) recognized a number of different types of discharge in terms of their chemistry, which was influenced by the hydrogeological conditions. They noted that drainage from flooded workings and pumped discharges were net alkaline whilst drainage from flooded and free-draining workings are either moderately net alkaline or net acidic. Drainage from spoil heaps usually had a pH value below 5 and net alkalinity values as low as $-2500\ \mathrm{mg\,l^{-1}}$ $CaCO_3$. Rees *et al.* further commented that iron was the major contaminant of concern but that many mine waters contained less than $30\ \mathrm{mg\,l^{-1}}$ and that iron to sulphate ratios were less than unity. Dey *et al.* (2003) contended that the best practice for the passive treatment of net alkaline ferruginous mine waters in Britain usually involves pre-treatment, through cascade aeration and settlement, followed by treatment in wetlands. The aim is to remove 30–50% of the iron in the settlement lagoons prior to the mine water entering the wetlands, hence allowing more effective sludge management and so prolonging the life of the wetland. Dey *et al.* also mentioned that an alternative for the pre-treatment stage could be to use autocatalytic oxidation and ochre accretion within an ochre bed without any supporting media. They suggested that such a system possibly could remove 90% of the dissolved iron from ferruginous mine water.

8.5. Case history 1

A tin mine in the North-West Province, South Africa, was forced to cease production because of the decline in the price of the metal (Bullock and Bell, 1995). However, before a mine in South Africa can close officially, a certificate of closure must be granted to the mine by the Department of Mineral and Energy Affairs. Section 2.11 of the Minerals Act (August, 1991) also states that mines that are to close, must implement an environmental management programme. The programme should include remedial measures to ensure that no pollution will occur when the mine is closed. Accordingly, a ground and surface water investigation was undertaken at the mine as part of the required environmental management programme. The objectives of the investigation were, first, to assess whether pollution of ground and surface water had occurred as a result of mining, second, to identify any sources of pollution and, third, to offer solutions to any problem of pollution found to exist.

The tin deposits are hosted by rocks of the Rooiberg Fragment, which is found within the western lobe of the Bushveld Complex. The local geology consists of the Leeuwpoort Formation, above which occurs the Smelterskop

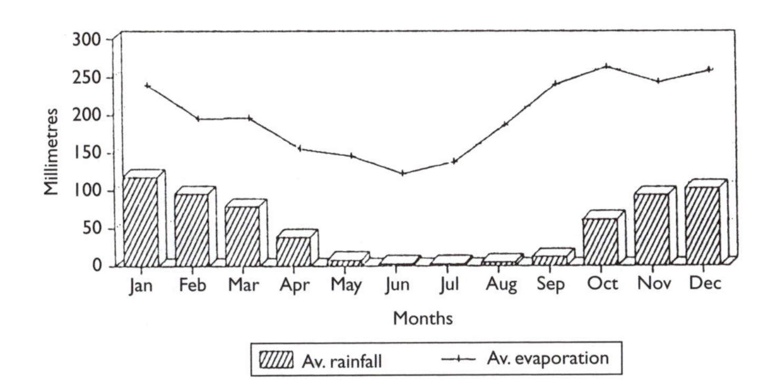

Figure 8.5 Average rainfall and evaporation in the region.

Quartzite Formation. The Leeuwpoort Formation consists of the Boschoffsberg Quartzite Member overlain by the Blaauwbank Shale Member. The tin occurs in Boschoffsberg Quartzite Member. The soils in the area are mostly residual, predominantly comprising sandy loams with lesser quantities of silty loams and silty clay loams.

The area is one of water shortage. The mean annual precipitation in the area is 620 mm with the average monthly rainfall varying from about 3–118 mm. The rainy season lasts from October to May, with most rain falling in January. Mean temperatures in the area reach a maximum in December–January and a minimum in June–July. For example, the maximum mean temperature in January is around 29°C–31°C, the minimum falling to 17°C. In July the maximum mean temperature is around 22°C, the minimum being between 3°C and 6°C. The amount of evapotranspiration affects the amount of surface run-off and groundwater recharge. It exceeds the precipitation every month of the year (Fig. 8.5). The potential water loss reaches a maximum during August, September and October. The water courses and watersheds in the area are shown on Figure 8.6. As the area is situated in a semi-arid region run-off is limited to very short periods of heavy rainfall. In order to catch surface run-off, two reservoirs were constructed in the mine lease area and are known as the Top and Bottom Dams.

8.5.1. Geohydrology and hydrochemistry

Almost all the groundwater in the region occurs in secondary aquifers associated with the Boschoffsberg Quartzite Member. These secondary aquifers consist of weathered and fractured rock, and lie directly beneath the soil surface. The occurrence of groundwater is controlled particularly by structural geology in that drillholes with moderate yields are associated with dykes, which have fractured the Boschoffsberg rocks. In fact, a series

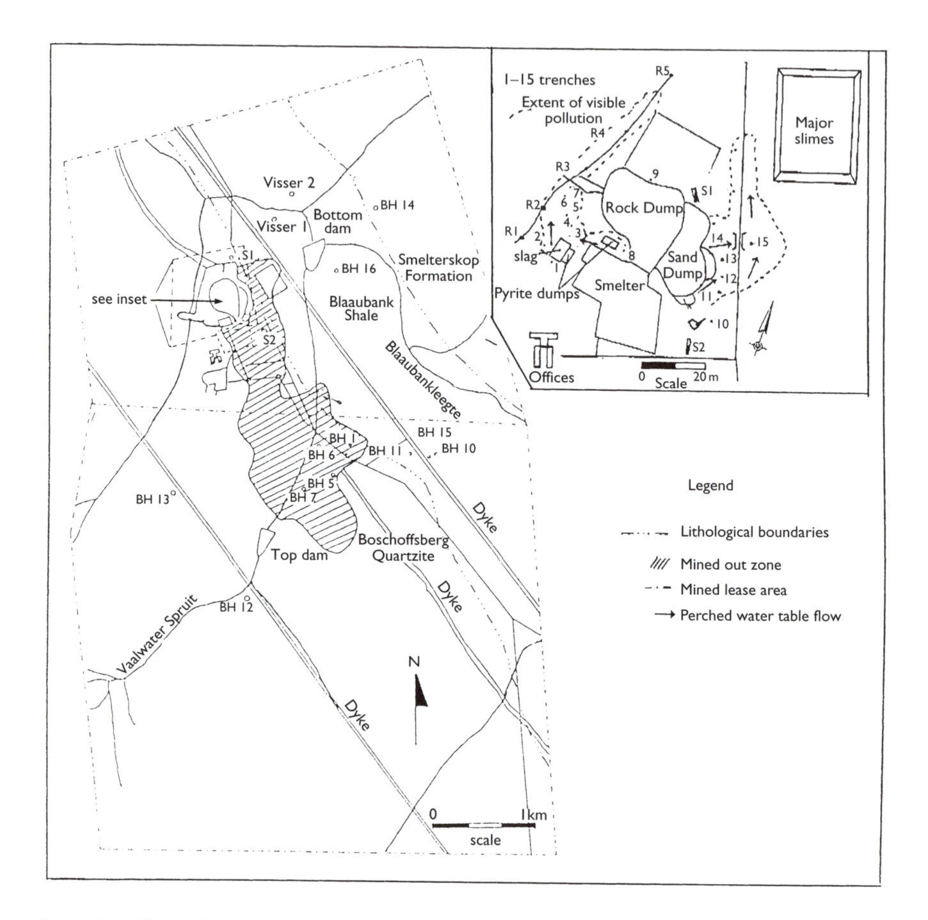

Figure 8.6 The mine lease area showing location of major dumps, boreholes, sampling points and trenches.

of cross-cutting dykes compartmentalizes the groundwater. Therefore, groundwater moves along fractures until it intersects a dyke where is it impounded. Lateral movement of groundwater then takes place along the dykes, flowing from south-east to north-west. The quantity of groundwater stored in the secondary aquifers is limited, while the permeability of such aquifers is generally low. Groundwater represents the primary supply of potable water for both domestic purposes and stock watering. Any possibility of pollution affecting these limited groundwater resources represents a significant threat to the community.

A perched water table exists around the major dumps at the mine and is impounded by a layer of ferricrete in the soil horizon. Water movement in the perched water table was determined by digging a series of exploratory trenches around the mine and observing from which direction water seeped into the trenches (Fig. 8.6). Movement of this water is topographically controlled and eventually finds its way into the Boschoffsberg Quartzite aquifer system. Although the quality of the water in the perched water table is poor, attenuation processes improve the quality as it percolates into the aquifer system proper. During the rainy season the sand and rock dumps at the mine site act as 'sponges' and water slowly percolates down through the dumps (Fig. 8.6). The sand and rock dumps, however, are underlain by impermeable tailings. Hence, the interface between the tailings (slimes) and dumps acts as a substrate along which water flows and slowly seeps into the surrounding soil horizon.

Extensive underground mine workings (hatched area on Fig. 8.6) further complicated the prediction of groundwater movement in the area. These workings were supplied by natural groundwater recharge. The underground tunnels overrode any structural control on groundwater movement in the vicinity of the mine. As water continues to flow into the mine workings, it will exert some influence on the groundwater regime until the workings are flooded.

In order to ascertain whether pollution of ground and surface water was occurring, samples of water and soil were taken for analysis. Samples also were taken from the dumps around the mine for analysis. Groundwater from all the boreholes on the mine lease area was sampled, as well as from the boreholes of local farmers. Groundwater samples were taken after a borehole had been pumped for a period so as to ensure that the groundwater sampled was actually from the aquifer and not groundwater that had been lying stagnant in the borehole. Further groundwater samples were obtained from the underground workings. As mentioned earlier, a series of trenches were excavated around the mine area and a number of these trenches contained groundwater that also was sampled. Surface water samples were taken from the reservoirs and streams. In the case of the latter, samples could only be taken after thunder showers as water flow in the streams only lasted a couple of hours after a rain storm. The soil samples were obtained from the trenches. They were all taken from a depth of 300 mm from the top of the trench. A number of sediment samples also were taken from the stream to the west of the smelter (Fig. 8.6).

Samples from the major dumps were taken from a depth greater than 500 mm below the surface. This was considered a suitable depth below which oxidation and leaching had not occurred. All the samples were taken from the upper parts of the dumps as both leaching and oxidation were more significant on the sides.

Table 8.6 lists the results obtained from chemical analyses carried out on surface water in the mine lease area. The high sulphate concentration and low pH in some of the surface samples was attributable to acid mine drainage.

Table 8.6 Analysis of surface water (After Bullock and Bell, 1995)

	Stagnant water	*River water*	*Bottom dam*	*Vaalwater*	*Top dam*	*South African limits (Anon., 1993)*	
						1	2
pH	2.4	4.1	7.4	7.1	7.3	6.0–9.0	5.5–9.5
COD	1410	38	—	83	—	*	*
EC (mS m^{-1})	865	57.2	65	82	138	70	300
Total hardness	32 409	219	249	523	663	300	650
Total N	38.4	2.5	<0.2	0.2	<0.2	6	10
Ca	163	64	82	66	93	150	200
Mg	74	12	32	30	22	70	100
Na	4	11	65	51	97	100	400
K	<1	9	8	23	10	200	400
SO_4	11 189	236	174	119	57	250	600
Cl	18	6	42	34	42	200	600
F	<0.1	<0.1	1.5	0.7	8.37	1.0	1.5
Fe	3580	2.4	—	0.2	—	0.1	1

Notes
All units expressed as mg l^{-1}, except pH and where stated.
* Chemical oxygen demand (COD) – general standards should not exceed 75 mg l^{-1} after applying the chloride correction.
1 Maximum limit for no risk.
2 Maximum permissible limit for insignificant risk.

Although the Piper diagram (Fig. 8.7) generally is used to classify groundwaters with different chemical compositions, it also was used to evaluate the surface waters in the mine area. This was to allow comparison of the chemical characteristics of the surface water with those of the groundwater. According to the classification introduced by Piper (1944), in all the water samples the alkaline earths exceeded the alkalies and the strong acids exceeded the weak acids. The fact that the strong acids exceeded the weak acids indicates that some surface water had been affected by acid mine drainage, presumably from the dumps. Non-carbonate hardness exceeded 50% in the stagnant water and the river water samples. In the water samples from the Vaalwater, Bottom Dam and Top Dam, not one of the cation-anion pairs exceeded 50%.

8.5.2. Dumps as sources of pollution

Analysis of surface water indicated that it had been affected by acid mine drainage. First, the material in the waste dumps was examined to determine whether they acted as sources of the pollution. This waste material had

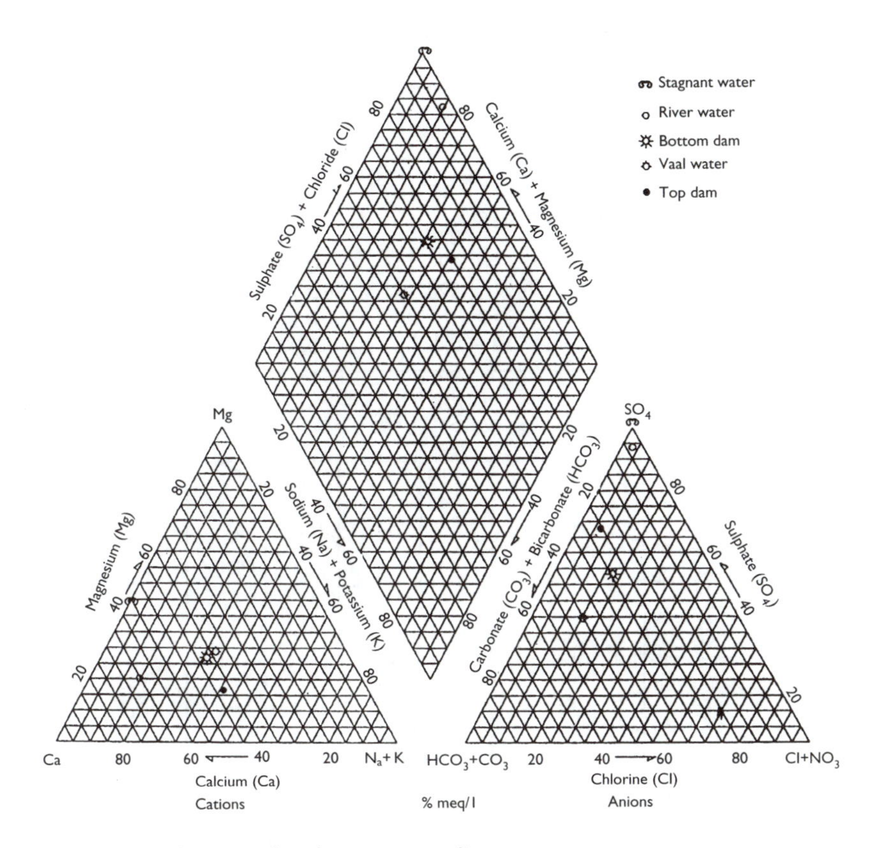

Figure 8.7 Piper diagram of surface water quality.

been disposed of in an area that covers approximately 1.2 km^2. The dumps included pyrite dumps, slimes dumps (tailings impoundments), a slag dump, a sand dump, and a rock dump.

An estimated 8125 tonnes of pyrite material had been deposited in pyrite dumps. An elemental analysis of material from the major pyrite dump showed that sulphur accounted for some 45% and iron for nearly 36% (Table 8.7). The particle size distribution of the material is illustrated in Figure 8.8. Because of the fine particulate nature of the material, the individual particles presented a very large surface area in relation to their volume that made them susceptible to leaching and other chemical reactions. The outer surfaces of the pyrite dumps were coated to a varying extent with a crust of sulphur material that was 20–70 mm in thickness. An aerobic zone could be distinguished from an anaerobic zone in this crust. The yellow and ochre coloured aerobic zone contained a greater than average concentration of iron and sulphur because of capillary action, which brought dilute salt

Table 8.7 Metals in the pyrite dumps

Element	%
S	45.14
Fe	35.57
Cu	2.20
Sn	1.02
Co	0.49
Ni	0.134
Pb	0.2
As	0.14
Sb	0.02

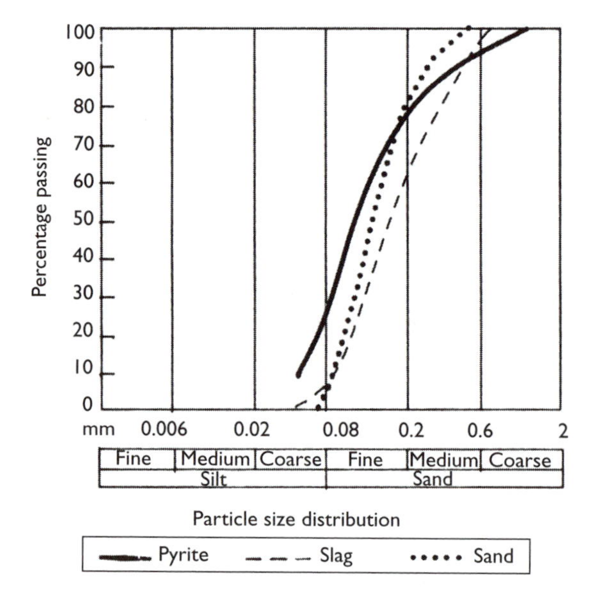

Figure 8.8 Particle size distribution of material in the pyrite, slag and sand dumps.

solutions from the interior of a dump to the surface. Subsequent loss of moisture at the surface due to evaporation resulted in a build-up of metallic salts. The surface temperature of the dumps was in excess of 40°C during the day in the summer months. The moisture content, 200 mm below the surface of the dumps, was approximately 12%. A paste pH value of 3.2 was obtained for the pyrite material. Brodie *et al.* (1989) maintained that a paste pH below 4 means the oxidation of sulphide minerals has occurred.

According to Halbert *et al.* (1983), the generation of sulphate with bacteria, at a pH of 3.0 and a temperature of 21°C is 1.17 moles kg^{-1} tailings per month. Taking into consideration the small particle size of this waste material, that the pH value of the paste is 3.2, that the surface temperature of the pyrite dumps exceeded 40°C in the summer months and 21°C in the winter, and the fact that *Thiobacillus ferrooxidans* cultures were present in the crust on the dumps, production rates of sulphate exceeded 1.17 moles kg^{-1} tailings per month. In fact, Halbert *et al.* maintained that the rate of generation of sulphate increased approximately threefold for every 10°C increase in temperature. Therefore, assuming an average annual surface temperature of 30°C, the sulphate generation rate on the pyrite dumps was estimated around 3.68 moles kg^{-1} tailings per month. This implies that in one month 353.5 mg of sulphate will form from 1 kg of waste. This rate of sulphate generation, however, probably was never reached because the yellow crust on the dumps inhibited the access of oxygen and so reduced the rate of oxidation. Nevertheless, after rain had fallen the yellow coating partly dissolves, although within two days the crust could reform, paying testimony to the initial rapid oxidation rates.

Further proof of the high generation rates of sulphate was obtained from the results of acid-base account tests on the pyrite material. The results gave an acid potential of 1410.625 g $CaCO_3$/kg, a neutralizing potential of −40.6 g $CaCO_3$/kg, and a net neutralizing potential of −1451.7 g $CaCO_3$/kg. According to Brodie *et al.* (1989), samples with a negative net neutralization potential and a ratio of neutralizing potential to acid potential of less than 1:1 have a high potential for acid generation. The acid-base accounting results therefore clearly demonstrated that the pyrite dump material had an extremely high potential to produce acidic waters.

The soil samples were scanned for a number of elements using an inductively coupled plasma mass spectrometer (I.C.P-MS). The results are given in Table 8.8. In addition, an attempt was made to ascertain the depth to which acidic waters had penetrated the substrate below the major pyrite dump. This was done by digging a trench to a depth of 2.5 m with a backactor. Soil samples were taken at different depths within the trench. The results of the analyses are given in Table 8.9 and suggest that acidic waters had penetrated to a depth greater than 2.5 m. There appeared to be a relationship between the pH value and the iron percentage, that is, as the iron content increased the pH decreased.

Metal ions and toxic salts are brought into solution more readily in acid conditions ($pH < 5.5$). Therefore, as acidic surface waters moved through the pyrite dumps they absorbed heavy metals and toxic salts. Once the waters infiltrated into the soil, however, the pH of the water increased as it came in contact with cations adsorbed on to clay minerals. For example, a predominance of cations such as Ca, Mg, Na and K tends to raise the pH when they are released from clay. As the water lost acidity, heavy metals and toxic salts were precipitated in the soil. The relatively high heavy metal concentrations observed in the soil accumulated in this fashion (Table 8.8).

Table 8.8 Semi-quantitative analysis of soil samples (After Bullock and Bell, 1995)

Element	*T1*	*T2*	*T3*	*T4*	*T5*	*T6*	*T7*	*T15*	*Sand*	*Slag*	*R1*	*R2*	*R3*	*R4*	*R5*
pH	4.36	7.78	5.04	4.62	2.9	7.1	3.73	3.04	8.24	8.05	6.54	6.25	2.84	4.37	6.67
EC	152	120	182	360	469	342	314	31 000	47	2000	35	2210	1228	279	47
Ca	871	95	8011	4424	3227	3982	4792	147	428	3836	141	1666	496	3997	374
Mg	378	25	434	1846	473	1577	894	47 296	130	19 419	50	2176	9949	643	122
Al	113	46	0	85	1399	2	860	30 782	0	8	8	17	15 829	167	8
Mn	98	0	91	421	345	46	262	46 202	6	14	4	380	759	89	8
Fe	15	3	2	10	454	9	10	24 731	1	16	1	19	3790	24	2
Co	20	0	3	43	91	1	67	3916	0	1	0	285	823	39	0
Ni	9	0	1	8	11	1	28	484	0	0	0	37	23	11	0
Cu^{2+}	178	1	0	35	450	0	225	14 820	0	5	0	12	3924	34	2
Cu^{3+}	179	1	0	35	440	0	231	15 192	0	5	0	12	3944	35	1
Zn	4	0	0	7	4	0	3	226	0	0	0	3	16	2	0

Notes

All values except pH and electrical conductivity ($mS\ m^{-1}$) are in $mg\ kg^{-1}$.

T1 etc. = trench samples; R1 etc. = stream samples.

Table 8.9 Results of soil analysis from a trench dug below the major pyrite dump

Depth (m)	*pH*	*Fe (%)*	*S (%)*
0.11	5.4	4.41	0.19
0.22	5.2	4.47	1.30
0.36	6.0	3.58	0.35
1.26	4.1	6.65	0.66
2.2	4.6	6.31	0.49

Samples also were taken of the sediment in the stream running alongside the dumps (Fig. 8.6; Table 8.8). Sediment sample R5 was taken from the stream bed just as it exited the mine lease area. Analysis of this sample suggested that pollution in the stream had not yet spread beyond the mine lease area.

The tailings or slimes material in the impoundments consisted of very fine rock powder ($<6\ \mu m$) that was left behind after crushing, milling and severance. Samples of this material analysed by X-ray fluorescence (XRF) gave the following results:

• Major elements (>1%)	Silica	Tin
	Aluminium	Iron
	Sodium	Calcium
	Potassium	Magnesium
• Minor elements (0.1–1%)	Rubidium	Zirconium
	Titanium	Phosphorus
• Trace elements (<0.1%)	Manganese	Chromium
	Sulphur	Barium
	Copper	Lead
	Cadmium	Strontium

The surface temperature of the old tailings impoundments varied, to some extent depending on the amount of vegetation cover, however, in summer it ranged from 20°C to 30°C. The moisture content of the old tailings also varied somewhat, with older impoundments showing a lower moisture content (3%) than the most recent impoundment (5%). The paste pH of the material was 7.8.

The results of acid-base accounting carried out on tailings material showed an acid potential of 5.0 g $CaCO_3$/kg, a neutralizing potential of 40 g $CaCO_3$/kg and a net neutralizing potential of 35 g $CaCO_3$/kg. Hence, the tailings material had a potentially acid consuming character and would not contribute to the acid mine drainage problem at the tin mine (Fig. 8.9). The fact that vegetation was well established on the flanks and upper surfaces of the tailings impoundments was indicative of their low pollution potential.

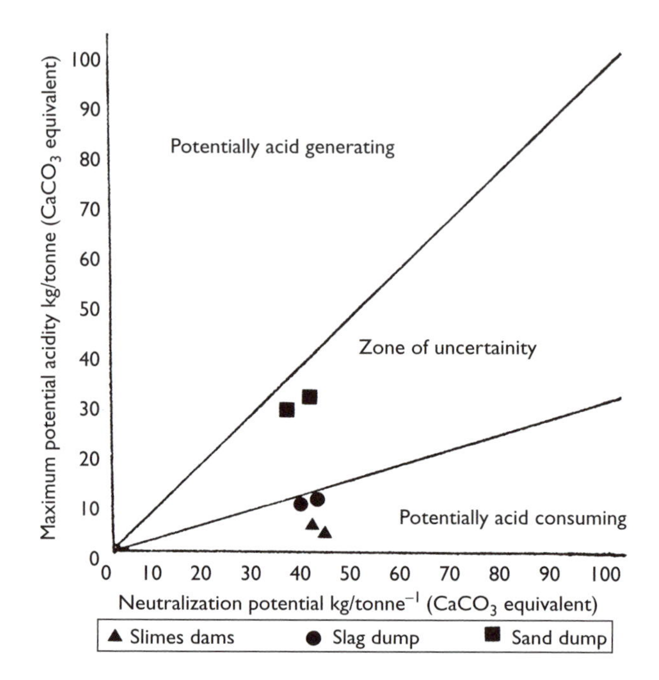

Figure 8.9 Acid-base account for slimes dams, and slag and sand dumps.

Table 8.10 Metals in the slag material

Element	%
Cu	0.12
As	0.009
Ni	0.008
Pb	0.01
Zn	52 ppm
S	0.35
Fe	3.63
Mg	5.01

The slag dump material consisted of crushed slag, char, ferrosilicon, limestone and fluorspar that remained after the secondary melt. An estimated 5000 tonnes of material existed in the dump. An analysis of the metals present in the slag dump is shown in Table 8.10 and a particle size distribution of the material is provided in Figure 8.8. The surface temperature of this dump exceeded 40°C during the summer months. The moisture content of the dump was not determined as active waterborne dumping was still in progress. However, the natural soil surrounding this dump had a moisture content of 15%. Salts also had precipitated on the surface of the soil. The paste pH of the material was 8.1. Grasses had started to inhabit the lower flanks of the dump but the upper surface remained barren.

The slag dump was not isolated from the natural drainage by cut-off berms or paddocks and therefore during heavy summer rains material was washed from the dump onto the surrounding flat land. If this material were to come into contact with acidic waters from the pyrite dumps, then the heavy metals present could be taken into solution. A small pyrite dump, east of slag dump, produced such acidic waters.

Trenches were dug in the soil around the slag dump (Fig. 8.6). Groundwater seeped into some of the trenches immediately but the water level tended to stabilize within 24 hours. Soil and water samples were taken from the trenches and the results are given in Tables 8.8 and 8.11, respectively.

The results of acid-base accounting tests done on the slag material showed an acid potential of 10 g $CaCO_3$/kg, a neutralizing potential of 40 g $CaCO_3$/kg and a net neutralizing potential of 30 g $CaCO_3$/kg. Consequently, the slag material had a potentially acid consuming character (Fig. 8.9). This is not necessarily surprising as sulphides had been extracted prior to dumping the slag.

The sand dump material was derived from the ore passing through a dense media separator. The separation was gravitational and therefore removed most of the sulphides along with the ore so that they were not disposed of with the sand material. An estimated 6000 tonnes of the sand dump material occurred in the area of the mine lease. This material had a dominant sand size fraction and consequently a high permeability (Fig. 8.8). The moisture content of the sand dump material was low, averaging 3%. The paste pH of the material was 8.2. Again the surface of the dump in the summer months was in excess of 40°C. The concentration of the major elements in this dump are given in Table 8.8.

Acid-base accounting tests carried out on the sand material indicated that the acid potential was 30 g $CaCO_3$/kg, the neutralizing potential was 38 g $CaCO_3$/kg and the net neutralizing potential was 8 g $CaCO_3$/kg. Accordingly, the sand material fell into the zone of 'uncertainty' (Fig. 8.9), that is, the material could either be acid consuming or acid generating depending on the local conditions.

The particles in the rock dump varied in size from 20 to 30 mm up to 300 mm. A mine rock classification developed by Brodie *et al.* (1989) was used to determine whether acidic waters seeped from this dump (Table 8.12). This classification uses visual, physical and geochemical characteristics to classify mine rock, in particular to identify relatively homogeneous rock units with respect to the quality of drainage water. The classification is based on six key properties, namely, particle size, sulphide type, sulphide surface exposure, alkali type, alkali surface exposure and slaking characteristics. The relative importance of each of these properties is taken into account by a weighting system. Since the neutralizing capacity of the alkaline material present prevents acid generation, these weighting factors are negative. Admittedly, the method does have shortcomings because it does not take account of a number of factors. Nevertheless, it does provide a relatively rapid means of assessment. The highest AMD or ARD potential value obtainable in the classification is 68 and the lowest ARD potential is −20. The rock dump material had a value of 2 and therefore probably would not produce any acidic seepage.

8.5.3. Chemistry of the groundwater

It can be seen from Table 8.11 that some of the groundwater samples had extremely high concentrations of sulphate. As with the surface water, these high values in the groundwater were associated with the acid mine drainage. Acidic waters, however, need not originate only at the surface. Oxidation of sulphides in old underground workings also can result in the formation of acidic water. The pH of this water, however, showed no evidence of acidity (see S1 and S2 in Table 8.11), as any acidic waters that did form were quickly neutralized by ankerite [$Ca\,(Mg, Fe^{+2}, Mn)(CO_3)_2$], which was present in relative abundance in the host rock. Therefore, the signs of mine water degradation were higher sulphate concentrations, higher electrical conductivity and total dissolved solids (TDS) values.

Figure 8.10(a) represents a Piper plot of water samples taken from farm and mine boreholes that appeared to be unpolluted. The alkaline earths exceeded the alkalies in these samples and, with the exception of those samples from BH12, Blaauwbank and Sleepwa, the weak acids exceeded the strong acids. In all samples the secondary alkalinity exceeded 50%, so that the chemical properties of the water were dominated by alkaline earths and weak acids. This can be compared with Figure 8.10(b) on which poor quality water samples from trenches and underground workings are plotted. In these samples the alkaline earths exceeded the alkalies, the strong acids exceeded the weak acids and the secondary salinity (non-carbonate hardness) exceeded 50%.

Piper (1944) maintained that in order to show that a certain water is a mixture of two others, one graphic criterion and one graphic-algebraic criterion must be satisfied. As far as the graphic criterion is concerned, in all three fields of the Piper diagram the apparent mixture must occur on straight lines between the points of its two inferred components. In addition, the area concentration of plots in the central field must conform to the principle that the concentration of a mixture is necessarily greater than the least, but less than the greatest, of the several concentrations of its

Table 8.11 Results of groundwater analysis

Sample	*pH*	*EC*	NO_3	*Ca*	*Mg*	*Na*	*K*	SO_4	*Cl*	HCO_3	CO_3	*TDS*	*Hardness*	*Alkaline*
T2	7.43	187	3.9	17.1	14.3	364	30.6	427	121.0	372.1	3	1167	102	310
T3	6.30	444	11.8	401	289	418	18.5	2937	26.7	24.4	0	4114	2171	20
T4	5.9	421	4.9	539	373	202	23.8	3275	73.5	18.3	0	4404	2866	15
T11	3.01	3130	29.4	493	3376	228	12.7	47 720	86.9	0.0	0	51 946	15 123	0
T12	2.9	2370	64.8	627	621	102	6.9	14 580	33.4	0.0	0	16 039	4121	0
T13	2.8	1680	19.6	509	561	70	0.7	27 070	53.5	0.0	0	28 284	3579	0
T15	3.2	857	7.4	614	309	103	12.5	2857	173.8	0.0	0	4078	2807	0
S1	7.54	497	4.7	475	430.8	367	20.1	3463	25.4	280.6	0	4933	2729	240
S2	7.74	190	1.0	153	123.5	131	5.2	805.5	62.2	353.8	9	1467	601	305
BH1	7.99	129	42.9	79	74.2	54	19.1	405.6	63.6	85.4	0	781	434	70
BH5	8.15	165	61.4	106	85.7	92	8.8	581.2	48.6	112.0	0	1057	520	103
BH6	7.73	182	57.5	129	99.5	85	10.0	683.4	49.4	122.0	0	1175	632	100
BH7	8.07	172	87.3	102	88.9	101	7.4	522.7	59.4	201.3	0	1070	457	165
BH10	7.2	69	0.2	70	3.1	69	9.0	12.0	4.3	374.4	0	448	317	375
BH11	7.3	71	0.4	64	3.2	68	6.0	15.0	12.6	374.3	0	461	334	375
BH12	7.8	112	< 0.2	76	3.7	138	4.0	263.0	12.3	296.2	0	728	452	298
BH13	7.2	58	< 0.2	53	4.5	54	3.0	62.0	17.4	263.4	0	377	226	265
BH14	7.1	51	0.4	59	3.4	20	3.0	7.0	10.3	308.6	0	332	286	309
BH15	7.5	75	0.2	75	3.3	91	9.0	7.0	4.6	407.0	0	488	314	409
BH16	7.3	87	0.3	102	3.7	54	17.0	148.0	24.5	297.4	0	566	452	298
Visser 1	7.65	81	2.6	22.8	57	71.7	2.1	23.0	6.4	536.8	0	435	439	440
Visser 2	7.58	78	1.1	55.7	85.5	19.5	3.3	4.1	4.6	610	0	478	490	500
Sleepwa	8.57	51	2.3	44.9	34.4	19.7	2.0	76.5	31.9	122.0	6	279	154	110
Knoppieskraal	7.74	45	20.8	32	23.6	37.6	2.4	5.6	18.5	189.1	0	235	22	155
Strydom	7.18	68	2.5	62.9	46.4	33.3	1.7	3.7	5.1	536.8	0	424	349	440
Blockdrift	7.32	77	0.0	76.5	54.0	30.0	1.4	2.0	6.8	494.1	0	439	405	—
Nieuwpoort	7.16	107	13.5	74.3	56.9	76.7	0.3	15.4	80.8	530.7	0	583	420	435
Blaauwbank	8.06	36	63.4	12.3	7.9	18.3	11.4	18.8	10.2	36.6	0	161	23	30
South 1.	6–9	70	6	150	70	100	200	200	250			450	300	
African 2. Limits (Anon., 1993)	6.5–9.5	300	10	200	100	400	400	600	600			2000	650	

Notes

Except for pH and EC ($mS\,m^{-1}$), all units are expressed in $mg\,l^{-1}$.

1 Maximum limit for no risk.

2 Maximum permissible limit for insignificant risk.

Table 8.12 Rock classification of geochemical and physical parameters for acid mine drainage (AMD) evaluations (After Brodie *et al.*, 1989)

Factor	*Highest chemical activity = 5*	*4*	*Intermediate chemical activity = 3*	*2*	*Lowest chemical activity = 1*	*Relative weight*	*ARD potential*
Sulphide type	Highly reactive, e.g. fine grained, $<5\ \mu$ pyrrhotite		Moderate reactivity, e.g. coarse grained (5–100 μ)		Massive pyrite crystals	6	6
Portion sulphide surface area exposed	Localized (zones) of concentration exposed on fracture surface, e.g. >50% of sulphides in stringers or veins		>20% of sulphides in zones, stringers or veins with some preferential exposure on fractures		Individual small crystals, uniformly distributed through host rock material	5	15
Alkali type	Highly reactive e.g. $CaCO_3$		Moderate reactivity		Slightly reactive	−4	−9
Alkali surface area exposure	Individual crystals uniformly distributed through the host rock material, or 50% alkali in bedding, fractures or veins which tends to become exposed on handling		>20% of alkali in bedding or veins with some preferential exposure fractures		Localized zones of concentration exposed on fracture surfaces e.g. >50% of alkali in bedding, fractures or veins which tend not to become exposed on handling	−3	−15
Grain size	>20% fines; e.g. > 20% sand size and finer		5% sand size and finer		Little fines; < 1% sand size and finer	2	2
Slaking	Highly slaking; > 20% of rock mass degrades to sand size		Slightly slaking; > 2% of rock degrades to sand size		Non-slaking	2	3
						Total	2

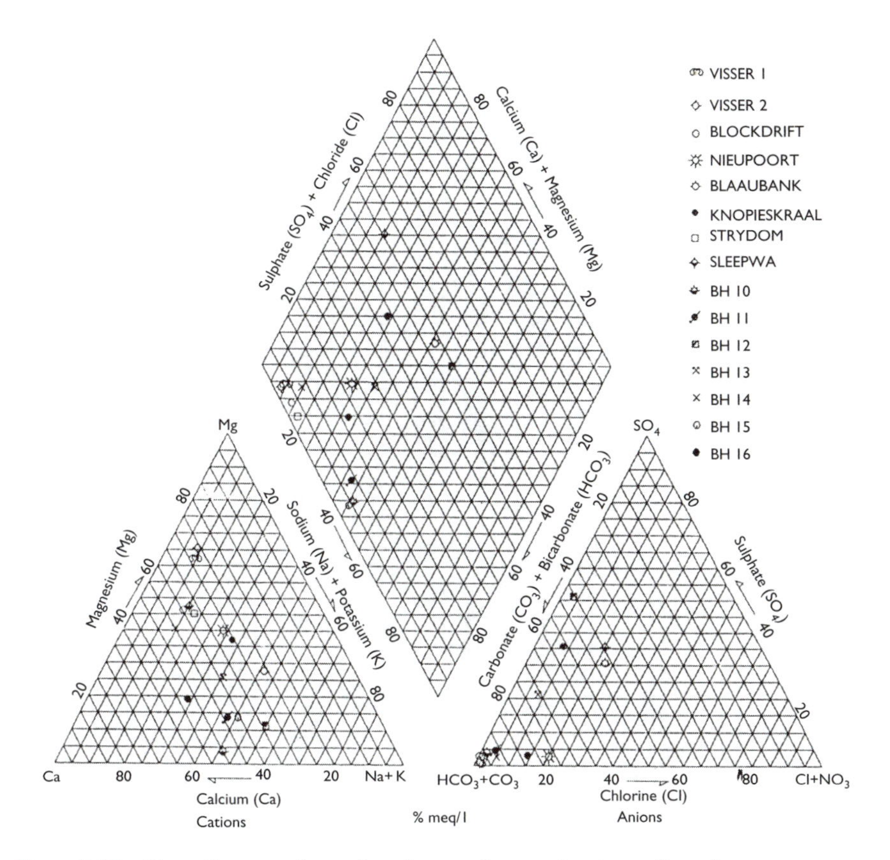

Figure 8.10a Piper diagram of samples of groundwater that is unaffected.

components. From examination of the Piper diagrams of the surface water, natural groundwater and affected groundwater, it is evident that the mixture (in this case the affected groundwater) between the natural groundwater and surface water does not plot on a straight line between these two components and therefore the graphic criterion is not met. Hence, the affected groundwater is not a simple mixture of surface water and natural groundwater. In such a situation the second criterion cannot be satisfied. This implies that the groundwater occurring in the tin field was subjected to different hydrochemical processes. Such processes could include ion exchange, mixing, dissolution and precipitation.

The character of the groundwater was examined with the aid of Stiff diagrams (Stiff, 1951). The advantage of the Stiff diagram is that it presents a better picture of the total salt concentration than other graphic methods. The effects of dilution and concentration are reduced to a minimum, this is

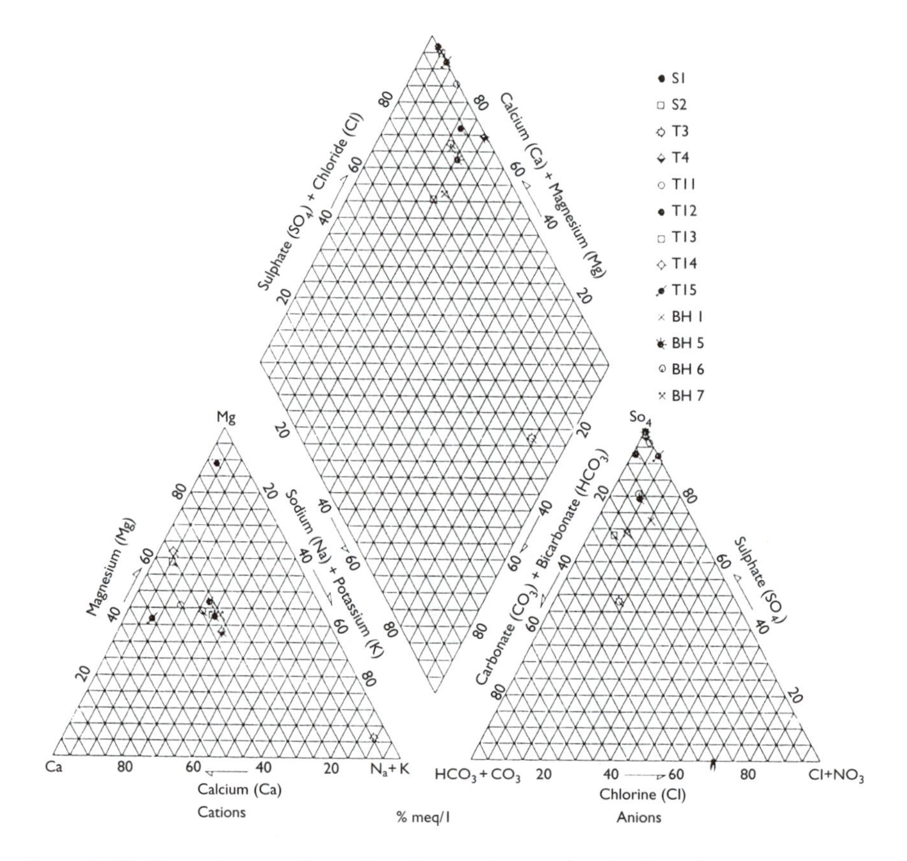

Figure 8.10b Piper diagram of samples of groundwater that is affected.

in contrast to the Piper diagram, which is directly affected by both dilution and concentration. In other words, the Stiff diagram offers a simple, practical means of characterizing and comparing groundwaters.

The use of Stiff diagrams allows a threefold classification of groundwater in the tin field. The natural groundwater has a 'kite' shaped tail with HCO_3 being the dominant anion. The Stiff diagrams for boreholes Blaauwbank, Blockdrift, Knoppieskraal, Nieuwport, Strydom, and Visser 1 and 2 offer examples (Fig. 8.11(a)). As the groundwater becomes affected by acid mine drainage, the HCO_3 percentage decreases, and the sulphate anion begins to dominate, leading to a change in the shape of the Stiff diagram. A 'flat' tail develops as the tail tends towards the sulphate ion and the HCO_3 percentage decreases. The groundwater from shafts S1 and S2, and from boreholes BH1, BH5 and BH6 provide 'flat' tailed Stiff diagrams (Fig. 8.11(b)). Water that is most affected by acid mine drainage has a low pH value, has a very

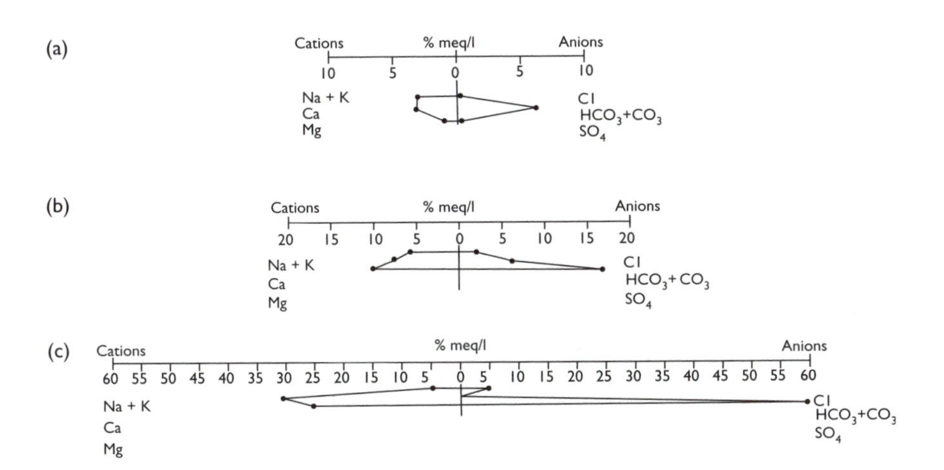

Figure 8.11 Examples of Stiff diagrams (a) borehole 11 (BH11), (b) shaft No 2 (S2), (c) trench No 4 (T4).

low or zero concentration of HCO_3 and an extended 'flat' tail towards a high concentration of sulphate ion. The groundwater sampled from trenches T3, T4, T11, T12, T13 and T15 provide examples of this type of Stiff diagram morphology (Fig. 8.11(c)).

In summary, Class 1 water is unaffected by acid mine drainage and has a 'kite' shaped tail; Class 2 has a 'flat' tail but HCO_3 is still present; and Class 3 water has an extended 'flat' tail, indicating a negligible concentration of HCO_3. Class 3 water has a low pH value and tended to occur close to the surface in the perched water table. Class 2 water, although affected by acid mine drainage, had a pH value near neutral. The neutralizing process was thought to be brought about by the predominance of the cations Ca, Mg, Na and K that were released from clay particles and raised the pH value. The second factor that increased the pH, so neutralizing acidic waters, was the ankerite present in the Boschoffsberg Quartzites.

The groundwater samples also were plotted on trilinear diagrams. Figure 8.12(a) shows the trilinear diagrams for groundwater sampled from the boreholes BH10 to BH16. The major anion diagram indicates that bicarbonate was the dominate ion with minor amounts of sulphate and chloride whereas the cation diagram suggests that calcium, magnesium and sodium were not dominant. Groundwater taken from farm boreholes had a lower concentration of chloride anion and magnesium cation than in the boreholes mentioned previously, but the calcium and sodium concentrations were greater (Fig. 8.12(b)). Trilinear diagrams of groundwater sampled from trenches and subsurface workings are shown in Figure 8.12(c). The anion triangle shows a distinct skewness towards the sulphate ion. The lower

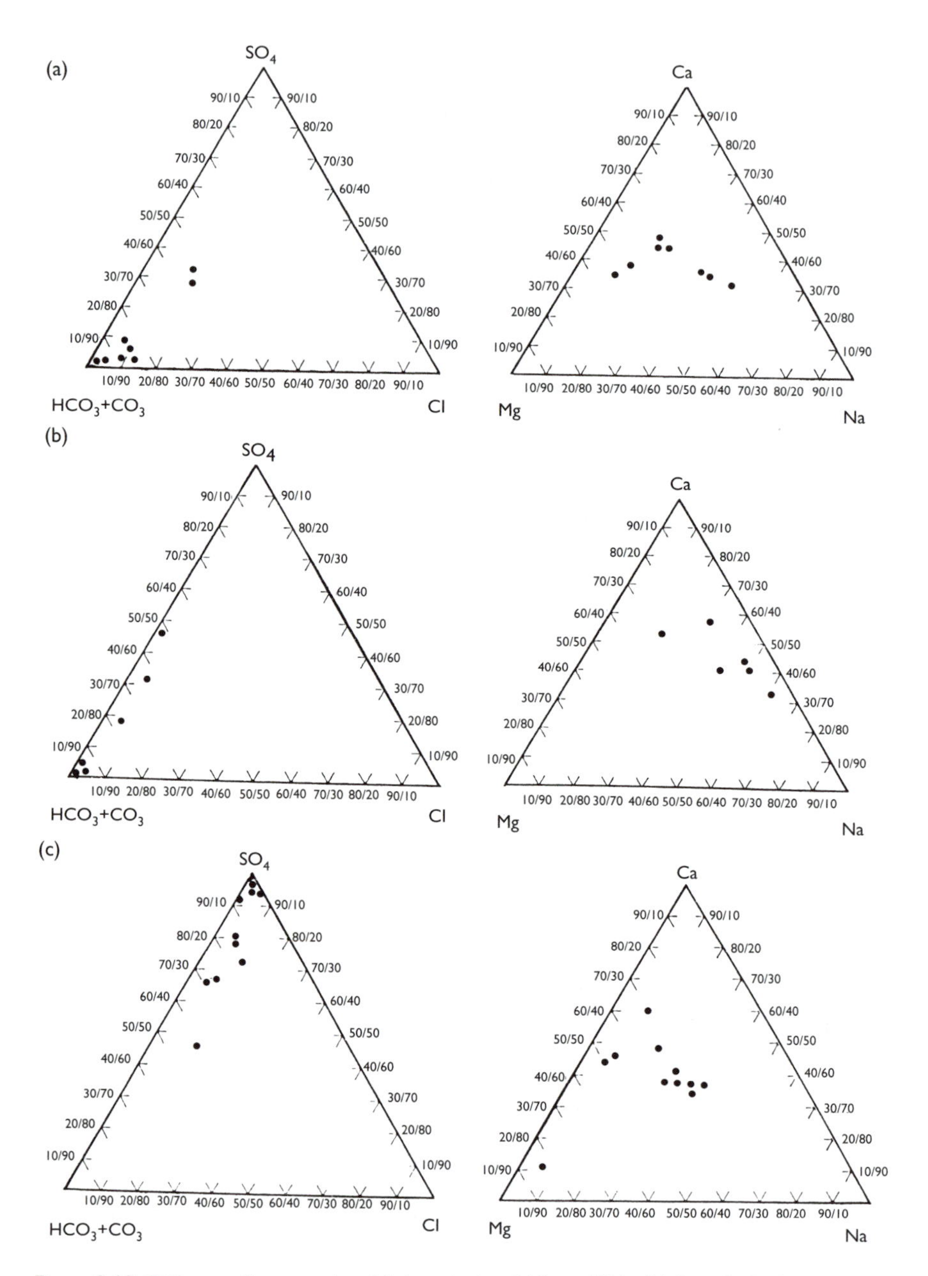

Figure 8.12 Trilinear diagrams for (a) boreholes B10 to B16, (b) boreholes Visser 1 to Blaauwbank, (c) trenches 2 to 15 and underground workings.

bicarbonate concentration and higher sulphate ion concentration was attributed to acid mine drainage. In the cation triangle some samples showed a tendency towards the magnesium ion. This also was attributed to acid mine drainage, magnesium being more soluble in waters with low pH and so was concentrated in the polluted water. Calcium values were relatively high due to cation exchange processes. Cation exchange occurs when subsurface waters containing solutes interact with organic and inorganic adsorbent processes in the soil. In other words, the cation exchange capacity measures the capacity of soils to hold positively charged elements such as calcium, and the ability to resist changes in pH.

Contours of pH value, electrical conductivity and the major cation and anion (Ca, Mg, Na, K, Cl, SO_4) concentrations, derived from groundwater samples, were drawn to provide an overall assessment of groundwater quality in the area around the mine (Fig. 8.13(a)–(h)). The anomalies that are present on the individual maps more or less corresponded with acid mine drainage from the pyrite dumps. In other words, the maps of sulphate concentration, pH values, TDS values, magnesium values and electrical conductivity show that plumes developed away from the dumps in an easterly direction, that is, in the direction of groundwater movement. On the other hand, there was a concentration of sodium on the western side of the dumps, the alkalinity of the groundwater declining eastwards. Calcium values tend to be concentrated beneath the sand dump and spread out fairly evenly from there.

8.5.4. Remedial treatment

The groundwater gradients in the area of the dumps were low, mainly due to the flatness of the surface topography. These low groundwater gradients, together with the presence of a dyke 500 m north of the major dumps and another about 1 km to the south-west, and the relatively impermeable nature of the Boschoffsberg strata, suggested that it would be unlikely that affected groundwater would migrate significant distances from the dumps. Indeed, migration of pollutants probably would be restricted to less than 1000 m from the dumps by these geological conditions.

Be that as it may, the pyrite was removed, it being sold for the manufacture of sulphuric acid. Inevitably, some pyrite was left behind. This remaining material was disposed of in an impoundment on site. The impoundment was designed so that it was isolated from the natural drainage by the construction of cut-off trenches and by the use of clay liners to prevent seepage into the groundwater. Once dumping within the impoundment was completed, it was capped with a clay layer and then completely vegetated. An impoundment also surrounds the slag dump. Part of the rock dump was placed in an abandoned opencast working. The remaining material was levelled. Both areas were covered with top soil and vegetated. The sand dump was stabilized by

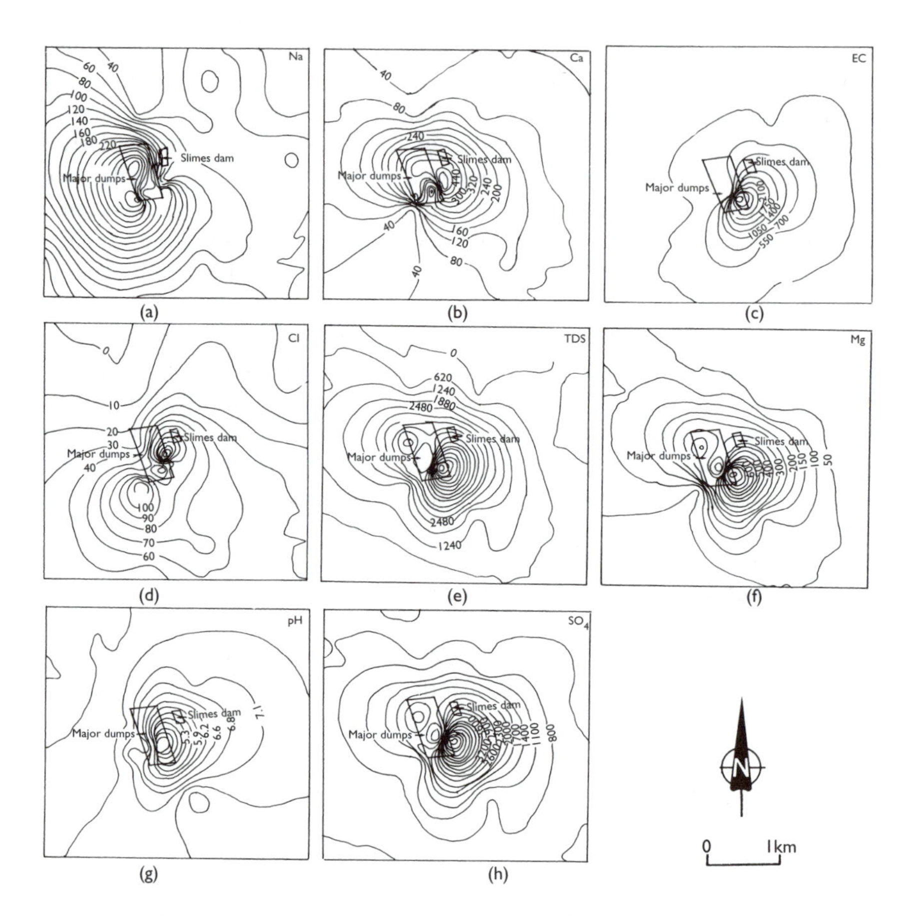

Figure 8.13 Maps showing contoured concentrations of (a) sodium ($mg\,l^{-1}$), (b) calcium ($mg\,l^{-1}$), (c) electrical conductivity ($mS\,m^{-1}$), (d) chloride ($mg\,l^{-1}$), (e) total dissolved solids ($mg\,l^{-1}$), (f) magnesium ($mg\,l^{-1}$), (g) pH value and (h) sulphate ($mg\,l^{-1}$).

covering with waste rock from the rock dump or by grassing. However, the latter proved difficult due to the low organic and nutrient content of the sand, its high permeability and excessive surface temperatures. The major tailings dam was relatively well-vegetated and apparently did not represent a principal source of pollution. The other tailings dam was vegetated and isolated from the natural drainage by embankments.

8.6. Case history 2

The Witbank Coalfield is located in the headwaters of the Olifants River in Mpumalanga Province, South Africa. Mining in Witbank Coalfield began in 1906, and there are currently over 50 operating or defunct coal mines. The coal and waste rock associated with the mining operations is pyrite bearing, and therefore acid mine drainage in the area has caused a deterioration in both surface and groundwater qualities. Opencast mining has the greatest impact on groundwater quality, in that the mines are acid generating and associated with heavy metals, which include iron, aluminium, manganese, copper and zinc. Investigations have shown that the average rate of sulphate generation in a backfilled opencast area is between 5 and 10 kg per hectare a day. On the basis of the present scale of opencast mining in the Witbank area this amounts to 70 tonnes per day of sulphate. Recharge to the groundwater in the opencast areas is as high as 20% of rainfall. Sulphate levels in the groundwater around these opencasts mines are typically between 2000 and 3000 mg l^{-1}. There are essentially four water management options at opencast mines. These are selective spoil handling, clay capping of backfilled opencast areas, flushing and containment. Shallow underground mining also has had an affect on regional groundwater quality. Shallow mining has given rise to subsidence leading to small troughs formed by pillar collapse, to open tension cracks and to crown holes at the surface formed by void migration, with the result that rainwater actively recharges the mines. Recharge in these areas has been found to range up to 15% of rainfall. A number of defunct mines west of Witbank are currently decanting extremely acid water to the surface. The decant water may have a pH as low as 1.5 and sulphate levels in excess of 3000 mg l^{-1}. Water quality deterioration is a two-phased phenomenon. First, base exchange occurs as groundwater seeps towards areas of high extraction, resulting in a sodium bicarbonate/chloride water entering these areas. The second phase occurs within the mine in that pyrite oxidation introduces sulphate, then releases calcium and magnesium from carbonates. Groundwater in high abstraction areas therefore tends to have a range of chemistries, depending on the evolutionary stage of the groundwater. Groundwater quality ranges from severely polluted in some instances to negligible in others.

Unfortunately, acid mine drainage is feeding from a number of abandoned mines into the waters of the Blesbokspruit, which is a tributary of the Olifants River. As a result, the water in the Blesbokspruit has a low pH and high total dissolved solids. This has become a matter of concern as far as the Olifants River is concerned, the latter ultimately flowing through Kruger National Park. The catchment area of the Olifants River is not only sensitive from the point of view of tourism and nature conservation but also because much of the river flows through areas of intensive agriculture. Water quality management and any environmental remediation needs to be considered within a regional context and requires cooperation between mining concerns in the area and downstream users. South Africa presently is moving towards regulation based on water quality receiving objectives that take account of the assimilative capacity of the receiving water, as well as the water requirements of the downstream user. A survey was undertaken at one particular abandoned mine (Middelburg Steam Colliery) by Bell *et al.* (2002) to determine the effect of acid mine drainage issuing from it on the Blesbokspruit.

The five recognized coal seams in the Witbank Coalfield occur within a succession some 70 m in thickness. Only one seam was mined during the life of the mine investigated. It occurs at a depth of approximately 18–23 m and ranges in thickness from 3.5 to 6 m. In the east of the mine property, the seam crops out approximately 100 m west of the Blesbokspruit.

The seam was mined primarily by the bord and pillar method from 1908 to 1947. Pillar robbing started in the late 1930s and resulted in the formation of crown holes at the surface due to void migration, and in discontinuous subsidence caused by multiple pillar failure with associated extensive surface fracturing (Bullock and Bell, 1997).

There are two spoil heaps on the site. One covers an area of approximately 56 250 m^2 and the other covers 66 000 m^2. It was assumed that the spoil material reflected the composition of the coal and shale in the old mine. Samples of shale and coal therefore were taken from the spoil heaps for analysis. Samples of coal also were obtained from an opencasted area. The shale was subjected to both chemical (XRF) and mineralogical analysis (XRD). The chemical and mineralogical composition of some samples of shale from the spoil heaps are given in Table 8.13, from which it can be seen that the two principal oxides, as expected, were silica and alumina. Calcium, magnesium, iron, sodium, potassium, titanium and phosphorus oxides are present in small concentrations. The sulphur content of this shale material averages approximately 1.5%. The mineralogical composition of the shale material in the coarse discard consists primarily of kaolinite, quartz and mica. Other minerals present included microcline, illite and jarosite. Table 8.14 indicates that the coal is low-rank bituminous coal. It also shows that the sulphide content, and therefore the pyrite content, frequently is in excess of 2%. The ash content of the coal averaged 24.5%, which is high. When analysed by XRF, silica and alumina were again the chief oxides, with Fe_2O_3 coming a lowly third, averaging 1.1%.

8.6.1. *Impact of mining on hydrogeology*

The surface subsidence and the associated underground fire at the mine have had impacts on both groundwater and surface water hydrology. These impacts include reduced surface run-off, increased groundwater recharge, and deterioration of water quality. Surface run-off is reduced as rainfall collects in collapsed areas after heavy summer rains. The ponded water percolates through subsidence related tension cracks and crown holes to the underground workings. The workings in the seam concerned act as an aquifer for the percolated water. An anticline in the mine area acts as a

Table 8.13 Analysis of shale (from spoil heaps). (a) Chemical composition; (b) Mineralogical composition (XRD) (After Bell *et al.*, 2002)

(a) Sample no.	SiO_2	Al_2O_3	Fe_2O_3	*MgO*	*CaO*	Na_2O	K_2O	TiO_2	P_2O_5	*Sulphur content*
1	63.4	32.7	0.96	0.24	0.04	—	1.01	1.43	0.08	0.94
2	80.1	15.7	0.47	0.03	0.03	0.18	0.22	2.64	0.47	1.72
3	51.8	44.0	0.48	0.13	0.02	0.32	0.70	2.19	0.08	1.25
4	65.2	31.0	0.78	0.11	0.07	0.2	0.43	1.59	0.14	2.01

(b) Sample no.	*Quartz*	*Kaolinite*	*Illite*	*Mica*	*Microcline*	*Jarosite*
1	10	68	0	14	0	6
2	18	53	2	21	2	3
3	14	72	4	6	1	2
4	23	66	1	6	0	3
5	29	58	2	4	2	1

water divide. Groundwater collecting on the western side of this regional anticlinal axis flows to the west and dams up against the boundary pillar with the adjacent mine. However, groundwater moving through the workings on the eastern side of the anticlinal axis flows towards the coal sub-outcrop in the vicinity of the Blesbokspruit. Since 1991, after the coal sub-outcrop pillar along the eastern boundary of the working was mined, groundwater, which previously had been retained behind the pillar, began to seep from the workings. This seepage water flows overland in a series of springs, before merging to enter the Blesbokspruit. A V-notch flume was installed to measure flow on the stream and weekly flow readings were recorded for a one-year period from December 1994 to November 1995. The total annual volume of water collecting over the mine site catchment area was derived from the rainfall data. The estimated recharge to the coal seam is around 50% of the volume of rain that falls. Figure 8.14 suggests that there is a lag time between the heaviest rainfall (January) and maximum flow over the V-notch (July).

Rainwater seeping into the old workings is affected by both the oxidation of pyrite and the presence of the underground fire. Table 8.15 includes some examples of water quality analyses of samples of water collected at the V-notch. Samples were obtained both during the summer (the wet season) and in winter (dry season) over a number of years, and are representative of the quality of water seeping from underground. Analyses of samples taken from drillholes sunk into the seam, along with analyses of water from drillholes sunk at adjacent collieries are given in Table 8.16 for comparison.

Inspection of the water quality data at the V-notch and from the mine shows that the waters are highly polluted and that the water in the

Table 8.14 Analysis of coal (a) from spoil heaps, (b) from open cast area. (c) Ash content (XRF) (After Bell *et al.*, 2002)

	Coal from spoil heap 1					*Coal from spoil heap 2*				
(a)	*Sample no.*	*Carbon content*	*Ash content*	*Sulphur content*	*Moisture content*	*Sample no.*	*Carbon content*	*Ash content*	*Sulphur content*	*Moisture content*
	1	64.2	22.1	2.25	5.5	1	66.8	22.7	2.35	2.3
	2	67.3	21.2	1.81	4.7	2	64.5	21.6	1.79	3.6
	3	63.2	24.3	2.31	4.1	3	63.2	29.6	1.85	1.9
	4	67.2	25.6	2.10	4.9	4	67.2	26.0	2.17	3.3
	5	64.6	27.1	2.21	5.2	5	65.7	27.3	2.28	3.9
(b)	*Sample no.*	*Carbon content*	*Ash content*	*Sulphur content*	*Moisture content*	*Sample no.*	*Carbon content*	*Ash content*	*Sulphur content*	*Moisture content*
	1	61.5	25.4	2.57	4.4	6	61.4	31.2	1.26	4.7
	2	66.5	23.9	1.92	3.8	7	58.6	33.0	2.74	3.6
	3	62.7	22.0	2.64	4.1	8	65.1	29.6	2.15	2.9
	4	64.5	26.9	2.79	4.1	9	66.7	27.9	1.98	3.1
	5	62.7	28.5	3.77	3.9	10	68.4	23.8	2.41	4.7
(c)	*Sample no.*	SiO_2	Al_2O_3	Fe_2O_3	MgO	CaO	Na_2O	K_2O	TiO_2	P_2O_5
	1	52.9	43.3	0.59	—	0.11	0.25	0.76	1.65	0.40
	2	40.6	35.7	1.99	0.06	10.2	0.23	0.35	1.75	8.84
	3	58.6	38.1	0.22	0.05	—	0.27	0.21	1.93	0.08
	4	52.8	43.4	1.48	0.12	0.09	0.12	0.33	1.01	0.21
	5	51.6	40.0	2.01	0.07	0.06	0.14	0.39	1.02	0.56
	6	52.6	41.4	1.90	0.10	1.26	0.22	0.54	1.11	0.32

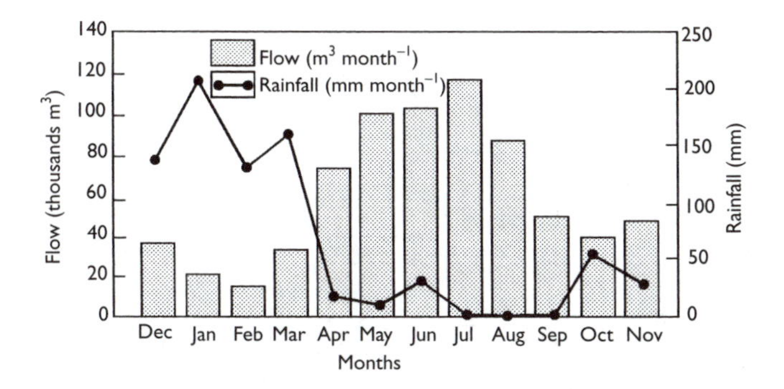

Figure 8.14 Monthly rainfall figures compared with seepage flow over the V-notch.

Blesbokspruit catchment is characterized by low pH and high total dissolved solids. As such, it is capable of mobilizing toxic metal concentrations. The pH values are well below and the total dissolved solids values significantly above the respective crisis limits recommended by the South African guidelines for domestic water (Anon., 1993). At such values there is a danger to health due to dissolved metal ions. Low pH values can be attributed to the formation of sulphuric acid as a product of reactions involving the oxidation of pyrite with the production of acid mine drainage. In this case, the oxidation of pyrite is enhanced by higher temperatures attributable to burning coal in the old mine. This is supported by the sulphate content, which in all the analyses is above the crisis limit guidelines and in some instances is twice that limit. The high sulphate content is not unexpected when compared with the high sulphur content in the coal and associated shale (Tables 8.13 and 8.14). Table 8.15 also shows that the concentrations of aluminium and iron far exceed the crisis limits for drinking water quality in South Africa. Most of the other constituents are around or exceed the maximum permissible limits.

The effect of the time lag between the period of maximum precipitation and maximum flow from the workings is reflected in TDS concentrations. Consequently, the salt concentrations for the wettest months, when flow is lowest due to the time lag mentioned, are generally greater than for the dry months when flow is highest. The increased volume of water therefore has a dilution effect on the concentration of dissolved salts in the water.

In an attempt to ameliorate the impact of underground water entering the Blesbokspruit, a series of four pollution control ponds were constructed (Fig. 8.15). A side stream of the seepage water was directed into the upper reservoir while the remaining flow entered the nearby stream below the ponds. Residence time in the ponds is not known but is likely to vary with

Table 8.15 Examples of the chemical composition of acid mine water collected from the V-notch (location W4 in Fig. 8.16) and South African guidelines for drinking water quality (Anon., 1993; after Bell *et al.*, 2002)

Determinand ($mg\,l^{-1}$)	*Sample 1 May 1990*	*Sample 2 June 1990*	*Sample 3 July 1990*	*Sample 4 Dec 1990*	*Sample 5 Jan 1991*	*Sample 6 Mar 1991*	*Sample 7 Dec 1993*	*Sample 8 Jan 1994*	*Sample 9 Feb 1994*	*Sample 10 Dec 1995*	*Sample 11 Aug 1996*	*Sample 12 Aug 1996*	*Sample 13 Aug 1996*	*Recommended limit (no risk)*	*Maximum permissible limit (insignificant risk)*	*Crisis limit (max limit for low risk)*
TDS	2749	2575	2575	2843	2760	2082	3376	3038	3575	4844	2968	3202	3604			
EC ($mS\,m^{-1}$)	293	327	465	421	379	298	424	463	418	471	430	443	340	70	300	400
pH value	2.3	2.4	3.0	1.8	1.9	2.0	2.8	6.6	2.8	1.9	2.4	2.95	2.8	6–9	5.5–9.5	>4 or <11
Nitrate NO_3 as N	0.05						0.04	0.02	0.18	0.1	0.1	0.1	0.1			
Chloride	91	84	186	120	106	124	179	174	170	310	431	406	611	250	600	1200
Fluoride										0.6	0.5	0.33	0.84	1	1.5	3
Sulphate as SO_4	2361	2239	2697	2462	2330	1692	2722	2378	2897	3250	1610	1730	1440	20	600	1200
Total hardness as $CaCO_3$											484	411	377			
Calcium hardness as $CaCO_3$											285	310		20–300	650	1300
Magnesium hardness as $CaCO_3$											199	101				
Calcium	135	115	176	76	132	98	162	179	186	173.8	114.0	124	84	150	200	400
Magnesium	56	50	90	41	55	40	84	90	83	89.4	48.4	49.5	31	70	100	200
Sodium	102	65	200	116	138	114	194	185	200	247.0	326.0	311	399	100	400	800
Potassium									9.4	7.3	9.4	8.9		200	400	800
Iron			140						128	248.3	128	140	193	0.1	1	2
Manganese			18						15	17.9	15	9.9	9.3	0.05	1.0	2.0
Aluminium			86						124		124		84	0.15	0.5	1.0

Table 8.16 Chemical composition of acid mine water from drillholes sunk at Middelburg Steam and adjacent collieries (After Bell *et al*., 2002)

Determinand ($mg\,l^{-1}$)	*Middelburg Steam Colliery (1)*	*Middelburg Steam Colliery (2)*	*Witbank Colliery (1)*	*Witbank Colliery (2)*	*Tavistock Colliery (1)*	*Tavistock Colliery (2)*
TDS	3604	5778	3048	3354	5778	7158
EC ($mS\,m^{-1}$)	340	355	389	403	355	368
pH value	2.8	2.8	2.65	2.7	2.8	2.9
Nitrate NO_3 as N	0.1	0.19	0.29	0.29	0.21	0.28
Chloride	611	184	951	989	18	4.8
Sulphate	1440	3233	910	1306	3253	3840
Total hardness as $CaCO_3$	377	214	106	161	2461	1977
Calcium	84	509	42	40	509	462
Magnesium	31	289	14.9	14.8	289	200
Sodium	399	47	620	775	47	32
Iron	193	198	122	99	198	726
Manganese	9.3	49	5.9	3.9	49	30
Aluminium	84	32	81	87	32	38

Figure 8.15 Pollution control pond, one of four.

season. Comparison of water quality data, especially of samples W7 and W8 (see Table 8.17; Fig. 8.16), show that the effect of cascading part of the seepage water through the decantation ponds is negligible. Indeed, it would appear that aluminium is being leached from soil particles in the ponds as aluminium concentrations leaving the lowest pond are about double those of water at the V-notch flume.

Table 8.17 Analyses of waters in the Blesbokspruit and its catchment (After Bell *et al.*, 2002)

Determinand (mg l^{-1})	*W1*	*W2*	*W3*	*W4*	*W5*	*W6*	*W7*	*W8*	*W9*	*W10*	*W11*	*W12*	*W13*	*W14*	*W15*	*W16*	*W17*	*W18*
TDS	578	2280	1748	2097	2356	2625	2295	2341	2504	2164	2526	2215	2225	2232	1070	360	230	180
EC (mS m^{-1})	84	357	371	373	329	390	363	364	378	308	438	316	330	330	166	126	33	23
pH	3.5	2.6	2.7	2.6	2.8	2.7	2.7	2.7	2.6	2.7	2.7	2.8	2.8	2.8	3.1	3.2	7.4	7.3
Chloride	111	468	220	397	486	572	363	380	552	366	425	319	299	271	127	89.6	50.5	48.5
Fluoride	1.1	1.2	1.2	1.2	2.0	1.3	1.3	1.2	1.3	1.3	0.04	1.4	1.5	2.1	1.7	1.2	0.1	0.08
Sulphate	284	1345	987	1226	1416	1512	1509	1511	1470	1417	1587	1487	1512	1510	716	128	45.8	22.9
Alkalinity	Nil	Nil	Nil	Nil	Nil	Nil	Nil	Nil	Nil	Nil	Nil	Nil	Nil	Nil	Nil	Nil	73.8	41.1
Calcium[a]	49.2	110	116	117	139	145	136	134	122	111	169	120	151	158	72.8	52.1	23.5	10.6
Magnesium[a]	24.7	43.1	47.1	45.4	56.7	56.9	51.9	51.3	47.0	44.6	64.5	51	64.4	81.4	36.4	25.7	11.6	7.4
Sodium[a]	101	296	360	294	248	302	229	228	298	213	266	228	188	199	111	56.4	23	18.2
Potassium[a]	4.5	11.5	13.4	11.2	4.85	24.5	2.34	33.0	10.5	5.01	8.9	6.7	5.9	7.4	4.3	4.5	3.3	1.9
Iron[a]	1.3	31.6	25.5	46.2	29.6	38.6	34.6	74.4	36.1	27.2	87.1	7.4	19.9	7.0	4.7	1.6	0.61	0.59
Manganese[a]	2.9	10.6	11.5	13.4	12.4	12.4	13.8	12.7	13.4	12	13.7	13.5	19.2	23.1	9.5	6.5	0.09	0.08
Aluminium[a]	7.9	89.4	115	101	96.1	153	190	189	111	132	233	137	161	152	60.3	39.6	0.33	0.27
Silicon[a]	7.8	44.8	44	42.2	31.7	59.9	54.3	55.5	42.2	41.2	64.8	39.4	39.8	36.8	17.1	13.0	0.75	2.67
Copper[a]	1.0	1.4	0.96	1.1	1.0	1.4	1.5	1.1	1.42	1.4	1.5	1.0	1.1	1.0	1.0	0.97	0.94	0.95
Nickel[a]	0.87	1.9	1.8	1.9	1.71	2.2	2.1	2.2	1.9	1.9	2.6	1.8	2.0	1.9	1.3	1.1	BDL	1.1
Lead[a]	1.2	1.6	1.5	1.4	1.4	1.6	1.4	1.4	1.4	1.6	1.6	1.6	1.4	1.7	1.1	1.3	0.96	1.1
Zinc[a]	0.74	2.4	2.7	2.6	1.9	2.3	3.4	2.8	2.6	2.3	4.3	2.2	2.7	2.8	1.4	1.1	0.4	0.4

Notes
See Figure 8.16 for sample locations. All samples from the Blesbokspruit (including W16) except those noted below.
W2 and W3 samples of seepage water.
W4 sample of water from V-notch.
W5, W6, W7 and W8 samples of water from the decantation ponds.
W15 sample of water from the wetland.
W17 and W18 samples of water from the uncontaminated Prison stream.
W16 sample of water from downstream of the confluence with Prison stream.
a Determined by ICP-AES.

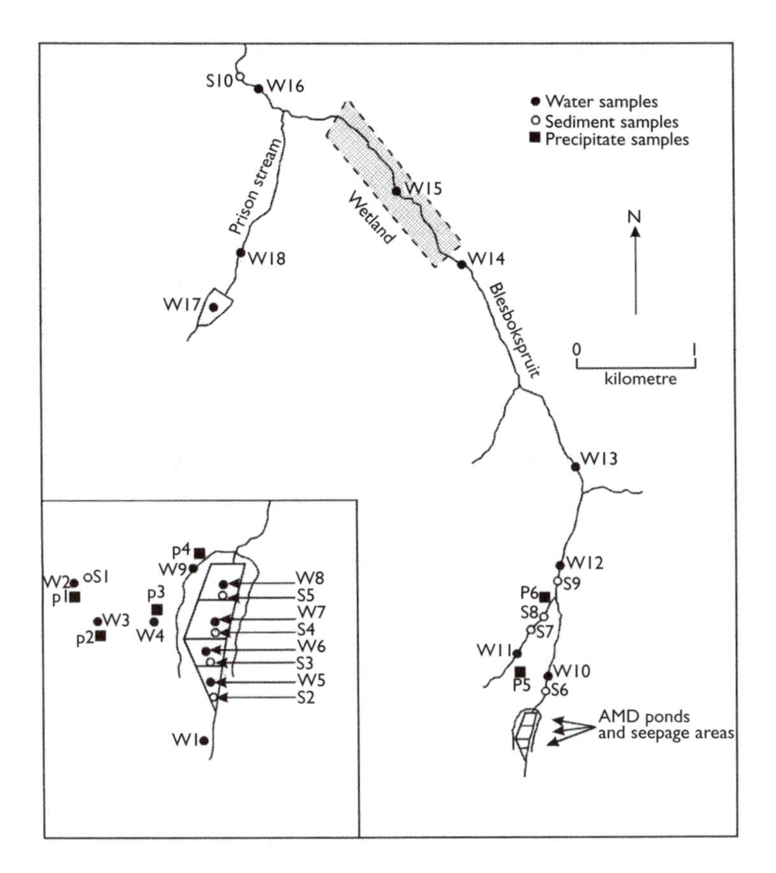

Figure 8.16 Location of water samples and wetland.

8.6.2. Aqueous geochemistry of the Blesbokspruit catchment

The pH value of the waters of the Blesbokspruit varied from pH 2.6 near where the water issued from the mine to pH 3.2 downstream of the wetland, which contained reed beds (Fig. 8.16; Table 8.17). By comparison, two samples taken from the unaffected tributary of the Blesbokspruit, namely, the Prison Stream, had values of pH above 7 (Table 8.17). The very low values of pH that characterized the rest of the Blesbokspruit catchment indicated a lack of neutralizing capacity in the mine workings and the catchment. The values of electrical conductivity were above 300 mS m^{-1} for water with a pH less than 3.0 (Fig. 8.17), whereas the unaffected water of the tributary stream had low electrical conductivity values (Table 8.17). This is a reflection of the TDS content.

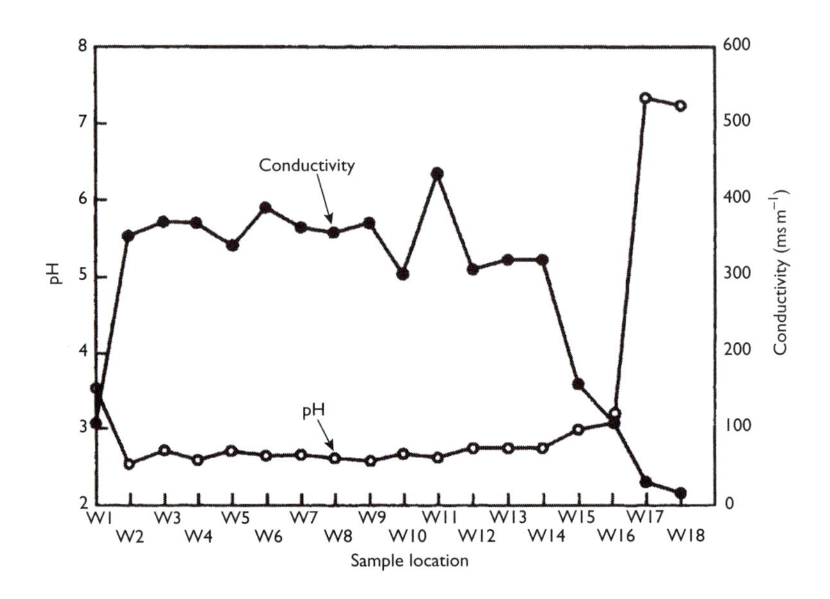

Figure 8.17 Relationship of electrical conductivity and pH value of water samples along the Blesbokspruit.

High concentrations of aluminium are typical of acid mine drainage water and presumably are derived from alumino-silicate minerals such as kaolinite and mica in shales associated with coal seams. The content of aluminium in the Blesbokspruit varied between 80 and 240 mg l^{-1} compared with 0.59–0.8 mg l^{-1} for stream water that was not contaminated. There was a twofold to fourfold decrease in Al concentration in the wetland.

The iron content of the Blesbokspruit was lower than that found in the waters obtained from drillholes, suggesting that much of the dissolved iron has been precipitated as a result of oxidation and hydrolysis. The equilibrium of iron, according to Karathanasis *et al.* (1988), is modified in an acid sulphate system. Although goethite ($HFeO_2$) and amorphous $Fe(OH)_3$ control Fe levels in most natural aquatic systems, the control of dissolved Fe in acid sulphate rich solutions appears to be more consistent with the solubility of the iron sulphate mineral jarosite [$KFe_3(OH)_6(SO_4)_2$].

Sulphate was the dominant anion in the Blesbokspruit, with concentrations up to 1587 mg l^{-1}. The high concentrations of sulphate are characteristic of water contaminated by acid mine drainage in this area. Once the sulphate rich acid mine drainage entered the Blesbokspruit the concentration remained fairly constant until it was diluted somewhat by uncontaminated water from the Prison Stream.

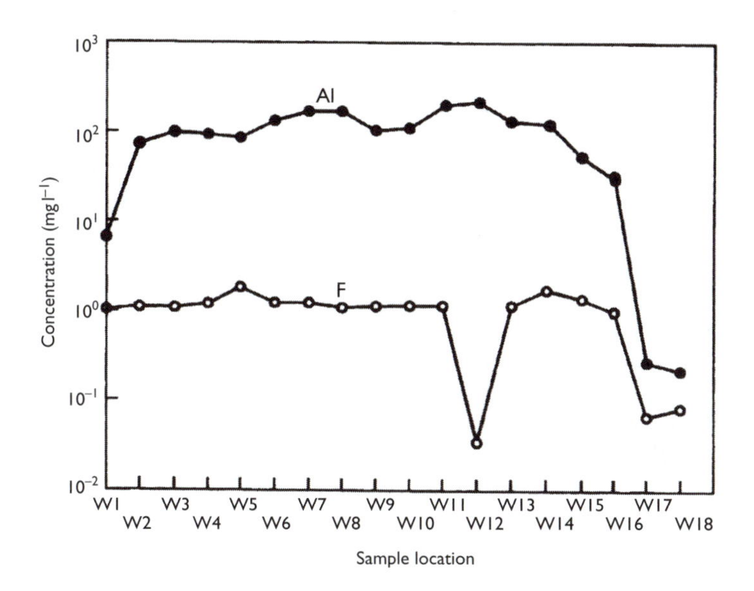

Figure 8.18 Changes in the concentration of fluoride compared with aluminium in water samples along the Blesbokspruit.

The chloride ion concentration up to about 580 mg l^{-1}. It, like sulphate, was derived from the coal and shale in the mine. It would be expected that the sodium and chloride present would have a similar origin. However, the higher values of chloride would suggest that the removal of dissolved sodium was occurring as a result of the precipitation of Na-jarosite. Calcium and magnesium had similar concentrations in the Blesbokspruit. The maximum concentration of manganese was 81 mg l^{-1} and the Blesbokspruit contained significant concentrations of lead and zinc (Table 8.17). As far as fluoride was concerned, its concentration fluctuated according to the changes in aluminium in the water (Fig. 8.18), aluminium-fluoride complexes presumably explain this fluctuation.

The waters of Blesbokspruit contained high concentrations of elements typical of acid mine drainage (Table 8.17). However, the concentration of elements decreased as the Blesbokspruit flowed through the small wetland. The reduction in the concentration of some elements can be attributed to element retention reactions involving adsorption, precipitation and co-precipitation. Dilution of the water exiting the wetland by uncontaminated water from the Prison Stream, further decreased the elemental concentrations of the Blesbokspruit. In particular, the precipitation of jarosite and goethite probably accounted for the removal of iron. However, the solubility of aluminium sulphate minerals under acidic conditions would seem to explain the high concentration of aluminium in the water downstream of the wetland.

If natural water is at saturation equilibrium, then the ion activity product (IAP) should be the same as the solubility product constant (K). The logarithm of this ratio is referred to as the saturation index (SI). A negative saturation index implies that the waters are unsaturated with respect to a particular mineral phase and the mineral should be expected to remain in solution. On the other hand, supersaturated waters have a positive saturation index, the mineral phase being expected to precipitate out of solution. Saturation indices of around zero indicate that the water is in equilibrium with a particular mineral phase. The MINTEQA2 program was used to determine the saturation indices (Allison *et al.*, 1991). Saturation indices of selected minerals were plotted as a function of pH in order to investigate the effects of dissolution or precipitation of minerals on the concentrations of metals in the Blesbokspruit. Water with a low pH is undersaturated with respect to kaolinite [$Al_2O_3 \cdot 2SiO_2 2H_2O$] and saturated with respect to quartz [SiO_2] (Fig. 8.19(a)). High concentrations of silica and aluminium correspond with low pH values where kaolinite is undersaturated. Most water samples with a pH value of less than 3 appear to be supersaturated with respect to jarosite [$KFe_3(OH)_6(SO_4)_2$] and goethite [$HFeO_2$], and undersaturated with respect to ferrihydrite [$Fe_2O_3 \cdot nH_2O$] (Fig. 8.19(b)). Gypsum [$CaSO_4 \cdot nH_2O$], alunite [$KAl_3(OH)_6(SO_4)_2$] and jurbanite [$AlOHSO_4$] may be at equilibrium or slightly undersaturated in water with a low pH (Fig. 8.19(c)). Waters with a pH exceeding 7 are undersaturated with respect to gypsum and jurbanite and supersaturated with respect to kaolinite, jarosite, ferrihydrite, goethite and alunite. The Blesbokspruit was slightly undersaturated with respect to anglesite ($PbSO_4$), indicating that lead remains in solution in these waters. In fact, the low pH of the water would counteract Pb adsorption to hydrous ferric oxides such as goethite and result in Pb concentrations remaining high (Schwertmann and Taylor, 1989).

8.6.3. Effect of acid mine drainage on vegetation and influence of algae

Most plants cannot tolerate low pH water because the high concentration of hydrogen ions causes inactivation of enzyme systems, restricting respiration and root uptake of mineral salts and water (Bradshaw *et al.*, 1982). A denuded area of approximately 3 ha exists in the coal sub-outcrop area between the eastern boundary of the mine and the decantation ponds. In this area, almost all vegetation has been killed and an algal mat has developed over part of the area (Fig. 8.20). Dissolved aluminium ions also are regarded as a major cause of plant toxicity in acid soils. As the total aluminium concentration in the seepage water is over 100 mg l^{-1}, aluminium is also likely to have played a significant role in the destruction of vegetation

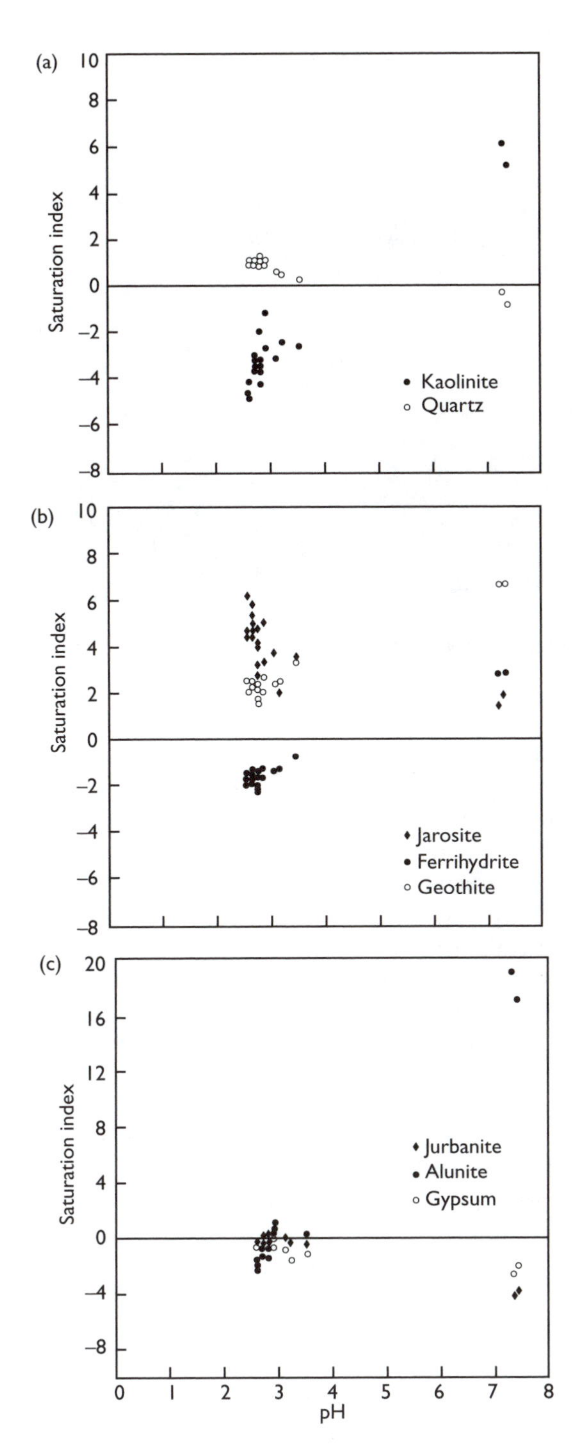

Figure 8.19 Saturation indices in relation to pH for (a) kaolinite and quartz, (b) ferrihydrite, goethite and jarosite, (c) gypsum, alunite and jurbanite in the waters of the Blesbokspruit.

Figure 8.20 Part of seepage area where vegetation has been decimated and an algal mat developed.

in this area. Only species of algae appear to exist in the seepage area, in the pollution control ponds and in the headwaters of the Blesbokspruit.

Many species of algae are known to tolerate acid mine drainage and they appear to play a role in metal attenuation. Intensive algal growth occurs in the seepage area. The green algae belongs to the genus Mongeotia and the red algae to the genus Microspora. There are several processes by which algae can remove metals from acid mine drainage water, including physical trapping of suspended metal particles and subsequent chemical binding (chelation) to the numerous anionic sites within the cell walls of the algae. The polymers that constitute the cell walls are rich in phosphoryl, carboxyl, hydroxyl and aromatic groups with cationic metals. In addition, the presence of micro-organisms that grow epiphytically on green algae filaments can result in mineral deposition on the cell walls of algae and algae can accumulate metals by intracellular uptake (Brady *et al.*, 1994).

High concentrations of Fe_2O_3 were detected by XRF in both types of algae, Mongeotia containing 77% by weight whilst Microspora contained 40% by weight (Table 8.18). This indicates that the algae take part in biomineralization of Fe from the iron rich acid water. The concentration of Al_2O_3 in the algae may be attributed to either the adsorption of aluminium to the precipitate crust or to the accumulation of aluminium due to the increased availability of the metal in low pH water. The amount of CaO and K_2O in Microspora is some six times higher than that of Mongeotia whereas the latter contained much more P_2O_5. There also is a significant difference in the chloride concentration.

Table 8.18 XRF analyses of algae (After Bell *et al.*, 2002)

Major oxides (weight %)	SiO_2	Al_2O_3	Fe_2O_3	*CaO*	*MgO*	Na_2O	K_2O	*Cl*	SO_3	TiO_2	P_2O_5
Mongeotia	3.5	1.9	77	0.68	0.35	0.75	0.51	0.98	9.8	0.02	2.3
Microspora	3.0	1.5	40	3.9	0.44	1.1	3.1	2.9	12	0.14	0.06
Precipitate (1)	2.1	4.3	79	0.06	0.02	0.12	0.06	0.22	9	0.02	1.2

Trace elements (mg kg^{-1})	*Zn*	*Cu*	*Ni*	*V*	*Cr*	*Mn*	*Co*	*Mo*	*Zr*	*Y*	*Sr*	*Rb*	*Th*	*Pb*
Mongeotia	91	52	9.0	797	81	221	13	0.4	7.1	17	4.9	14	2.0	4.3
Microspora	230	14	63	79	337	789	33	0.5	17.1	22	33	165	2.3	6.5
Precipitate (1)	17	9.4	<0.8	800	69	37	6.4	0.3	7	4.5	1.4	3.5	3.7	19

The contents of zinc, copper, nickel, manganese and lead in both types of algae exceeded those found in the acid drainage water seeping from the mine. Hence, both algae would appear to have the ability to abstract and concentrate these trace metals. Furthermore, these algae, especially Mongeotia, presumably play an important role in the formation of precipitate material. Iron rich encrustations occurred on dead algae, notably ferrihydrite, and can result in the co-precipitation of heavy metals.

High concentrations of zinc were detected in the algae, with Microspora containing the higher concentration (Table 8.18). Zinc can be adsorbed to negatively charged cell walls of algae or may be co-precipitated with iron minerals (Kiekens, 1995). An antagonistic zinc–copper relationship would appear to occur in the case of Microspora where the ratio of the two metals is 16.4, which compares with a Zn:Cu ratio of 1.7 in Mongeotia. This suggests that the presence of zinc inhibits the uptake of copper in Microspora, which may indicate that the latter algae have the same carrier sites in adsorption mechanisms for both metals.

Microspora contained significantly higher concentrations of chromium than Mongeotia. On the other hand, Mongeotia possessed more or less 10 times a much vanadium as did Microspora (Table 8.18). In fact, Mongeotia contained similar concentrations to that of precipitate material from near the seepage point of the acid mine drainage. Arnon and Wessels (1953) noted that high values of vanadium in acidic conditions were attributable to it being an essential micro-nutrient for certain green algae and that the presence of aluminium and ferric ions caused precipitation of vanadium. Subsequently, Edwards *et al.* (1995) mentioned that V^{3+} has a similar ionic radius to that of Fe^{3+} and that it therefore may replace Fe^{3+} in iron minerals, resulting in vanadium enrichment.

According to Heier and Billings (1970), rubidium is easily taken up by algae and may substitute for potassium sites. Mongeotia had a K/Rb ratio of 320 and that of Microspora was 155. However, potassium was present in the cytoplasm of the latter. These differences in the ratios could indicate different metal accumulation mechanisms and/or different rates of metal accumulation between the two types of algae.

References

Adam, K., Kourtis, A., Gazea, B. and Kontopoulis, A. 1997. Evaluation of static tests used to predict the potential for acid drainage generation at sulphide mine sites. *Transactions of the Institution of Mining and Metallurgy*, Section A, Mining Industry, **106**, A1–A8.

Allison, J.D., Brown, D.S. and Novo-gradac, K.J. 1991. *MINTEQA2/PRODEFA2, A Geochemical Assessment for Environmental Systems (ERA/600/3-91/021).* United States Environmental Protection Agency, Athens, GA.

Anon. 1993. *South African Water Quality Guidelines; Volume 1, Domestic Use.* Department of Water Affairs and Forestry, Pretoria.

Arnon, D.J. and Wessels, G. 1953. Vanadium as an essential element for green plants. *Nature*, **172**, 1039–1040.

Banks, D., Parnachev, V.P., Frengstad, B., Holden, W., Vedernikov, A.A. and Karnachuk, O.V. 2002. Alkaline mine drainage from metal sulphide and coal mines: examples from Svalbard and Siberia. In: *Mine Water Hydrogeology and Geochemistry*, Special Publication 198, Younger, P.L. and Robins, N.S. (eds), Geological Society, London, 287–296.

Bell, F.G. and Kerr, A. 1993. Coal mining and water quality with illustrations from Britain. *Proceeding of the International Conference on Environmental Management, Geowater and Engineering Aspects*, Wollongong, Chowdhury, R.N. and Sivakumar, S. (eds), Balkema, Rotterdam, 607–614.

Bell, F.G., Hälbich, T.F.J. and Bullock, S.E.T. 2002. The effects of acid mine drainage from an old mine in the Witbank Coalfield, South Africa. *Quarterly Journal of Engineering Geology and Hydrogeology*, **35**, 265–278.

Best, G.T. and Aikman, D.T. 1983. The treatment of ferruginous groundwater from an abandoned colliery. *Water Pollution Control*, **82**, 557–566.

Bodurtha, P. and Brassard, P. 2000. Neutralization of acid by steel-making slags. *Environmental Technology*, **21**, 1271–1281.

Bradshaw, A.D., Williamson, M.S. and Johnson, M.S. 1982. Mine wastes reclamation. *Mining Journal*, **299**, 75–80.

Brady, D., Letebele, B., Duncan, J.R. and Rose, P.D. 1994. Bioaccumulation of metals by Scenedesmus, Selenastrum and Chlorella algae. *Water South Africa*, **20**, 213–218.

Brodie, M.J., Broughton, L.M. and Robertson, A. 1989. A conceptional rock classification system for waste management and a laboratory method for ARD prediction from rock piles. *British Columbia Acid Mine Drainage Task Force, Draft Technical Guide*, **1**, 130–135.

Bullock, S.E.T. and Bell, F.G. 1995. An investigation of surface and groundwater quality at a mine in the north west Transvaal, South Africa. *Transactions of the Institution of Mining and Metallurgy*, **104**, Section A, Mining Industry, A125–A133.

Bullock, S.E.T. and Bell, F.G. 1997. Some problems associated with past mining in the Witbank Coalfield, South Africa. *Environmental Geology*, **32**, 233–242.

Burke, S.P. and Younger, P.L. 2000. Groundwater rebound in the South Yorkshire coalfield: a first approximation using the GRAM model. *Quarterly Journal of Engineering Geology and Hydrogeology*, **33**, 149–160.

Bussiere, B., Aubertin, M. and Chapuis, R.P. 2003. The behavior of inclined covers used as oxygen barriers. *Canadian Geotechnical Journal*, **40**, 512–535.

Bussiere, B., Benzaazoua, M., Aubertin, M. and Mbonimpa, M. 2004. A laboratory study of covers made of low sulphide tailings to prevent acid mine drainage. *Environmental Geology*, **45**, 609–622.

Cambridge, M. 1995. Use of passive systems for treatment of mine outflows and seepages. *Minerals Industry International, Bulletin Institution of Mining and Metallurgy*, **1024**, 35–42.

Choi, J.-C. and West, T.R. 1995. Evaluation of phosphate pebble as a precipitant for acid mine drainage treatment. *Environmental and Engineering Geoscience*, **1**, 163–171.

Connelly, R.J., Harcourt, K.J., Chapman, J. and Williams, D. 1995. Approach to remediation of ferruginous discharge in the South Wales Coalfield and its application to closure planning. *Minerals Industry International, Bulletin Institution of Mining and Metallurgy*, **1024**, 43–48.

Coulton, R., Bullen, C. and Hallett, C. 2003. The design and optimization of active mine water treatment plants. *Land Contamination and Reclamation*, **11**, 273–280.

Davison, W. 1990. Treatment of acid waters by inorganic bases, fertilizers and organic material. *Transactions of the Institution of Mining and Metallurgy*, Section A, Mining Industry, **99**, A153–A157.

Dey, M., Sadler, P.J.K. and Williams, K.P. 2003. A novel approach to mine water treatment. *Land Contamination and Reclamation*, **11**, 253–258.

Dumpleton, S., Robins, N.S., Walker, J.A. and Merrin, P.D. 2001. Mine water rebound in south Nottinghamshire: risk evaluation using 3-D visualization and predictive modelling. *Quarterly Journal of Engineering Geology and Hydrogeology*, **34**, 307–319.

Edwards, R., Lepp, N.W. and Jones, K.C. 1995. Other less abundant elements of potential environmental significance. In: *Heavy Metals in Soils*, Alloway, B.J. (ed.), Blackie, Glasgow, 306–351.

Fytas, K. and Bousquet, P. 2002. Silicate micro-encapsulation of pyrite to prevent acid mine drainage. *CIM Bulletin*, **95**, 96–99.

Gandy, C.J. and Younger, P.L. 2003. Effect of a clay cap on oxidation of pyrite within mine spoil. *Quarterly Journal of Engineering Geology and Hydrogeology*, **36**, 207–215.

Geldenhuis, S. and Bell, F.G. 1997. Acid mine drainage at a coal mine in the Eastern Transvaal, South Africa. *Environmental Geology*, **33**, 233–242.

Halbert, B.E., Scharer, J.M., Knapp, R.A. and Gorber, D.M. 1983. Determination of acid generation rates in pyritic mine tailings. *Proceedings of the 56th Annual Conference of Water Pollution and Control Federation*, Atlanta (Offprint 6p).

Hallberg, K.B. and Johnson, D.B. 2003. Passive mine water treatment at the former Wheal Jane tin mine, Cornwall: important biogeochemical and microbiological lessons. *Land Contamination and Reclamation*, **11**, 213–220.

Hedin, R.S., Nairn, R.W. and Kleinmann, R.L.P. 1994. *Passive Treatment of Coal Mine Drainage*. Information Circular 9389, United States Bureau of Mines, Department of the Interior, Government Printing Office, Washington, DC.

Heier, K.S. and Billings, G.K. 1970. Rubidium. In: *Handbook of Geochemistry*, Wedepohl, K. (ed.), Springer-Verlag, Berlin, 37/B/1–37/N/1.

Jarvis, A., England, A. and Mee, S. 2003. Mine water treatment at Six Bells Colliery, South Wales: problems and solutions, from conception to completion. *Land Contamination and Reclamation*, **11**, 153–160.

Karathanasis, A.D., Evangelou, V.P. and Thompson, Y.L. 1988. Aluminium and iron equilibria in soil solutions and surface waters of acid mine watersheds. *Journal Environmental Quality*, **17**, 534–542.

Kiekens, L. 1995. Zinc. In: *Heavy Metals in Soils*, Alloway, B.J. (ed.), Blackie, Glasgow, 284–303.

Lindsay, P., Bell, F.G. and Mowatt, C. 2002. Contaminated sites in New Zealand and their clean-up using natural zeolites. *Proceedings of the Ninth Congress International Association of Engineering Geology and the Environment*, Durban. Van Rooy, L. and Jermy, C.A. (eds), Published on CD-Rom.

Lundgren, D.G. and Silver, D. 1980. Ore leaching by bacteria. *Annals Reviews Microbiology*, **34**, 263–283.

Nuttall, C.A. 2003. Testing and performance of a newly constructed full-scale passive treatment system at Whittle Colliery, Northumberland. *Land Contamination and Reclamation*, **11**, 105–112.

Nuttall, C.A. and Younger, P.L. 2000. Zinc removal from hard circum-neutral mine waters using a novel closed-bed limestone reactor. *Water Research*, **34**, 1262–1268.

Piper, A.M. 1944. A graphic procedure in the geochemical interpretation of water analyses. *Transactions of American Geophysical Union*, **25**, 914–923.

Pugh, C.E., Hossner, L.R. and Dixon, J.B. 1984. Oxidation rate of iron sulphides as affected by surface area, morphology, oxygen concentration and autotrophic bacteria. *Soil Science*, **137**, 309–314.

Rees, S.B., Bowell, R.J. and Wiseman, I. 2002. Influence of mine hydrogeology on mine water discharge chemistry. In: *Mine Water Hydrogeology and Geochemistry*, Special Publication 198, Younger, P.L. and Robins, N.S. (eds), Geological Society, London, 379–390.

Schwertmann, U. and Taylor, R.M. 1989. Iron oxides. In: *Minerals in Soil Environments*, Second Edition, Dixon, J.B. and Weed, S.B. (eds), Soil Society of America, Madison, WI, 379–438.

Sracek, O., Choquette, M. and Gelinas, P. 2004. Geochemical characetrization of acid mine drainage from a waste rock pile, Mine Doyon, Quebec, Canada. *Journal of Contaminant Hydrology*, **69**, 45–71.

Stiff, H.A. 1951. The interpretation of chemical water analysis by means of patterns. *Journal of Petroleum Technology*, **3**, 11–15.

Woodward, G.M. and Selby, K. 1981. The effect of coal mining on water quality. *Proceedings of the Symposium on Mining and Water Pollution*, Nottingham, Institution of water pollution, Nottingham, Institution of Water Engineers and Scientists, 11–19.

Younger, P.L. 1995. Hydrogeochemistry of waters flowing from abandoned coal workings in the Durham Coalfield. *Quarterly Journal of Engineering Geology*, **28**, Supplement 2, S101–S113.

Younger, P.L. 2000. Holistic remedial strategies for short- and long-term water pollution from abandoned mines. *Transactions of the Institution of Mining and Metallurgy*, Section A, Mining Industry, **109**, A210–A218.

Younger, P.L. and Robins, N.S. 2002. Challenges in the charcterization and prediction of the hydrogeology and geochemistry of mined ground. In: *Mine Water Hydrogeology and Geochemistry*, Special Publication 198, Younger, P.L. and Robins, N.S. (eds), Geological Society, London, 1–16.

Younger, P.L. and Rose, P.D. 2000. Using one waste stream to cancel out another: towards holistic management of industrial waste waters and solid wastes in the UK and South Africa. *Proceedings of the Chartered Institution Water, Environment and Management, Millenium Conference, Wastewater Treatment Standards and Technologies to Meet the Challenges of the 21st Century*, Leeds, **1**, 349–356.

Zinck, J.M. and Aube, B.C. 2000. Optimization of lime treatment processes. *CIM Bulletin*, **93**, 98–105.

Chapter 9

Dereliction and contamination associated with mining and related industries

Mining, and associated mineral processing and beneficiation have a notable impact on the environment. Such impacts depend on many factors, especially the type of mining and the size of the operation. Mining leads to land being disturbed and hydrogeological conditions are affected. In the past, the mining industry frequently showed a lack of concern for the environment. In particular, the disposal of waste led to unsightly spoils being left to disfigure the landscape, and to surface streams and groundwater being polluted. Some urban areas suffered subsidence damage due to undermining. In addition, some of the worst environmental dereliction has been associated with past mineral workings, mining activities and associated industries (Fig. 9.1).

Mineral processing may involve grinding the ore, adding various chemicals and possibly several physical separation processes. These processes result in tailings, which contain numerous metal and non-metal residues from the minerals worked, along with concentrations from the process chemicals. These compounds may include petroleum based or organic compounds, organic acids, cyanide and related compounds, and various other acids. Washing and beneficiation of coal also results in the production of sludges that have to be disposed of (Fig. 7.8(b)). Coal may have been coked at mine sites.

9.1. Dereliction

Derelict land can be regarded as land that has been damaged by mining/industrial exploitation to an extent that it has to undergo remedial treatment before it can be of beneficial use. Indeed, as remarked, some of the worst dereliction has been associated with past mineral workings and mining activities (see Chapters 6 and 7). Such land frequently has been abandoned in an unsightly condition and often is located in urban areas where land for development may be scarce. Consequently, this land, if it is not reclaimed, is not only a wasted resource but has a blighting effect on the surrounding area. Its reclamation therefore is highly desirable, not only

Figure 9.1 Derelict land left behind after the closure of Woolley Colliery and Opencast Mine, South Yorkshire, England.

by improving the appearance of the area but also by making a significant contribution to its economy by bringing derelict land back into worth while use. The use to which derelict land is put should suit the needs of the surrounding area and be compatible with other forms of land use. Any type of reclamation must take account of safety (e.g. in terms of derelict buildings, mine shafts etc) and potential problems of pollution on and off-site.

Land recycling in mining/industrial areas also can be advantageous since the infrastructure generally is still in place. Moreover, its regeneration should help prevent the exploitation of greenfield sites. The regeneration of derelict sites is linked with the process of sustainable development. Hence, the potential for redevelopment of a derelict site needs to be assessed in terms of economic, environmental and social factors that contribute to the overall concept of sustainable development, as well as site-based factors. The environmental geotechnics of a derelict site influence its subsequent reuse as well as affecting the costs of bringing the site into a state suitable for redevelopment.

Any project involving the reclamation of derelict land requires a feasibility study. This needs to consider accessibility to the site; land use and market value; land ownership and legal issues; topography and geological conditions; site history and contamination potential; and the local environment and existing infrastructure. The results of the feasibility study allow an initial assessment to be made of possible ways to develop a site and an estimation of the costs involved. This is followed by a site investigation. The investigation provides essential input for the design of remedial measures. Derelict land may present hazards, for example, disposal of mining or associated industrial waste may have contaminated land. Contaminated land may emit gases or may represent a fire hazard. Details relating to such hazards should be determined during the site investigation. Site hazards result in constraints on the freedom of action, necessitate following stringent safety requirements, may involve time-consuming and costly working procedures, and affect the type of development. For instance, Leach and Goodyear (1991) noted that constraints may mean that the development plan has to be changed so that the more sensitive land uses are located in areas of reduced hazard. Alternatively, where notable hazards exist, then a change to a less sensitive end use may be advisable. Settlement is another problem that frequently has to be faced when derelict land, which consists of a substantial thickness of fill, is to be built on. If this is not contaminated, then dynamic compaction or vibro-compaction can be used to minimize the amount of settlement. On the other hand, derelict sites may require varying amounts of filling, levelling and regrading. Once regrading has been completed the actual surface needs rehabilitating. This is not so important if the area is to be built over as it is, if it is to be used for amenity or recreational purposes. In the case where buildings are to be erected, however, the ground must be adequately compacted so that they are not subjected to adverse settlement. On the other hand, where the land is to be used for amenity or recreational purposes, then soil fertility must be restored so that the land can be grassed and trees planted.

9.2. Contamination, mining and associated industries

Many abandoned sites associated with mining and mineral processing are heavily contaminated. Indeed, one of the legacies in most of the mining areas of the world is that land has been contaminated. Hence, when such sites are cleared for redevelopment they can pose problems. Contamination can take many forms and can be variable in nature across a mining/industrial site, each site having its own characteristics. In some instances, only a single previous use of a site may be identified, that is, mining, which may have a characteristic pattern of contamination. On the other hand, at those sites where there were associated industries such as smelting or the manufacture of coal gas, the pattern of contamination may be much more complicated. The types of contaminants that may be encountered include metals, sulphates, asbestos, various organic compounds, toxic and flammable gases, combustible materials and radioactive materials. Nonetheless, the presence of potentially harmful substances at a site may not necessarily require remedial action, if it can be demonstrated that they are inaccessible to living things or materials that may be detrimentally affected. However, consideration always must be given to the migration of soluble substances.

The large amounts of waste associated with mining frequently represent a source of contamination of land, and both ground and surface water.

These wastes normally possess chemical and physical characteristics that prevent the re-establishment of plants without some form of prior remedial treatment. As noted, consideration must always be given to the migration of soluble substances. The migration of soil borne contaminants is associated primarily with groundwater movement, and the effectiveness of groundwater to transport contaminants is dependent mainly upon their solubility. The quality of groundwater can provide an indication of the mobility of contamination and the rate of dispersal. In an alkaline environment, the solubility of heavy metals becomes mainly neutral due to the formation of insoluble hydroxides. Providing groundwater conditions remain substantially unchanged during the development of a site, then the principal agent likely to bring about migration will be percolating surface water. On many sites the risk of migration off-site is of a very low order because the compounds have low solubility, and frequently most of their potential for leaching has been exhausted. However, in the longer term, the possible effects of groundwater rebound may have to be considered. Liquid and gas contaminants, of course, may be mobile. Obviously, care must be taken on-site during working operations to avoid the release of contained contaminants (e.g. liquors in buried tanks) into the soil.

Soil may be contaminated by mining waste from old workings. Davies (1972) quoted an example of lead contamination from the Tamar Valley in south-west England where some garden soils contained as much as 522 mg kg^{-1} of lead. Another notable example of lead contamination is provided by Parc Mine in North Wales. This mine produced lead and zinc from the beginning of the nineteenth century until the mid-twentieth century. A tailings dam, containing some 250 000 tonnes of fine waste, was located in a valley head on a tributary of the River Conwy. Storms during the winter of 1963–64 eroded the tailings dam spreading tailings over a distance of approximately 1 km, which effectively destroyed or seriously damaged 11 ha of agricultural land. It was estimated that several hundred tonnes of lead and zinc was deposited in the area affected and subsequently a number of instances of lead poisoning in cattle were detected, together with zinc toxicity in cereals. More recently, agricultural land in Swaledale and the Vale of York, England, was contaminated by floods in the autumn of 2000. Soil samples showed heavy metals from disused mines in the Yorkshire Dales had been washed down by the flooded River Swale for a total distance of some 80 km. The flood waters had dredged up old contaminated deposits associated with nineteenth century ore refining at old lead and copper mines. In some instances the levels of lead in fields that had been submerged by the floods were 10 times higher than the accepted safe limits.

Gold mining in South Africa has meant that huge amounts of tailings have been impounded within tailings dams. Poor construction and management, in particular, of some of the older tailings dams resulted in seepage loss that adversely affected both surface water and groundwater. Rosner and van Schalkwyk (2000) noted that some tailings dams have been partially or totally reclaimed thereby leaving behind contaminated footprints. They found that the topsoil in such areas has been highly acidified and contains heavy metals. As such, it poses a serious threat to the underlying dolomitic aquifers. Rosner and van Schalkwyk recommended that soil management measures such as liming could be used to prevent the migration of contaminants from the topsoil into the subsoil and groundwater, and would aid the establishment of vegetation. The removal of contaminated soil, because of the cost, can only be undertaken in situations where small volumes are involved.

Aucamp and van Schalkwyk (2003) referred to fine grained material that had been eroded by wind and water from five tailings dams at the abandoned Machavie Gold Mine near Potchefstroom in North West Province, South Africa, that was deposited over an area of more than 1 km^2. Sulphate salts were precipitated on the surface of this material, as well as on the surface of the tailings in the impoundments. These salts showed high water soluble concentrations of cobalt, chromium, copper, nickel, silver, lead and zinc, and represented a potential source of contamination. Furthermore, acidic leachate had penetrated into the soils beneath the tailings dams and contaminated them. Aucamp and van Schalkwyk maintained that soil pH could be used as a first indication of contamination. For example, the uppermost soil beneath both the deposited material and the tailings dams usually had a pH value of less than 4.5. In addition, Aucamp and van Schalkwyk determined the mobilities of these trace elements by comparing total trace element concentration with extractable trace element concentration. They found that the average trace element mobilities normally were higher in the material from the tailings than the soils beneath and that they increased where the pH values were less than 3.5 and 5 respectively.

Acid mine drainage is responsible for contamination of ground, and ground and surface water in coal and in metal mining areas around the world (see Chapter 8). Acid generation gives rise to elevated levels of heavy metals in the drainage water, which pollutes natural waters and can be precipitated on or in sediments (Tables 9.1 and 9.2). Bell *et al.* (2002) suggested that the accumulation of zinc, copper, nickel, manganese, chromium and lead in material precipitated from a stream polluted by acid mine drainage in the Witbank Coalfield, South Africa, could be due to adsorption to iron and sulphate minerals, which are characteristically associated with acid mine drainage. For example, jarosite [$KFe_3(OH)_6(SO_4)_2$] acted as an important trace element accumulator with high concentrations of rubidium being present in precipitates containing this mineral. The high concentrations of lead in some of the sediments concerned may be the result of sorption to goethite. High concentrations of heavy metals in surface

Table 9.1 XRF analysis of precipitates sampled in the headwater area of the Blesbokspruit, Witbank Coalfield, South Africa (After Bell *et al.*, 2002)

Oxides/elements	*Sample number*					
	P1	*P2*	*P3*	*P4*	*P5*	*P6*
Major oxides (in weight %)						
SiO_2	2.1	35	20	6.5	62	46
TiO_2	0.02	0.42	0.28	0.12	0.33	0.18
Al_2O_3	4.3	7.4	6.7	2.9	3.1	1.6
Fe_2O_3	79	37	65	81	20	40
MgO	0.02	0.09	0.18	0.01	0.12	0.01
CaO	0.06	0.12	0.06	0.02	0.4	0.02
Na_2O	0.12	1.0	0.31	0.25	0.34	0.03
K_2O	0.06	2.01	0.61	0.83	0.59	0.1
BaO	0.003	0.03	0.02	0.01	0.01	0.01
P_2O_5	1.2	0.06	0.16	0.05	0.03	0.03
SO_3	9	11	6.5	8.5	5.3	6.3
Cl	0.22	0.08	0.1	0.1	0.05	0.08
Trace elements (mg kg^{-1})						
Zn	17	44	53	34	18	8.5
Cu	9.4	20	50	39	14	12
Ni	<0.8	19	11	<0.8	15	7.1
V	800	62	123	263	198	178
Cr	69	112	570	104	93	81
Mn	37	149	276	34	178	54
Co	6.4	7.3	10	4.0	4.0	2.4
Mo	0.3	0.9	3.3	0.8	2.4	0.9
Rb	3.5	96	24	31	37	5.3
Pb	19	19	24	76	20	4.1
U	1.1	2.4	3.6	4.2	0.7	<0.9

Note
P1 and P2 samples from seepages points.

water and soils can decimate aquatic life and vegetation, and can seriously affect the health of animals.

According to Hatheway (2002) abandoned manufactured gas sites represent the most common and difficult of all contaminated waste sites. Former manufactured gas plants (FMGPs) roasted coal to drive off gas and in the process produced toxic wastes, notably tar. Unfortunately, most tar residuals and gas oils are highly resistant to natural degradation or attenuation in the environment. Therefore, potential problems associated with tar could persist for centuries. Tar residuals and gas oils are composed of complex mixtures of hundreds of aliphatic and aromatic organic hydrocarbons. The polycyclic aromatic hydrocarbons (PAHs) are of particular concern as they

Table 9.2 XRF analysis of sediments in the Blesbokspruit and catchment, Witbank Coalfield, South Africa (After Bell *et al.*, 2002)

Oxides/elements	*Sample number*										
	S1	*S2*	*S3*	*S4*	*S5*	*S6*	*S7*	*S8*	*S9*	*S10*	*S11*
Major oxides (in weight %)											
SiO_2	52	80	83	67	46	84	87	76	87	64	102
TiO_2	0.91	0.53	0.42	0.56	0.49	0.27	0.51	0.59	0.43	0.83	0.59
Al_2O_3	22	7.7	3.8	8.4	7.9	3.9	4.6	8.1	2.0	8.4	7.4
Fe_2O_3	17	5.5	10	15	36	6.1	4.4	4.2	6.8	8.2	3.2
MgO	0.18	0.07	0.03	0.10	0.08	0.05	0.06	0.11	0.02	0.24	0.28
CaO	0.04	0.02	0.02	0.03	0.04	0.08	0.02	0.03	0.02	0.11	0.04
Na_2O	0.08	0.04	0.05	0.13	0.16	0.51	0.04	0.06	0.04	0.05	0.04
K_2O	0.93	0.34	0.26	0.78	0.97	0.16	0.18	0.35	0.12	0.68	0.73
P_2O_4	0.35	0.02	0.03	0.04	0.03	0.05	0.02	0.03	0.03	0.21	0.10
SO_3	1.6	0.4	0.85	1.9	4.0	0.85	0.35	0.83	0.48	1.0	0.83
Cl	0.04	0.01	0.02	0.02	0.03	0.02	0.04	0.01	0.01	0.05	0.02
Organic carbon (%)	nd	0.33	0.48	0.47	0.61	1.76	0.48	0.94	0.68	9.5	3.9

Trace elements (mg kg^{-1})											
Zn	67	13	12	25	38	140	13	16	14	73	43
Cu	39	13	16	21	30	24	26	26	27	56	21
Ni	29	14	20	16	11	41	26	28	35	63	40
V	181	49	47	92	109	56	71	148	68	107	47
Cr	157	118	165	305	225	169	179	186	204	244	138
Mn	141	82	112	108	108	181	147	106	209	243	158
Co	9.7	3.9	4.1	3.8	5.0	14	3.9	5.7	3.9	25	19
Zr	271	617	624	514	383	294	655	442	920	240	345
Y	42	19	15	25	23	14	17	24	13	77	45
Sr	152	14	11	29	30	17	8.5	14	9.4	14	8.0
Rb	26	19	16	38	44	10	15	25	7.2	36	36
Pb	34	7.8	10	23	36	21	21	64	21	26	10
U	6.0	2.6	1.9	4.1	5.3	3.2	1.8	3.9	2.9	11	3.3

Notes
S1 sample from seepage point.
S2, S3, S4 and S5 samples from decantation ponds.
S10 sample from wetland.

are suspected of being carcinogenic. Hatheway further pointed out that most of the broad advances in dealing with toxic and persistent groundwater contaminants have been focused on dealing with solvents, pesticides and heat dissipation oils. However, solvents are volatile organic compounds (VOCs) that are very different in character from the predominant semi-volatile organic compounds (SVOCs) associated with gas manufacture. Gas manufacturing sites produced significant amounts of solid, as well as liquid waste. For instance, Hatheway estimated that three tons of brick were removed from, and replaced, at each generator set per year. Other waste solids included ash, clinker, slag, scurf (hard carbon deposits formed on the interior surfaces of retorts and generators), spent lime, spent wood chips, spent iron spirals (for capturing sulphur), and retort and bench fragments. The solid material, some of which may be contaminated, was disposed of in dumps, which usually were located around the periphery of the plant, along an adjacent stream or occupying topographic hollows. Because dumps had high void ratios, toxic wastes and sludges may have been disposed of within them. Adjacent low land often was used as unlined tar ponds. Once in the ground most manufactured gas wastes become immobile. Semi-volatile organic compounds are highly viscous and have a low solubility in water, and so come to rest in the vadose zone.

9.3. Investigation of contaminated sites

If any investigation of a site that is suspected of being contaminated is going to achieve its purpose, then its objectives must be defined and the level of data required determined, the investigation being designed to meet the specific needs of the project concerned (Anon., 1988). According to Johnson (1994) all investigations of potentially contaminated sites should be approached in a staged manner. This allows for communication between interested parties, and helps minimize costs and delays by facilitating planning and progress of the investigation. After the completion of each stage, an assessment should be made of the degree of uncertainty and of acceptable risk in relation to the proposed new development. Such an assessment should be used to determine the necessity for, and type of further investigation.

The first stage in any investigation of a site suspected of being contaminated is a desk study that provides data for the design of the subsequent ground investigation. The desk study should identify past and present uses of the site, and the surrounding area, and the potential for and likely forms of contamination. The objectives of the desk study are to identify any hazards and the primary targets likely to be at risk; to provide data for health and safety precautions for the site investigation; and to identify any other factors that may act as constraints on development. Hence, the desk study should provide, wherever possible, information on the layout of the site, including structures below ground; its physical features; the geology and hydrogeology of the site; the previous history of the site; the nature and quantities of materials handled; the processing involved; health and safety records; and methods of waste disposal. It should allow a preliminary risk assessment to be made and the need for further investigation to be established. Genske *et al.* (1994) indicated that the interpretation of historical data could be done with the aid of CAD and GIS programmes. The evaluation of the data obtained by the desk study aids the planning of the site investigation programme.

Geochemical maps can provide basic data for investigations in areas contaminated by past and present mining and smelting operations. Remote sensing techniques such as thermal hyperspectral scanners may be effective in providing information on the nature of the ground. Also, aerial photographs, especially infra-red colour photographs, can help detect contaminated ground. Vegetation affected by chemical contamination of water, or by methane does not reflect infra-red waves as well as healthy vegetation. Consequently, this can be reflected in contrasts on infra-red photographs, thereby identifying areas for further investigation.

The preliminary reconnaissance, which is often referred to as a land quality appraisal, provides the data for planning the site exploration, which includes the personal and equipment needs, the sampling and analytical requirements and the health and safety requirements. In other words, it is a fact-finding stage that should confirm the chief hazards and identify any additional ones so that the site exploration can be carried out effectively. It thereby supplements the desk study. The preliminary reconnaissance also should refer to any short-term or emergency measures that are required at the site before the commencement of full-scale operations and should formulate objectives so that the work is cost effective. A number of factors should be noted during the preliminary reconnaissance. These include the state of the site and its topography; the location of any buildings and any evidence of previous buildings; the location of disposed materials, storage tanks, ponds and pits; man-made and natural drainage; unusual colours, fumes and odours; site hazards and signs of contamination; presence or absence of vegetation, its type and condition; and site access and adjacent land use. Simple on-site testing and sampling of soil materials and water can be undertaken. Gases can be tested, for example, by hand-held photo-ionization detectors.

Contaminated sites may pose a health hazard to the personnel involved in an investigation. Clark *et al.* (1994) recommended that a safety plan, including an initial risk assessment, should be prepared prior to commencement of work on-site. The plan should include the known or suspected hazards, provide guidance as to how to recognize and deal with the contaminants involved and outline the effects that they can have on personnel. An assessment should be made as to whether protective clothing and equipment (e.g. hard hat, face shield, overalls, industrial boots, respiratory equipment) should be worn by operatives. A decontamination unit should be provided on hazardous sites.

Just as in a conventional site investigation, one that is involved with the exploration for contamination needs to determine the nature of the ground (Anon., 1999). In addition, it needs to assess the ability of the ground to transmit any contaminants either laterally, or upward by capillary action. Permeability testing therefore is required. The investigation also must establish the location of perched water tables and aquifers, and any linkages between them, as well as determining the chemistry of the water on-site. The exploratory methods used in the site exploration can include manual excavation, trenching and the use of trial pits, light cable percussion boring, power auger drilling, rotary drilling, and water and gas surveys (Bell *et al.*, 1996). Visually different materials should be placed in different stockpiles as a trench or pit is dug to facilitate sampling. Of the geophysical methods, resistivity and electromagnetic techniques have the more general application (Jewell *et al.*, 1993). They can be used for the detection of cavities and subsurface structures, and for contaminant mapping, in addition to assessment of soil type and distribution (Tables 9.3 and 9.4). If thermal activity is suspected at a site, then temperature surveys should be carried out. Thermocouples can be set out on a grid pattern, initially at 25–100 m centres, and subsequently at closer centres at suspected or discovered hot spots. Subsurface temperatures can be recorded by lowering thermocouples down boreholes, but movement of air or groundwater may affect results, or by probes driven into the ground. A sampling plan should specify the objectives of the investigation, the history of the site, analyses of any existing data, types of samples to be used, sample locations and frequency, analytical procedures and operational plan. If unidentified substances are encountered, then the investigation should be suspended until a specialist can visit the site to determine the nature of the substance and whether it poses a hazard.

Some materials can change as a result of being disturbed when they are obtained or during handling. Hence, sampling procedure should take account of the areas of a site that require sampling; the pattern, depth, types and numbers of samples to be collected; their handling, transport and storage; as well as sample preparation and analytical methods. One of the factors that should be avoided is cross contamination. This is the transfer of materials by the exploratory technique from one depth into a sample taken at a different depth. Consequently, cleaning requirements should be considered, and ideally a high specification of cleaning operation should be carried out on equipment between both sampling and borehole locations. Bell *et al.* (1996) indicated that on-site chemical analysis in mobile laboratories is becoming increasingly important in relation to the investigation of contaminated sites. Such on-site investigation for characterization of toxic

Table 9.3 Geophysical methods frequently used in environmental site assessment

Method	*Dependent physical property*	*Major applications*
Seismic refraction	Elastic moduli	Type of lithology and depth. Overburden characteristics. Degree of saturation. Location of waste boundaries
Vertical seismic profiling (surface to downhole)	Elastic moduli	Vertical and lateral changes in lithology properties. Anisotropy of linear features such as faults and major discontinuities
Resistivity	Electrical conductivity	Type of lithology and depth. Degree of saturation. Location of water table. Contaminant mapping. Cavity detection
Frequency domain electromagnetic (EM) techniques	Electrical conductivity	Type of lithology. Degree of saturation. Contaminant mapping. Cavity detection. Location of buried objects
Transient electromagnetic (TEM) techniques	Electrical conductivity	Type of lithology and depth. Linear features such as faults and major discontinuities. Contaminant mapping
Magnetometry	Magnetic susceptibility	Type of lithology. Linear features such as faults and major discontinuities. Cavity detection. Location of buried objects
Ground probing radar	Dielectric permittivity	Location of water table. Contaminant mapping. Cavity detection. Location of buried objects

chemicals has many advantages over conventional laboratories since transportation time is more or less eliminated and sample integrity is maintained. Volatile contaminants or gas producing material can be determined by sampling the soil atmosphere by using a hollow gas probe, inserted to the required depth. The probe is connected to a small vacuum pump and a flow of soil gas induced. A sample is recovered by using a syringe. Care must be taken to determine the presence of gas at different horizons by the use of sealed response zones. Standpipes can be used to monitor gases during the exploration work.

As an investigation proceeds, it may become apparent that the distribution of material about the site is not as predicted by the desk study or preliminary investigation. Hence, a site investigation programme needs to be flexible. The testing programme should identify the types, distribution and concentration (severity) of contaminants, and any significant variations or local

Table 9.4 Geophysical method: applicability of method

Problem	*Geophysical method*				
	Seismic	*Resistivity*	*EM*	*Radiowave*	*Radar*
Dimensions and depth of fill	Yes	Yes	Yes	Yes	No
Large heterogeneities	Yes	Yes	Yes	No	No
Metal drums containers and tanks	No	Yes	Yes	Yes	Yes
Cavities	Possible	Possible	Yes	Yes	Yes
Buried services	No	Possible	Yes	Yes	No
Pollution plumes, ionic	No	Yes	Yes	Yes	No
Pollution plumes, organic	No	No	No	Yes	Possible
Groundwater conditions	No	Yes	Yes	No	No

anomalies. The data obtained on the distribution of the contaminants may need to be analysed in order to estimate any possible degree of error (see Case history 1).

Investigation of contaminated sites frequently requires the use of a team of specialists and without expert interpretation many of the benefits of site investigation may be lost. Once completed, the site characterization process, when considered in conjunction with the development proposals, will enable the constraints on development to be identified. These constraints, however, cannot be based solely on the data obtained from the site investigation but must take account of financial and legal considerations. If hazard potential and associated risk are regarded as too high, then the development proposals will need to be reviewed. When the physical constraints and hazards have been assessed, then a remediation programme can be designed, which allows the site to be economically and safely developed. It is at this stage when clean-up standards are specified in conjunction with the assessment of the contaminative regime of the surrounding area.

9.4. Remediation of contaminated sites

A wide range of technologies are available for the remediation of contaminated sites and the applicability of a particular method depends on the site conditions, the nature and extent of contamination, and the extent of the remediation required. In some situations it may be possible to rely upon natural decay or dispersion of the contaminants. When the acceptance criteria are demanding and the degree of contamination complex, it is important that the feasibility of the remediation technology is tested in order to ensure that the design objectives can be satisfied (Swane *et al.*, 1993). This can necessitate that a thorough laboratory testing programme be undertaken, along with field trials. In some cases the remedial operation requires the employment of more than one method.

It obviously is important that the remedial works do not give rise to unacceptable levels of pollution either on-site or in the immediate surroundings. Hence, the design of the remedial works should include measures to control pollution during the operation. The effectiveness of the pollution control measures need to be monitored throughout the remediation programme.

In order to verify that the remedial operation has complied with the clean-up acceptance criteria for the site, a further sampling and testing programme is required (Anon., 1989). Swane *et al.* (1993) recommended that it is advisable to check that the clean-up standards are being attained as the site is being rehabilitated. In this way any parts of the site that fail to meet the criteria can be dealt with there and then, so improving the construction schedule.

9.4.1. Soil remediation

Most of the remediation technologies differ in their applicability to treat particular contaminants in the soil. Landfill disposal and containment are the exceptions in that they are capable of dealing with most soil contamination problems. Removal of contaminants from a site for disposal in an approved disposal facility frequently has been used. However, the costs involved in removal are increasing and can be extremely high, especially if an approved disposal site is not within a reasonable transport distance. On-site burial can be carried out by the provision of an acceptable surface barrier or an approved depth of clean surface filling over the contaminated material. This may require considerable earthworks. Clean covers are most appropriate for sites with a number of previous uses and associated contamination. They should not be used to contain oily or gaseous contamination. It frequently is argued that the removal of contaminated material from a site for disposal in a special landfill facility transfers the problem from one location to another. Hence, the in place treatment of sites, where feasible, is the better course of action.

Another method is to isolate the contaminants by *in situ* containment by using, for example, cut-off barriers. In addition, containment may include a cover placed over the contaminated zone(s) to reduce infiltration of surface water and to act as a separation layer between land users and the contaminated ground. Low grade contaminated materials can be diluted below threshold levels by mixing with clean soil. However, there are possible problems associated with dilution. The need for quality assurance is high (unacceptable materials cannot be reincorporated into the site). The dilution process must not make the contaminants more leachable.

Bell and Genske (2000) described the rehabilitation of Prosper III Coal Mine, and associated industries, site in the Ruhr district of Germany (Fig. 9.2). The mine was established in 1906 and together with associated industries covered an area of about 290 000 m^2. Most of the site was occupied by a coking plant and chemical factories. The mine closed in 1986, leaving a derelict site, part of which was identified as highly contaminated. Since this former mine site is located in the centre of Bottrop, it was decided that it should be rehabilitated in order to allow new industries and residential areas to be developed. Rehabilitation of the site involved excavation of contaminated areas including massive foundation remnants to a depth of about 2 m. The ground that was excavated was replaced by coarse cohesionless material, the soil properties of which had to be good enough to allow the construction of the industrial and residential buildings. The contaminated soil and foundation remnants were stored within a sector of the site that was identified as highly contaminated whereas the less contaminated material was distributed over the remaining sectors. This involved the movement of some 180 000 m^3 of soil and foundation remnants. Such excavation work, however, could lead to the release of contaminated dust into the air. Consequently, special protective measures were taken by the work force so that they were not affected. As the degree of the contamination present in the excavated soil involved a risk, the soil was continuously tested and if a given contamination threshold was exceeded, then the soil affected was removed to a special waste site. Fortunately, this was necessary only occasionally. The remediation procedure is shown in Figure 9.2. The excavated material was used to form an undulating topography, part of which is now a recreation area. The highly contaminated sectors were covered with a drain and seal system, consisting of geosynthetics and clay cover, to avoid immediate contact with contaminated soil, and to prevent the percolation of meteoric water through the contaminated ground that would cause migration of contaminants into the saturated zone beneath the water table. In addition, a number of observation wells were installed. These are continuously sampled in order to detect any possible contamination. If increased values are detected, then the observation wells will be used as recovery wells to extract and clean the contaminated groundwater.

Fixation or solidification processes reduce the availability of contaminants to the environment by chemical reactions with additives (fixation) or by changing the phase (e.g. liquid to solid) of the contaminant by mixing with another medium. Such processes usually are applied to concentrated sources of contaminated wastes. Various cementing materials such as Portland cement and quicklime can be used to immobilize heavy metals (Harris *et al.*, 1995). Cementitious stabilization permits the use of site

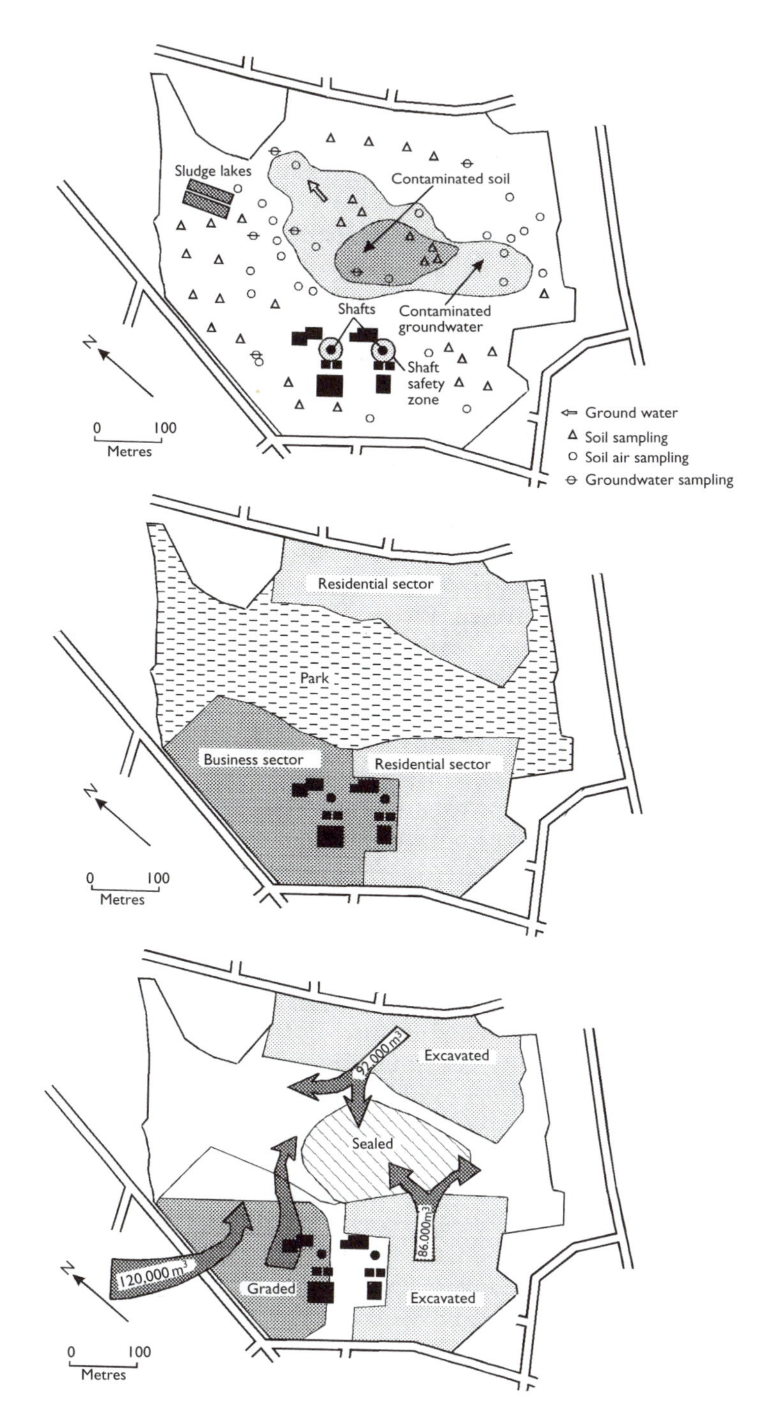

Figure 9.2 Contamination, projected land-use and remediation strategy of Prosper III site. (After Bell and Genske, 2000.)

specific mixtures and appropriate methods of mixing for a wide range of situations. For example, Al-Tabbaa *et al.* (1998) used *in situ* auger mixing, with seven different soil-grout mixes to treat contaminated ground. The mixes were made up of varying proportions of cement, lime and pulverized fuel ash and were injected to form overlapping stabilized soil columns. The auger consisted of cutting flights and mixing rods, and was designed to produce homogeneous mixing of soil and grout with minimal exposure of contaminated material. Contaminants are held by entrapment in the cementitious compounds rather than by the formation of binding chemical compounds. Accordingly, there is the possibility that contaminants could leach out in the long-term if the cementitious compounds begin to break down. In order to investigate the possibility of break down, Reid and Brookes (1999) undertook a test programme to assess the durability of lime stabilized contaminated materials. Leaching tests showed an initial concentration of mobile ions such as Na and Cl, with concentrations of Cu, Ni and phenols exhibiting a similar pattern, that is, high initial concentrations that decreased as testing progressed. Overall, however, the test results showed that the stabilized material maintained its long-term integrity and the concentrations of most metals in the leachate were below detection limits. Reid and Brookes therefore concluded that the mobility of the individual contaminants appears to depend upon their speciation. They supported the use of cementitious stabilization as a method of dealing with contaminated ground but pointed out that it would be necessary to carry out a site specific risk assessment to show that both people and the environment would be unaffected by such treatment.

Soil washing involves using particle size fractionation; aqueous-based systems employing some type of mechanical and/or chemical process; or counter current decantation with solvents for organic contaminants, and acids/bases or chelating agents for inorganic contaminants, to remove contaminants from excavated soils (Trost, 1993). Particle size fractionation is based on the premise that contaminants generally are more likely to be associated with finer particle sizes. Hence, in particle size fractionation the finer material is subjected to high pressure water washing, which can include the use of additives. Aqueous-based soil wash systems generally make use of the froth flotation process to separate contaminants. Basically, counter current decantation uses a series of thickeners in tanks into which the contaminated soil is introduced and allowed to settle. The contaminated soil is pumped from one tank to the next and the solvent, acid/base or chelating agent is in the last tank. Variation in the type of soil and contaminants at a site mean that soil washing is more difficult, and that a testing programme needs to be carried out to determine its effectiveness. Welsh and Burke (1991) described the use of high pressure soil washing to treat soil contaminated with phenol. In this method, a drillhole is advanced to the required depth, then water and air are jetted from closely spaced nozzles at the base of a triple fluid phase pipe, as in jet grouting, to displace the contaminated soil to the surface. At the same time bentonite slurry is introduced to stabilize the ground. The removed soil is cleaned by oxidation in a self-contained decontamination unit, then mixed with cement and returned in place.

Steam injection and stripping can be used to treat soils in the vadose zone contaminated with volatile compounds. The steam provides the heat that volatizes the contaminants, which then are extracted (i.e. stripped). The efficiency of the process depends upon the ease with which steam can be injected and recovered. As steam is used in near-surface soils, since much of it can be lost due to gas pressure fracturing the soil, then an impermeable cover may be used to impede the escape of steam. One of the disadvantages is that some steam turns to water on cooling that means that some contaminated water remains in the soil.

Solvents can be used to remove contaminants from the soil. For example, soil flushing makes use of water, water-surfactant mixtures, acids, bases, chelating agents, oxidizing agents and reducing agents to extract semi-volatile organics, heavy metals and cyanide salts from the vadose zone of the soil. The concept of soil flushing is based on the premise that a selective liquid extractant that is introduced into the soil will concentrate and remove contaminants as it moves downwards (Dawson and Gilman, 2001). Ultimately, hydraulic capture such as pump and treat is required to recover the contaminated liquid. The technique is used in soils that are sufficiently permeable (ideally not less than 10^{-5} m s^{-1}) to allow the solvent to permeate and the more homogeneous the soil is, the better. However, prefabricated vertical drains can be used to facilitate the flushing operation by reducing the drainage path and the travel times between injection and extraction points in less permeable ground. Every effort should be made to prevent the contaminated extractant from invading the groundwater. Other disadvantages include the difficulty in achieving good coverage and mixing, and the incomplete capture of the contaminated liquid. In fact, due to the possibility of soil flushing having an adverse impact on the environment, the United States Environmental Protection Agency (Anon., 1990) recommended that the technique should be used only when others with lower impacts on the environment are not applicable. Solvents also can be used to treat soils that have been excavated.

Some contaminants can be removed from the ground by heating. For instance, soil can be heated to between 400°C and 600°C to drive off or decompose organic contaminants such as hydrocarbons and volatile organic compounds. Mobile units can be used on site, the soil being removed, treated and then returned as backfill. Norris *et al.* (1999) referred to the use of low temperature thermal desorption (LTTD) to treat soil on part of a site affected by polychlorinated biphenyl (PCB) contamination. Low temperature thermal desorption at around 400°C is suitable for dealing with soils on site that are contaminated with low to middle distillate organic compounds such as petrol, diesel fuel and lubricating oils. The contaminated soil is fed continuously through a rotary kiln where it is heated to temperatures sufficient to evaporate or combust the contaminants, so stripping them from the soil. The exhaust gases and any non-combusted vapours pass through dust filters into a thermal oxidizer unit or after burner, where a minimum temperature of 850°C ensures their destruction. The treated soil passes out of the plant and is available for reuse.

Incineration, whereby wastes are heated to between 1500°C and 2000°C, is used for dealing with contaminated wastes containing organic compounds such as polychlorinated biphenyl (PCBs) that are difficult to remove by other techniques. Incineration involves removal of the soil and it is then usually crushed and screened to provide fine material for firing. The ash that remains may require additional treatment since heavy metal contamination may not have been removed by incineration. It then is disposed of in a landfill.

In situ vitrification transforms contaminated soil into a glassy mass. It involves electrodes being inserted around the contaminated area and sufficient electric current being supplied to melt the soil (the required temperatures can vary from 1600°C to 2000°C). According to Dawson and Gilman (2001), a layer of graphite conducting material and glass frit (a fusable ceramic mixture) is added to the surface of the soil in order to facilitate the development of the high temperatures required. The volatile contaminants are either driven off or destroyed, whilst the non-volatile contaminants are encapsulated in the glassy mass when it solidifies. It may be used in soils that contain heavy metals.

Bioremediation involves the use of microbial organisms to bring about the degradation or transformation of contaminants so that they become harmless (Loehr, 1993). The micro-organisms involved in the process either occur naturally or are artificially introduced. In the case of the former, the microbial action is stimulated by optimizing the conditions necessary for growth (Singleton and Burke, 1994). The principal use of bioremediation is in the degradation and destruction of organic contaminants, although it also has been used to convert some heavy metal compounds into less toxic states. The rate of bioremediation may be slow in some cases because of the physical and/or chemical nature of the contaminants. Physical behaviour may influence the availability of micro-organisms due to low solubility or strong adsorption to soil particles. Biodegradation also may be inhibited if special organisms are required or if different conditions are needed for various stages of the degradation process. Bioremediation can be carried out on the ground *in situ* or ground can be removed for treatment. *In situ* bioremediation depends upon the amenability of the organic compounds to biodegradation, the ease with which oxygen and nutrients can reach the contaminated area, the permeability of the soil, its temperature and the pH value. Air circulation brought about by soil vacuum extraction enhances the supply of oxygen to micro-organisms in the vadose zone of the soil. Other systems used for *in situ* bioremediation in the vadoze zone include infiltration galleries or injection wells for the delivery of water carrying oxygen and nutrients. In the phreatic zone water can be extracted from the contaminated area that then is treated at the surface with oxygen and nutrients, and subsequently injected back into the contaminated ground. This allows groundwater to be treated. However, the *in situ* method of treatment when applied to soil or water alone may prove unsatisfactory due to recontamination by the remaining untreated soil or untreated groundwater. For example, contaminants in the groundwater, notably oils and solvents that are immiscible, recontaminate soil as groundwater levels fluctuate. On the other hand, contamination in the soil may be washed out continuously over many years so contaminating groundwater. Therefore, as Adams and Holroyd (1992) recommended, it frequently is worthwhile cleaning both soil and groundwater simultaneously. *Ex situ* bioremediation involves the excavation of contaminated ground, placing it in beds where it is treated and then returning it as cleaned backfill to from where it was removed. There are a number of different types of *ex situ* bioremediation that include land farming, composting, purpose built treatment facilities and biological reactors.

9.4.2. Groundwater remediation and soil vapour extraction

Contaminated groundwater either can be treated *in situ* or can be abstracted and treated. The solubility in water and volatility of contaminants influence the selection of the remedial technique used. Some organic liquids are only slightly soluble in water and are immiscible. These are known as non-aqueous phase liquids (NAPLs), when dense they are referred to as DNAPLs or 'sinkers' and when light as LNAPLs or 'floaters'. Examples of the former include many chlorinated hydrocarbons, whilst petrol, diesel oil and paraffin provide examples of the latter. The permeability of the ground influences the rate at which contaminated groundwater moves and therefore the ease and rate at which it can be extracted.

The pump and treat method is the most widely used means of remediation of contaminated groundwater (Haley *et al.*, 1991). The latter is abstracted from the ground concerned by wells, trenches or pits and treated at the surface. It then is injected back into the ground. The pump and treat method proves most successful when the contaminants are highly soluble and are not readily adsorbed by clay minerals in the ground. On the other hand, LNAPLs can be separated from the groundwater either by using a skimming pump in a well or at the surface using oil–water separators. It usually does not prove possible to remove all light oil in this way so that other techniques may be required to treat the residual hydrocarbons. Oily substances and synthetic organic compounds normally are much more difficult to remove from the ground. In fact, the successful removal of DNAPLs is impossible at present. As such, they can be dealt with by containment. Methods of treatment that are used to remove contaminants that are dissolved in water, once it has been abstracted, include standard water treatment techniques, air stripping of volatiles, carbon adsorption, microfiltration and bioremediation (Swane *et al.*, 1993).

Active containment refers to the isolation or hydrodynamic control of contaminated groundwater. The process makes use of pumping and recharge systems to develop zones of stagnation or to alter the flow pattern of the groundwater. Cut-off walls are used in passive containment to isolate the contaminated groundwater. Additives capable of capturing and holding or degrading contaminants can be placed in permeable barriers that are located in the flow path of a plume. Such barriers must be wide enough for the residence time of the captured liquid to be sufficient to complete its task.

In situ bioremediation makes use of microbial activity to degrade organic contaminants in the groundwater so that they become non-toxic. Oxygen and nutrients are introduced into the ground to stimulate activity in aerobic bioremediation whereas methane and nutrients may be introduced in anaerobic bioremediation.

Electrokinetic remediation involves using electro-osmosis and electromigration to remove contaminants (Taha, 1997). A direct current is sent between a series of electrodes located across the area to be treated, causing charged particles in the pore fluid to move towards the electrodes. This movement, in turn, causes water and associated non-ionic contaminants to flow in the same direction. Sorbents or other media are placed in the path of induced flow between the electrodes to recover the contaminants. Groundwater also is pumped from the zones around the electrodes.

Vacuum extraction involves the removal of the gaseous and liquid phases of contaminants by the use of vacuum extraction wells. It can be applied to volatile organic compounds (VOCs) residing in the unsaturated soil or to volatile LNAPLs resting on the water table. The method can be used in most types of soils, although its efficiency declines in heterogeneous and high permeability soils. The VOCs are transferred from the liquid to the vapour stage during extraction and vapour from the exhaust pipes can be treated when necessary.

The applicability of vacuum systems can be extended to less volatile constituents and to clay soils where the moisture content inhibits the flow of soil gases during the extraction process by heat enhanced soil vapour extraction. Heat may be introduced into the ground by way of electrodes, by injection of hot air or steam, or by the application of radio frequency waves. The heat increases the vapour pressure of the contaminants and evaporates pore water thereby opening additional conduits for vapour movement. Dawson and Gilman (2001) indicated that in some cases enough heat can be added to dewater saturated zones and provide access to DNAPL that might otherwise be inaccessible.

Air sparging is a type of *in situ* air stripping in which air is forced under pressure through the ground in order to remove volatile organic contaminants. It also enhances desorption and bioremediation of contaminants in saturated soils. The air is removed from the ground by soil venting systems. The injection points, especially where contamination occurs at shallow depth, are located beneath the area affected. According to Waters (1994) the key to effective air sparging is the contact between the injected air and the contaminated soil and groundwater. Because air permeability is higher than that of water, greater volumes of air have to be used to treat sites. Nevertheless, air sparging can be used to treat sites when pump and treat methods are ineffective. The air that is vented may have to be collected for further treatment as it could be hazardous.

Biosparging is a closed loop system used to remove gases from a contaminated plume, treat the vapour in a bioadsorption tank, enrich the vapour stream with oxygen and heat, and re-inject it into screened wells beneath groundwater level at the source of the plume for complete degradation to occur (Madhav *et al.*, 1997). However, disadvantages include the difficulty in achieving good coverage and mixing, and the incomplete capture of the contaminated liquid. Extraction wells are located around the perimeter of the plume to induce the migration of contaminants from the more concentrated zone towards the wells. In this way, the concentration is reduced to the point where natural degradation by micro-organisms can take over.

9.5. Some metals and environmental health

Plants and animals, including humans, essentially are made up of 11 elements, namely, H, O, C, N, Ca, Mg, K, Na, P, S and Cl. In addition to these elements certain others are required in trace amount. Two main groups of trace elements are of particular importance as far as health is concerned. According to Mills (1996), those that are essential to animals are Fe, Mn, Ni, Co, Cr, Cu, Zn, V, Mo, Sn, Se, I and F. By contrast, potentially harmful elements, which have adverse physiological significance at relatively low levels include Pb, As, Cd and Hg, and some of the daughter products of U (Table 9.5).

9.5.1. Lead

Although lead (Pb) is a major constituent of more than 200 minerals, most of these are very rare and only three are commonly found in sufficient abundance to form mineable lead deposits. These are galena (PbS), anglesite

Table 9.5 Some trace elements and man[a] (From Bowen, 1979; WHO, 1993; reproduced by kind permission of the World Health Organization)

Element	*Symbol*	*Normal intake (mg day^{-1})*	*Intake level above which toxicity occurs (mg day^{-1})*	*Lethal intake (mg day^{-1})*	*WHO (1993) drinking water guideline maxima (mg l^{-1})*
Arsenic	As	0.1–0.3	5–50	100–300	0.01 (P)
Boron	B	10–30	4000		0.3
Cadmium	Cd	0.5	3		0.003
Chromium	Cr	0.05	200	3000	0.05 (P)
Cobalt	Co	0.0002	500		2 (P)
Copper	Cu	2–5	250–500		0.01
Lead	Pb	0.3–0.4		10 000	
Manganese	Mn	3–9			0.5(P)
Mercury	Hg	0.005–0.02		150–300	0.001
Molybdenum	Mo	0.5			0.07
Selenium	Se	0.03–0.075	3.0		0.01
Silver	Ag	0.06–0.08	60	1300	

Notes
a Mean body weight = 70 kg; weight of dry diet = 750 g day^{-1}.
WHO = World Health Organization (Anon., 1993); P = provisional.

($PbSO_4$) and cerrusite ($PbCO_3$), a lead sulphide, sulphate and carbonate respectively. Galena is a common primary constituent of sulphide ore deposits whereas anglesite and cerrusite normally form by the oxidation of galena close to the surface. Ores of lead and zinc are often closely associated in deposits formed by replacement of limestone or dolostone. Also, lead ore frequently is present together with ores of copper, zinc, silver, arsenic and antimony in complex vein deposits that are related genetically to granitic intrusions, but it may occur in a variety of host rocks.

Soil may be contaminated by mining waste from old lead workings. As mentioned earlier, Davies (1972) noted an example from the Tamar Valley in south-west England where some garden soils contained as much as 522 mg kg^{-1} lead. Radishes were used to measure lead uptake and they contained up to 74 mg kg^{-1} in their dry ash. However, a number of factors decrease the solubility of lead from mine waste (Davis *et al.*, 1994). These include the mineral composition, the degree of encapsulation in pyrite of silicate matrices, the nature of the alteration rinds and the particle size. As such they decrease the bioavailability of lead in soils derived from or contaminated by mining wastes, especially when compared with smelter derived or urban soils (Fig. 9.3). Although guidelines on acceptable levels of Pb in urban areas generally have not been established, Calder *et al.* (1994) suggested the thresholds given in Table 9.6 for the Port Pirie lead abatement programme, South Australia.

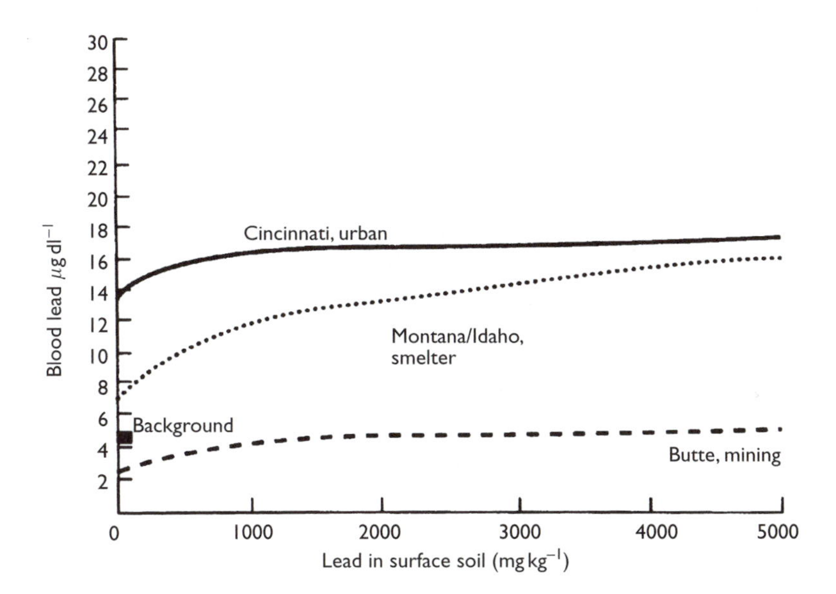

Figure 9.3 Comparison of the blood lead dose response at mining smelter and urban sites in the United States. (After Davis *et al.*, 1994; reproduced by kind permission of Springer.)

Table 9.6 Port Pirie lead soil abatement protocols 1984–94 (After Calder *et al.*, 1994; reproduced by kind permission of Springer)

Category	*Pb level (ppm)*	*Action*
A	<250	No action
B	250–1250	Home owners are advised to maintain a barrier between soil and children and advice given to parents about behaviour
C	1250–5000	As for category B plus assistance to cover contaminated soil with 50 mm of gravel or soil
D	>5000	Excavate contaminated soil and replace with clean fill

Lead in animals, including man, is concentrated largely in bone. Lead interferes with normal maturation of erythroid elements in the bone marrow and inhibits haemoglobin synthesis in precursor cells. Studies conducted on animals have shown that lead is an immuno-suppressive agent at levels well below those causing overt toxicity. It also affects porphyrim metabolism and interferes with the activity of several enzymes (Cannon, 1976). Overt manifestations of acute lead poisoning, plumbism, in children differ from those of adults, and their mortality rate is relatively high. Indeed, children less than 6 years old are those at highest risk from exposure to environmental lead (Lutz *et al.*, 1994). Symptoms include anaemia, gastrointestinal distress and encephalopathy. The latter may result in early death, permanent symptoms of brain damage or recovery may take place. Adult or chronic lead poisoning requires years to develop to a critical level, which is recognizable by symptoms such as headache, irritability, behavioural changes, constipation, impairment of mental ability, abdominal tenderness, colic, weight loss, fatigue and neuro-muscular problems and muscle pains (Tebbutt, 1983). Renal lesions, which may be caused by lead ingestion, can give rise to hypertension and may predispose to gout as a result of defective urate secretion. Chronic exposures to lead in water can result in increased miscarriages and stillbirths.

Yet another health problem that appears to be associated with lead was mentioned by Anderson and Davies (1980). They reported the results of surveys of the prevalence of dental caries in children in the Tamar Valley, England, and in Ceredigion, Wales. Both are areas where past base metal mining has left a legacy of extensive heavy metal contamination of agricultural and garden soils. In both areas a higher level of dental caries was associated with high levels of lead in soils, which was available to plants (Fig. 9.4).

9.5.2. Arsenic

Arsenic (As) is the main constituent of more than 200 mineral species of which 60% are arsenates, 20% sulphides and sulphosalts, and the remaining 20% include arsenides, arsenites, oxides, silicates and elemental arsenic (Thornton, 1996). It is associated with many types of mineral deposits, especially those containing sulphide minerals. The common arsenic minerals are arsenopyrite (FeAsS), orpiment (As_2S_3), realgar (As_2S_2) and enargite (Cu_3AsS_4). Arsenic usually is present in chalcopyrite ($CuFeS_2$), iron pyrite (FeS_2) and galena (PbS), and more rarely in sphalerite (Zn, Fe)S.

Weathering may mobilize arsenic as salts of arsenous acid and arsenic acid. Under oxidizing conditions arsenates are stable species. Under reducing conditions arsenites are the predominant arsenic compounds. Inorganic arsenic compounds can be converted to methylated arsenic species by micro-organisms, by plants, by animals and by man. The oxidative methylation reactions act on trivalent arsenic compounds and produce methylarsonic acid, dimethylarsinic acid and trimethylarsine oxide. Under reducing conditions these arsenic compounds can be reduced to volatile and easily oxidized methylarsines (Thornton, 1996). Arsenic species in aqueous systems are mainly of arsenite and arsenate oxyanions and are highly soluble over a wide range of pH and Eh conditions. However, under reducing conditions in the presence of sulphide, the mobility of arsenic is reduced due to precipitation as orpiment, realgar or arsenopyrite (although at low pH, the aqueous species $HAsS_2$ may be present). Biomethylation of arsenic also may occur giving rise to monomethylarsonic acid (MMAA) and dimethylarsinic acid (DMAA). These usually are rare in natural waters compared to

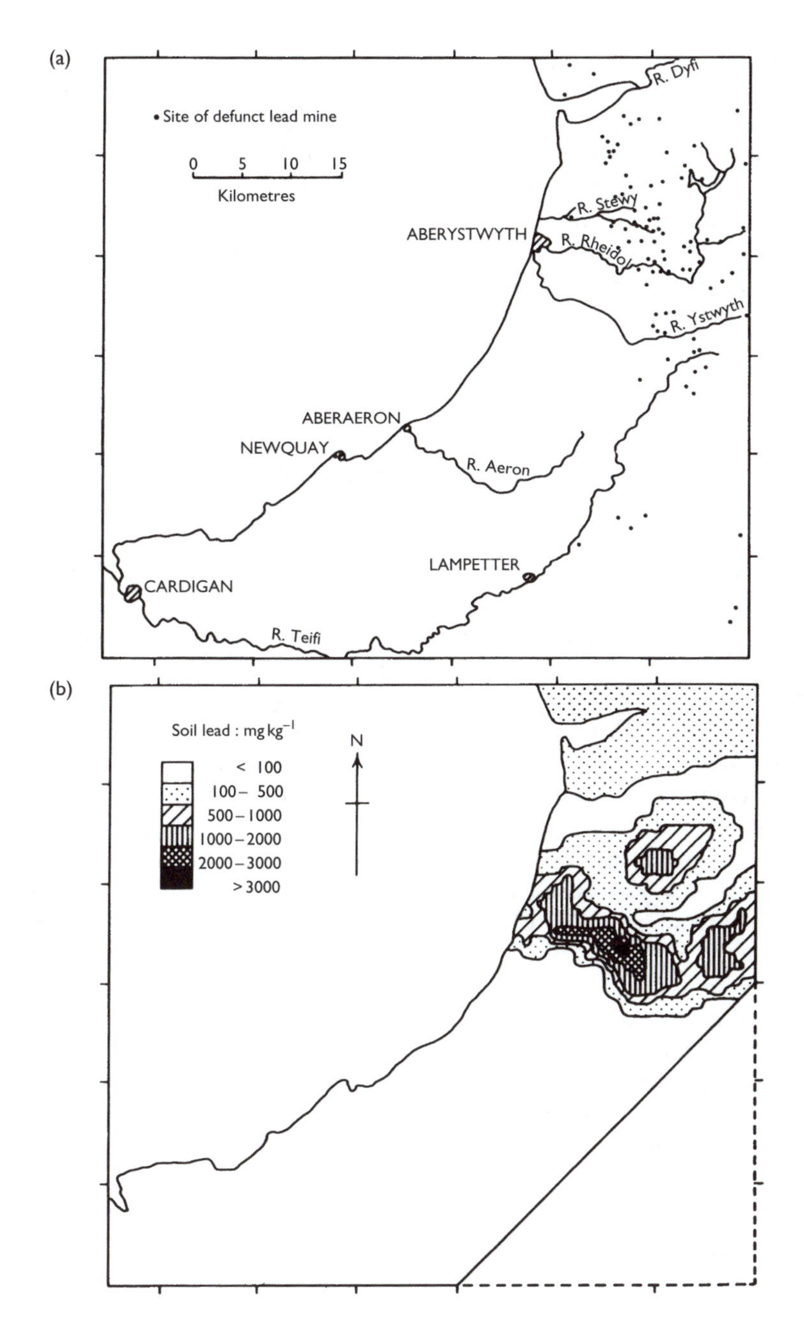

Figure 9.4 Ceredigion district of Wales (a) positions of old lead mines, (b) map of soil lead levels. (After Anderson and Davies, 1980; reproduced by kind permission of the Geological Society of London.)

the organic forms but may be present in relatively high concentrations in organic rich waters.

The average concentration of arsenic in soil is around 56 mg kg^{-1} and so generally is higher than that found in rocks. However, soils close to or derived from sulphide ore deposits may contain up to 8000 mg kg^{-1}. The concentration of arsenic in unpolluted fresh waters typically ranges from 1 to 10 $\mu g\ l^{-1}$, but can rise to 100–5000 $\mu g\ l^{-1}$ in areas of sulphide mineralization such as pyrite and arsenopyrite, and of mining whereby high arsenic may be produced by processing and by disposal of mine wastes.

Arsenic contamination of the environment has arisen as a result of mining and smelting activities in several countries. However, a survey of the effects of arsenic mining wastes and soils in south-west England by Mitchell and Barr (1995) suggested that the additional number of deaths arising from widespread contamination was small. On the other hand, arsenic toxicity in cattle was positively identified in the form of dysentry and respiratory distress. In fact, it was suggested that uptake of arsenic by affected cattle could be as high as 50 mg day^{-1} due to grazing on contaminated pastures, with 60–70% of the arsenic being present in accidentally ingested soil. Drainage from disused metalliferous mines and coal mines is frequently a source of both soluble and particulate arsenic in surface river systems. The roasting of sulphide ores containing arsenic and burning of arsenic rich coal releases arsenic trioxide, which may react in air with basic oxides, such as alkaline earth oxides, to form arsenates. These inorganic arsenic compounds then can be deposited onto soils and may be leached into surface and groundwaters.

Arsenic is toxic and as little as 0.1 g of arsenic trioxide may be lethal to man (Jarup, 1992). Also, at high doses arsenic is a human carcinogen. Skin cancer and a number of internal cancers such as cancers of the bladder, liver, lung and kidneys, can result (Morton and Dunette, 1994). Hyperpigmentation, depigmentation, keratosis and peripheral vascular disorders are the most commonly reported symptoms of chronic arsenic exposure. Neurological effects include tingling, numbness and pheripheral neuropathy. Inhalation of arsenic may give rise to respiratory cancer. As illustration, Tseng (1977) showed that a relationship existed between high concentrations of arsenic in drinking water and skin cancer, keratosis and Blackfoot Disease (a type of gangrene) in south-west Taiwan. Subsequent investigations by Chen *et al.* (1992) established relationships between high arsenic exposure and the internal cancers mentioned previously.

Toxicity depends on the form of arsenic ingested, notably the state of oxidation and whether in organic or inorganic form. Reduced forms of arsenic are more toxic than oxidized forms, with the order of toxicity decreasing from arsine, through organo-arsine compounds, arsenite and oxides, arsenate, arsonium to native arsenic (Welsh *et al.*, 1988). Arsenic intake by man probably is greater from food (e.g. seafood) than from drinking water, however, that present in fish occurs as organic forms of low toxicity. Drinking water therefore represents by far the greatest hazard since the species present in groundwater are predominantly the more toxic in inorganic forms (Smedley *et al.*, 1996). In specific situations exposure may be added to through the inhalation of atmospheric particulates derived from smelter emission or from arsenic rich soils and waste materials. Like that of lead, the World Health Organization (Anon., 1993) guideline maxima for arsenic in drinking water is 0.01 mg l^{-1} (the latter is provisional).

9.5.3. Cadmium

Cadmium (Cd) is more volatile than most heavy metals, with a boiling point of 790°C. Consequently, appreciable amounts of cadmium, primarily in gaseous form, are released to the atmosphere when zinc and lead ores are processed. For instance, the cadmium content of sphalerite may be as high as 5000 ppm. The released gas is oxidized rapidly and then is deposited as fine particles over the surrounding areas. As such, this represents a major source of cadmium to the environment. The concentration of cadmium in natural waters also can be affected by mine waters. The World Health Organization (Anon., 1993) recommended a guideline with a maximum for cadmium as 0.003 mg l^{-1} for drinking water.

Cadmium is an acute toxin, producing symptoms such as giddiness, vomiting, respiratory difficulties, cramps and loss of consciousness at high doses. Chronic exposure to the metal can lead to kidney disorders, anaemia, emphysema, anosmia (loss of sense of smell), cardiovascular diseases, renal problems and hypertension (Robards and Warsford, 1991). Furthermore, there is evidence that increased ingestion of cadmium can promote copper and zinc deficiency in man and it also may be a carcinogen (Tebbutt, 1983). Schroeder (1965) suggested that cadmium may be implicated in hypertension, in that humans exhibiting arterial hypertension pass appreciably more cadmium in their urine than do normal individuals. Once cadmium is in the blood stream, it is transported to the kidneys and liver, where about two-thirds of the total body cadmium is retained. In particular, cadmium accumulates in the kidneys during childhood and adolescence where it is bound to a protein (metallothiomein) of unknown function. Cadmium accumulates slowly over the course of a lifetime and there is no known mechanism for ridding the body of it.

Itai-itai appears to be a cadmium related disease, which is a very painful disease that causes wastage and embrittlement of bones. The disease occurred along the River Zintsu in Japan in 1945. Mining wastes associated with the processing of lead and zinc, and containing cadmium, were disposed of into the river system. The contaminated water was used for agricultural and domestic purposes (Pettyjohn, 1972). Although samples of water generally contained less than 1 mg l^{-1} of cadmium and 50 mg l^{-1} of zinc, these elements were selectively concentrated in sediment and yet more highly in plants. For example, the cadmium content in rice averaged 125 mg kg^{-1} and in the roots it was 1250 mg kg^{-1}.

9.5.4. Mercury

Although there are more than one dozen mercury (Hg) bearing minerals, only a few are abundant in nature, cinnabar (HgS) being the most important. It generally is found in mineral veins (e.g. quartz-gold veins), as impregnations or having replaced quartz in rocks near recent volcanic or hot spring areas.

Because of the tendency of mercury to vapourize, it enters the atmosphere in both gaseous and particulate forms. Gaseous and particulate mercury commonly occur in the fumes given off by various smelting processes. Mercury has been used to recover gold, for example, in placer mining. Indeed, one of the legacies of the Californian gold rush of the mid-nineteenth century is mercury contamination. It is estimated that 11.9 million kg of elemental mercury was used in the gold mining industry between 1850 and 1900 and much of that mercury is still in soils and sediments. Basically, rock particles containing gold are crushed to free the gold. Mercury then is added that combines with the gold dust to form an amalgam (the process requires about 2 kg of mercury for each kilogram of gold recovered). The amalgam is heated to drive off the mercury as vapour leaving a residue of gold. If the heating is done in a retort, then mercury vapour can be condensed and recovered. However, not all the mercury is recovered in this process and this, together with mercury spilled, gets into the environment. In the case of artisanal or small-scale gold mining, heating the amalgam exposes miners to elemental mercury vapour. Inhalation of the mercury vapour is the commonest route of elemental mercury poisoning. Acute symptoms include cough, chest tightness, difficulty breathing, confusion, weakness, mouth sores, kidney damage and even pneumonia. The discharge of this mercury into rivers means that it enters the food chain.

In 1983, several alluvial and epigenetic gold occurrences were discovered by artisanal miners on Mindanao Island, in the Philippines. This was followed by uncontrolled gold rushes and the development of several mining communities. The most significant of these was Diwalwal, which had 100 000 inhabitants at its peak in the mid-1980s. It was estimated that 50 tonnes of mercury was discharged into the local river system from the mineral processing that involved amalgamation of the gold ore. Most of the gold now is extracted with the aid of cyanide but amalgamation with mercury is still used to process about 25% of the ore. The rudimentary mining and processing of the ore caused severe environmental contamination, so that in the 1980s and 1990s there was a decline of 50% in the local yields of rice, the death of a high proportion of cattle, and widespread skin disease in the local population. Investigations carried out at the Naboc River showed that mercury in solution was 40 to 3000 times greater than the recommended concentrations for drinking water. The World Health Organization (Anon., 1993) suggests a guideline with a maximum value of 0.001 mg l^{-1} for mercury in drinking water. Mercury contaminated water from the river was used to irrigate rice in paddy fields. Consequently, mercury in soil samples from the paddy fields averages 24 and reaches a maximum of 96 mg kg^{-1} (Appleton, 2001). In Britain the maximum permissible concentration of mercury in agricultural soils is 1 mg kg^{-1}.

Mercury in living tissues is largely organic and primarily methyl mercury (CH_3Hg), the latter being soluble in water. It tends to concentrate in living tissue and at critical concentrations can be extremely toxic. The toxic effects of water borne mercury on man were emphasized in 1953 when over 100 people in the vicinity of Minamata, Kyushu, Japan, developed strange symptoms, including tottering, jerky gaits, loss of manual dexterity, impairment of speech, and commonly, deafness and blindness (Minamata disease). Eventually 50 of those afflicted died. Extensive investigations revealed that the deaths were caused by the consumption of fish and shellfish contaminated by mercury from Minamata Bay. The bay had received large amounts of methyl mercury compounds in the waste effluents from a factory (Kurland *et al.*, 1960). Methyl mercury can penetrate and erode brain tissue, particularly the areas that control sight, hearing and equilibrium.

9.5.5. Uranium

Naturally occurring uranium occurs as three isotopes, ^{238}U, ^{235}U and ^{234}U, all of which are radioactive. ^{238}U and ^{235}U are the parent nuclides of two independent decay series whilst ^{234}U is a decay product of the ^{238}U series. However, uranium is not found native but occurs in a number of minerals such as pitchblende (uraninite, UO_2), carnotite [$K_2(UO_2)_2(VO_4)_2.3H_2O$], and minerals of the uranite group such as autunite [$Ca(UO_2)_2(PO_4)_2.10H_2O$] and torbernite [$Cu(UO_2)_2(PO_4)_2.12H_2O$]. Radioactivity is the spontaneous emission of radiation. Emission of radiation from a radioactive isotope involves the loss of small particles from the nuclei of its atoms and by losing particles, the nuclei are transformed into nuclei of a different element. This second element also may be radioactive so that its atoms decay to a third element and so on until ultimately a nucleus is formed that is stable (Table 9.7). The time taken for half of a given mass of an element to decay to the next element in the series is referred to as the half-life. Two kinds of particles account for the bulk of those emitted, namely, alpha particles and beta particles. Alpha particles consist of 2 protons and 2 neutrons (nuclei of helium atoms) and beta particles consist of electrons. The electromagnetic waves that may accompany alpha or beta particles are

Table 9.7 Radioactive decay series of ^{238}U

Radioactive isotopes	*Half-life*	*Type of radiation*[a]
^{238}U	4.5×10^9 years	Alpha
^{234}Th	24 days	Beta
^{234}Pa	6.8 hr	Beta
^{234}U	2.5×10^5 years	Alpha
^{230}Th	80 000 years	Alpha
^{226}Ra	1600 years	Alpha
^{222}Rn	3.8 days	Alpha
^{218}Po	3 min	Alpha
^{214}Pb	29 min	Beta
^{214}Bi	20 min	Beta
^{214}Po	64 microseconds	Alpha
^{210}Pb	22 years	Alpha
^{210}Bi	5 days	Beta
^{210}Po	138 days	Alpha
^{206}Pb	Stable	

Notes
a All decays emit gamma radiation.
Bi = bismuth, Pa = Protactinium, Pb = lead, Po = polonium, Ra = radium, Rn = radon, Th = thorium, U = uranium.

called gamma rays and are the most penetrating of the three types of radiation, and so are the most damaging to the body. According to Krauskopf (1988), beta particles are absorbed by a few metres of air but if skin is unprotected they can penetrate the flesh sufficiently to cause burns and can do serious harm if a beta particle source is taken into the body. Alpha particles do no damage from an external source, however, if even traces of an alpha emitting isotope enter the body by ingestion or inhalation they can cause serious disruption of nearby cells.

Uranium remains practically immobile as long as conditions remain reducing, measured concentrations in groundwater are around 1 ppb. Where there is evidence of migration of groundwater into the rock surrounding the ore, then the movement of uranium is restricted to several metres. On the other hand, if oxidizing groundwater has been in contact with uranium ore the uranium is oxidized to the hexavalent state in which it is mobile. The oxide may be reprecipitated from solutions of hexavalent uranium if the solutions move from an oxidizing to a reducing environment.

Uranium is mined in open-pit or underground mines or by the *in situ* leach process. The latter involves pumping a leaching liquid through drillholes into the uranium deposits and then pumping out the uranium bearing liquid (see Chapter 10). This method only can be used when the uranium deposits occur in a permeable rock mass confined between impermeable rock masses. The uranium content of ore that is mined is often between 0.1 and 0.2% so that large amounts of ore have to be mined to obtain the uranium, meaning that there are large amounts of waste produced. However, some uranium deposits discovered in Canada have uranium grades of several percent. Spoil heaps of waste rock produced during mining often contain elevated concentrations of radioactive isotopes so that they can release dust, radon and seepage water containing radioactive and toxic materials. Ore mined in open-pits or underground mines is crushed and leached in a uranium mill. In most cases sulphuric acid is used as the leaching agent, extracting not only uranium from the ore but other constituents such as arsenic, lead, iron, molybdenum, selenium and vanadium. Hence, the uranium must be separated out of the leaching solution and in so doing yellow cake is produced (U_3O_8 with impurities). This needs further refining before it can be used as nuclear fuel. As can be inferred from above, because of the low grades large amounts of tailings are produced when the uranium ore is milled and processed. All the uranium is not extracted from the ore so that the tailings may contain between 5% and 10% residual uranium and as long-lived decay products such as ^{230}Th and ^{226}Ra are not removed, then the tailings may contain around 85% of the initial radioactivity of the ore, as well as heavy metals and chemical reagents. Various reactions can take place within the tailings that cause additional hazards to the environment. For instance, in arid areas salts containing contaminants can migrate to the surface of the tailings where they are subject to wind erosion. The presence of pyrite within the tailings can lead to the generation of sulphuric acid that causes leaching of contaminants. Radon (Rn) gas also is emitted by the tailings. Although ^{222}Rn only has a half-life of 3.8 days it is continuously produced by the decay of ^{226}Ra, the half-life of which is 1600 years, so that radon presents a long-term hazard. In addition, radionuclides in uranium tailings emit 20–100 times as much gamma radiation as natural background levels, although gamma radiation decreases rapidly with distance from the tailings. Of the decay products the one posing the greatest hazard is ^{226}Ra, which may cause damage if ingested because of its intense alpha radiation and, as mentioned, its decay produces ^{222}Rn gas. When the latter decays it produces solid radioactive products, mainly ^{210}Pb, as suspended particles, which on being inhaled can lodge in lung tissue causing cell damage that may lead to cancer. The United States Environmental Protection Agency (Anon., 1986) has taken 4 pCi l^{-1} (1 pCi l^{-1} = 37 Bq m^{-3}) as the limit beyond which radon is considered a health hazard.

In most cases uranium tailings are disposed of in such a way as to limit the release of contaminants in to environment. For instance, a cover of 2 or 3 m of clayey material should be placed on top of the tailings to protect against the escape of radon gas and gamma radiation, wind erosion of tailings particles and infiltration of precipitation. Seepage from tailings impoundments can pose the risk of contamination to ground and surface water supplies. This is especially the case when tailings are acidic as the radionuclides concerned are more mobile under acidic conditions. A liner must be placed below the tailings in order to prevent seepage of contaminant water if no impermeable rock is present. It may be necessary to collect and treat seepage water.

Unfortunately, there are cases on record where tailings have been released into the environment. A notable example was at the Bukhovo Mill in Bulgaria where tailings were deposited in a valley between 1947 and 1958, the finer particles moving into the river (Diehl, 2004). Heavy precipitation spread the tailings over a wider area, contaminating 120 ha of agricultural land. Rates of gamma radiation detected at the surface were about one hundred-fold those of background levels. The most severely contaminated areas were fenced off after 1957. However, the fences deteriorated later and some of the land was re-used for farming. Excessive radium concentrations of up to 1077 Bq kg^{-1} were found in cereals grown in these areas.

In some cases uranium is removed from low grade ore by heap leaching. Again, the leaching agent usually is sulphuric acid. The heaps present a hazard during leaching since they release dust, radon and leaching liquid. A long-term problem may result after leaching if the heap contains pyrite, the breakdown of which can produce sulphuric acid, which results in further leaching of uranium and other contaminants that can adversely affect groundwater. For example, this has occurred in Thuringia, Germany.

9.6. Case history I

The former Graf Moltke coal mine and coking plant was located in the Ruhr district near Essen, Germany (Bell *et al.*, 2000). It covered an area of some 230 000 m^2. The first shaft was sunk in 1873 and three more followed within the next 30 years. A coking plant was built in 1903–04, followed by benzol and ammonia factories. Over the next 50 years further industries were established, turning the site into a multi-use industrial complex. The mine was closed in 1971. In the succeeding years the Graf Moltke site became a typical wasteland. However, because of the existing infrastructure it eventually was decided to rehabilitate the site and establish an industrial park. Funding was provided by the European Fund for Regional Development, leaving about 50% of the costs involved in the remediation to be found by the owners of the former mining site.

The site consisted of waste rubble fill that contained large remnants of foundations. This fill varied from 2 to 9 m in thickness beneath which there were sediments of Quaternary age that, in turn, overlay fractured chalky marl of Cretaceous age. The water table occurred at a depth of about 5 m (Fig. 9.5).

A historical analysis of the site was carried out in the first phase of the rehabilitation process and its evaluation facilitated the planning of the site investigation programme. After the analysis of samples to determine the

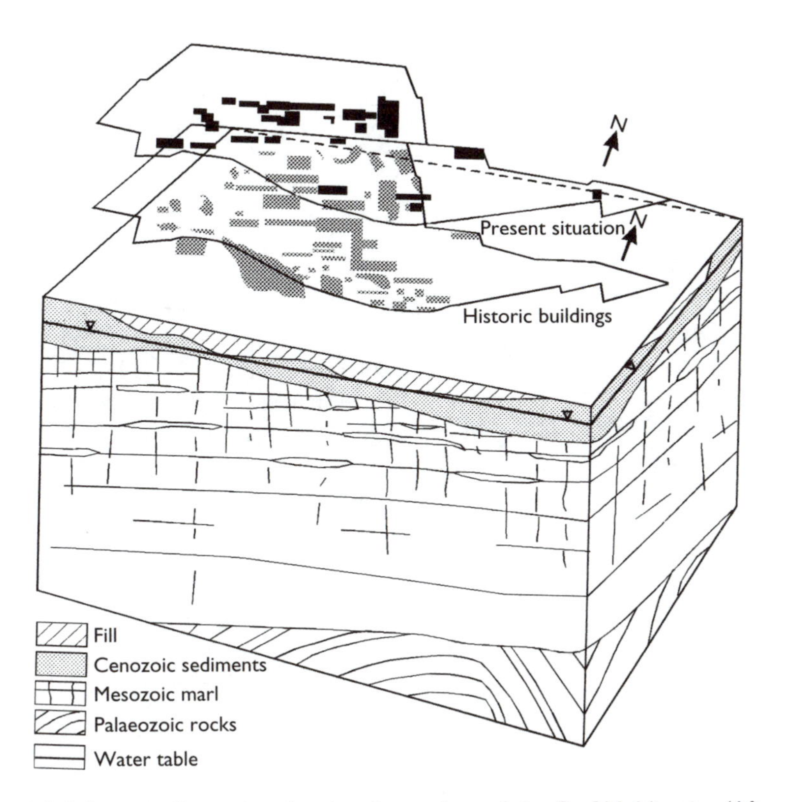

Figure 9.5 Schematic illustration showing the geology of the Graf Moltke site. (After Bell *et al.*, 2000.)

nature of the contamination on site, the grades of contamination of the sampling points were interpreted as regionalized variables and a block kriging routine was applied. In addition to a map depicting the contaminated zones, the prediction error was quantified on a second map (Fig. 9.6). The site investigation programme was further optimized in relation to the error map, zones of high contamination and of large error indicating where additional investigation was necessary. In particular, serious hydrocarbon contamination was detected in the vicinity of the former coking plant. Pollutants had migrated through the sediments into the fractured rock beneath, percolating along the network of discontinuities.

In order to help visualize the complex task of remediating the site, a videotape was prepared for viewing by all parties involved in the project. This consisted of three parts, namely, demonstration of the historical analysis, three-dimensional visualization of the geology and the migration of contaminants into the substrata, and explanation of the remediation concept. The first part, that is, the historical analysis was accomplished by merging

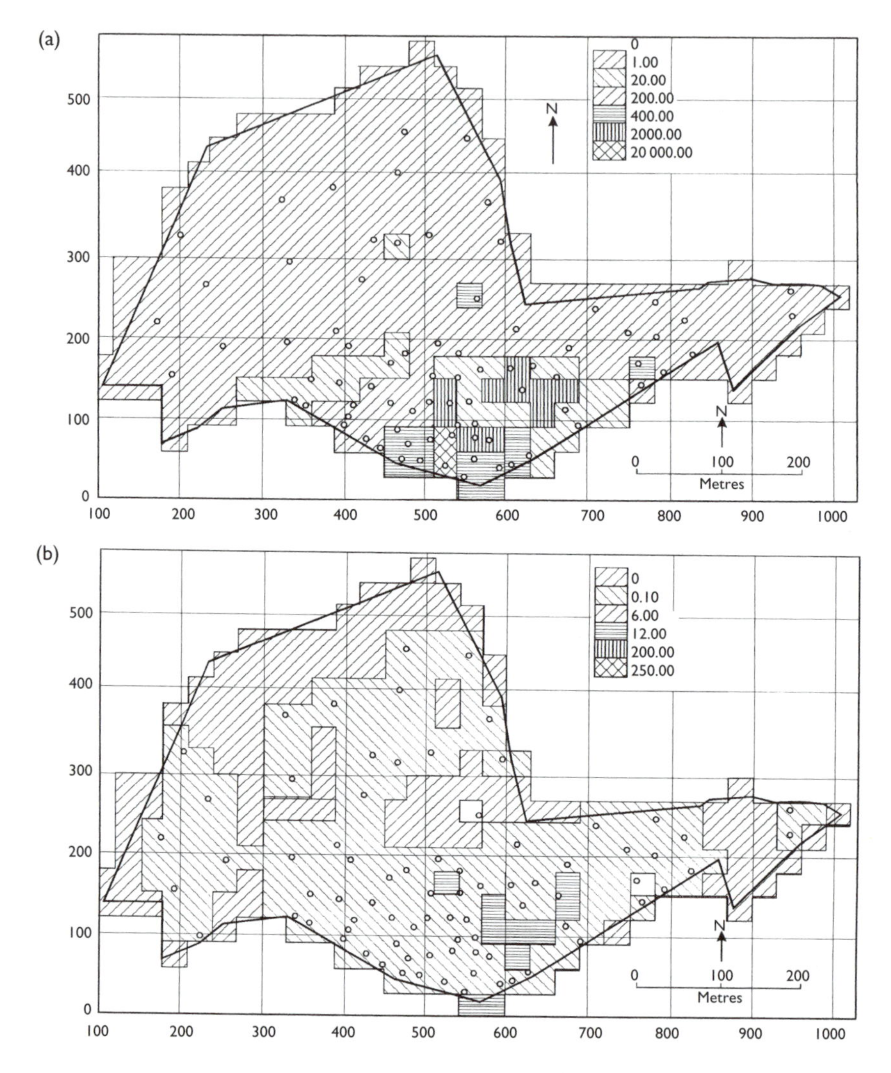

Figure 9.6 (a) Distribution of contamination (in mg kg^{-1}). (b) Estimation of error in the distribution of contamination (in mg kg^{-1}). (After Bell *et al.*, 2000.)

the map of the present position of the site with evidence from historical building documents and aerial photographs. The second part, three-dimensional visualization, was produced as an animated idealization of the geological situation that could be viewed by means of a virtual camera performing circuits around the three-dimensional image. Subsequently, the migration of contaminants into the ground was animated. As far as the rehabilitation process was concerned, the virtual camera intersected the three-dimensional model and zoomed in on the rehabilitation system. The latter consisted of a reinforced cover design accompanied by flushing wells.

The Environmental Protection Agency required that excavation was kept to a minimum at the Graf Moltke site. Hence, a reinforced cover design, using geotextiles, was chosen for the heavily contaminated zones. The surface confinement of the contaminated ground had to meet three goals. First, it had to be impermeable in order to prevent the penetration of precipitation into the affected zones. Second, it had to be gas-proof to stop the migration of any toxic gas to the surface. Third, it had to be stiff enough to allow any construction to proceed. The third aspect was particularly important in relation to the Graf Moltke site as large massive remnant foundations occurred within loose fill that could have led to differential settlement beneath structures erected on the site. A composite soil-geosynthetic system was designed to take account of the three factors mentioned (Fig. 9.7). Basically, the system consisted of a lower reinforced supporting layer, a draining and sealing layer incorporating geotextiles and geomembranes, and an upper reinforced foundation layer to accommodate the structural loading.

As the technique was somewhat novel, it was decided to design and construct a full-scale test on site. The aims of the field test were to study the feasibility of using geosynthetics, to establish quality control tests and measures to investigate the potential for damage occurring during construction, and to develop new design criteria for such structures. The

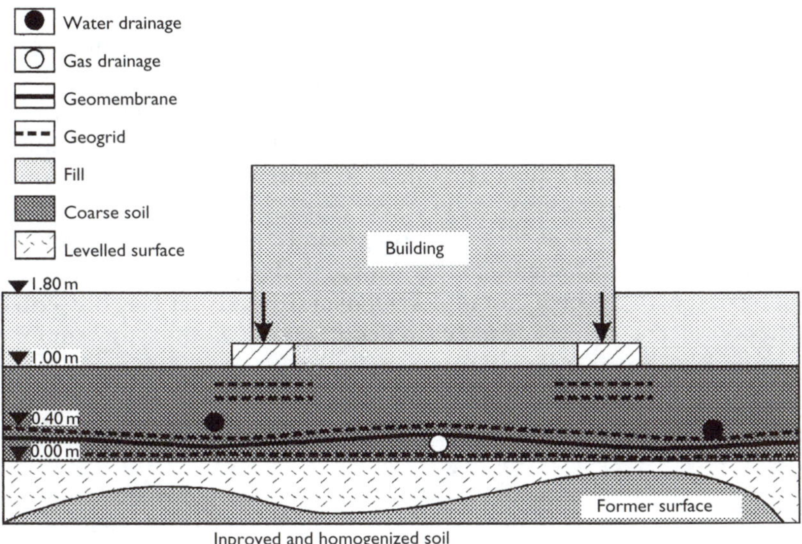

Figure 9.7 A geotextile reinforced sandwich system for the Graf Moltke site. (After Bell *et al.* 2000.)

location of the field test on site was chosen at an area where the subsoil conditions were representative of the whole site. An area 24 m × 19 m was levelled and densified for the test area. The lower reinforced layer was constructed of two lifts of waste material ($c_u = 3.8$ kPa; $C_c = 1.7$; 10% $d > 12$ mm) reinforced with a geogrid. Above this a non-woven sheet was installed as a protective layer over which a geomembrane, 2.5 mm in thickness, was placed as a sealing layer. Then came the drainage layer (Secudrain) over which a further geogrid was laid. The upper layer was subdivided into two parts, an unreinforced layer and a reinforced foundation layer using geogrids as reinforcing elements. The area then was loaded up to a maximum load of 2013 kN using steel plates.

In order to monitor the stress-strain behaviour of the reinforced foundation layers and the performance of the waterproofing geomembrane a number of earth pressure cells, displacement transducers and horizontally installed inclinometer rods for measurement of settlement were incorporated in the test set up. The results of the strains developed in the geogrids with loading are illustrated in Figure 9.8, which shows the effect of placing reinforcement above the sealing geomembrane. The geomembrane developed very small strains (<0.05%) as compared with the geogrid (maximum of 1.3%) in the five day period of loading. Nonetheless, the geosynthetics remained more or less undamaged. Furthermore, on unloading there was little recovery of strain, it falling to 0.9%. Settlement recorded over the reinforced area only amounted to some 4 mm. Hence, the results indicated that geosynthetic material could be used effectively in the rehabilitation process.

Those areas of the site that were identified as slightly contaminated were covered with approximately 0.5 m of granular soil. A number of observation wells were installed that are not only used for groundwater monitoring but also can be used as recovery wells if a given degree of contamination is detected in the groundwater. The tolerable concentrations were negotiated with the Environmental Protection Agency.

It was estimated that this type of ground rehabilitation would cost approximately $60 per square metre. Although this appeared high, it had to be compared with the high cost of development land in that part of Germany. Furthermore, the cost of disposal of contaminated soil was minimized. Therefore, the system adopted was by no means cost prohibitive.

9.7. Case history 2

The Mont Cenis coal mine in the Ruhr district of Germany started in 1872 and gave rise to the community of Herne-Sodingen. In 1893 a washery was built and 20 years later the first coke oven facility was constructed (Bell *et al.*, 2000). The life of the mine extended over 100 years, it closing in 1978 when the 26 ha site was abandoned. Not only were contamination and massive subsurface structures associated with the site but so were the

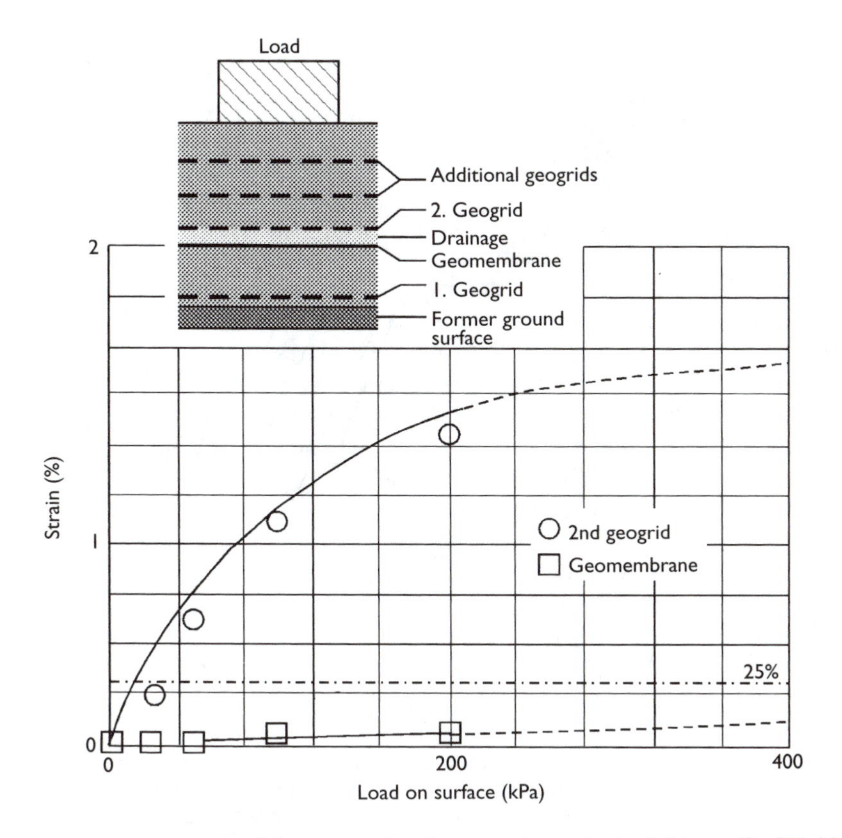

Figure 9.8 Strain versus load for a geogrid and geomembrane in the field test, Graf Moltke site. (After Bell *et al.*, 2000.)

effects of acid mine drainage and subsidence. In the northern part of the site, in particular, the soil was contaminated (Fig. 9.9). The authorities would not allow excavation and removal of the contaminated soil since this would risk release of contaminants, represent an expensive off-site treatment procedure, and storage of non-degradable wastes in landfills simply would occupy space elsewhere. The solution chosen was to line the contaminated area with geomembrane or clay liner, over which a gravel and sand filter was placed to intercept infiltrating precipitation. Finally, topsoil was laid over the gravel and sand. Reinforcement of the system with geogrids would permit the construction of buildings at some future date.

In 1990 Mont Cenis was one of the largest and most ambitious remediation projects in North Rhine-Westfalia, with new industry to be established on the site. The plan also had to incorporate a state building for the Ministry of the Interior, civic administration buildings, a library and a multi-purpose meeting hall. In addition, there was to be a new shopping centre, 250 housing units and a recreational park, with associated

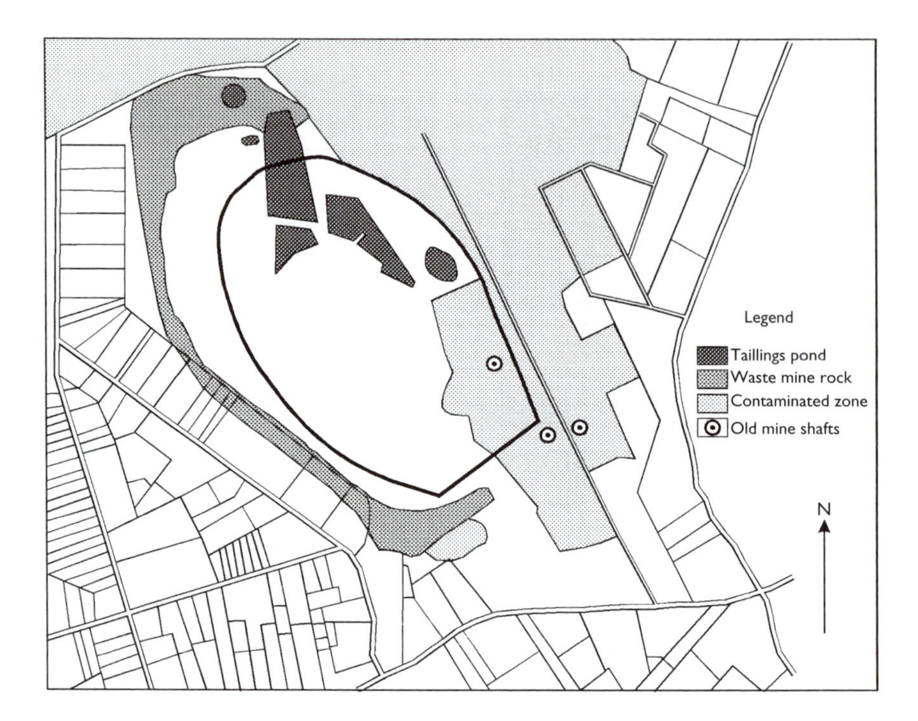

Figure 9.9 Schematic illustration showing Mont Cenis site showing distribution of waste mine rock, tailings ponds, contaminated zone and three abandoned shafts, together with the position of the planned oval vista. (After Bell and Genske, 2000.)

infrastructure. Short distances for communication and transport in order to save time and resources were central aspects of the project, as was the incorporation of alternative energy concepts.

A tent-like structure constructed of glass and timber was proposed to house the Ministry of the Interior, the library, a hotel, the meeting hall, service and sports facilities, and housing. This structure saves some 23% of energy costs and hence a net reduction of 18% in CO_2 emissions. In other words, the microclimate created in the building enables its users to reduce energy consumption, compared with conventional buildings, to 32 kWh per year per square metre. The largest roof-integrated solar power plant in the world was installed to cover the structure. It consists of some 10 000 m^2 of solar cells that produce 1 MW, that is, more than twice the energy needed to operate the centre. In addition to the use of solar energy, methane gas that is emitted from the three old shafts associated with the coal mine is utilized. This is the first time in the Ruhr area that this has happened even though it is estimated that 120 million m^3 of methane escape annually from decommissioned coal mines in the area, which is equivalent to 100 000

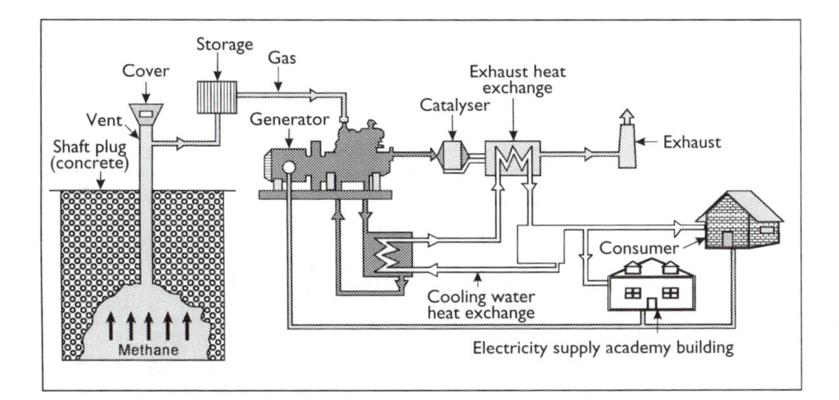

Figure 9.10 Collection and storage of methane for use in the generation of electricity. (After Bell and Genske, 2000.)

tonnes of fuel oil. Approximately 1 million m^3 of gas is available for use from Mont Cenis that produces around 2 million kWh of electrical energy and 3 million kWh of heat for new buildings. However, the emission of gas is not constant, fluctuating according to atmospheric pressure. Hence, gas is collected in sealed chimney-like structures constructed at the top of each of the three shafts from which it is piped to a storage tank and from the tank to the generator (Fig. 9.10). In this way a constant supply of gas is available to the electricity generating plant. Nonetheless, a supplementary natural gas plant of 1800 kWh plus a hot water storage tank was constructed to make sure of a secure source of energy and heat supply. To buffer peak consumption the installation also is connected to the municipal supply, the connection also allowing discharge of surplus energy produced from mine gas. In fact, the Mont Cenis project combines all aspects of modern site rehabilitation in Europe, namely, the sustainable management of land, the introduction of innovative technology to achieve high efficiencies as far as energy and resource consumption are concerned, and alternative architectural concepts to attract consumers and investors.

9.8. Case history 3

Coal mining in Lünen in the east of the Ruhr district took place for over 100 years and the Minister Achenbach was one of the largest mine complexes in Lünen, being founded in 1897. It closed in 1992 when a programme was initiated for new industries to be developed on the mine site, which covers approximately 54 ha. A good system of communications, including autobahn, regional roads, rail and canal offer access to the site (Fig. 9.11).

A thorough site investigation was conducted to indicate the extent to which the soil and groundwater in the area had been contaminated. In fact, it was revealed that the site was heavily contaminated, contaminants having been leached into the ground continuously since the mine started.

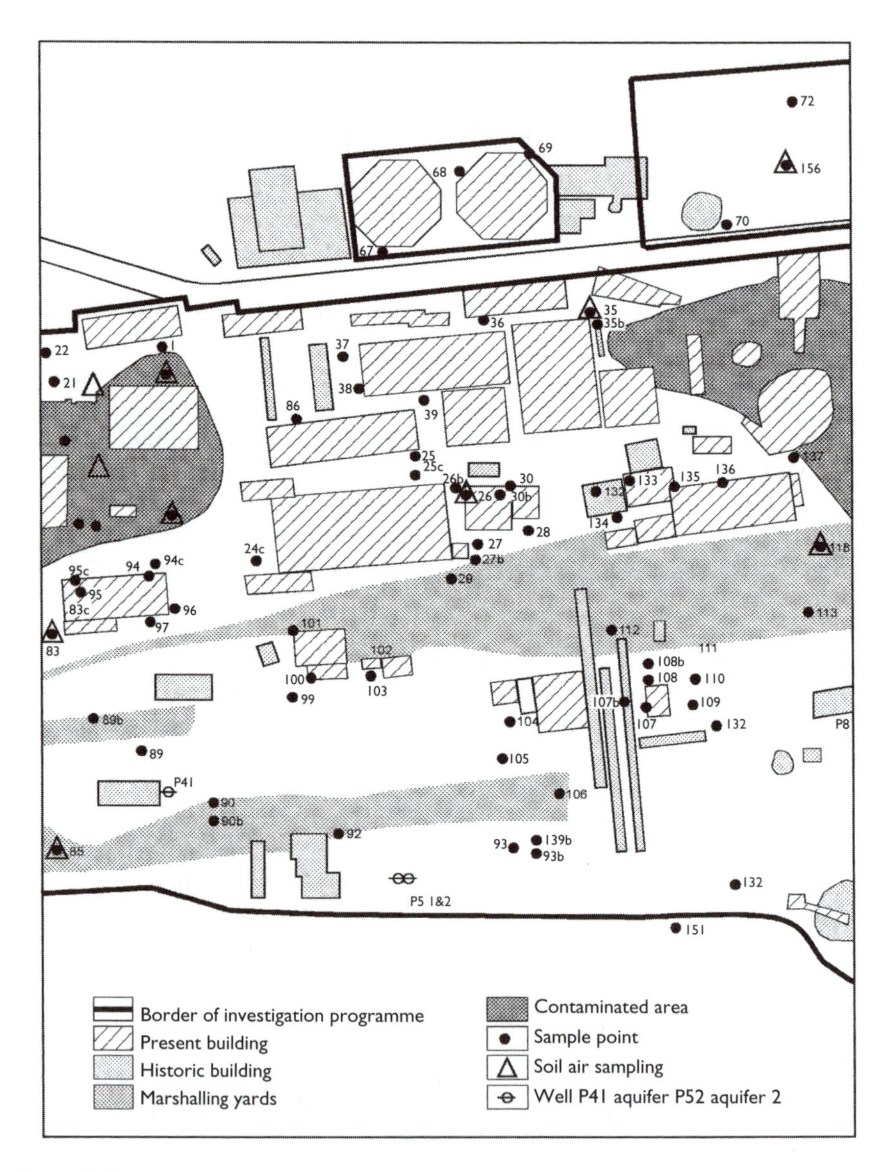

Figuge 9.11 A schematic section of a risk map of the Minister Achenbach depicting the former and present use of the site, the field sampling programme and the sectors identified as contaminated. (After Bell *et al.*, 2000.)

Aromatic hydrocarbons, especially those that are non-volatile, were widespread in the unsaturated zone. Contaminants associated with the coking processes such as BTX (benzene, toluene, xylene) were detected in the immediate vicinity of the former coking plants. Some contaminants, notably from coal tar, had migrated through the overlying Quaternary sediments into the fractured chalk marl beneath. The coking process also had been responsible for elevated concentrations of arsenic present in the waste. Furthermore, shallow contamination with heavy metals such as mercury, zinc, copper and cadmium were found locally across the site. Yet another factor contributing to contamination was the destruction of mining facilities during the Second World War. In particular, bomb craters had been filled quickly with all kinds of material. At that time no one paid any attention to possible environmental hazards that subsequently may be caused by the fill. In fact, these fills often represented serious sources of contamination.

The site investigation programme was based primarily upon a historical analysis of the records, maps and aerial photographs of the site over its 95 years of existence. This indicated that two coking plants and a large number of chemical factories had been present on the site. An important part of the historical analysis was the interpretation of aerial photographs that were compared with old building documents and maps with the aid of CAD and GIS programs (Genske *et al.*, 1994). Data gained from the historical analysis allowed the field investigation to be more cost effective, with special attention being paid to the critical sectors.

The results obtained from the field investigation were embodied in a risk assessment report. The latter included a plan of the critical sectors

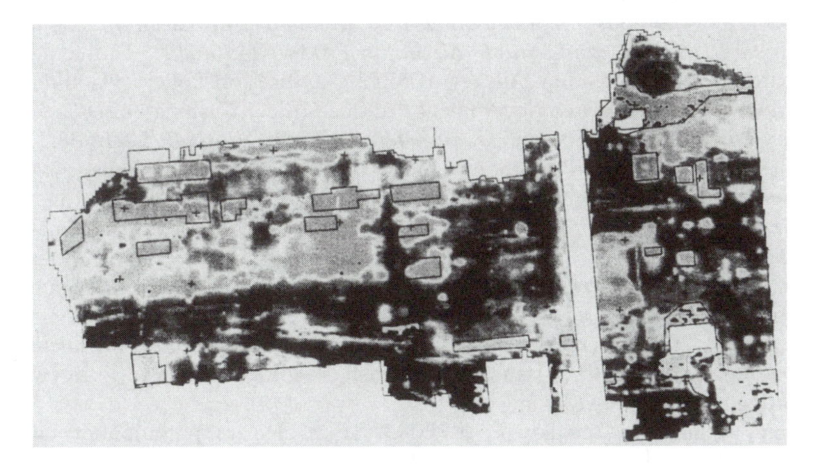

Figure 9.12 Suspected anomalies from analysis of an electromagnetic survey of the Minister Achenbach, darker zones are associated with foundations, pipes and tunnels. (After Bell *et al.*, 2000.)

(Fig. 9.11). This report served as a guideline for the city planners and a master plan for the rehabilitation of the Minister Achenbach was produced. During the rehabilitation programme it became necessary to detect derelict foundations and subsurface structures that remained in the ground after the superstructures had been dismantled. As a consequence, it was decided to use an electromagnetic survey in a certain part of the site for this purpose. The survey penetrated the ground to a depth of about 3 m and detected remnants that were larger than 10 m^3. The resulting map revealed zones of low conductivity that reflected the natural ground conditions together with zones of high conductivity suggesting anomalies associated with foundations, pipes and tunnels (Fig. 9.12).

References

Adams, D. and Holroyd, M. 1992. *In situ* soil bioremediation. *Proceedings of the Third International Conference on Construction on Polluted and Marginal Land.* London, Forde, M.C. (ed.), Engineering Technics Press, Edinburgh, 291–294.

Al-Tabbaa, A., Evans, C.W. and Wallace, C.J. 1998. Pilot *in situ* auger mixing treatment of a contaminated site, Part 2: site trial. *Proceedings of the Institution of Civil Engineers, Geotechnical Engineering*, **131**, 89–95.

Anderson, R.J. and Davies, B.E. 1980. Dental caries prevalence and trace elements in soil with special reference to lead. *Journal of Geological Society*, **137**, 547–558.

Anon. 1986. *Citizens Guide to Radon, EPA-86–004.* United States Environmental Protection Agency (USEPA), Office of Policy, Planning and Evaluation, United States Government Printing Office, Washington, DC.

Anon. 1988, *Draft for Development, DD175:1988, Code of Practice for the Identification of Potentially Contaminated Land and Its Investigation.* British Standards Institution, London.

Anon. 1989. *Methods for Evaluating the Attainment of Clean-up Standards, Volume 1: Soils and Solid Media, EPA/540/2–90/011.* United States Environmental Protection Agency (USEPA), Office of Policy, Planning and Evaluation, United States Government Printing Office, Washington, DC.

Anon. 1990. *Subsurface Contamination Reference Guide, EPA/540/5–91/008.* United States Environmental Protection Agency (USEPA), Office of Emergency and Remedial Response, United States Government Printing Office, Washington, DC.

Anon. 1993. *Guidelines for Drinking Water Quality.* World Health Organization, Geneva.

Anon. 1999. *Code of Practice on Site Investigations, BS: 5930.* British Standards Institution, London.

Appleton, D. 2001. Mercury pollution from artisanal gold mines. In: *Earthwise, Geology and Health*, Issue 17, British Geological Survey, Keyworth, Nottinghamshire, 12–13.

Aucamp, P. and van Schalkwyk, A. 2003. Trace element pollution of soils by abandoned gold mine tailings, near Potchefstroom, South Africa. *Bulletin of Engineering Geology and the Environment*, **62**, 123–134.

Bell, F.G. and Genske, D.D. 2000. Restoration of derelict mining sites and mineral workings. *Bulletin of Engineering Geology and the Environment*, **59**, 173–185.

Bell, F.G., Bell, A.W., Duane, M.J. and Hytiris, N. 1996. Contaminated land: the British position and some case histories. *Environmental and Engineering Geoscience*, **2**, 355–368.

Bell, F.G., Genske, D.D. and Bell, A.W. 2000. Rehabilitation of industrial areas: case histories from England and Germany. *Environmental Geology*, **40**, 121–134.

Bell, F.G., Halbich, T.F.J. and Bullock, S.E.T. 2002. The effects of acid mine drainage from an old mine in the Witbank Coalfield, South Africa. *Quarterly Journal of Engineering Geology and Hydrogeology*, **35**, 265–278

Bowen, H.J.M. 1979. *Trace Elements in Biochemistry*. Academic Press, New York.

Calder, I.C., Maynard, E.J. and Heyworth, J.S. 1994. Port Pirie Lead Abatement Program, 1992. *Environmental Geochemistry and Health*, **16**, 137–145.

Cannon, H.L. 1976. Lead in the atmosphere, natural and artificially occurring lead, and the effects of lead on health. In: *Lead in the Environment*, Lovering, T.G. (ed.), United States Geological Survey, Professional Paper 957, 75–80.

Chen, C.J., Chen, C.W., Wu, M.M. and Kuo, T.L. 1992. Cancer potential in liver, lung, bladder and kidney due to ingested inorganic arsenic in drinking water. *British Journal of Cancer*, **66**, 888–892.

Clark, R.G., Scarrow, J.A. and Skinner, R.W. 1994. Safety considerations specific to the investigation of landfills and contaminated land. *Proceedings of the First International Congress on Environmental Geotechnics*, Edmonton, Carrier, W.D. (ed.), Bi-Tech Publications Ltd, Richmond, British Columbia, 167–172.

Davies, B.E., 1972. Occurrence and distribution of lead and other metals in two areas of unusual disease incidence in Britain. *Proceedings of the International Symposium on Environmental Health Aspects of Lead*, Amsterdam, 125–134.

Davis, A., Ruby, M.V. and Bergstrom, P.D. 1994. Factors controlling lead bioavailability in the Butte mining district, Montana, USA. *Environmental Geochemistry and Health*, **16**. 147–157.

Dawson, G.W. and Gilman, J. 2001. Land reclamation technology – expanding the geotechnical engineering envelope. *Proceedings of the Institution of Civil Engineers, Geotechnical Engineering*, **149**, 49–61.

Diehl, P. 2004. Uranium mining and milling wastes: an introduction. http://www.antenna.nl/wise/uranium/uwai.html

Genske, D.D., Kappernagel, T. and Noll, P. 1994. Computer aided remediation of contaminated sites. *Proceedings of the Seventh International Congress of International Association of Engineering Geology*, Lisbon, Oliveira, R., Rodriques, L.F., Coelho, A.G. and Cunha, A. (eds), A.A. Balkema, Rotterdam, **4**, 4557–4562.

Haley, J.L., Hanson, B., Enfield, C. and Glass, J. 1991. Evaluating the effectiveness of groundwater extraction systems. *Ground Water Monitoring Review*, **12**, 119–124.

Harris, M.R., Herbert, S.M. and Smith, M.A. 1995. *Remedial Treatiment for Contaminated Land: In Situ Methods of Remediation*. Construction Industry Research and Information Association (CIRIA), Special Publication No. 109, London.

Hatheway, A.W. 2002. Geoenvironmental protocol for site and waste characterization of former manufactured gas plants; worldwide remediation challenge in semi-volatile organic wastes. *Engineering Geology*, **64**, 317–338.

Jarup, L. 1992. *Dose-Response Relations for Occupational Exposure to Arsenic and Cadmium*. National Institute for Occupational Health, Stockholm.

Jewell, C.M., Hensley, P.J., Barry, D.A. and Acworth, I. 1993. Site investigation and monitoring techniques for contaminated sites and potential waste disposal sites. *Proceedings of the Conference on Geotechnical Management of Waste and Contamination*, Sydney, Fell, R., Phillips, A. and Gerrard, C. (ed.), A.A.Balkema, Rotterdam, 3–38.

Johnson, A.C. 1994. Site investigation for development on contaminated sites – how, why and when? *Proceedings of the Third International Conference on Re-use of Contaminated Land and Landfills*, London, Forde, M.C. (ed.), Engineering Technics Press, Edinburgh, 3–7.

Krauskopf, K.B. 1988. *Radioactive Waste Disposal and Geology*. Chapman and Hall, London.

Kurland, L.T., Faro, S.N. and Siedler, H.S. 1960. Minimata disease. *World Neurologist*, **1**, 320–325.

Leach, B.A. and Goodyear, H.K. 1991. *Building on Derelict Land*. Special Publication 78, Construction Industry Reseach and Information Association (CIRIA), London.

Loehr, R.C. 1993. Bioremediation of soils. In: *Geotechnical Practice for Waste Disposal*, Daniel, D.E. (ed.), Chapman and Hall, London, 520–550.

Lutz, P.M., Jayachandran, C., Gale, N.L., Hewett, J., Phillips, P.E., Looney, F.M. and Bengsch, H. 1994. Immunity in children with exposure to environmental lead: 1. Effects on cell numbers and cell-mediated immunity. *Environmental Geochemistry and Health*, **16**, 167–177.

Madhav, M.R., Bouazza, A. and Van Impe, W.F. 1997. Reclamation of landfills and contaminated ground – a review. In: *Environmental Geotechnics – Geoenvironment 97*, Bouazza, A., Kodikara, J. and Parker, R. (eds), A.A. Balkema, Rotterdam, 505–510.

Mills, C.F. 1996. Geochemistry and trace element diseases. In: *Environmental Geochemistry and Health with Special Reference to Developing Countries*, Special Publication No. 113, Appleton, J.P., Fuge, R. and McCall, G.J.H. (eds), Geological Society, London, 1–5.

Mitchell, P. and Barr, D. 1995. The nature and significance of public exposure to arsenic: a review of its relevance to south west England. *Environmental Geochemistry and Health*, **17**, 57–82.

Morton, W.E. and Dunette, D.A. 1994. Health effects of environmental arsenic. In: *Arsenic in the Environment, Part II: Human Health and Ecosystem Effects*, Nriagu, J.O. (ed.), Wiley, New York, 159–170.

Norris, G., Al-Dhahir, Z., Birnstingl, J., Plant, S.J., Cui, S. and Mayell, P. 1999. A case study of the management and remediation of soil contaminated with polychlorinated biphenyls. *Engineering Geology*, **53**, 177–185.

Pettyjohn, W.A. 1972. No thing is without poison. In: *Man and His Physical Environment*, McKenzie, G.D. and Utgard, R.O. (eds), Burgess Publishing, Minneapolis, MN, 109–110.

Reid, J.M. and Brookes, A.H. 1999. Investigation of lime stabilised contaminated material. *Engineering Geology*, **53**, 217–231.

Robards, K. and Warsford, P. 1991. Cadmium toxicology and analysis: a review. *Analyst*, **16**, 549–568.

Rosner, T. and van Schalkwyk, A. 2000. The environmental impact of gold mine tailings footprints in the Johannesburg region, South Africa. *Bulletin of Engineering Geology and the Environment*, **59**, 137–148.

Schroeder, H.A. 1965. Cadmium as a factor in hypertension. *Journal of Chronic Disease*, **18**, 647–656.

Singleton, M. and Burke, G.K. 1994. Treatment of contaminated soil through multiple bioremediation technologies and geotechnical engineering. *Proceedings of the Third International Conference on Reuse of Contaminated Land and Landfills*, London, Forde, M.C. (ed.), Engineering Technics Press, Edinburgh, 97–107.

Smedley, P.L., Edmunds, W.M. and Pelig-Ba, K. 1996. Mobility of arsenic in groundwater in the Obuasi gold mining area of Ghana: some indications for human health. In: *Environmental Geochemistry and Health with Special Reference to Developing Countries*, Appleton, J.P., Fuge, R. and McCall, G.J.H. (eds), Geological Society Special Publication No. 113, London, 223–230.

Swane, I.C., Dunbavan, M. and Riddell, P. 1993. Remediation of contaminated sites in Australia. *Proceedings of the Conference on Geotechnical Management of Waste and Contamination*, Sydney, Fell, R., Phillips, A. and Gerrard, C. (eds), A.A. Balkema, Rotterdam, 127–163.

Taha, M.R. 1997. Some aspects of electrokinetic remediation of soil. In: *Environmental Geotechnics – Geoenvironment 97*, Bouazza, A., Kodikara, J. and Parker, R. (eds), A.A. Balkema, Rotterdam, 511–516.

Tebbutt, T.H.Y. 1983. *Relationship between Natural Water Quality and Health*, UNESCO, Paris.

Thornton, I. 1996. Sources and pathways of arsenic in the geochemical environment: health implications. In: *Environmental Geochemistry and Health with Special Reference to Developing Countries*, Appleton, J.P., Fuge, R. and McCall, G.J.H. (eds), Geological Society Special Publication No. 113, London, 153–161.

Trost, P.B. 1993. Soil washing. In: *Geotechnical Practice for Waste Disposal*, Daniel, D.E. (ed.), Chapman and Hall, London, 585–603.

Tseng, W.P. 1977. Effects of dose response relationships on skin cancer and Blackfoot disease with arsenic. *Environmental Health Perspectives*, **19**, 109–119.

Waters, J. 1994. *In Situ* remediation using air sparging and soil venting. *Proceedings of the Third International Conference on Re-use of Contaminated Land and Landfills*, London, Forde, M.C. (ed.), Engineering Technics Press, Edinburgh, 109–112.

Welsh, A.H., Lico, M.S. and Hughes, J.L. 1988. Arsenic in groundwater of the western United States. *Ground Water*, **26**, 333–347.

Welsh, J.P. and Burke, G.K. 1991. Jet grouting for soil improvement. In: *Geotechnical Engineering*, Geotechnical Special Publication No. 27, American Society Civil Engineers, New York, 334–345.

Chapter 10

Other problems associated with mining

10.1. Spontaneous combustion

Most of the major coalfields of the world have experienced spontaneous combustion. Spontaneous combustion can be regarded as an atmospheric oxidation process in which self-heating occurs (i.e. an exothermic reaction emitting between 5 and 10 $kcal\,g^{-1}$ of coal). It is the most common cause of fires in coal mines (Fig. 10.1). Spontaneous combustion of colliery spoil is dealt with in Chapter 7.

Fire remains one of the principal hazards facing all stages of coal mining, storage and transportation (Walker, 1999). Coal fires can give off noxious gases and burning coal in mines can lead to ground instability. The cost of controlling fires can be high, and in some instances this has led to the abandonment of a mine, loss of lives and the loss or sterilization of coal or reserves. All coals will spontaneously ignite if the right conditions exist for the particular coal, although some coals ignite more easily than others (Anon., 1999). The factors that aid the spontaneous combustion of coal include the temperature of the surroundings, the rank of the coal, the supply of air, the moisture content of the coal, and the surface area of coal exposed (i.e. the larger the surface area exposed, the greater is the opportunity for oxidation of the coal).

Coal and carbonaceous materials may be oxidized in the presence of air at temperatures below their ignition points. If heat is lost to the atmosphere, then the ignition temperature for coal is between 420°C and 480°C. However, where the heat of reaction is retained, the ignition point of coal falls appreciably and can be somewhere between 35°C and 140°C. If the heat generated cannot be dissipated, then the temperature rises that, in turn, increases the rate of oxidation so that the reaction becomes self-sustaining. In fact, the rate of reaction increases exponentially with increasing temperature. For example, Mohan (1996) maintained that the rate of oxidation is slow below 40°C but subsequently accelerates by a factor of 1.8 and becomes self-sustaining for lignites above 50°C and for bituminous coals above 70° to 80°C. Generally, the rate of oxidation of coal increases tenfold as the temperature rises from 30°C to 100°C. On the other hand, the rate

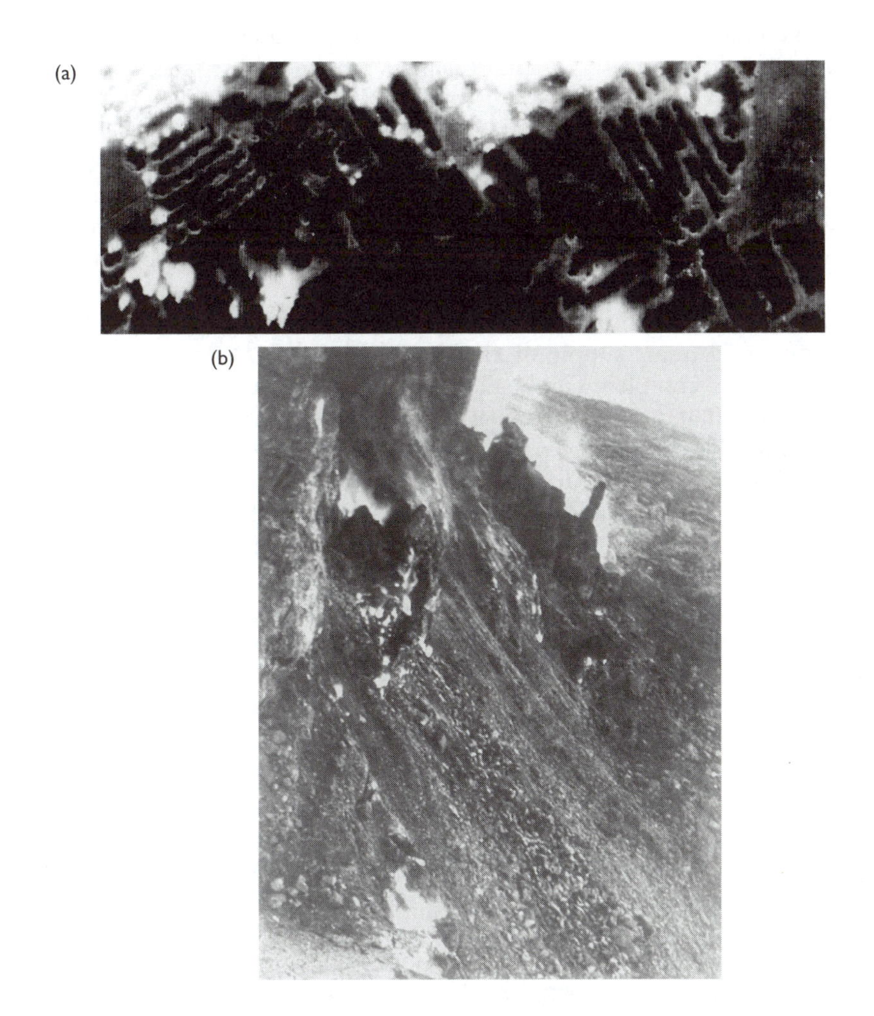

Figure 10.1 (a) Airborne pre-dawn thermal imagery showing near-surface old room and pillar workings affected by combustion in South Africa. The bright (white) areas are trees and vegetation. (Courtesy of Anglo-American Plc & Infoterra Limited.) (b) East fire area of Beishan Coalfield, Qitai County, China, coal seam burning and collapsing down slope.

of oxidation is reduced as oxidation of coal proceeds due to the development of oxidation products on the surface of the coal through which oxygen must penetrate if oxidation is to continue. According to Mohan, at constant temperature, the rate of oxidation is reduced by a tenth of its value for each hour up to 100 hours from the start of oxidation. Depending on the reactivity of the coal or carbonaceous material involved, a time comes when the temperature has risen sufficiently to cause the material to combust. As the likelihood of spontaneous combustion occurring increases as the temperature increases, so with increasing depth, and consequently increasing geothermal gradient, coal has a greater tendency to ignite.

Generally, lower rank and high volatile coals are more reactive and accordingly more susceptible to self-heating than coals of higher rank (Michalski *et al.*, 1990). In addition, low rank coals are more porous and possess chain-like molecular structures that are more susceptible to oxidation than structures in high rank coals. A high ash content in coal tends to reduce the risk of self-heating.

Obviously, the continuation of spontaneous combustion depends on a continuing adequate supply of air. Van Vuuren (1995), for instance, indicated that for complete combustion of 1 tonne of run-of-mine coal, then between 6000 and 7000 m^3 of fresh air is needed, with proportionally less to sustain self-heating. The velocity of air flow is also important in that if it is high enough it can carry heat away, while stagnant air soon becomes deoxygenated. An optimum air flow velocity that will permit oxidation to continue, according to Voracek (1997), is 0.3–0.4 $m\,min^{-1}$ for air that has a minimum oxygen content of 7%.

The moisture content also is an important factor in spontaneous combustion, especially sudden wetting or drying. For instance, the surface temperature of dry coal can be raised by as much as 30°C by wetting and so the rate of oxidation is accelerated. Therefore, if a coal seam has a lower than normal moisture content and then if the moisture content rises, this leads to the liberation of heat. In low rank coals at low moisture contents the heat generated by wetting can exceed the heat released by condensation appreciably.

In combustible material of large size the movement of air can cause heat to be dissipated while in fine material the air remains stagnant and this means that burning ceases when the supply of oxygen is consumed. Accordingly, ideal conditions for spontaneous combustion exist when the grading is intermediate between these two extremes and hot spots may develop in spoils under such conditions. Furthermore, the rate of oxidation generally increases as the specific surfaces of particles increases.

Oxidation of pyrite at ambient temperature in moist air leads to the formation of ferric and ferrous sulphate, and sulphuric acid. This reaction also is exothermic. When present in sufficient amounts, and especially when in finely divided form, pyrite associated with coaly material, increases the likelihood of spontaneous combustion. For instance, Michalski *et al.* (1990) mentioned that if the pyrite content in coal exceeds 2%, then this helps spontaneous combustion. Moreover, oxidation of pyrite leads to a significant increase in volume that can cause micro-fracturing in coal, thereby exposing a larger surface area to oxidation.

When air gains access to shallow abandoned coal workings via surface fissures or due to partial collapse of the workings, conditions conducive to self-ignition may exist. The mine itself provides pathways through which air can be carried to coal. As mentioned, the retention of heat by the coal is largely dependent on the air flow, in that there is a critical velocity below which the coal is oxidized but the air flow is not capable of removing the heat generated. The flow of air through partially collapsed workings is unlikely to have a high enough velocity to convey away the heat generated. Furthermore, in old abandoned workings, the sides of the pillars normally are fractured and coal commonly is strewn in the roadways. Therefore, a large surface area of coal is available for oxidation and the exothermic reaction produces a rise in temperature that eventually becomes self-generating. If such occurrences are not detected early and controlled adequately, large areas of coal can be destroyed by self-combustion and surface areas can be seriously affected if a mine is at shallow depth. Partially burnt pillars can collapse leading to subsidence and ground fissuring, which can further accentuate the problem by allowing greater access of air to the workings. Air also can gain access to workings by the development of crown holes at the surface or via poorly sealed shafts. Gases such as steam, carbon dioxide, carbon monoxide and sulphur dioxide may escape from the fissures or crown holes (Fig. 10.2).

If a coal seam affected by spontaneous combustion is at shallow depth, then the spread of a fire may be limited by excavating a trench ahead of the fire into the coal seam and then backfilling. Obviously, depth imposes a limitation on trenching. Old workings sometimes have been flooded to extinguish fires. This means that the water pumped into the area of the mine

Figure 10.2 Steam emanating from a crown hole above workings in the No. 2 seam Witbank Coalfield, South Africa, due to seam burning.

concerned must be impounded by dams or pumped into the mine more quickly than it can be discharged. The water must remain in place for a sufficient length of time to cool the coal and surrounding strata otherwise re-ignition will occur when the water level is lowered. However, neither of these techniques always proves successful. Furthermore, pillar stability may be affected in workings that have been flooded when the workings drain as the pore water pressure within fractured pillars may cause sidewall scaling that can result in a reduction of the pillar dimensions.

In a review of the problems created by mine fires in the Jharia and Singrauli Coalfields in India, Michalski and Gray (2001) maintained that an aggregate surface area occupying some 10 km^2 was degraded by underground mine fires in the Jharia Coalfield, which have severely scarred the landscape and virtually destroyed its productivity. They mentioned that 65 fires continue to burn making this the largest complex of above ground and underground coal fires in the world. The fires also have resulted in a legacy of uncontrolled subsidence, as well as adversely affecting the health of the people living in the region, even leading to death at times. Fires reduce the dimensions of pillars meaning that eventually they are unable to support the load they carry so that subsidence results. Void migration in shallow mines, together with fractures created by subsidence provide access for air to mines, thereby exacerbating the problem of spontaneous combustion. Michalski and Gray suggested that pulverized fuel ash from local thermal power stations could be placed in opencast and underground mines for disposal and at the same time contribute to the abatement of mine fires. From the point of view of mine fire abatement, the Jharia Coalfield produces some 30 million tonnes of fuel ash per year so that there is a plentiful supply for use in the control of mine fires.

Adamus (2002) noted that unreactive gases were first used to fight mine fires in the latter half of the nineteenth century. Previously, Morris (1987) had indicated that probably the earliest recorded case of the atmosphere in a deep mine being rendered inert was in the 1850s at the Clackmannan Mine, Scotland. There a mixture of CO_2, N_2, SO_2 and steam was pumped into the mine for about a month to extinguish the fire. In the second half of the twentieth century nitrogen began to be used to smother fires in deep mines. According to Adamus, nitrogen helps to protect rescuers from fires and explosions and provides the opportunity to open sealed fires earlier (in some instances allowing a fire to be extinguished directly), as well as helping to control spontaneous combustion in wastes. Adamus also provided a review of the use of nitrogen in a number of countries. For example, he described the use of nitrogen at the Doubrava Mine in the Czech Republic. There the fire was exacerbated by explosions of coal dust and methane. Both downcast and upcast shafts were sealed by airtight plugs and covered with clay and sand at the surface. Between 16 000 and 17 000 m^3 of nitrogen were pumped into the mine daily for 35 days.

When grouting is used to extinguish a burning coal seam, drillholes initially are sunk to intercept the workings at the lowest area where burning occurs. Holes are drilled at closely spaced intervals and systematic filling takes place moving towards the higher levels of the mine. Foams, again introduced into the mine via drillholes, also have been used in attempts to extinguish mine fires. Any fissures or crown holes at the surface that allow entry of air to a mine should be sealed prior to treatment. Thermocouples are used to determine the extent of the fire, its subsequent movement and the success of the treatment.

Coal mines in China that were burning have been collapsed by controlled explosions (Fig. 10.3). However, the technique has not always proved successful in extinguishing fires, presumably because the fragmented rock material may allow access for air.

Spontaneous combustion of coal frequently takes place in opencast coal mines (Fig. 10.4). For instance, some opencast coal mines in India generate the largest fires in the world. These fires burn the coal reserves, damage the land and buildings and structures, and pose danger to miners working in the mines, as well as to the local residents. The spontaneous combustion of the coal transforms it into an unconsolidated, weak, residual ash. This frequently results in local instability and failure where the residual ash accumulates in moderate thicknesses (Fig. 10.5). However, since the zone of burning usually is restricted to the outer layers of the coal, penetrating to about 1 m behind the face, small rock falls and topples tend to be more frequent in occurrence.

Figure 10.3 Extinguishing mine fire by directional blasting in the south fire area of Beishan Coalfield, Qitai County, China.

Figure 10.4 Fires caused by spontaneous combustion of coal in an opencast mine, Singrauli Coalfield, India. (Reproduced by kind permission of International Mining Consultants Limited.)

Many of the large opencast coal mines in India are worked by draglines. Individual faces often reach at least 2 km in length. Extraction of the coal by the draglines creates a working face from 60 to 100 m high. The dragline operations leave behind pillars of coal 6–10 m wide along their base, and 18–20 m high and around 2 m wide at the top. These pillars retain the dragline tips, that is, material that has been dumped by the draglines above the overburden. The coal in the pillars contains a high density of discontinuities consisting of small faults, joints (cleat), bedding planes and mining induced fractures. These become dilated and upon exposure to the atmosphere begin to smoulder, with ignition occurring within a few hours to a few months of exposure. However, the residual ash in the pillars that are affected by spontaneous combustion is considerably weaker than the coal itself, and this therefore aids the eventual retrogressive rotational failure of the tipped material into the workings. Some examples of slope failures at an opencast mine in India associated with spontaneous combustion during 1997 and 1998 are given in Table 10.1. Where the seams are dipping into the high wall face instability problems are less likely. Rainwater during the monsoon season accumulates within the 'valleys' in the overburden tips. These generate high pore water pressure within the groundwater in the tipped material and in particular behind the pillars, which further facilitates instability. Visual inspections of the pillar fires, when undertaken after dusk, provide the opportunity to investigate the mechanisms of combustion of the coal. The zones of burning are controlled largely by the density and orientation of the joints, bedding planes and mining induced fractures.

Figure 10.5 Slope failures in an opencast coal mine due to spontaneous combustion, Singrauli Coalfield, India.

These frequently induce small-scale toppling and wedge failures of blocks of coal up to 1 m^3. Violent combustion of methane gas occurs in the accumulated scree material at the foot of the coal pillars, and the gas can be recognized by the presence of a deep blue coloured flame. This contrasts considerably with the deep orange-red coloured fires on the face of the pillars. Fewer fires occur along the high wall face, where the rock mass

Table 10.1 Examples of slope failures induced by spontaneous combustion at an opencast mine in Singrauli Coalfield, India

Date of failure	*Description of events*
10 February 1997	A slope failure affected the lower section of the steepest part of the dragline dumps, to a height of 48 m. The total height of the tip before failure was 83 m and was inclined at 54°. The rotational slide was 140 m wide and the coal underwent translational shift by 5 m along the floor of the seam, in the dip direction. The collapsed material accumulated up to 15 m in the mined out section of the workings
28 May 1997	A slope failure affected the lower 52 m of the dump material. The total height of the tip before failure was 82 m and was inclined at 51°. The width of the slide was 51 m, and the tip material accumulated 18 m into the workings
3 October 1997	The total pre-slide dump height was 106 m and the overall slope was 48°, steepening to 76° in the lower 50 m of the slope. Failure affected the dragline dump 50 m high and 220 m long. The collapsed material accumulated on the floor of the mine where the Turra seam was being extracted. The coal pillar underwent translational shift by 7 m towards the high wall face
19 September 1998	A slope failure involved the failure of a 94 m high, 47°, dragline tip. This created a scar 75 m high and 250 m long and caused the down-dip translational shift of the coal pillar by 65 m, stopping just short of the high wall

Figure 10.6 Extinguishing fires by water jetting and then spraying face with a bitumen compound at high pressure, Singrauli Coalfield, India.

discontinuities are tight or covered with a veneer of ochreous groundwater that percolates from within the coal seams. Specialist fire fighting units cool the coal with jets of high pressure water (Fig. 10.6). This also removes the loose rocks that have formed by fissuring in the coal. The fissures then are sprayed with a high-pressure bitumen based compound that temporarily seals the fissures on the working face. The latter is left to stand for about 12 months before excavation. Although this technique has been used with some success, it is not a completely satisfactory process. Furthermore, the technique generates large quantities of steam and ash, which fill the mine.

10.2. Gases

10.2.1 Coal mine gases

The hazards associated with mine gas emissions have been known since the earliest mines were sunk and continue to be a major cause of deaths, and damage to mine and surface infrastructure in operating and abandoned mines around the world. A variety of gases can accumulate in coal mines, especially abandoned mines as ventilation becomes inadequate. These include methane (referred to as fire damp), which is the most common mine gas, carbon monoxide (white damp), carbon dioxide, hydrogen sulphide (stink damp) and stythe gas (black damp, air depleted of oxygen). These gases are dangerous and may be combustible, explosive or toxic (Table 10.2). Admittedly, gas problems are not present in every mine. Also, some of these gases can be found in mines other than coal mines.

Methane not only is toxic but it also is combustible and explosive when mixed with air at concentrations between 5% and 15%. Also, it is lighter than air, having a density of 0.7 $kg\,m^{-3}$ at 10°C. As methane is less dense than air it tends to migrate to zones of higher elevation in workings or into fractured strata. Methane gas in coal mines may contain small amounts of ethane, propane, carbon dioxide, carbon monoxide and nitrogen. What is more, methane can be oxidized during migration to form carbon dioxide.

Methane, like carbon dioxide, is generated by the breakdown of organic matter. In other words, methane is a by-product formed during coalification, that is, the process that changes peat into coal. Biodegradation takes place in the early stages of accumulation of plant detritus with the evolution of some methane and carbon dioxide. The major phase of methane production, however, probably takes place at a later stage in the coalification process after the deposits have been buried. During this process approximately 140 m^3 of methane is produced per tonne of coal. As a consequence, the quantity of methane produced during coalification exceeds the holding capacity of the coal, resulting in excess methane migrating into reservoir rocks that surround or overlie the coal deposit. In addition, coal has a very large internal surface area (93 000 000 m^2 $tonne^{-1}$ of coal) and,

Table 10.2 Effects of noxious gases (After Anon., 1973; reproduced by kind permission of the Coal Authority)

Gas	*Concentration by volume in air ppm*	*Effect*
Carbon monoxide	100	Threshold limit value (TLV) under which it is believed nearly all workers may be repeatedly exposed day after day without adverse effect (TLV)
	200	Headache after about 7 hr if resting or after 2 hr if working
	400	Headache and discomfort, with possibility of collapse, after 2 hr at rest or 45 min exertion
	1200	Palpitation after 30 min at rest or 10 min exertion
	2000	Unconsciousness after 30 min at rest or 10 min exertion
Carbon dioxide	5000	TLV. Lung ventilation slightly increased
	50 000	Breathing is laboured
	90 000	Depression of breathing commences
Hydrogen sulphide	10	TLV
	100	Irritation to eyes and throat; headaches
	200	Maximum concentration tolerable for 1 hr
	1000	Immediate unconsciousness
Sulphur dioxide	1–5	Can be detected by taste at the lower level and by smell at the upper level
	5	TLV. Onset of irritation to the nose and throat
	20	Irritation to the eyes
	400	Immediately dangerous to life

Notes

1 Some gases have a synergic effect, that is, they augment the effects of others and cause a lowering of the concentration at which the symptoms shown in the above Table occur. Further, a gas that is not itself toxic may increase the toxicity of one of the toxic gases, for example, by increasing the rate of respiration; strenuous work will have a similar effect.

2 Of the gases listed carbon monoxide is the one most likely to prove a danger to life. The others become intolerably unpleasant at concentrations far below the danger level.

as methane can exist as a tightly packed monomolecular layer adsorbed on the internal surfaces of the coal matrix, it is able to hold 2 to 3 times more gas than conventional reservoirs. Measurements of methane contents have ranged from trace amounts to 25 $m^3 tonne^{-1}$, the highest values being associated with anthracites. In the bituminous coals of Britain the volumes generally are between 5 and 6 m^3 per tonne, the variation depending upon the depth of burial and the geological history. Methane in the Coal Measures is adsorbed on coal, may be trapped in gas pockets or dissolved in groundwater. For instance, sandstones in Coal Measures may act as reservoirs, methane having migrated there from coal seams over geological time. Creedy (1991) mentioned that degassing rates of methane appeared to depend largely upon the amount of fusinite in coal, the greater the fusinite content the faster a coal degassed. He explained that this relationship is due to the ready fracturing of friable maceral on microscopic planes to form channels or partings along which the gas flows. However, because of the low permeability of coal this only occurs when coal is destressed by mining activity.

During modern coal mining operations, methane is liberated routinely into the mine ventilation system and vented to the atmosphere (Creedy *et al.*, 1997). For example, methane drainage above longwall panels by inclined underground drillholes is common practice at such working mines. However, not all the methane need be removed by these processes and unknown but probably substantial volumes can accumulate in goaf areas, in fractured strata, pores and voids, and in waste and stowed material.

Gases may be dissolved in groundwater depending on the pressure, temperature or concentration of other gases or minerals in water. Dissolved gases may be advected by groundwater, and only when the pressure is reduced and the solubility limit of the gas in groundwater exceeded, do they come out of solution and form a separate gaseous phase. Such pressure release occurs in the strata affected by the removal of coal. It is essential that such degassing of methane is not allowed to occur in confined spaces where an explosive mixture could develop.

Methane can move through coal by diffusion, which is relatively slow but as coal seams are fractured the diffusion rate is increased. Gas also migrates through rocks via intergranular permeability or, more particularly, along discontinuities. Where strata has been disturbed by mining subsidence gas permeability is enhanced, as is that of groundwater holding gas.

Gas may be liberated from mine workings at the ground surface or into the built environment by more than one factor. For example, the reservoirs of gas that exist in certain formations in coal mines expand and contract with changes in atmospheric pressure. The idea that the efflux of methane gas from mines is associated with lower barometric pressure has long been known and has been referred to as the 'breathing effect' (Donnelly and McCann, 2000). Other factors that influence the escape of gas include changes in or termination of mine ventilation. During underground coal mining operations, ventilation is essential to prevent gas accumulation, it being controlled by fans, screens, air locks and doors.

The rate of fall of barometric pressure is a factor in determining the quantity of gas emissions into mine atmospheres (Fig. 10.7). The differential pressure created may reach 40 mb in severe cases, which can cause gas hazards at the ground surface. If the barometric pressure falls gradually, the gas is diluted in a working mine as fast as it overflows into the ventilating currents. On the other hand, if the fall is rapid and the ventilation poor, then the volume of gas that escapes from the walls and/or waste may be sufficiently large to cause dangerous fouling of the airways. When the mine ventilation is stopped, under falling barometric pressure, mine gases begin to accumulate in the mine workings and eventually may find their way to the ground surface via shafts, adits or the overlying strata (if fractured and

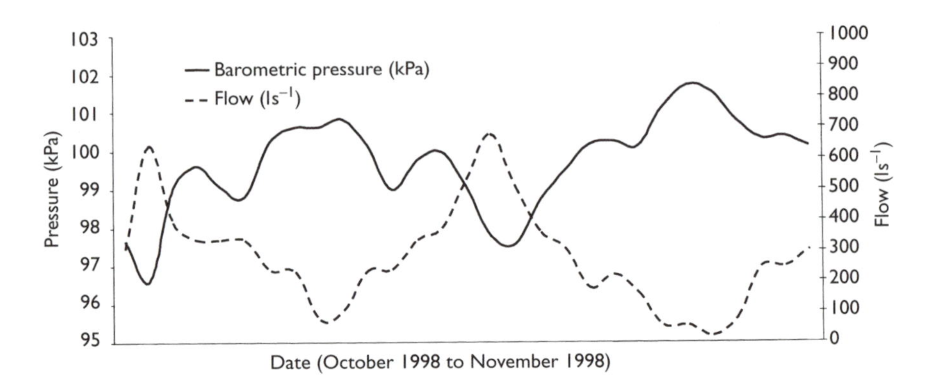

Figure 10.7 Changes in barometric pressure related to methane gas emission from a shaft, Donbass Coalfield, Ukraine. (Reproduced by kind permission of White Young Green Consulting Limited and International Mining Consultants Limited.)

permeable, and at shallow depth). Furthermore, the hydrostatic head imposed by mine water rebound, that is, by rising mine water caused by the cessation of large-scale groundwater pumping once a mine ceases operation, can drive mine gases to be released at the ground surface. However, once mine water levels have recovered, then the problem of mine gas at the surface will be reduced.

Methane detectors can record concentrations as low as 1 ppm and so have been used for the detection of old coal mine shafts (see Chapter 2). Anomalies associated with old shafts tend to range between 10 and 100 ppm. A detector is used in a series of traverses across the site near the ground but should not be used on a windy day. A contoured map is produced showing methane concentration.

When methane is detected at the ground surface it is necessary to determine its source so that the most appropriate remedial action can be taken to minimize any possible danger. Obviously, methane can originate in more than one way, it need not be associated with Coal Measures strata, it could be from a landfill. In major construction operations or in domestic dwellings, the sources of methane need to be considered and measures taken to minimize any risk posed. For instance, gas may accumulate in abandoned mines and methane, in particular, since it is lighter than air may escape from old workings via shafts and via crown holes where the workings are at shallow depth. Joints and fractures, probably enhanced by mining subsidence also may act as gas migration pathways. The escape of methane at the surface can lead to the accumulation of methane in buildings. Because, as noted, when air is mixed with 5% to 15% methane it becomes explosive, this can give rise to disastrous consequences, that is, a building being destroyed and/or lives being lost. There were over 70 recorded incidents of gas emission at the surface from abandoned coal mine workings in Britain in the 50 years following 1945. According to Sizer *et al.* (1996), over two-thirds of these were of methane, the remainder being mainly stythe gas (see later). To date, these emissions have been at locations on exposed coalfield areas and in some instances appeared to be related to the occurrence of sandstone at the surface. By contrast, areas with a surface cover of argillaceous rocks or till seemed to be at lower risk. Robinson (2000) referred to more than a dozen incidents in the Northumberland and Durham coalfield between 1950 and 1995, in which four people were killed, others injured and families had to be re-housed. Raybould and Anderson (1987) described grouting of old mine workings that had acted as conduits of methane to residential properties. Although methane is not toxic to plant life, the generation of significant quantities can displace oxygen from the root zone and so suffocate plant roots (Fig. 10.8). Large concentrations of carbon dioxide, carbon monoxide or hydrogen sulphide can produce the same result.

Generally, the connection between a source of methane and the location where it is detected can be verified by detecting a component of the gas that is specific to the source or by establishing the existence of a migration pathway from the source to the location where the gas is detected. An analysis of the gas obviously helps identify the source. For example, methane from landfill gas contains a larger proportion of carbon dioxide (16–57%) than

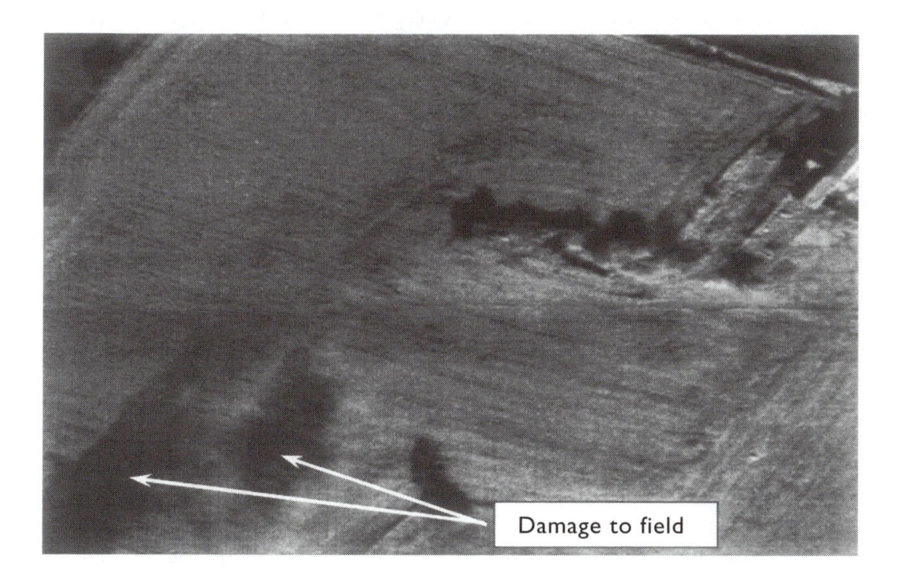

Figure 10.8 Damaged ground in a field partly affected by emission of methane gas (two patches bottom-centre of field). (Reproduced by kind permission of White Young Green Consulting Limited and International Mining Consultants Limited.)

does coal gas (up to 6%). Analysis also may involve trace components or isotopic characterization using stable isotope ratios ($^{13}C/^{12}C$; $^{2}H/^{1}H$) or the radio-isotope ^{14}C. This again allows distinction between coal gas and landfill gas as in the former all the radiocarbon had decayed (Williams and Aitkenhead, 1991).

Determination of the migration pathway of gas may involve a hydrogeological investigation. Groundwater flow needs to be determined if gases are dissolved in groundwater, particular note being taken of areas of discharge. Hydrogeological assessments require accurate measurement of piezometric pressure, and sampling and chemical analysis of water. If methane is dissolved in groundwater, then samples should be obtained at *in situ* pressure and they must not be allowed to degas as it equilibrates to atmospheric pressure.

Under favourable economic, environmental and geological conditions, methane accumulations in active and abandoned mine workings, or methane driven ahead of rising mine water may be exploited as an energy resource (Gayer and Harris, 1996). Methane drainage in an active mine is brought about by degasification and by the use of vertical and horizontal drillholes into the coal, as well as by mine ventilation systems. This methane is referred to as coal mine methane (CMM). Coal bed methane (CBM) on the other hand, involves the recovery and utilization of methane, from a drillhole, or an array of wells drilled into coal seams, which have not been disturbed by mining. Underground gasification of coal seams (UCG) is a chemical process that involves converting the coal into a combustible gas. This gas then is utilized as an energy resource or to produce heat. Underground coal gasification involves the in-seam gasification of strata-bound coal reserves, by drilling into the coal horizons, injecting air or oxygen, igniting the seam and thereby gasifying the coal. The gaseous products then are transported to the surface where they undergo processing. Following the abandonment of mining, the recovery of gas for energy purposes is referred to as abandoned mine methane (AMM) (Fig. 10.9). For example, Bell *et al.* (2000) described the utilization of methane gas emitted from three old shafts at an old coal mine site in North Rhine-Westphalia, Germany, which was scheduled for restoration (see Case history 2, Section 9.7).

As remarked, stythe gas is deoxygenated air, which occurs in abandoned coal mines. The oxygen is removed to a varying degree by the oxidation process that takes place between oxygen and coal, rotting timber supports or minerals such as pyrite. Hence, stythe gas consists mainly of nitrogen and carbon dioxide. For instance, the concentration of carbon dioxide in stythe gas varies from 0.1% to 20% compared with a concentration of 0.03% in normal air. Because of its carbon dioxide content, stythe gas is heavier than air so it tends to accumulate on the floors of mines and at the bottom of mine shafts. Stythe gas has been known to accumulate in houses, where it usually tends to occupy the area above the floor to a depth of 0.5 m or

Figure 10.9 Collection and storage of methane for use in the generation of electricity at the former Shirebrook Colliery site, West Midlands, England. (Reproduced by kind permission of the Coal Authority.)

more. Because stythe gas is deficient in oxygen, it is a suffocating gas and can give rise to fatal consequences. For example, one of the fatalities in the Northumberland and Durham Coalfield referred to by Robinson (2000) was from stythe gas. The eight hour occupational exposure limit is 0.5%.

Both carbon dioxide and carbon monoxide are toxic (Table 10.2). The former is heavier than air and so hangs about the floor of a mine excavation or may form a layer on the floor of a home, or can concentrate in cellars or confined spaces. Carbon monoxide is lighter than air. Hydrogen sulphide is heavier than air and is highly toxic (Table 10.2). It also is explosive when mixed with air. Hydrogen sulphide may be absorbed by water, which then becomes injurious as far as concrete is concerned. Sulphur dioxide is a colourless, pungent, asphyxiating gas that dissolves readily in water to form sulphuric acid. It may be formed by the breakdown of pyrite.

Carbon dioxide sequestration is aimed at reducing the volumes of CO_2 emitted into the atmosphere. This involves the capturing of the gasses after combustion and injecting them into the ground where they can remain for large periods of geological time. Potential storage sites for CO_2 include old mines, brine formations and salt domes (below land and sea), and abandoned hydrocarbon reservoirs. One possible solution is that CO_2 sequestration could be undertaken in conjunction with UCG. A continuous system of UCG would result in highly fractured and porous strata. Abandoned cavities could be penetrated by drillholes and the CO_2 could be injected under high pressure for storage.

10.2.2 Radon

Radon is a naturally occurring radioactive gas that is produced by the radioactive decay of uranium (U) and thorium (Th). It is the only radioactive gas, and is colourless, odourless and tasteless. Uranium and thorium are the parents of a series of radioactive daughter products that ultimately decay to stable lead isotopes. Three isotopes of radon are members of these series, namely, ^{219}Rn (actinon), ^{220}Rn (thoron) and ^{222}Rn (radon). The half-lives of the two former are only a matter of seconds whilst that of the latter is 3.82 days. However, the immediate parent of ^{222}Rn is radium, ^{226}Ra. Its half-life is 1622 years. Radon decays to the solid daughter product ^{218}Po (polonium).

Although radon frequently is present in notable amounts in areas where there is no uranium mineralization, it normally is associated with rocks with high concentrations of uranium (Ball *et al.*, 1991). The range of concentration of ^{238}U in rocks and soil varies significantly. Rocks such as sandstone generally contain less than 1 mg kg^{-1}, whilst some carbonaceous shales, some rocks rich in phosphates and some granites may contain more than 3 mg kg^{-1}. The amount of radon that is emitted at the Earth's surface is related to the concentration of uranium in the rock and soil, as well as the efficiency of the transfer processes from the rock or soil to soil–water and soil–gas. In fact, most of the radon produced in a mineral remains entrapped but some may escape into voids in rock or soil. How much radon is released depends upon the surface area, shape, degree of fracturing and other imperfections in the host mineral. The amount of radon that escapes is greater in soils than rocks. Movement of radon in the pores depends upon fluid flow through the rock and soil in that most radon is transported by carrier gases or liquids, the movements of which are governed by the permeability of the rocks or soils. As radon is moderately soluble in water it can be transported over considerable distances, and so anomalous concentrations can occur far from the original sources of uranium or thorium. Transport by fluids is especially rapid in limestones and along faults.

Radon represents a health hazard since it emits alpha particles (Howes, 1990). Outside the body, these do not present a problem because their large size and relatively high charge mean that they cannot pass through the skin. However, when alpha particles are ingested or inhaled, they can damage tissue because they are not penetrative and therefore release energy over a relatively small volume of tissue. Indeed, the inhalation of radon and its daughter products accounts for about one half of the annual average exposure to ionizing radiation in Britain and the United States. Although radon is the principal way alpha particles enter the human body, most is breathed out. The solid daughter products (e.g. ^{218}Po that adheres to dust), however, are also alpha particle emitters but are more dangerous because they often are retained in lung tissue where they increase the risk of lung cancer. The basis of radon as an aetiological factor in lung cancer derived from the increased incidence of lung cancer in uranium miners. Jones (1995) found an association between dispersed uranium and the incidence of lung cancer in males and females in Illinois (Fig. 10.10). Some radon may be dissolved in body fats and its daughter products transferred to the bone marrow. The accumulated dose in older people can be high and may give

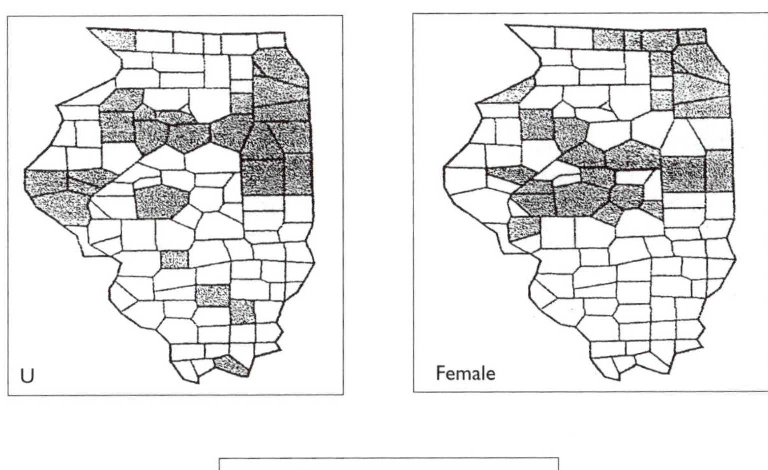

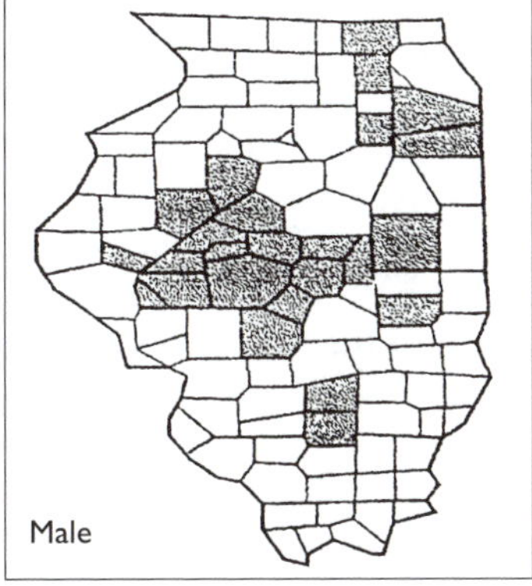

Figure 10.10 Relationship between dispersed U and average annual incidence of lung cancer in females and males for counties in Illinois. (After Jones, 1995; reproduced by kind permission of Springer.)

rise to leukaemia (Henshaw *et al.*, 1990). Radon also has been linked with melanoma, cancer of the kidney and some childhood cancers. The United States Environmental Protection Agency (Anon., 1986) has taken 4 $pCi\,l^{-1}$ as the limit beyond which radon is considered a hazard. The average concentration of radon outdoors is around 0.2 $pCi\,l^{-1}$ as compared with approximately 1.0 $pCi\,l^{-1}$ for indoors.

Radon and its daughter products accumulate in confined spaces such as buildings. Soil gas is drawn into a building by the slight underpressure indoors that is attributable to warmer air rising. A relatively small contribution is made by building materials. Radon can seep through concrete floors and foundations, drains, small cracks or joints in walls below ground level, or cavities in walls. Emission of radon from the ground can vary, for example, according to barometric pressure and the moisture content of the soil. Accumulation of radon also is affected by how well a building is ventilated. It also can enter a building via the water supply, particularly if a building is supplied by a private well. Public supplies of water, however, usually have relatively low radon contents as the time the water is stored helps radon release.

Reduction of the potential hazard from radon in homes involves locating and sealing the points of entry of radon and improved ventilation by keeping more windows open or using extraction fans. Ventilation systems can be built into a house during its construction, for example, a system can be installed beneath the house.

Ball *et al.* (1991) stated that radon can be detected by zinc sulphide scintillation counting, liquid scintillation counting, alpha track registration, semi-conductor detectors and absorbers (e.g. activated charcoal, silica gel or charged metal plates). When alpha particles interact with zinc sulphide, they give off pulses of light that may be counted electronically. It is possible to calculate the activities of both radon and thoron because of the different half-lives of radon, thoron and their immediate daughter products. The technique frequently is used for the measurement of radon in soil gases. In liquid scintillation counting an organic fluorescence agent is dissolved in an organic solvent, such as xylene or toluene. The solvent extracts radon from gas or liquid phases or absorbers. The ionization of the solvent by alpha and beta particles, and its subsequent de-excitation results in light emission that is proportional to the radon extracted. Alpha particles may be registered or detected by cellulose nitrate film that is damaged by the particles but is insensitive to light. The track density that the alpha particles make is proportional to the radon concentration. Due to the lack of penetration of alpha particles, surface barrier semi-conductor detectors are suitable for measuring radon and its daughter products. Absorbers can concentrate radon and/or its decay products. The build-up of the radioactivity of the daughter products may be determined, after a suitable time, by extraction of the radon into a liquid scintillator.

Radon in soil gas is measured by means of a hollow spike hammered into the ground and linked to a gas pump and detection unit. Detection of the radon normally is by the zinc sulphide scintillation method. Account must be taken of the weather conditions and the permeability of the soil as both can have a profound effect on radon levels in soil gas. Alternative, passive methods employ detectors that are buried in the soil and recovered some time later, often up to a month. This procedure is used when long-term monitoring is required to overcome the problems of variation in concentration of radon due to changing weather conditions. Etched track methods also are used for long-term monitoring of soil radon. Occasionally semi-conductor or absorber methods are employed.

Abandoned mine shafts provide a place for the accumulation of radon and its daughter products. Similarly, since faults represents zones of mechanical disintegration that may contain fissures and/or voids, they also may provide a suitable location for the accumulation and storage of radon gas. Radon gas anomalies therefore may occur above abandoned mine shafts and fault zones. However, abandoned mine shafts and faults occur in a variety of geological settings and therefore anomalous levels of gas are not likely to be present in every instance. Moreover, positive radon gas anomalies over shafts and faults are not always repeatable, the quantity of radon gas emission varying from hour to hour, as well as from day to day. This is due to a number of reasons such as variations in ground permeability and local weather conditions, along with the sensitivity of the radon gas detectors. Being a noble gas, radon tends to remain in the ground or the mine air until the air above it is discharged. The emission of radon gas from the ground to the surface therefore is dependent upon the local atmospheric pressure, being emitted during periods when it is lower. Local wind speed also plays an important role in the ability to detect the radon and hence all measurements ideally should be taken below the ground surface in a small borehole (made, for example, with hand held auger). Also, the weather conditions at the time the survey was undertaken should be recorded. The residence time of air in a fault zone or mine shaft will be variable. The half-life of radon as mentioned, is around 3.82 days and therefore the concentration of radon in a fault zone or mine shaft is dependent on its rate of emanation, and its travel time within the ground surface (Donnelly and Bamford, 1996).

10.3. Induced seismicity

Induced seismicity occurs where changes in the local stress conditions give rise to changes in strain in a rock mass. The sudden release of strain energy due to deformation and failure within a rock mass results in detectable earth movements (McCann, 1988). Induced seismicity has been associated with many mining operations where changes in local stress conditions have given rise to corresponding changes in strain and deformation in the rock masses concerned. These changes have been responsible for movements that took place along discontinuities, which may be on a macroscopic or microscopic scale. In the former case they generate earth movements that are detectable at the ground surface. Such movements can cause damage to buildings, but generally this is minor, although very occasionally high local intensities are generated. Obviously, the extent of damage is related to the magnitude of the earth tremor and the distance from the source, as well as to the nature of the surface rock and the strength of the structure. Nonetheless, such seismic events are a cause of concern to the general public.

Seismicity related to mining activity may occur directly as the shock waves from rock blasting, or as a result of the effects of mining, where stresses in the rock may result in the occurrence of rockbursts. Cook (1976) suggested that the changes in stress consequent on mining were responsible for a sudden loss of stored strain energy that results in brittle fracture in the rock mass in the excavation. These changes in strain energy may be related to a decrease in potential energy of the rock mass as the rock is mined. On the other hand, it may be that mining triggers latent seismic events in rock masses that are in a near unstable condition.

The incidence of seismic events and their magnitudes can be predicted statistically from the calculated values of the spatial rate of energy release, where it is associated with extension of mining excavations. As a result of making these excavations, the potential energy of the system is changed by an amount equal to the product of the weight of rock mined and the depth from which it is mined. At most, half this energy can be stored as elastic strain energy in stress concentrations and the remainder must be released. The amount of energy released is determined by the volumetric closure of the excavation and the value of the virgin rock stress acting on its surfaces before mining (Cook, 1976). Mining creates voids in the strata. Failure of the rock occurs in regions of maximum stress concentrations near the edges of the excavations. This gives rise to new fracture planes closely parallel to the working faces and to continuous seismic activity. Foci of seismic events (magnitude usually less than 3) occur mainly within tens of metres of advancing working faces.

The phenomena that constitute rockbursting are very complex and cover a wide range of magnitudes, as well as modes of origin. However, although all rockbursts generate seismic events, all seismic events associated with mining activity are not rockbursts. Their are a number of factors that influence the incidence of rockbursts. These include the stoping width and span, the rate of face advance, the presence of faults or dykes and shapes of the abutments (Cook *et al.*, 1966). Rockbursts are capricious and it is therefore very difficult to make objective measurements of their causes, mechanisms and effects that are necessary preliminaries to finding solutions to the problem. Often, much more can be gleaned by studying good photographs than from on-site examination. Early efforts to record rockburst damage reliably were hampered by a lack of suitable photographic equipment. After major rockbursts it often is impossible to gain more than a few isolated, peripheral glimpses of the devastated scene. Still, reliable descriptions of the damage and circumstances prevailing before the events are necessary.

Kinetic energy is made available as a result of a change in the energy level of that portion of a rock mass that is in a state of unstable equilibrium. As an example, surficial, violent failure can occur at the face of a tunnel at moderate depth. Usually, very little energy is involved and the fracture is restricted to the immediate surface of the tunnel, where it is dictated by the stress concentrations resulting from the detailed shape of the opening (Ortlepp, 1984). Such failures may be termed strainbursts and involve splitting, flaking, scaling, or slabbing of rock at the face (Morris and Vorster, 1984). There is a substantial increase in the number of these small events directly after blasting. On the other hand, large seismic events can occur along geological discontinuities so distant from excavations that the shape of even extensive mined areas have no influence, and the instability is promoted by low intensity, widespread stress changes resulting from the diffused effect of regional mining. These larger events tend also to be independent of blasting and may have magnitudes of up to 5.5, but often produce no significant damage to excavations other than widespread falls of loose ground. Events resulting in seismicity in the intermediate range of magnitude (in the region of 3 on the Richter Scale) usually occur a few to tens of metres distant from mined areas. These events are often the most hazardous and can be considered true rockbursts. In such cases, the shape of the excavation plays a major role in determining the magnitude and distribution of stress changes in the zone of potential instability (Ortlepp, 1984). Instability can involve movement along an existing discontinuity or it can involve extensive fresh shear fracture through previously intact massive rock.

The mine tremor phenomenon in southern Africa is unparalleled in its extent and intensity, and adds a hazard to mining. The tremors are, it seems, only hazardous in the immediate vicinity of mines (Shapira *et al.*, 1989). Damage to structures on the surface can be caused partly by the permanent displacement or fracture of the rock but mostly it is caused by the elastic waves that cause ground shaking. The degree and extent of any damage depends on the amount of energy released at the focus (magnitude), the focal depth, the focal mechanism, the transmitting medium (especially the properties of the upper layers) and the distance from the source. High frequency energy is rapidly attenuated whereas low frequency (less than 10 Hz) energy may travel long distances. Where seismicity is mining induced, the type and quality of support, the strength and conditions of the surrounding rock, and the shape and size of the excavations are, in addition, important in governing the extent of any damage.

Many seismic events, admittedly of small magnitude, have occurred within the Johannesburg region, South Africa (see later). Studies of these events indicate that their origin is due to extensive mining operations because there is an intimate relationship between a large percentage of them and the incidence of blasting in mines (Gibowicz, 1984).

Some damage has occurred to surface structures. For example, in 1976, a tremor of magnitude 5.1 in the centre of Welkom, South Africa, damaged many buildings and culminated in the collapse of a reinforced concrete framed block of flats. An enquiry into the cause of the disaster did not establish with certainty whether release of stress in the mine workings at a depth of 1600 m below the surface contributed to the event. The event did, however, prompt the installation of a permanent regional seismic network, consisting of 24 geophones, between 1978 and 1980 (Lawrence, 1984). In fact, a number of networks in the Rand and Free State monitor the seismicity occurring in the vicinity of all the deep gold mines (Shapira *et al.*, 1989). Neff *et al.* (1992) referred to severe cracking of some houses in a gold mine township in the North West Province, some houses having to be evacuated. The magnitudes of the seismic tremors ranged up to 3.4. A seismic investigation was carried out that revealed that the response of the houses was dependent on the type of their foundation. In other words, the maximum peak particle velocity for a house on conventional spread footings was 40 $mm\,s^{-1}$, whereas a house on a light raft foundation with a compacted gravel mattress was 10 $mm\,s^{-1}$. These values compared with 20 $mm\,s^{-1}$ for the maximum peak particle velocity of the ground surface. Hence, the rafts with gravel mattresses experienced an attenuation of ground vibration and were not damaged.

The Coalbrook disaster took place in South Africa in 1962. This disaster was the most notable example of pillar failure occurring in a working mine. Collapse occurred in minutes over a whole pillar district some 300 ha in area at 140 m average depth. The ground surface suffered substantial rapid subsidence, severe local strain and tilt damage, and a fracture zone developed. The associated seismic activity was recorded 720 km from Coalbrook. It is significant to note that the sudden release of stored strain energy usually is associated with mining areas with high stress levels in strong rocks. This combination is not immediately associated with conditions in coal mines at depths of only 140 m, and Bryan *et al.* (1964) concluded that the peculiar geological conditions had had a significant influence on the collapse. A 40 m strong thick dolerite sill immediately above the workings was expected not to collapse but the pillar support proved inadequate.

Seismic events have been associated with deep longwall mining of coal in the United Kingdom. In 1975–77 earth tremors were recorded in Stoke-on-Trent, in the North Staffordshire coalfield, having magnitudes between 3 and 5. One of the largest tremors occurred at 7.15 on the morning of 13 May 1976. It lasted for 2 seconds and was centred on Trent Vale, affecting a wide area of the Lyme Valley between Penkhull and Clayton. Exaggerated claims have been made regarding the 'violence' of this tremor, in fact one report of damage was recorded – a cracked kitchen ceiling. Nonetheless, continued tremors could weaken old or poorly constructed buildings. Over 100 low magnitude earth tremors, but less than a dozen large enough to be felt by the public, were recorded. At times as many as 12 tremors per week were recorded. It was suggested that there was a link between these earth tremors and the mining activity at Hem Heath Colliery. The foci of the tremors affecting the Hanford and Knutton districts generally were located above the level of these mine workings. Retreat mining began at Hem Heath Colliery in the late 1960s and the area is interrupted by two major faults. Further earth tremors occurred in 1980–81 (Kusznir *et al.*, 1982).

Subsequently, Redmayne (1988) catalogued earthquake events in Britain and showed that approximately 25% of small to moderate earthquakes ($M_L \leq 3$) were associated with coalfields. He suggested that it seems likely that all deep coal mining was accompanied by such activity of varying magnitude, depending on geological conditions, local faulting, the local stress field, the rate and type of extraction, seam thickness, past mining in the area and the nature of the overburden. Earth tremors were associated particularly with the South Wales, Staffordshire, Midlothian and Nottinghamshire coalfields.

The Midlothian Coalfield comprises a north-east to south-west trending synclinal basin, with a number of subsidiary basins, the deepest of which underlies Rosewell. During 1985 and 1988 the Great Seam and the Parrot Coal seam were mined at depths of 600 and 900 m, respectively, beneath the village of Rosewell in a series of three longwall panels. The east to west striking Roslin Fault, a normal fault with a northerly direction of dip of 45° bounded the take in the south. A swarm of over 200 earthquakes with epicentres in the Rosewell area reached a peak in October 1986, with 41 events of less than 2.8 M_L (Redmayne, 1988). A large number of tremors were felt in the village. Minor structural damage was caused by events as low as 1.7 M_L. For instance, slight damage occurred to Rosslyn Chapel that was protected against potential mining subsidence by a support pillar of 300 m radius around the church. The computed epicentre for the shocks presumed responsible was located 1.5 km to the east of the chapel, but a sand and gravel filled erosion channel beneath the church may have been responsible for the amplification of the seismic waves. In August 1986 a seismic event coincided with the appearance of a crown hole that formed above the intersection of two roadways in abandoned mine workings 20 m below the surface. Initially, the crown hole was 3 m in diameter but increased to 5 m as a result of subsequent seismic events. The largest event on 9 October 1986, coincided with the closest approach of two working panels, the intensity being 5 M_{SK}. All the epicentres were within 1 km of the working face. The seismograph signal was characteristic of mining induced events, being dominated by surface waves, and the S- and P-waves were poorly developed and of a low frequency (Fig. 10.11). Redmayne was able to show a correlation between earth tremors and mining activity, the tremors showing a marked decrease during holidays and weekends (Fig. 10.12). In other words, histogram plots of the total number of events with time clearly illustrated a link between seismicity and mining activity. The continuation of the miners strike in early 1985 and then three weeks for holidays in July 1985 meant that production was halted, and account for the lack of seismicity. Only one single event occurred during the 1984 miners strike. The higher seismicity rates reflect periods of increased coal output and longwall face advance. On a daily basis, the greater number of events can be related to the week-day periods of mining activity. The increase in seismicity on a Saturday may reflect the time lag between extraction and seismic induction. Earthquake locations migrated with the mining and the focal depths of recorded events were within 1.4 km of the surface, with many being concentrated between 500 and 600 m, the average depth being 620 m. Redmayne *et al.* (1998) noted a rapid decay in earthquake activity following pit closure, further indicating a mining induced cause. They also stated that residual stress from past mining appeared to have been an important factor in generating seismicity and observed that limiting the width of the workings or the rate of extraction may reduce or eliminate mining induced earthquake activity.

Bishop *et al.* (1993) investigated the earth tremors associated with a number of collieries in north Nottinghamshire, England. Like Redmayne (1988), they found that maximum activity occurred mid-week while the

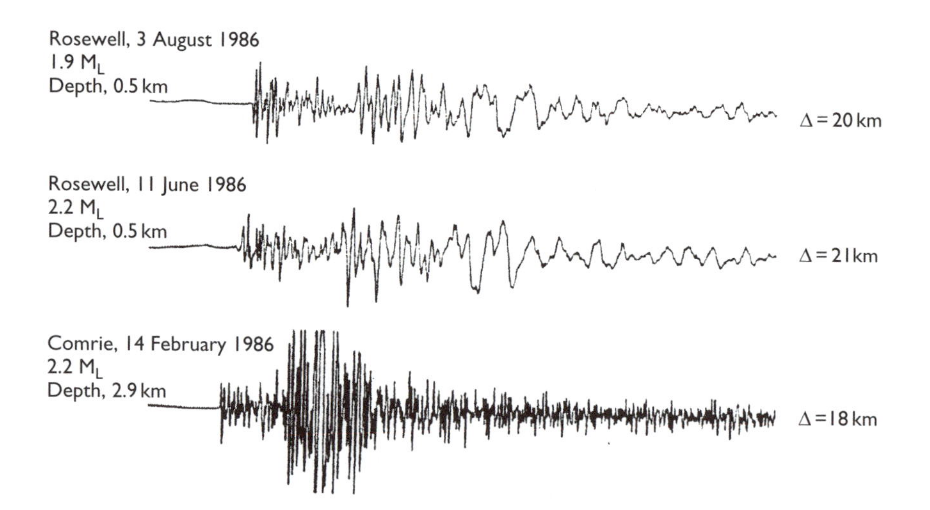

Figure 10.11 A comparison of the velocity seismograms of two Rosewell Coalfield earthquakes with a natural earthquake near Comrie in Perthshire, Scotland, epicentral distances (Δ) being similar. (From Redmayne, 1988; reproduced by kind permission of the Geological Society of London.)

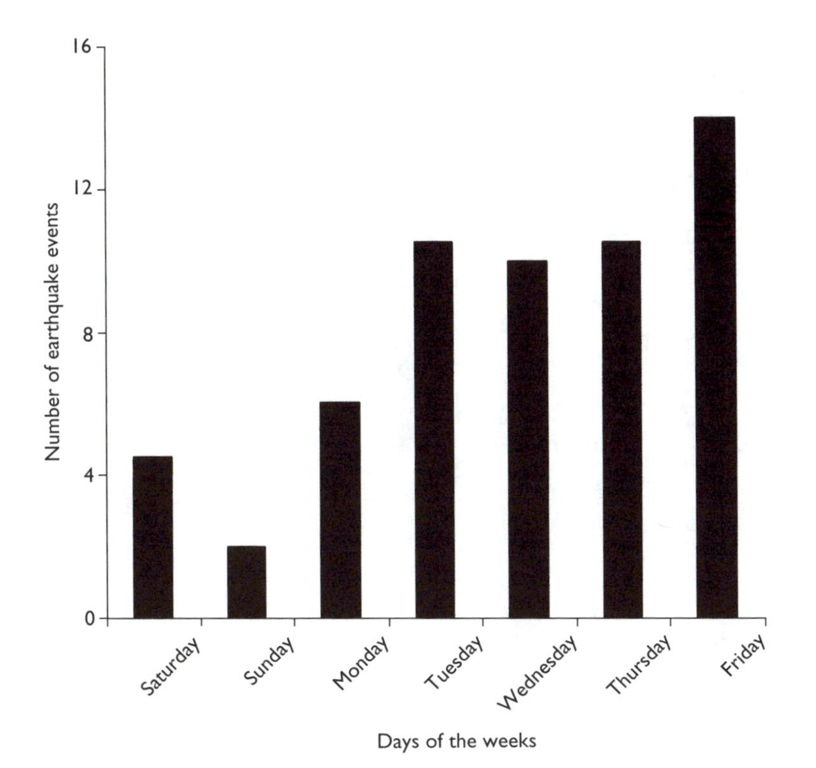

Figure 10.12 Histogram of Rosewell earthquakes versus day of the week. (From Redmayne, 1988; reproduced by kind permission of the Geological Society of London.)

minimum activity occurred at the weekends and during holidays. Bishop *et al.* noted that events occurred within days of a face going into full production and ceased when production came to an end. They also noted a good correlation between face advance and epicentral position (Fig. 10.13). Be that as it may, the seismic hazard and risk to property was low as the maximum magnitude recorded was less than 2.5. It was suggested that the potential magnitude of a tremor was related to the nature of the overburden, the local fault population, the local stress field, the rate of extraction, the seam thickness and past mining.

Induced seismicity also may be associated with post-mining activity. For example, at Easington in the Durham Coalfield, England, the formation of graben and fault scarps, over 300 m long, caused severe damage to rows of terraced houses and roads in the vicinity of the former Easington Colliery, and were accompanied by seismicity (Donnelly, 1998). Similarly, fault reactivation caused slight damage to houses in the Ryhope area, 6 km south of Sunderland, County Durham. From December 1998 to March 1999 earth

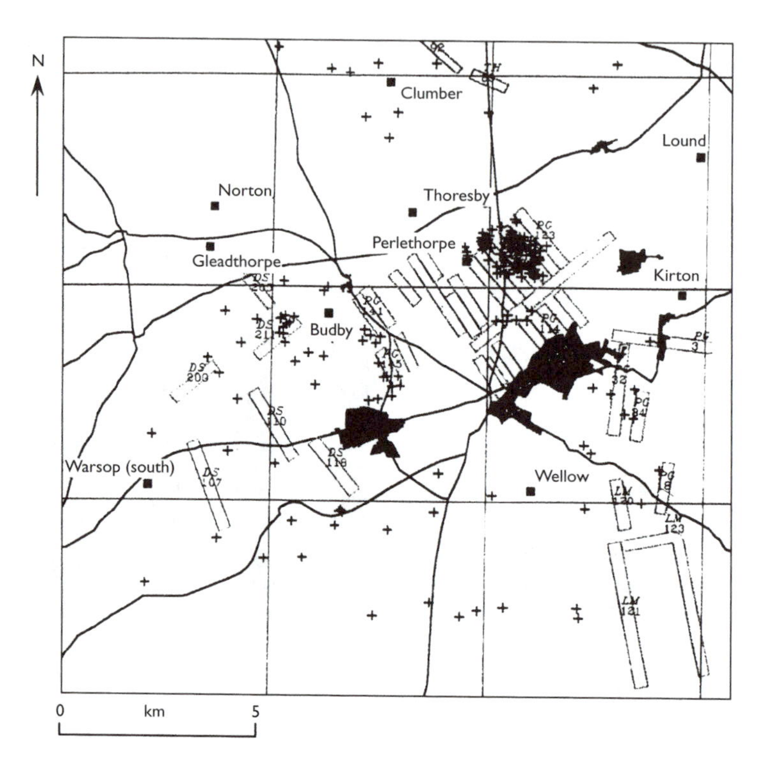

Figure 10.13 Epicentre location map for all the possible mining induced events from the Edwinstowe area, Nottinghamshire, England. (After Bishop *et al.*, 1993; reproduced by kind permission of the Geological Society of London.)

tremors were felt by a number of residents in some 14 separate streets. Twenty tremors were felt in six of these streets in a 12 month period. Subsequent investigations were carried out to try to find the source and origin of the seismicity. The possible causes of the earth tremors were neotectonic movements on faults, solution of the Magnesian Limestone (Permian) bedrock and the collapse of subsurface cavities, mining induced fault reactivation, rising mine water along faults and the displacement of mine gases along fault zones. Although the actual origin of the tremors was not identified conclusively, they were thought to be related to past mining.

The use of water injection to maximize the yield of oil can give rise to induced seismicity, presumably due to the injection of water leading to a reduction in the resistance to faulting consequent on the increase in pore water pressure. However, this would suggest that the rock masses involved were stressed to near failure stress prior to the injection of water. Gibbs *et al.* (1973) referred to seismic activity in the Rangely area, Colorado, associated with water injection in a nearby oilfield. They recorded 976 seismic events between 1962 and 1972, of these 320 exceeded magnitude 1 on the Richter Scale. They described an apparent correlation between the number of earth tremors recorded and the quantity of water injected. Ten years later, Rothe and Lui (1983) recorded 31 earth tremors between March 1979 and March 1980 at the Sleepy Hollow oilfield in Nebraska. Again water had been injected to maximize the yield of oil. The range of magnitude varied between 0.6 and 2.9 on the Richter Scale. The source of the tremors coincided remarkably with the area of the oilfield.

According to Maury *et al.* (1992), between 1957, when the Lacq gas field in south-west France began production, and 1967, some 25 mm of subsidence occurred due to a resultant depletion of pore pressure of around 30 MPa. Over the next 22 years the pressure was reduced by a further 25 MPa that brought about another 30 mm of subsidence. More than 1000 earthquakes were recorded during this latter period, 44 with magnitudes in excess of 3 and four that were greater than magnitude 4. Maury *et al.* found that between 1969 and 1979 the strain energy release rate by seismic events increased but subsequently there has been an annual decline. It appears that gas production and induced stresses have reactivated small movements along pre-existing faults.

Ground and loading conditions at any mine are practically unique. Therefore, a microseismic monitoring system to be effective should be matched to the characteristics of the microseismic activity being generated at a particular mine. A test programme to determine these characteristics, as well as those of mine and background noise is necessary. In addition, the goal of monitoring and the problem of mine stability must be kept in mind in order to use the microseismic data effectively. Four main problems that complicate field studies of microseismic activity can be identified. These are the presence of environmental noise; the attenuation of stress waves associated with the microseismic activity as they pass through the rock mass; the difficulty in location of the source position because of lack of information on the acoustic properties of the rock mass; and the basic complexity of the geological structure of the rock mass.

Various attempts have been made to distinguish natural seismic events from those related to mining. The criteria for discrimination are based largely on the characteristics of the seismic energy. Seismic mining events are typically long period in character and dominated by surface waves. On the other hand, natural seismic events are of higher frequency and lack the extensive development of surface waves due to their greater focal depth. High intensities at or near the epicentre, decreasing rapidly outwards imply a shallow focal depth and a possible mining induced cause. Also, the stress release mechanism can be expected to reflect the stress pattern created by the removal of a volume of mineral and the subsidence of the overburden in response to gravity. According to Redmayne (1988), vertical rather than horizontal movements are more likely during the failure of rock around a mine opening, with variations due to the orientation of the original plane of weakness and the ambient stress direction. Implosion type events or rockbursts are likely to give dilational first motions of the P-waves (Westbrook *et al.*, 1980). If a seismic event occurs in a mining area and if the focus is located within the upper kilometre or so of the surface, then this would suggest a mining origin. By contrast, naturally occurring earthquakes typically have the depths of their foci below 2.5 km. Mining induced activity is believed to occur just above or below the working horizon according to Westbrook *et al.* Finally, correlation of mining activity with seismic events and reduced seismicity during weekends or holidays is indicative of a mining induced origin.

Geophysical observations have the ability to provide warnings long before any effects are directly observable at the surface or underground. Particular contributions of seismology are the definition of zones of weakness and failure by a three-dimensional determination of rockburst locations. Statistical rockburst distributions with regard to magnitude and time assist in making predictions. In recent years, acoustic emission techniques have been in use or under evaluation for stability monitoring of underground structures. The technique was originally developed in an attempt to predict and reduce the incidence of rockbursts in hard rock mines.

Bath (1984), investigating a four-year rockburst sequence at an iron ore mine in central Sweden concluded that for supervision of rockbursts, sites of interest should be equipped with 3-D geophone networks, containing at least a minimum number of geophones in an optimum configuration. Both the piezoelectric accelerometer and geophone have sufficient particle velocity range to reproduce events between magnitude 0–3.5 in the distance range 200 m–5 km. For magnitudes greater than 3.5, at distances of less than 200 m from the source, the accelerometer is more suitable. The geophone is capable of yielding data above 1 kHz with a better signal to noise ratio between 30 Hz and 3.0 kHz than the accelerometer. The geophone also has a much lower threshold limit (Green, 1984).

10.4. Heap leaching

Heap leaching involves low grade ores, such as finely disseminated gold deposits, from which the metal cannot be extracted by conventional methods, being placed on bases that have low permeability and then being sprayed with a solvent to extract the metal. For example, cyanide commonly is used to dissolve gold. As the complex formed between gold and cyanide is very strong, only a relatively weak solution is required for cyanidation. The solvent should percolate uniformly through the heap so that it comes in contact with all the sources of metal in the ore. The solution and dissolved metal are collected in a plastic lined pond and then treated to recover the metal.

The ores used in heap leaching can be run-of-mine material, the blasted ore being placed on the base for leaching without any prior preparation. Alternatively, the ore may be crushed. If there are excessive fines in the crushed ore, these can be bound together with Portland cement to form coarser particles.

Heap-leach projects can be developed on permanent bases, that is, the spent ore is left on the base after leaching, the latter acting as a liner. Conversely, if the leached ore is disposed of after the metal is extracted, then the base can be used again. A single composite liner can provide environmental protection for permanent or reusable bases where the hydraulic head is low. A double composite liner may be necessary in the case of valley leach facilities because of the higher hydraulic head that is maintained in the heap.

In order to reclaim the heaps after leaching, they are rinsed. Generally, a heap is regarded as having been successfully rinsed when, for instance, in the case of gold extraction the monitoring shows that the weak acid dissociable cyanide content is 0.5 $mg\,l^{-1}$ or less (Van Zyl, 1993).

The chemistry of cyanide is complex and consequently numerous forms of cyanide are present in the leached heaps and associated tailings dams at gold mines. Toxicity primarily arises from free cyanide and generally metal–cyanide compounds are less toxic. Toxicity also is dependent upon the degree to which these compounds dissociate to release free cyanide. The different forms of cyanide vary widely in their rates of decay and their potential toxicity. For example, free cyanide exists in solution as hydrocyanic acid and the cyanide anion. As the former has a low boiling point and high vapour pressure it is reasonably volatile at atmospheric temperature-pressure conditions. Cyanide leaching of gold from ore usually is undertaken at a pH value of 10.3 or above so that most of the free cyanide in solution exists in the stable anion form, in this way reducing to a minimum the loss of cyanide by volatilization. When the waste is disposed of in a tailings lagoon, the pH value of the decant decreases to around 7, at which point most of the free cyanide occurs as hydrocyanic acid and is volatile. The greater the depth of water in the impoundment, the slower is the rate of loss of free cyanide by volatilization.

Cyanide forms compounds with many metals including cadmium, cobalt, copper, iron, mercury, nickel, silver and zinc. As such, these cyanide compounds occur in the waste and effluents at gold mines that use the heap-leach process. Generally, the toxicity of these cyanide–metal compounds is related to their stability, in that the more stable the compound is, the less toxic it is, especially to aquatic life. For instance, zinc cyanide is weak, copper cyanide and nickel cyanide are moderately strong and iron cyanide is very strong, the latter being more or less non-toxic. On the other hand, the weak and moderately stable metal-cyanide compounds can break down to form highly toxic free cyanide forms.

In addition to copper cyanide and zinc cyanide, the principal weak acid dissociable metal-cyanide species present in cyanide leach material are nickel, cadmium and silver cyanide. The breakdown of these metal-cyanide species in an impoundment, first, involves dissociation of the cyanide ion and free metal ion. Hydrolysis of the cyanide ions then leads to the formation of hydrocyanic acid, which subsequently is lost by volatilization from the water in the impoundment. The general rate of decay of cyanide in the water in an impoundment therefore depends upon the rate at which metal-cyanide species dissociate and the rate of volatilization of free cyanide (Simovic *et al.*, 1984).

Residual cyanide in the spent ore represents a potential source of cyanide release to the environment. However, the extent to which this represents a threat to the environment is debatable. For instance, Smith *et al.* (1984) found that cyanide concentrations at various depths within an impoundment in South Africa were very low. Subsequently, Miller *et al.* (1991) showed that natural decay and transformation processes within a tailings impoundment removed soluble cyanide from pore water at depth and that as a result the seepage that took place from the impoundment contained very low cyanide.

10.5. *In situ* leaching

In situ leaching involves using leaching liquids (e.g. ammonium carbonate, sodium bicarbonate or weak sulphuric acid) that are pumped into the ground to recover the mineral from the ore. Consequently, there is little surface disturbance and no tailings or waste rock produced. The orebody, however, needs to be permeable to the leaching liquid, and confined beneath and above by impermeable formations so that the leaching liquid does not contaminate groundwater. For example, uranium deposits have been worked by *in situ* leach methods in the United States since 1974. The time of production of an *in situ* leach wellfield in the United States usually is less than 3 years, typically 6 to 10 months. Initial concentration from a well soon peaks at values typically between 300 and 600 $mg\,l^{-1}$ and then declines rapidly. The decline slows down as the concentration reaches 30–50 $mg\,l^{-1}$ and a well usually is closed down when the concentration reaches 10–20 $mg\,l^{-1}$. Average uranium concentrations are commonly 40–70 $mg\,l^{-1}$ (Anon., 1989).

The uranium ores are often in sands or sandstones that are more or less horizontal. The leaching liquid is pumped into and out of the ground via wells that are cased to ensure that the liquid only flows to and from the ore zone. Wellfield design usually consists of a grid of alternating injection and production wells with a spacing between them ranging from 30 to 50 m (Fig. 10.14). A series of monitoring wells is positioned around the wellfield to ensure that contaminated groundwater does not move outside the mining

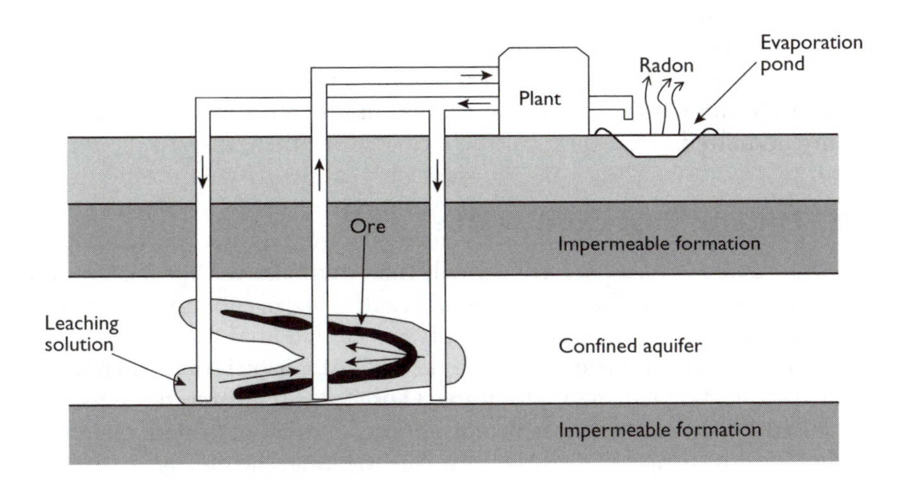

Figure 10.14 Diagram of *in situ* heap leaching process.

area. The uranium bearing solution goes to a treatment plant for the uranium to be recovered. Most of the leaching liquid is returned to the injection wells but a little is bled off and treated as waste water. It contains various dissolved elements such as arsenic, radium and iron. Barium chloride is added to precipitate the radium. The purpose of the bleed is to ensure that there is a steady flow of groundwater from the surrounding rock mass into the wellfield rather than having any leach liquid move in the other direction. Oxygen is injected into the leach liquid as it is returned to the injection wells, and if necessary the liquid is recharged with leach material. Wells are sealed or capped upon decommissioning.

Although the *in situ* leach method of mining reduces the risk of radiation and the need for large tailings impoundments, there are still a number of disadvantages associated with the process. First, there is still the risk of leaching liquid excursions beyond the mining area with subsequent contamination of groundwater. Second, the effects of the leaching liquid on the host rock of the uranium ore are unpredictable. Third, some amounts of waste water and waste sludge are produced when recovering the leaching liquid. Fourth, various decay products of uranium also are leached and can reach considerable activities in the leaching liquid, depending on the leaching agent used. Large amounts of radon may escape into the atmosphere during processing of the solution, while other decay products are transferred to the waste solutions. These solutions usually are disposed of in evaporating ponds, resulting in a concentrated waste slurry. Fifth, after completion of *in situ* leach mining the ore zone should be restored to its pre-leaching conditions. Groundwater restoration is a very protracted process that is not yet understood completely. It still proves impossible to establish pre-leach conditions for all parameters concerned. Of course, groundwater quality may be very low to start with such as at some of the proposed sites in South Australia where *in situ* leach mining has been considered. There the groundwater in the ore bodies is very saline and too high in radionuclides for any permitted use.

10.6. Mineral dusts and health

As far as mineral dusts are concerned, they can take the form of fumes, particulates or fibres. The diseases that result from occupational exposure to dust are referred as pneumoconioses.

Wagner (1980) described how the inhalation of mineral dusts can damage lung tissue, leading to illness and death. He showed that excessive exposure to mineral dusts, such as those produced in coal mining, in slate quarrying and in the processing of china clay, damage the lung, the damage consisting of little nodules of scar tissue. Once these nodules begin to coalesce, the disease becomes progressive and is independent of further exposure. It is at this stage that breathlessness and ill health become apparent. This is associated with disturbance of the exchange of gases, which ultimately leads to respiratory failure. In addition, the lungs become more sensitive to infections. These diseases are caused by the accumulation of isometric dust particles in the lung. Particles are retained in the lung if they are small enough not to fall onto the wall of the airway before they reach the alveolar (gas exchange) part of the lung. Consequently, isometric particles are retained if they are smaller than about 5 μm in diameter and larger than about 0.5 μm. Isometric particles of the size that are retained in the lung are removed by scavenger cells known as macrophages. Unless the particles damage the cells, as happens in the case of quartz dust, the dust is removed either to the lymphatic system or up the airways leaving the lung undamaged.

Although exposure to quartz dust, that is, silicosis, produces severe scaring of the lungs, it rarely causes death by itself. The most common cause of death among individuals who contact silicosis is tuberculosis. An outline of the pathogenesis of silicosis has been provided by Wagner (1980).

Unlike quartz dust, coal dust has very little effect on the macrophages (macrophages engulf the dust in the air spaces of the lungs, quartz is able to kill these cells without any damage to itself). Hence, macrophages can ingest large quantities of coal dust with very little deleterious effect. It is considered that other materials in coal dust such as mica, kaolinite and small amounts of quartz, may play a part in lungs that are so full of dust and engorged macrophages that the normal methods of clearance are overwhelmed.

A considerable quantity of dust is required in the lung to cause disease. For example, Elmes (1980) suggested that a coal miner needed to retain 100 g or more of dust, a slate quarryman around 10–15 g and a worker exposed to pure quartz sand around 5 g. The quantity of dust required depends partly on the amount of crystalline quartz it contains, on the surface area of the particles, and on the presence of other compounds such as iron oxides, which may modify the effect of the quartz. However, dust related diseases can be prevented by measures that suppress dust (wet processing), by adequate ventilation, or by the use of masks that filter out dust.

Elongated particles and fibres have different aerodynamic properties from isometric particles, and tend to fall at a speed that is related to their cross-sectional diameter and is independent of their length. Length, however, becomes a limiting factor when the particles are longer than the diameter of the airways. Consequently, fibres are retained if they are less than 3 μm in diameter and less than 50–100 μm long. The lower limit of retention may be around 0.1 μm in diameter and 5 μm in length. Unfortunately, elongated particles cannot be engulfed and removed by the macrophages. Fibres exceeding 10–12 μm in length remain in the alveolar part of the lung. Smaller fibres move into the lung tissue and, although some may enter the lymphatic system, others reach the surface of the lung and the pleural space. In fact, the way in which particle shape influences clearance from the lung determines relative particle potency as regards causing disease and the location at which the disease develops.

Wagner (1980) also described how fibrous minerals like asbestos can cause extensive lung scarring (asbestosis), primary lung cancer and a cancer of the pleura or peritoneum called a mesothelioma. The difference between the effect of isometric and fibrous dust is qualitative in that lung scarring is diffuse and not nodular in the case of the latter, and both types of cancer can occur, whereas isometric dusts (unless radioactive or contaminated with chemical carcinogens) do not cause cancer. There also is a notable quantitative difference in that severe disease can be associated with the retention of less than 1 g of asbestos, and mesothelioma with less than 1 mg.

Inhalation of asbestos dusts is associated with a variety of pathological changes. Some of these depend upon the type of asbestos to which the individual is exposed. Amphibole fibres are straight and stiff, and can split along their longitudinal cleavage. They can have an extremely fine diameter. Of the common commercial types of asbestos, crocidolite (a fibrous form of riebeckite $[Na_2Fe^{2+}{}_3Fe^{3+}{}_2(OH)_2Si_8O_{22}]$ forming blue asbestos, which has a diameter of <0.1 μm) can penetrate the lung parenchyma through the conducting airways. Amphiboles do not undergo leaching or disintegrate in the tissue. The longer amphibole fibres become coated with an iron containing protein complex to form asbestos bodies that probably are inert. It is probably the smaller uncoated fibres that cause damage and may continue to do so for a lifetime. By contrast, although chrysotile $[Mg_3(OH)_4Si_2O_5]$ has a diameter that is less than that of crocidolite, because of its coiled configuration, it presents a far larger aerodynamic profile and so most of these fibres tend to become caught and immobilized, and do not penetrate the smaller conducting airways. Consequently, only the shortest fibres can migrate through the lung and pleura. Once in the tissue, the magnesium in the fibres is leached out and the fibres eventually disintegrate.

In asbestosis the scarring progresses until the alveoli along the respiratory bronchioles are replaced with a layer of scar tissue. The fibrous tissue then spreads further down the walls of the air sacs, so that more and more lung becomes involved, with scar tissue surrounding collapsed and useless air spaces. Breathlessness becomes evident once the scarring includes a third of the lung parenchyma. This increases as the disease progresses until the heart cannot cope and goes into muscle fatigue and eventual failure. Excessive exposure to all types of asbestos fibre produces asbestosis.

The incidence of lung cancer is dose related and becomes frequent when exposure causes fibrosis. Elmes and Simpson (1977) showed that the risk of lung cancer developing in workers exposed to asbestos was enhanced significantly if they smoked. According to Elmes (1980), it would appear that mesothelioma is more likely to develop when a person is exposed to crocidolite. This is a rare cancer and the latent period between first exposure and development of cancer averages about 40 years. In other words, a significant load of amphibole asbestos fibres is required to induce mesothelioma.

10.7. Mining and landslides

The high frequency occurrence of landslides, or of a single major destructive landslide, on slopes in active and abandoned mining areas, raises concerns as to whether mining has been and is a significant contributory factor in landslide generation in such areas. However, mining induced landslides may be difficult to prove since the occurrence of landslides tends to be influenced by several other interrelated factors. Indeed, the number of factors that influence slope stability may be numerous and varied, and interact in complex and often subtle ways. Frequently, the final factor is simply a trigger mechanism that sets in motion a mass that already was near the point of failure. Basically, landslides occur because the combination of forces creating movement exceed those resisting movement. Mining can be one of those forces and has the potential for triggering landslides where unfavourable ground conditions exist. Nonetheless, an analysis of mining subsidence in relation to slopes is complex and involves both mining and geological factors. The mining factors include the method of working, the width and depth of the extraction, and the height of the mineral worked. Geological factors include the nature of rock types involved, the geological structure and the surface topography. Franks (1985), for example, showed that in the case of longwall working of coal that sloping ground appeared to affect horizontal movements and ground strains to a much greater degree than vertical movements and tilts.

One of the most notable examples of landslides occurred on 29 April 1903, at the small town of Frank, in southern Alberta, Canada, when Turtle

Figure 10.15 The Frank Slide, Alberta, Canada.

Mountain collapsed (Fig. 10.15). This resulted in one of the greatest landslides recorded in North America. At least 70 people were killed when approximately 90 million tonnes of limestone fell some 750 m from the peak of the mountain and rose 145 m up the slope on the opposite side of the valley. The movement lasted 100 seconds, which suggests an average velocity of 30 $m s^{-1}$. Three quarters of the homes in Frank were completely destroyed along with over 1.5 km of railroad. The initial report into the landslide pointed to the role of coal mining as one of the causes of the catastrophe. This subsequently was confirmed by the investigation carried out by Daly *et al.* (1912). Excavation near the foot of the mountain slope began in 1901 and a drift mine was developed deep into the mountain to work thick deposits of coal. The drift was excavated through steeply dipping Devonian and Carboniferous limestones that had been thrust over near-vertical Cretaceous shales, sandstones and coals. The coal was mined from a series of stopes opened out along the strike of the coal seams. These stopes were about 40 m long by 4 m wide and were separated from their neighbours by pillars some 12 m long. Six months before the landslide, the progressive loosening and collapse of the coal, and squeezing in the stopes were observed. This prompted strengthening of the mine workings but unfortunately did not prevent the collapse.

There are a few published examples of landslides where past mining has been cited as the principal cause of the movements. For instance, Pomeroy (1982) documented two examples where shallow failures occurred on fill slopes that were underlain by mine workings in the Pittsburgh coal seam,

Pennsylvania. It was the collapse of the workings that influenced the stability of the slopes. Also, Malgot *et al.* (1986) reported several damaging landslides that occurred in the Vtacnik Mountains, Handlova, Czechoslovkia, which they maintained were caused by mining 3–9 m thick brown coal seams of Jurassic age, overlain by strong andesistic lavas and tuffs. Heslop (1974) described an example of toppling failure that developed adjacent to caved ground at the Havelock asbestos mine in Swaziland. This resulted in the generation of a series of uphill facing (antislope) scarps on the hillslope immediately uphill of the failure.

The dilation of joints in rock masses as a result of mining beneath slopes, thereby enabling increased water penetration and accompanying increased hydrostatic pressures that, in turn, influence slope failure has been considered by Thompson and Tiedemann (1982). Donnelly *et al.* (2000) considered a number of cases where mining subsidence and associated fault reactivation was thought to have contributed to deep-seated rotational landslides in South Wales. These were envisaged to have reactivated pre-existing landslides, as well as causing first time failures of slopes, long after normal ground movement caused by mining had ceased. Jones and Siddle (2000) also investigated the potentially destabilizing effects of shallow mining close to outcrop beneath steep hillslopes in South Wales. Physical and finite element modelling of mining beneath the slopes enabled the principal stress regime, fracturing and subsidence to be predicted. In this way, for example, an explanation of the mechanisms responsible for the East Pentwyn Landslide, which occurred in 1954, was provided. Figure 10.16 shows the development of the stress regime and potential fracture orientations associated with shallow mining in the Brithdir Seam at East Pentwyn. Steep stress gradients suggest a high probability that strata had been weakened and zones of more or less continuous fracturing occurred above remnant pillars of coal, whilst punching of pillars into the floor gave rise to lateral displacements beneath the seam. The deeper zone of fracturing appears to be similar to the failure surface of the 1954 landslide.

In early 1991 fissuring of the ground surface was observed along the crest of an escarpment at Bolsover, Derbyshire, England. This subsequently developed over the following two years into a large compound landslide, causing severe damage to 18 properties located at the toe of the slope and to 9 buildings on the crest of the escarpment (Fig. 10.17). Ten buildings subsequently were demolished. The crest of the slope is underlain by approximately 10 m of Magnesian Limestone (Permian), comprising thinly bedded, moderately strong, dolomitic limestone that rests on a about 20 m of weak calcareous mudstone forming the Lower Permian Marl. This overlies a basal Permian breccia approximately 0.2 m thick. These beds dip gently to the east and unconformably overlie Middle Coal Measures (Upper Carboniferous). The solid strata are covered by up to 4 m of colluvium. At least six seams of coal were extracted beneath the escarpment from 1902 to 1984. The seams were extracted by narrow longwall panels oriented

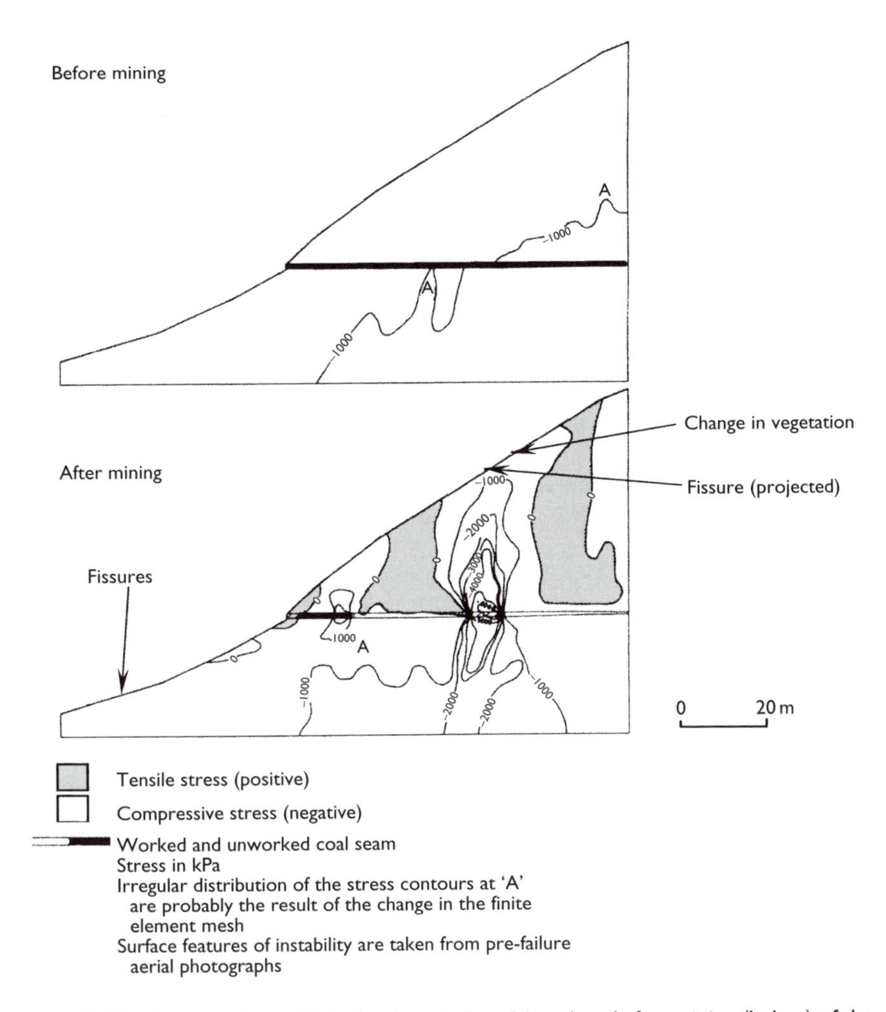

Figure 10.16a Stress regime within the slope before (above) and after mining (below) of the Brithdir Seam at East Pentwyn. (After Jones and Siddle, 2000; Halcrow Group Limited, 1989; reproduced by kind permission of Halcrow Group Limited.)

sub-parallel to the faulting in the area. Subsidence monitoring and a comparison of Ordnance Survey benchmarks established in the latter part of the 1800s indicated that there had been up to 0.45 m of subsidence in Bolsover town. These investigations showed that mining had a significant effect on the strata. This had caused the development of the initial failure and its subsequent extension in 1991. These movements, however, occurred several years after the expected cessation of normal surface subsidence due to longwall mining (Cobb *et al.*, 2000).

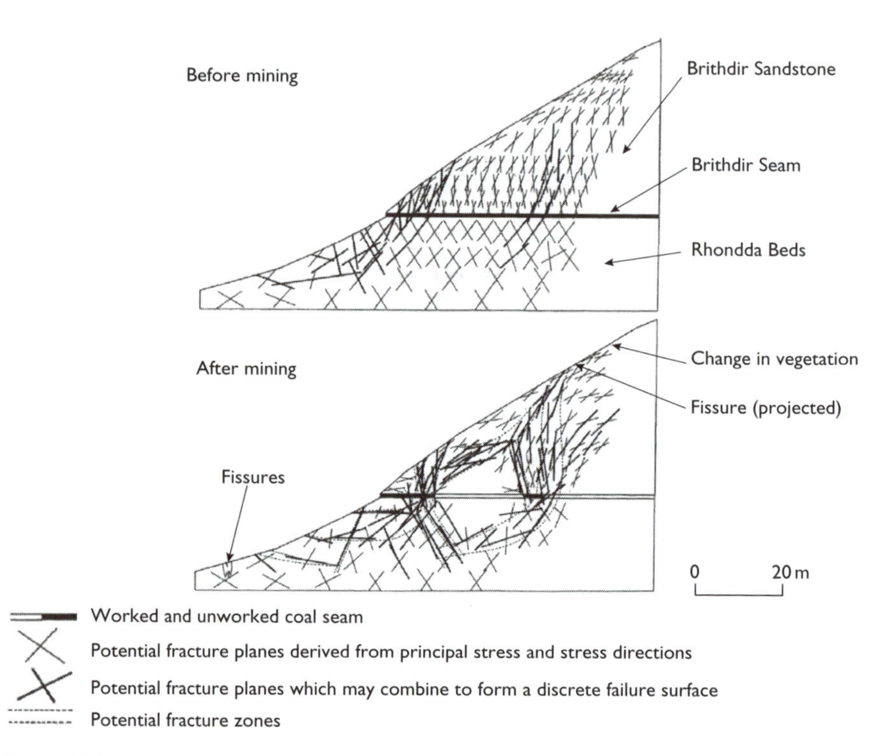

Figure 10.16b Potential fracture orientations before (above) and after mining (below) of the Brithdir Seam at East Pentwyn. (After Jones and Siddle, 2000; Halcrow Group Limited, 1989; reproduced by kind permission of Halcrow Group Limited.)

The destabilizing influence of mining at opencast coal sites was discussed by Walton and Taylor (1977). They quoted two examples of failure of the high wall in two active opencast sites, in which the back walls coincided with the tensional zones associated with the rib side of longwall workings. The authors concluded that opencast excavations advancing towards and parallel to rib side positions are particularly prone to failure, especially if advancing up dip. The St Aiden's extension opencast coal site is located approximately 10 km south-east of Leeds, on the flood plain of the River Aire, Yorkshire, England. The bedrock succession consists of Upper Carboniferous (Westphalian) Coal Measures. Several seams had been worked in the area by underground mining and at the site by opencast mining. A massive failure occurred in the wall of the opencast workings in March 1985, causing the displacement approximately 600 000 m^3 of Coal Measures strata into the void, the failure measuring about 350 m long, 120 m wide and 50 m high. This caused a breach in the riverbanks and flood control levees along the River Aire for at least three days, which

Figure 10.17 (a) The backscarp of the mining-induced landslide at Bolsover, Derbyshire, England. (b) The upheaved toe of the landslide at Bolsover, now landscaped. (c) Structural damage to a perimeter garden wall and a greenhouse at the crest of the Bolsover landslide.

resulted in flooding the opencast void. Around 17 million m^3 of water flowed into the opencast workings forming a lake up to 70 m deep and covering an area of about 100 ha. Coal mining operations were suspended for ten years, sterilizing approximately 2 million tonnes of coal reserves. Remediation involved the re-routing of the river and a canal. The pumping of the flood waters from the workings and the re-establishment of new mining operations were estimated to have cost approximately £36 million (Hughes and Clarke, 2001).

The effects of mining on spoil heaps was investigated by Forrester and Whittaker (1976). They found that vertical subsidence of a spoil heap was greater than predicted in both magnitude and extent. In addition, where mining had taken place at shallow depth, then the measured strains were about twice those predicted. Tensile strains greater than 3 $mm\,m^{-1}$ gave rise to tension cracks and compressive strains that exceeded 10 $mm\,m^{-1}$ resulted in compression humps. Excessive subsidence at the crest, heave at the toe and downslope displacement were noticed in some cases, and in one case resulted in a rotational failure. It therefore was concluded that the shear strength of spoil along a plane subjected to cracking and differential subsidence could be reduced thereby having a detrimental influence on spoil heap stability. These findings subsequently were supported by Siddle *et al.* (1985) in their investigation of a spoil heap at Elsecarr, South Yorkshire, England. They also found that the effects of mining were most notable at the flanks of the heap. Furthermore, where mining displacement coincided with the slope of the spoil heap, then excessive downslope displacements took place. On the other hand, where mining displacements were in the opposite direction to that of the spoil heap, then excessive subsidence occurred at the crest and the toe bulged.

10.8. Case history 1

Mining operations at Middelberg Steam Colliery in Mpumalanga Province, South Africa, began in 1908, the coal being worked by the bord and pillar method (Bullock and Bell, 1997). The extraction ratio was approximately 60% and the average mining depth was 18–20 m, with a mining height of 2.5 m. From 1908 to 1947, when operations at the mine ceased, a total area of some 1700 ha was undermined. Pillar robbing started in the late 1930s and continued until 1946. This meant that the extraction ratio increased from 60% to about 90%. Pillar robbing occurred over an area of approximately 215 ha. This resulted in multiple pillar failure with associated surface subsidence. Generally, areas of pillar collapse were several hundred square metres in extent and the collapsed areas often were bounded by near vertical sides. Surface tension cracks occur around the outer edges of the collapsed areas and were typically 200–800 mm in width. They may extend in length for up to 100 m. In addition, high compressive stresses acting in the vicinity of pillars can cause failure of roof strata. Stresses tend to develop at an angle to the maximum compressive stress. In the incompetent roof strata this tends to be at a high angle (75°). These stress fractures combined with inter-pillar tensional areas, caused collapse, resulting in upward void migration through overlying strata, until the weathered zone, which varies between 5 and 10 m in thickness, is reached. Usually, the weathered material has subsided by 2 to 3 m, but in some cases it has collapsed totally into the old workings leaving voids of 15 to 20 m deep. The resultant crown holes at the surface have diameters between 5 and 10 m, and often are spaced regularly. Tension cracks, varying in width from 200 to 500 mm, are found around the perimeter of individual crown holes.

In 1947 when the mine was being decommissioned, spontaneous combustion within the workings was noticed for the first time. Despite efforts to extinguish the fire, it is still burning, emitting steam, and noxious NO_x and SO_x gases into the atmosphere from fractures and crown holes. Accordingly, the coal in the mine has been undergoing spontaneous combustion for over fifty years and it is estimated that the area affected by burning is between 150 and 200 ha (Fig. 10.18).

All entries to the mine were sealed on closure in 1947. However, a shaft was opened for access when a small area was mined in 1982–83. This was sealed when mining ceased. Hence, except for this limited period, air and water have not gained access to the mine by mine entries. However, the crown holes and tension cracks that were caused by subsidence, and which extend to mine level, allow free passage of air into the mine workings, thereby leading to the oxidation of coal and its spontaneous combustion. Most of the factors that aid spontaneous combustion of coal, which have been outlined earlier, exist at this particular mine. Some burning pillars have collapsed and weakened the integrity of the remaining pillars, leading to further multiple pillar failure and related subsidence, thus further aggravating the problem. These coals are low rank bituminous coals with notable sulphide content, and therefore the pyrite content frequently is greater than 2% and so presumably enhances the process of spontaneous combustion. Although the natural moisture content of the coal prior to the abandonment of the mine is unknown, it is assumed that it is now higher since the mine is partially flooded in places. Certainly, sulphurous steam emanates from some crown holes almost continuously, as well as from some tension cracks (Fig. 10.2). The partially collapsed nature of the workings impedes the rapid flow of air and so heat is not readily conducted away from hot spots. Indeed, the existing conditions appear almost ideal for the spontaneous combustion of coal.

Various attempts have been made to control or extinguish the burning coal in the mine. These include the construction of cut-off trenches around burning zones, which then were backfilled with earth (Fig. 10.18), and

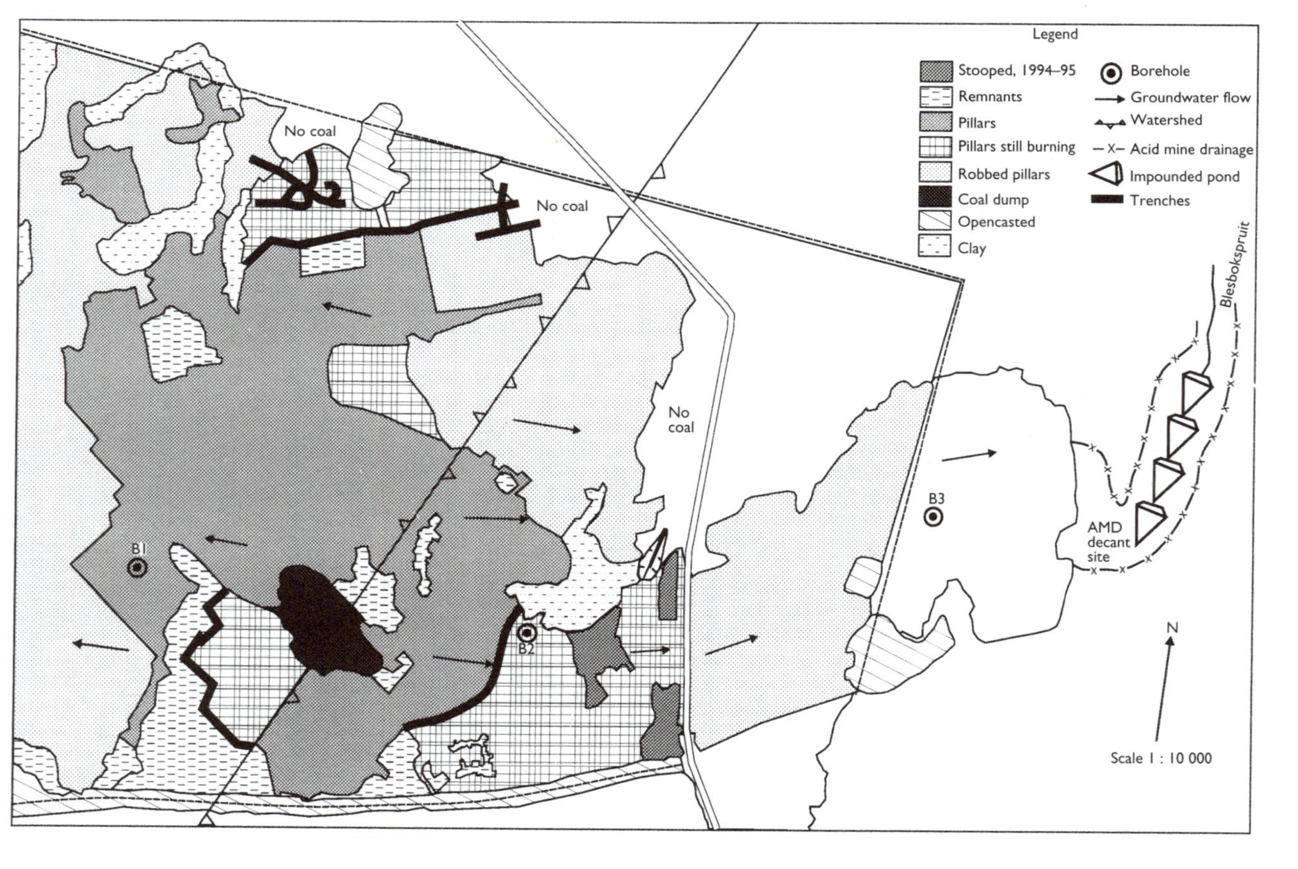

Figure 10.18 Distribution of the various types of mining used at an abandoned mine in the Witbank Coalfield, South Africa, showing the areas where underground fires still burn. Trenches (marked in black) were dug in an attempt to stop the fires spreading. (After Bullock and Bell, 1997.)

injection of water into the workings. In addition, collapsing the workings by subsurface explosions to try to inhibit the access of air, has been suggested. As this is just as likely to increase the access of air, the suggestion has not been put into effect. None of the efforts have had any real success. Burning appears to have by-passed many of the trenches simply because they were not long enough or wide enough and water injection has made no impression on the burning coal. The latter probably is because no barriers were placed to stop water draining away. In addition, water injection may take considerable time, sometimes years, to be successful. The programme was not continued for a long enough time, a matter of weeks being insufficient. In fact, there can be potential problems associated with water injection. First, carbon monoxide and hydrogen may be produced from watergas reactions with burning coal and these then would be emitted at the surface. Second, the release of heat of condensation causes the temperature to rise in the mine.

10.9. Case history 2

Gold was discovered on the Witwatersrand, South Africa, in the mid-1880s and mining commenced in 1886, with the first earth tremor being felt in Johannesburg in 1894 (De Bruyn and Bell, 1997). Prior to 1908, one or two tremors per year were recorded in the Johannesburg mining area. In July 1910 the installation of a Wiechert seismograph took place at the Union Observatory 6.5 km north of the mining area in Johannesburg, and from that year until 1937 some 14 830 seismic events were recorded (Gane, 1939). The number and the intensity of tremors in Johannesburg increased at a slow but relatively steady rate, and 29 669 events were recorded between 1938 and 1949 (Finsen, 1950). The increase in the rate of occurrence, together with that of average intensity followed the increases in tonnage mined and average depth of mining (Gane *et al.*, 1946). An analysis of recorded seismic events conducted by Grobbelaar (1970) revealed that they reached a maximum in 1953 and only decreased significantly after 1965. This decrease both in number and intensity was ascribed to the decrease in mining activity immediately south of the city. The total production of ore on the mines of the central Witwatersrand decreased between 1950 and 1968 from 16 741 000 to 5 761 000 tonnes, a reduction of 65%. Seismic events decreased from 4504 in 1953 to 2851 in 1968, a reduction of 37%. This would suggest that seismic activity would continue on the central Witwatersrand even after all mining activity ceases. A statistical analysis of seismographic records of the Union Observatory published in 1939 proved conclusively that blasting in the mines was by far the most effective trigger in initiating a tremor (Gane, 1939). The daily records showed a sudden increase in the relative frequency of tremors immediately after 14.00 hours on weekdays. This was explained by the general custom of the mines to blast at about this time.

In a routine investigation of more than a thousand earth tremors in the Witwatersrand region, not once was the position of an epicentre found, with confidence, to be outside the area where mining was taking place. In addition, the focal depths from which the tremors originated were shown to be in close proximity to the level of active mining (Gane *et al.*, 1952). From 1971 onwards, the location and magnitude of events were recorded in the Earthquake Catalogue of the Geological Survey of South Africa. The catalogue indicates that most seismic events in South Africa below magnitude 5 originate in the mining areas, with most of the tremors from the Witwatersrand and Far West Rand gold mining areas. Since these tremors generate excellent P and S phases, it usually is possible to evaluate the epicentral distance from the recording station at Pretoria, and consequently their magnitudes. These epicentres are obtained by using the arrival times of the first P-waves at the South African Seismological stations, supplemented by the available P readings as listed in the bulletins of neighbouring countries. However, the magnitudes of the majority of these events are such that they are not recorded by a sufficient number of stations to enable them to be located exactly.

Ockleston (1968), observing the main characteristics of Johannesburg tremors, found that they are essentially transverse seismic waves of which the horizontal component of ground movement predominates. The tremors are of short duration, seldom lasting much longer than a second. Ground oscillation is of higher frequency, smaller amplitude and shorter duration than is the case with deep-seated tectonic earthquakes.

In the 1960s it became possible to quantify the effects of mining on the stresses in the rock mass surrounding narrow mined out areas in deep level, hard rock mines. Empirical relations between the rockburst hazard, and the stress and energy changes induced in a rock mass by extracting the gold-bearing reefs could be established. This development also made possible the identification of seismically hazardous areas or structures, providing mine management with the means of devising mining layouts and sequences that minimized the risk of rockburst occurrence, and also to improve the density or quality of support, thus minimizing the destructive effects of seismic events. New types of stope support were developed as a result, notably rapid-yielding hydraulic props and barrier chocks.

Also in the 1960s, the introduction of underground seismic monitoring networks yielded valuable insight into the origin and nature of seismic events, enabling more accurate estimates of the energy contents of these events to be made (Blake, 1984). A strong relationship between the frequency of rockbursts and the energy release rate associated with mining was identified. The practical conclusion was that, in order to minimize rockburst hazard, it was necessary to design mine layouts in which the energy release rate was kept as low as possible. This is best achieved by utilizing straight longwall faces in preference to scattered mining, by backfilling with waste rock or slime, or by leaving stabilizing pillars. There appears, however, to be some disagreement on the first of these points. For instance, the lower incidence of rockbursts in the Free State gold mining district has been attributed to low energy release rates due to the rarity of longwall mining in this region, in preference to scattered mining (Lawrence, 1984). The establishment of stabilizing pillars has been shown to be the most effective method of controlling the energy release rate and thus reducing the rockburst hazard. A disadvantage of this method, however, is that up to 20% of the reef has to be left intact to form pillars. By 1984, at least four deeper gold mines had successfully adopted this method of rockburst control. In one of these mines it was shown that large, widely spaced stabilizing pillars in a section of the mine reduced the seismicity accompanying stoping by 60%.

References

Adamus, A. 2002. Review of the use of nitrogen in mine fires. *Transactions of the Institution of Mining and Metallurgy*, Section A, Mining Technology, **111**, A89–A98.

Anon. 1973. *Spoil Heaps and Lagoons, Technical Handbook*. National Coal Board, London.

Anon. 1986. *Citizen's Guide to Radon*. EPA-86-004 United States Environmental Protection Agency, Washington, DC.

Anon. 1989. *In Situ Leaching of Uranium: Technical, Environmental and Economic Aspects*. IAEA-TECDOC-492, International Atomic Energy Agency, Vienna.

Anon. 1999. *Uncontrolled Fires in Coal and Coal Wastes*. International Energy Agency (IAE) Coal Research, London.

Ball, T.K., Cameron, D.G., Colman, T.B. and Roberts, P.D. 1991. Behaviour of radon in the geological environment: a review. *Quarterly Journal of Engineering Geology*, **24**, 169–182.

Bath, M. 1984. Rockburst seismology. *Proceedings of First International Congress on Rockbursts and Seismicity in Mines*, Gay, N.C. and Wainwright, E.H. (eds), Johannesburg, South African Institution of Mining and Metallurgy, Symposium Series No. 6, Johannesburg, 7–15.

Bell, F.G., Genske, D.D. and Bell, A.W. 2000. Rehabilitation of industrial areas: case histories from England and Germany. *Environmental Geology*, **40**, 121–134.

Bishop, I., Styles, P. and Allen, M. 1993. Mining induced seismicity in the Nottingham Coalfield. *Quarterly Journal of Engineering Geology*, **26**, 253–279.

Blake, W. 1984. Design considerations for seismic monitoring systems. *Proceedings of the First International Congress on Rockbursts and Seismicity in Mines*, Johannesburg, Gay, N.C. and Wainwright, E.H. (eds), South African Institution of Mining and Metallurgy, 79–82.

Bryan, Sir A., Bryan, J.C. and Fouche, J. 1964. Some problems of strata control in pillar workings. *Mining Engineer*, **123**, 238–266.

Bullock, S.E.T. and Bell, F.G. 1997. Some problems associated with past mining at a mine in the Witbank coalfield, South Africa. *Environmental Geology*, **23**, 61–71.

Cobb, E.A., Jones, H.J. and Siddle, H.J. 2000. The influence of mining on the Bolsover landslide. *Proceedings of the Eighth International Symposium on*

Landslides: Landslides in Research, Theory and Practice, Cardiff, Bromhead, E., Dixon, N. and Ibsen, M.L. (eds), Thomas Telford Press, London, 287–292.

Cook, N.G.W. 1976. Seismicity associated with mining. *Engineering Geology*, **10**, 99–122.

Cook, N.G.W., Hoek, E., Pretorius, J.P.G., Ortlepp, W.D. and Salamon, M.D.G. 1966. Rock mechanics applied to the study of rockbursts. *Journal of South African Institute of Mining and Metallurgy*, **66**, 435–528.

Creedy, D.P. 1991. An introduction to geological aspects of methane occurrence and control in British deep coal mines. *Quarterly Journal of Engineering Geology*, **24**, 209–220.

Creedy, D.P., Saghafi, R. and Lama, R. 1997. *Gas Control in Underground Coal Mining*. International Energy Agency (IAE), Coal Research, London.

Daly, R.A., Miller, W.G. and Rice, G.S. 1912. *Report on the Commission Appointed to Investigate Turtle Mountain, Frank, Alberta*. Canadian Geological Survey, Memoir 27, Canadian Department of Mines, Ottawa.

De Bruyn, I.A. and Bell, F.G. 1997. Mining induced seismicity in South Africa: a survey. *Proceedings of the International Symposium on Engineering Geology and the Environment*, Athens, Marinos, P.G., Koukis, G.C., Tsiamboas, G.C. and Stournaras, G.C. (eds), Balkema, Rotterdam, **3**, 2321–2326.

Donnelly, L.J. 1998. *Fault Reactivation and Ground Deformation Investigation, Easington Colliery, County Durham*. Report No. WN/98/9, British Geological Survey, Nottingham.

Donnelly, L.J. and Bamford, L.V. 1996. *Exploration for Abandoned Mine Shafts Using Radon Soil Gas Geochemistry*. Report No. WN/96/38, British Geological Survey, Keyworth, Nottingham.

Donnelly, L.J. and McCann, D.M. 2000. The location of abandoned mine workings using thermal techniques. *Engineering Geology*, **57**, 39–52.

Donnelly, L.J., Northmore, K.J. and Jermy, C.A. 2000. Fault reactivation in the vicinity of landslides in the South Wales Coalfield. *Proceedings of the Eighth International Symposium on Landslides, Landslides in Research, Theory and Practice*, Cardiff, Bromhead, E.N., Dixon, N. and Ibsen, M.-L. (eds), Thomas Telford Press, London, **1**, 481–486.

Elmes, P.C. 1980. Fibrous minerals and health. *Journal of Geological Society*, **137**, 525–535.

Elmes, P.C. and Simpson, M.J.C. 1977. Insulation workers in Belfast: a further study of mortality due to asbestos exposure (1940–1975). *British Journal of Industrial Medicine*, **34**, 174–180.

Finsen, W.S. 1950. The statistics of Witwatersrand earth tremors, 1933.5 to 1949.0. *Circular 110 of Union Observatory*, Johannesburg.

Forrester, D.J. and Whittaker, B.N. 1976. Effect of mining subsidence on colliery spoil heaps – II. Deformational behaviour of spoil heaps during undermining. *International Journal of Rock Mechanics and Mining Science and Geomechanical Abstracts*, **13**, 121–133.

Franks, C.A.M. 1985. Mining subsidence and landslips in South Wales. *Proceedings of the Symposium on Landslides in South Wales*. Morgan, C.S. (ed.), Polytechnic of Wales, Pontypridd, 225–230.

Gane, P.G. 1939. A statistical study of the Witwatersrand earth tremors. *Journal of the Chemical, Metallurgy and Mining Society of South Africa*, **40**, 155–164.

Gane, P.G., Hales, A.L. and Oliver, H.A. 1946. A seismic investigation of the Witwatersrand earth tremors. *Bulletin of Seismological Society of America*, **36**, 129–142.

Gane, P.G., Seligman, P. and Stephen, J.H. 1952. Focal depths of Witwatersrand tremors. *Bulletin of Seismological Society of America*, **42**, 239–250.

Gayer, R. and Harris, I. (eds) 1996. *Coalbed Methane and Coal Geology*. Special Publication 109, Geological Society, London.

Gibbs, J.F., Healy, J.H., Raleigh, G.B. and Cookley, J. 1973. Seismicity in Rangely, Colorado, Area: 1962–70. *Bulletin of Seismological Society of America*, **63**, 1557–1570.

Gibowicz, S.J. 1984. The mechanism of large mining tremors in Poland. *Proceedings of the First International Congress on Rockbursts and Seismicity in Mines*, Johannesburg, Gay, N.C. and Wainwright, E.H. (eds), South African Institution of Mining and Metallurgy, 17–28.

Green, R.W.E. 1984. Design considertions for an underground seismic network. *Proceedings of the First International Congress on Rockbursts and Seismicity in Mines*, Johannesburg, Gay, N.C. and Wainwright, E.H. (eds), South African Institution of Mining and Metallurgy, 67–74.

Grobbelaar, C. 1970. Statistical study into the influence of dykes, faults and raises on the incidence of rockbursts. Association of Mine Managers of South Africa, Chamber of Mines, Johannesburg, 1033–1052.

Halcrow Group Limited. 1989. Landslides and Undermining, Research Project. Joint project undertaken by Sir William Halcrow, Consultant Engineers and Nottingham University, Department of Mining Engineering on behalf of the Department of the Environment.

Henshaw, D.L., Eatough, J.P. and Richardson, R.B. 1990. Radon: a causative factor in the induction of myeloid leukaemia and other cancers in adults and children? *Lancet*, **335**, 1008–1012.

Heslop, F.G. 1974. Failure by overturning in ground adjacent to cave mining, Havelock Mine, Swaziland. *Proceedings of the Third Congress of International Society Rock Mechanics*, Denver, **2B**, 1085–1089.

Howes, M.J. 1990. Exposure to radon daughters in Cornish tin mines. *Transactions of the Institution Mining and Metallurgy*, Section A, Mining Industry, **99**, A85–A90.

Hughes, D. B and Clarke, B. G. 2001. The River Aire slope failure at the St. Aidans extension opencast coal site, West Yorkshire, United Kingdom. *Canadian Geotechnical Journal*, **38**, 339–359.

Jones, D.B. and Siddle, H.J. 2000. Effect of mining on hillslope stability. In: *Landslides and Landslide Management in South Wales*, Siddle, H.J., Bromhead, E.N. and Bassett, M.G. (eds), National Museum of Wales, Geological Series No. 18, Cardiff, 40–42.

Jones, R.L. 1995. Soil uranium, basement radon and lung cancer in Illinois, USA. *Environmental Geochemistry and Health*, **17**, 21–24.

Kusznir, N.J., Al-Saigh, N.H. and Farmer, I.W. 1982. Induced seismicity resulting from roof caving and pillar failure in longwall mining. In: *Strata Mechanics*, Farmer, I.W. (ed.), Elsevier, Amsterdam, 7–12.

Lawrence, D.A. 1984. Seismicity in the Orange Free State gold mining district. *Proceedings of the First International Congress on Rockbursts and Seismicity in Mines*, Johannesburg, Gay, N.C. and Wainwright, E.H. (eds), South African Institution of Mining and Metallurgy, 121–130.

McCann, D.M. 1988. Induced seismicity in engineering. In: *Engineering Geology of Underground Movements*, Engineering Geology Special Publication No. 5, Bell, F.G., Culshaw, M.G., Cripps, J.C. and Lovell, M.A. (eds), Geological Society, London, 405–413.

Malgot, J., Baliak, F. and Mahr, T. 1986. Prediction of the influence of underground coal mining on slope instability in the Vtacnik Mountains. *Bulletin of International Association of Engineering Geology*, **33**, 57–65.

Maury, V.M.R., Grasso, J.-R. and Wittlinger, G. 1992. Monitoring of subsidence and induced seismicity in the Lacq gas field (France): the consequences on gas production and field operation. *Engineering Geology*, **32**, 123–135.

Michalski, S.R. and Gray, R.E. 2001. Ash disposal – mine fires – environment: an Indian dilemma. *Bulletin of Engineering Geology and the Environment*, **60**, 23–29.

Michalski, S.R., Winschel, L.J. and Gray, R.E. 1990. Fires in abandoned mines. *Bulletin of Association of Engineering Geologists*, **27**, 479–495.

Miller, S.D., Jeffery, J.J. and Wong, J.W.C. 1991. In-pit identification and management of acid forming rock waste at Golden Cross Gold Mine, New Zealand. *Proceedings of the Second International Conference on the Abatement of Acid Drainage*, Montreal, 125–132.

Mohan, S. 1996. Genesis of mine fires. *Journal of Mines, Metals and Fuels*, **44**, 195–198.

Morris, D.M. and Vorster, F.A.L. 1984. A record of fall of ground accidents on gold mines, 1975 to 1981. *Proceedings of the First International Congress on Rockbursts and Seismicity in Mines*, Johannesburg, Gay, N.C. and Wainwright, E.H. (eds), South African Institution of Mining and Metallurgy, 179–182.

Morris, R.A. 1987. A review of experiences of the use of inert gases in mine fires. *Mining Science and Technology*, **6**, 37–69.

Neff, P.A., Wagener, F. von M. and Green, R.W.E. 1992. Houses damaged by mine tremor distress, repair and design. *Proceedings of the Symposium on Construction over Mined Areas*, Pretoria, South African Institution Civil Engineers, Yeoville, 133–137.

Ockleston, A.J. 1968. The effect of earth tremors on high-rise buildings in Johannesburg. *South African Institution of Civil Engineers Forum on High-rise Buildings*, Johannesburg.

Ortlepp, W.D. 1984. Rockbursts in South African gold mines: a phenomenological view. *Proceedings of the First International Congress on Rockbursts and Seismicity in Mines*, Johannesburg, Gay, N.C. and Wainwright, E.H. (eds), South African Institution of Mining and Metallurgy, 121–130.

Pomeroy, J.S. 1982. *Landslides in the Greater Pittsburgh Region, Pennsylvania.* United States Geological Survey, Professional Paper 1229, Washington DC.

Raybould, J.G. and Anderson, J.G. 1987. Migration of landfill gas and its control by grouting – a case history. *Quarterly Journal of Engineering Geology*, **20**, 78–83.

Redmayne, D.W. 1988. Mining induced seismicity in U.K. coalfields identified on the BGS National Seismograph Network. In: *Engineering Geology of Underground Movements*, Engineering Geology Special Publication No. 5, Bell, F.G., Culshaw, M.G., Cripps, J.C. and Lovell, M.A. (eds), Geological Society, London, 405–413.

Redmayne, D.W., Richards, J.A. and Wild, P.W. 1998. Mining-induced earthquakes monitored during pit closure in the Midlothian Coalfield. *Quarterly Journal of Engineering Geology*, **31**, 21–36.

Robinson, R. 2000. Mine gas hazards in the surface environment. *Transactions of the Institution of Mining and Metallurgy*, Section A, Mining Technology, **109**, A228–A236.

Rothe, G.H. and Lui, C. 1983. Possibility of induced seismicity in the vicinity of Sleepy Hollow oilfield, south western Nesbraska. *Bulletin of Seismological Society of America*, **73**, 1357–1367.

Shapira, A., Fernandez, L.M. and du Plessis, A. 1989. Frequency-magnitude relationships of southern African seismicity. *Tectonophysics*, **167**, 261–271.

Siddle, H.J., Oliver, M.E. and Ansell, P. 1985. The effect of mining subsidence on spoil heap stability, a case history. *Proceedings of the Third International Conference on Ground Movements and Structures*, Cardiff, Geddes, J.D. (ed.), Pentech Press, London, 264–278.

Simovic, L., Snodgrass, W.J., Murphy, K.L. and Schmidt, J.W. 1984. Development of a model to describe the natural degradation of cyanide in gold with effluents. *Proceedings of the Conference on Cyanide and the Environment*, Tucson, 413–430.

Sizer, K., Creedy, D. and Sceal, J. 1996. *Methane and other Gases from Disused Coal Mines: The Planning Response*. Department of the Environment, Technical Report, Her Majesty's Stationery Office, London.

Smith, A., Dehomann, A. and Pullen, R. 1984. The effects of cyanide bearing gold tailings disposal on water qualities in the Witwatersrand, South Africa. *Proceedings of the Conference on Cyanide in the Environment*, Tucson, 221–229.

Thompson, S. and Tiedemann, A. 1982. A review of the factors affecting landslides in urban areas. *Bulletin of International Association of Engineering Geology*, **19**, 55–65.

Van Vuuren, M.C.J. 1995. *Guidelines for the Prevention of Spontaneous Combustion of Coal During Storage and Transport*. Report No. ES9307, Ministry of Minerals and Energy Affairs, Pretoria.

Van Zyl, D.J.A. 1993. Mine waste disposal. In: *Geotechnical Practice for Waste Disposal*, Daniel, D.E. (ed.), Chapman and Hall, London, 269–287.

Voracek, V. 1997. Current planning procedures and mine practice in the field of prevention and suppression of spontaneous combustion in deep coal mines in the Czech part of the Upper Silesian Coalfield. *Proceedings of the 27th International Conference Safety in Mines Research Institutes*, New Delhi, Dhar, B.B. and Bhowmick, B.C. (eds), Oxford and IBH Publishing Company, New Delhi, 437–441.

Wagner, J.C. 1980. The pneumoconioses due to mineral dusts. *Journal of Geological Society*, **137**, 537–545.

Walker, S. 1999. *Uncontrolled Fires in Coal and Coal Wastes*. International Energy Agency (IEA), Coal Research, London.

Walton, G. and Taylor, R.K. 1977. Likely constraints on the stability of excavated slopes due to underground coal workings. *Proceedings of the Conference on Rock Engineering*, Newcastle upon Tyne, British Geotechnical Society, London, 329–349

Westbrook, G.K., Kuznir, N.J., Browit, C.W.A. and Holdsworth, B.K. 1980. Seismicity induced by coal mining in Stoke-on-Trent (UK). *Engineering Geology*, 16, 225–241.

Williams, G.M. and Aitkenhead, N. 1991. Lessons from Loscoe: the uncontrolled migration of landfill gas. *Quarterly Journal of Engineering Geology*, **24**, 191–208.

Index